HANDBOOK OF COMPARATIVE GENOMICS

HANDBOOK OF COMPARATIVE GENOMICS
Principles and Methodology

CECILIA SACCONE
Department of Biochemistry and Molecular Biology
University of Bari
Italy

GRAZIANO PESOLE
Department of Physiology and General Biochemistry
University of Milan
Italy

WILEY-LISS

A JOHN WILEY & SONS PUBLICATION

2003

Published by John Wiley & Sons, Inc., Hoboken, New Jersey.
Published simultaneously in Canada.

For general information on our other products and services please contact our Customer Care Department within the U.S. at 877-762-2974, outside the U.S. at 317-572-3993 or fax 317-572-4002.

Wiley also publishes its books in a variety of electronic formats. Some content that appears in print, however, may not be available in electronic format.

Library of Congress Cataloging-in-Publication Data:

Saccone, Cecilia.
   Handbook of comparative genomics: Principles and methodology /
Cecilia Saccone, Graziano Pesole.
      p. cm.
   Includes bibliographical references and index.
   ISBN 0-471-39128-X (cloth : alk. paper)
   1. Genomics—Handbooks, manuals, etc.   2. Evolutionary ge-
netics—Handbooks, manuals, etc.   I. Pesole, Graziano.
II. Title.
   [DNLM: 1. Genomics.   2. Computational Biology.   3. Evolu-
tion, Molecular.   4. Sequence Analysis, DNA. QU 58.5 S119h
2003]
QH447.S23 2003
572.8—dc21   2002011158

Printed in the United States of America

10 9 8 7 6 5 4 3 2

To
Ernesto Quagliariello

CONTENTS

**6   COMPUTATIONAL METHODS FOR THE ANALYSIS OF
GENOME SEQUENCE DATA**   **187**

PREFACE

No sooner had the genomics era begun when a post-genomics era was declared. Since the beginning of large-scale sequencing, there has been a need for new approaches and tools for the study of living matter at the molecular level, and particularly of new hypotheses for progress in biology. There are tremendous expectations about future benefits of sequencing complete genomes; realistically, however, much still remains to be done. Indeed, the exploitation of genome information is still in its infancy and new methodologies may be required to make the most of them.

Some fundamental questions are: To what extent has large-scale sequencing increased our knowledge of the properties and functions of species and, in general, of the structure and function of the genome? What value-adding notions have emerged since sequencing? How has our knowledge of the molecular basis of biological processes improved, if at all? What is our capability to perform comparative genomic studies? What general rules can we derive?

It is well known that advances in comparative biology are based largely on the concepts of analogy and homology. For this reason, comparative analyses of complete genomes rather than single genes, gene families, or specific genome regions must be performed. The sequencing of complete genomes from prokaryotes, eukaryotes, and organelles has aided research on the structure and evolution of the genome as a unit, as opposed to the previous focus on its components. Hence, the new discipline of evolutionary genomics is emerging and will make a revolutionary impact in terms of helping to disclose the linear structure of nucleic acids and proteins, their three-dimensional folding, cytogenetics, gene expression, and regulatory pathways.

We are persuaded that comparative evolutionary genomics is the key to unraveling the hidden messages of living matter. Our use of *genomics* includes its expression and the regulatory mechanisms that are at the basis of all biological processes. In other words, genomics also includes transcriptomics, proteomics, and other "omics," concepts that interconnect and overlap. "Omics" research needs broad databases and dynamic technologies capable of facilitating collaborative efforts across many disciplines, such as biology, chemistry, informatics, mathematics, and physics.

The ambitious goal of this book is to offer a tool for trained biologists who want to tackle the new genomic dimension of modern biology. It will also be useful for technology developers, managers, industries, and funding agencies—in essence anyone interested in the exciting new applications in this new dimension. We are aware that the book is a personal vision and because of its extent and complexity, cannot cover all the literature. Nonetheless, we have tried to highlight milestones, emerging principles, key methods used, and the most urgent needs of this new field. Beginning with a description of complete genomes sequenced from living organisms, this handbook pinpoints new concepts emerging from available data. At the same time we describe the leading methods used to study complete genomes and their evolution.

In summary, we have written this book as a guide for students and researchers not necessarily specialized in genomics. It is written for anyone who approaches, with theoretical and practical aims, this intriguing new chapter of biology—perhaps already a new discipline. We have organized the book in three parts.

- In Part I we describe the state-of-the-art in molecular knowledge of the main biological processes achieved with modern biotechnological tools. We introduce recent insights brought about by genome sequencing. In particular, we summarize the major features of the genomes completely sequenced in prokaryotes, eukaryotes, and organelles.

- In Part II we illustrate the most recent methodologies used in genomics. We describe the available experimental and bioinformatics tools with particular emphasis on molecular biology techniques, biological databases, and computational methods for the analysis of sequence data.

- Part III contains data derived from comparative studies. We discuss fundamental, cutting-edge topics such as the evolution of genome size, base compositional constraints, and the structure and origin of organisms at the molecular level. We conclude by addressing recent advances in molecular phylogenetics.

ACKNOWLEDGMENTS

We thank our dearly missed friend and colleague Giuliano Preparata, a distinguished theoretical physicist who introduced us to the multidisciplinary approach in the study of molecular evolution. Warm thanks also to several colleagues and friends who offered advice and suggestions on several topics covered in the book; they include Marcella Attimonelli, Giorgio Bernardi, Rita Casadio, Nicla Cataldo, Victor De Lorenzo, Annamaria D'Erchia, Ilenia D'Errico, Carmela Gissi, Alessandro Minelli, Aurelio Reyes, Teresa M. R. Regina, Elisabetta Sbisà, and Apollonia Tullo. Finally, special thanks to Alessandra Larizza, for recovering the bibliography and organizing the material, and to Marilina Lonigro, for assistance in English.

PART I

GENOME FEATURES

CHAPTER 1

PROKARYOTES

1.1 INTRODUCTION

While this book is being written, complete sequences of bacterial genomes are being produced at a rate of about two genomes per month, and the National Center for Biotechnology Information (NCBI) Web site (see the URL in Table 5.1) reports about 60 completely sequenced prokaryotic genomes. Data reported in this chapter refer to the status of completely sequenced genomes, summarized in Table 1.1. Obviously, by the time you read this book, many more will have been sequenced and perhaps some of the aspects dealt with could be viewed differently, although we do not expect dramatic changes in our knowledge unless technology speeds its pace considerably.

Table 1.1 reports the prokaryotic genomes completely sequenced up to now and includes such features as species name, EMBL data library accession number, size, shape, presence of extrachromosomal elements, and bibliographic references. From a look at this list, one can gain an appreciation of the diverse reasons for promoting the sequencing of one species rather than another. Bacterial species are sequenced according to their research interest in basic or applied science: their importance for phylogenetic investigations, to shed light into the metabolic machinery (mainly Archaea) as well as for their importance as human and/or animal pathogens, and for their role as a source of industrial enzymes. In other words, priority has been given to species already well known or species presenting attractive opportunities in applied fields; thus from a phylogenetic point of view, the choice turns out to be very random.

We know we are at the infancy of the genomic era; despite the fact that completely sequenced organisms are still tiny in number, they have already turned out to be full of surprises. In this chapter we summarize the principal sequencing achievements that have improved our knowledge of the prokaryotic genomes and have contributed to outlining methods and approaches to be used in such studies.

TABLE 1.1. Prokaryotic Genomes Completely Sequenced

Species	Main Chromosome		Extrachromosomal Elements		References
	Accession Number	Size (bp)	Accession Number	Size (bp)	
Archaea					
Aeropyrum pernix	BA000002	1,669,695			Kawarabayasi, Hino et al. (1999)
Archaeoglobus fulgidus	AE000782	2,178,400			Klenk, Clayton et al. (1997)
Halobacterium sp. NRC-1 (3 chromosomes)	AE004437	2,014,239			Ng, Kennedy et al. (2000)
	AE004438	365,425	AF016485	191,346	
	AF016485	191,346	AE004438	365,425	
Methanobacterium thermoautotrophicum	AE000666	1,751,377			Smith, Doucette-Stamm et al. (1997)
Methanococcus jannaschii	L77117	1,664,970	L77118	58,407	Bult, White et al. (1996)
			L77119	16,550	
Methanococcus kandleri AV19	AE009439	1,694,969			Slesarev, Mezhevaya et al. (2002)
Methanosarcina acetivorans str. C2A	AE010299	5,751,492			Galagan, Nusbaum et al. (2002)
Methanosarcina mazei Goe1	AE008384	4,096,345			Deppenmeier, Johann et al. (2002)
Pyrobaculum aerophilum	AE009441	2,222,430			Fitz-Gibbon, Ladner et al. (2002)
Pyrococcus abyssi	AL096836	1,765,118			Lecompte, Ripp et al. (2001)
Pyrococcus furiosus DSM 3638	AE009950	1,908,256			Robb, Maeder et al. (2001)
Pyrococcus horikoshii	AP000001–AP000007	1,738,505			Kawarabayasi, Sawada et al. (1998)
Sulfolobus solfataricus	AE006641	2,992,245			She, Singh et al. (2001)
Sulfolobus tokodaii	BA000023	2,694,765			Kawarabayasi, Hino et al. (2001)
Thermoplasma acidophilum	AL445063–AL445067	1,564,905	AJ010405	41,229	Ruepp, Graml et al. (2000)
Thermoplasma volcanium	AP000991–AP000996	1,584,799			Kawashima, Amano et al. (2000)
Bacteria					
Agrobacterium tumefaciens str.C58(Cereon)	AE007869	2,841,581			Goodner, Hinkle et al. (2001)
Agrobacterium tumefaciens str.C58(U.Washington)	AE008688	2,841,490	AE008687	542,780	Wood, Setubal et al. (2001)
			AE008690	214,234	

Organism	Accession	Size (bp)	Plasmid	Size (bp)	Reference
Aquifex aeolicus	AE000657	1,551,335	AE000667	39,456	Deckert, Warren et al. (1998)
Bacillus halodurans	BA000004	4,202,353			Takami, Nakasone et al. (2000)
Bacillus subtilis	AL009126	4,214,814			Kunst, Ogasawara et al. (1997)
Borrelia burgdorferi[a]	AE000783	910,725			Fraser, Casjens et al. (1997); Casjens, Palmer et al. (2000)
			AE000791	9,386	
			AE000792	26,498	
			AE001575	30,750	
			AE001576	30,223	
			AE001577	30,299	
			AE001578	29,838	
			AE001579	30,800	
			AE001580	30,885	
			AE001581	30,651	
			AE001583[a]	5,228	
			AE000793[a]	16,823	
			AE001582[a]	18,753	
			AE000785[a]	24,177	
			AE000794[a]	26,921	
			AE000786[a]	29,766	
			AE000784[a]	28,601	
			AE000789[a]	27,323	
			AE000788[a]	36,849	
			AE000787[a]	38,829	
			AE000790[a]	53,561	
			AE001584[a]	52,971	
Brucella melitensis	AE008917	2,117,144	AP001070	7,258	DelVecchio, Kapatral et al. (2002)
Buchnera sp. APS	AP000398	640,681	AP001071	7,786	Shigenobu, Watanabe et al. (2000)
Campylobacter jejuni	AL111168	1,641,481			Parkhill, Wren et al. (2000)
Caulobacter crescentus	AE005673	4,016,947			Nierman, Feldblyum et al. (2001)
Chlamydia pneumoniae AR39	AE002161	1,229,853		4,524	Read, Brunham et al. (2000)
Chlamydia pneumoniae CWL029	AE001363	1,230,230			Kalman, Mitchell et al. (1999)
Chlamydia trachomatis MoPn	AE002160	1,069,412	AE002162	7,501	Read, Brunham et al. (2000)
Chlamydia trachomatis serovar D	AE001273	1,042,519			Stephens, Kalman et al. (1998)
Chlamydophila pneumoniae J138	BA000008	1,226,565			Shirai, Hirakawa et al. (2000)
Clostridium acetobutylicum	AE001437	3,940,880	NC_001988	192,000	Nolling, Breton et al. (2001)

(Continued)

TABLE 1.1. *Continued*

Species	Main Chromosome		Extrachromosomal Elements		References
	Accession Number	Size (bp)	Accession Number	Size (bp)	
Clostridium perfringens	BA000016	3,031,430			Shimizu, Ohtani et al. (2002)
Corynebacterium glutamicum	AX114121	3,309,400			Tauch, Homann et al. (2002)
Deinococcus radiodurans R1 (2 chromosomes)	AE000513	2,648,638	AE001826	177,466	White, Eisen et al. (1999)
	AE001825	412,348	AE001827	45,704	
Escherichia coli K-12	U00096	4,639,221			Blattner, Plunkett et al. (1997)
Escherichia coli O157:H7 EDL933	AE005174	5,528,970			Perna, Plunkett et al. (2001)
Escherichia coli O157:H7 Sakai	BA000007	5,498,450	AB011549	92,721	Hayashi, Makino et al. (2001)
			AB011548	3,306	
Fusobacterium nucleatum subsp.nucleatum ATCC25586	AE009951	2,174,500			Kapatral, Anderson et al. (2002)
Haemophilus influenzae Rd	L42023	1,830,138			Fleischmann, Adams et al. (1995)
Helicobacter pylori 26695	AE000511	1,667,867			Tomb, White et al. (1997)
Helicobacter pylori J99	AE001439	1,643,831			Alm, Ling et al. (1999)
Lactococcus lactis subsp. *lactis*	AE005176	2,365,589			Bolotin, Wincker et al. (2001)
Listeria innocua	AL592022	3,011,208	AL592102	81,900	Glaser, Frangeul et al. (2001)
Listeria monocytogenes EGD-e	NC_003210	2,944,528			Glaser, Frangeul et al. (2001)
Mesorhizobium loti	BA000012	7,036,074	AP003015–16	351,911	Kaneko, Nakamura et al. (2000)
			AP003017	208,315	
Mycobacterium leprae	AL450380	3,268,203			Cole, Eiglmeier et al. (2001)
Mycobacterium tuberculosis	AL123456	4,411,529			Cole, Brosch et al. (1998)
Mycoplasma genitalium	L43967	580,074			Fraser, Gocayne et al. (1995)
Mycoplasma pneumoniae	U00089	816,394			Himmelreich, Hilbert et al. (1996)
Mycoplasma pulmonis	AL445566	963,879			Chambaud, Heilig et al. (2001)
Neisseria meningitidis MC58	AE002098	2,272,351			Tettelin, Saunders et al. (2000)
Neisseria meningitidis Z2491	AL157959	2,184,406			Parkhill, Achtman et al. (2000)
Nostoc sp. PCC 7120	NC_003272	6,413,771	NC_003240	186,614	Kaneko, Nakamura et al. (2001)
			NC_003267	101,965	
			NC_003273	55,414	
			NC_003270	40,340	
			NC_003241	5,584	

Organism	Accession	Size (bp)	Reference
Pasteurella multocida	AE004439	2,257,487	May, Zhang et al. (2001)
Porphyromonas gingivalis	NC_002950	2,343,478	Unpublished
Pseudomonas aeruginosa PA01	AE004091	6,264,403	Stover, Pham et al. (2000)
Ralstonia solanacearum	AL646052 AL646053	3,716,413 2,094,509	Salanoubat, Genin et al. (2002)
Rickettsia conorii	AE006914	1,268,755	Ogata, Audic et al. (2001)
Rickettsia prowazekii	AJ235269	1,111,523	Andersson, Zomorodipour et al. (1998)
Salmonella enterica subsp. *enterica* serovar Typhi	AL513382 AL513383 AL513384	4,809,037 218,150 106,516	Parkhill, Dougan et al. (2001)
Salmonella enterica serovar *typhimurium* LT2	AE006468 NC_003277	4,857,432 93,939	McClelland, Sanderson et al. (2001)
Sinorhizobium meliloti	AL591688 AE006469 AL911985	3,654,135 1,354,226 1,683,333	Galibert, Finan et al. (2001)
Staphylococcus aureus	BA000017	2,878,040	Kuroda, Ohta et al. (2001)
Streptococcus pneumoniae R6	AE007317	2,038,615	Hoskins, Alborn et al. (2001)
Streptococcus pneumoniae TIGR4	AE005672	2,160,837	Tettelin, Nelson et al. (2001)
Streptococcus pyogenes	AE004092	1,852,441	Ferretti, McShan et al. (2001)
Streptomyces coelicolor A3(2)	AL645882 NC_003903 NC_003904	8,667,507 356,023 31,317	Bentley, Chater et al. (2002)
Synechocystis PCC6803	AB001339	3,573,470	Kaneko, Sato et al. (1996)
Thermoanaerobacter tengcongensis	AE008691	2,689,445	Bao, Tian et al. (2002)
Thermotoga maritima	AE000512	1,860,725	Nelson, Clayton et al. (1999)
Treponema pallidum	AE000520	1,138,011	Fraser, Norris et al. (1998)
Ureaplasma urealyticum	AF222894	751,719	Glass, Lefkowitz et al. (2000)
Vibrio cholerae (2 chromosomes)	AE003852 AE003853	2,961,151 1,072,914	Heidelberg, Eisen et al. (2000)
Xanthomonas axonopodis pv.citri str. 306	NC_003919		da Silva, Ferro et al. (2002)
Xanthomonas campestris pv.campestris str.ATCC33913	NC_003902		da Silva, Ferro et al. (2002)
Xylella fastidiosa	AE003849 AE003851 AE003850	2,679,305 51,158 1,285	Simpson, Reinach et al. (2000)
Yersinia pestis	AL590842 NC_003132 NC_003131 NC_003134	4,653,728 9,612 70,305 96,210	Parkhill, Wren et al. (2001)

[a] Linear chromosome.

We include some basic knowledge of prokaryotes, such as morphology, classification, and main features regarding the organization, replication, and expression of genetic material. Such descriptions are far from exhaustive. Since we focus our attention on the main aspects that emerged from knowledge of complete genome sequencing, we ask the reader to refer to more in-depth studies by specialists in the field and to the numerous reviews and papers available in the literature.

1.2   MORPHOLOGY AND CLASSIFICATION

Prokaryotes are unicellular organisms and are the most numerous organisms on Earth (4 to 6×10^{30} cells, 3 to 5×10^{17} g of C; Whitman, Coleman et al. 1998); commonly known as bacteria, they include both Archaea and Bacteria. Their morphology is quite simple: the prokaryotic cell (ranging from 0.2 to 10 μm in diameter) can be considered as one unit, a single compartment with no membrane-bound organelles inside.

Typically, prokaryotes have a cell wall containing peptidoglycan (except mycoplasmas and Archaea) which surrounds the cell and confers rigidity and protection. The way that peptidoglycan is arranged is the basis for the identification of bacterial organisms; two distinct cell wall types are revealed by a commonly used staining procedure, the Gram stain (see below), according to the relative content of peptidoglycan. Many pathogenic bacteria also have capsules, structures made of polysaccharides or proteins that are external to the cell wall. Such coatings are useful for both adhesion and resistance to host immune response.

The cytoplasm of prokaryotic cells is enclosed in the plasma membrane, a phospholipid bilayer, whose function is not only to control what enters and leaves the cell but also to provide a site for protein attachment and enzyme activity. Indeed, the size limit of prokaryotes is greatly influenced by the surface-to-volume ratio, since there are no internal membranes. The cytoplasm is diffuse and granular, due to the presence of many ribosomes—sites for protein synthesis, smaller than those of eukaryotes. Very often, there are inclusion bodies whose function is material storage. The genetic material is confined to the nucleoid, a region not delimited by a membrane, but visibly distinct from the rest of the cell by electron transmission microscopy. Sometimes, one or more additional circular DNA molecules (plasmids) are also present.

Even though only about 4000 species are described, it is estimated that the true number could range between 400,000 and 4,000,000. It is evident that a definition of *species* in small unicellular organisms such as prokaryotes is not easy. In eukaryotes, individuals belong to the same species if they are capable of fertile interbreeding, share specific morphological traits, and form a monophyletic group. In prokaryotes, instead, the classic species concept, based primarily on morphological traits, cannot be used reliably (Rossello-Mora and Amann 2001). For this reason, novel approaches have been adopted for microbial classification based on a variety of evidence, including chemotaxonomic markers (e.g., cell wall, polyamines, quinones, etc.), DNA properties (e.g., G + C content, extent of DNA hybridization, etc.), and rRNA sequences. In particular, classification based on comparative analysis of rRNA sequences, mostly 16S rRNA, is the one used most at present. Furthermore, in Bacteria there are three taxonomic ranks below the species: (1) the type/group, a group of isolates or strains sharing a single characteristic (e.g., *Chlamydia trachomatis* serovar D); (2) the isolated pure culture, a clonal population from a single cell or isolate strain having a known set of characteristics (e.g., *Neisseria meningiditis* MC58); and (3) the single cell, that is, a single individual.

In general, two main types of classification can be adopted: the phylogenetic classification, grouping Bacteria according to their evolutionary relationships; and the phenetic classification, based on similarity in bacterial features, without considering origin or evolution. The main characteristics that define the major bacterial groups are the nature of the cell wall (gram-positive, gram-negative, no cell wall), cell shape (cocci, rods, helical), physiology (aerobes, thermophiles, chemiolitotrophs, intracellular parasites, etc.), and motility (presence/absence of flagella, corkscrew motion).

There are many different classifications of Bacteria, based on one or more than one of the above-mentioned features. No official classification is available, but bacteriologists have fixed some rules for naming new and old species of bacteria. These rules are collected in the *International Code of Nomenclature of Bacteria (Bacteriological Code)*. The 1980 Approved Lists of Bacterial Names [see Pittman, Walczak et al. (1991) for an update] contained 2212 names of genera, species, or subspecies, and 124 names of higher taxa. A continuously updated list of validly published bacterial names, including more than 5500 taxa, can be found at the LBSN Web site (see the URL in the Appendix; Euzeby 1997).

Following a classic taxonomy scheme the bacterial species can be grouped into four divisions:

1. *Firmicutes* (Gibbons and Murray 1978; see also Pittman, Walczak et al. 1991) are gram-positive Bacteria with thick cell walls.
2. *Gracilicutes* (Gibbons and Murray 1978; see also Pittman, Walczak et al. 1991) are gram-negative Bacteria with thin cell walls.
3. *Mendosicutes* (Murray 1984; see also Woese and Fox 1977) enclose the single class Archaeobacteria, which represents the Archaea domain in phylogenetic classification (Woese, Kandler et al. 1990).
4. *Tenericutes* (Murray 1984) comprise bacterial species lacking the cell wall.

A remarkable breakthrough in the classification of bacteria was achieved with the advent of molecular data, particularly the sequence analysis of 16S rRNA. The classification based on molecular features divides prokaryotes into two domains, Bacteria and Archaea. This classification dates back to as early as the mid-1970s, when Woese and Fox (1977), using small subunit ribosomal RNA (SSU rRNA) data, first described a number of features characterizing a distinct group of unicellular organisms, the Archaebacteria renamed Archaea in 1990 by Woese, Kandler et al. (1990), thus abandoning the classical bipartite division of living organisms into prokaryotes and eukaryotes. Ever since, the classical *tree of life* is usually illustrated with three main branches: Bacteria, Archaea, and Eukarya. There are, however, recent conflicting claims regarding this classification, and several fundamental questions regarding the time and mode of evolution, as well as the phylogeny of the three domains, are discussed in Sec. 8.5.

Although the rRNA-based classification of prokaryotes may have a limited resolving power in some cases and can be questioned for the peculiar evolutionary dinamics of rRNA genes (see Sec. 8.1), it remains a stable and operationally satisfactory framework for prokaryotic classification. Furthermore, recent evolutionary analyses based on differences in gene content (Snel, Bork et al. 1999) or on protein sequence comparison (Brown, Douady et al. 2001; Brochier, Bapteste et al. 2002) were remarkably consistent with rRNA trees. This indicates that processes of lateral gene transfer (see Sec. 7.4), although frequent in bacterial evolution, have not completely erased the phylogenetic signal.

rRNA-derived phylogenetic trees as well as aligned and annotated rRNA sequences are provided by the Ribosomal Database Project (RDP; see the URL in the Appendix). The availability of completely sequenced genomes represents a powerful tool for shedding light on these fundamental questions and will provide the opportunity not only to better understand prokaryotic organisms at the molecular level, but also to discover new and unexpected features of their evolution. Figure 1.1 shows the classification of bacterial species whose complete genome has been fully sequenced, provided by the NCBI Taxonomy Browser (see the URL in Table 5.1).

In this chapter we describe the main features of the prokaryotic genome, treating the two domains, Archaea and Bacteria, separately whenever possible.

1.3 GENOME SHAPE AND SIZE

Until recently, bacterial genomes were believed always to be circular. If this is true for most bacteria, there are several species whose chromosomes are linear; actually, there may be a natural interchange between the linear and the circular geometry (Volff and Altenbuchner 1998), such as in *Streptomyces*. In the list of completely sequenced organisms reported in Table 1.1, only one species, *Borrelia burgdorferi*, has a linear chromosome.

Telomeres in linear replicons have been described in only a few species. In *Borrelia*, telomeres are covalently closed hairpins, where one DNA strand loops around, becoming the other (Hinnebusch, Bergstrom et al. 1990; Hinnebusch and Barbour 1991; Casjens, Murphy et al. 1997). In *Streptomyces*, instead, the chromosome is open and ends with specific proteins covalently bound to the 5' ends of the DNA (Sakaguchi 1990; Chen 1996). Linear replicons have been described on several branches of the bacterial phylogenetic tree, which would suggest that linearity arose more than once from circular progenitors (Casjens 1998).

The presence of a single large chromosome in prokaryotes also appears to be a general rule, although several bacterial genomes contain two or three large replicons (chromosomes) several hundred kilobase pairs long. This feature is a stable property of some genera [e.g., *Borrelia* (Barbour 1988) and *Rhizobium* (Honeycutt, McClelland et al. 1993)], which suggests these replicons are essential for their lifestyles. In addition, extrachromosomal DNA elements are present in many species.

The haploidy of bacteria is indeed an oversimplification (Casjens 1998). Many fast-growing species (such as *Azotobacter vinelandii* and *Borrelia hermsii*) contain more than one complete chromosome copy per cell during the exponential growth phase; *Deinococcus radiodurans* also has four copies of its replicon in the stationary phase. *M. jannaschi* has one to five chromosome copies per cell during the stationary growth phase but can harbor more than 10 copies in the exponential

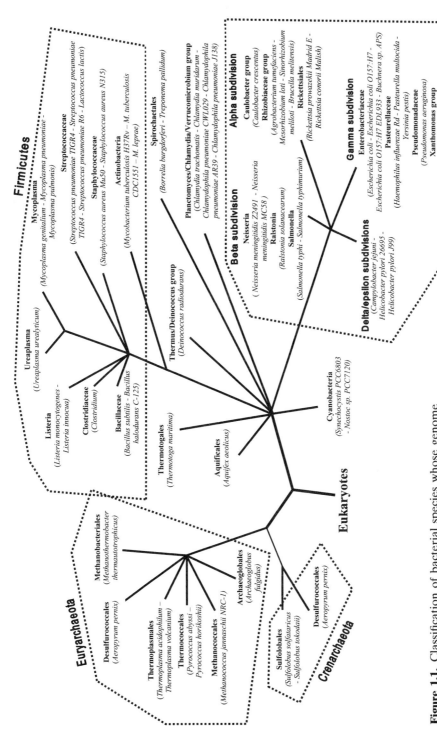

Figure 1.1. Classification of bacterial species whose genome has been sequenced completely according to the NCBI Taxonomy Browser (data updated June 2002).

11

phase, in contrast with what occurs in other archaeal species, such as *Sulpholobus* (Bernander 2000). As far as the Archaea are concerned, the genomes sequenced to date are circular in shape, and only few species have extrachromosomal elements.

The size of bacterial chromosomes is notably variable even within a genus, as in *Treponema* (from 1040 to 3000 kbp), *Mycoplasma* (from 580 to 1350 kbp), or *Streptomyces* (from 6400 to 8400 kbp) (Casjens 1998). In the bacterial genomes completely sequenced at present, dimensions range from 580,074 bp (*Mycoplasma genitalium*) to 7,036,074 bp (*Mesorhizobium loti*) (see Table 1.1). However, bacterial genomes can be much larger: restriction analyses have shown that the largest known genome is *Myxococcus xanthus* (9200 kbp; He, Chen et al. 1994). Most sequenced bacterial genomes have genome size around 1 to 1.8 Mbp; the largest, beside *M. loti*, being, *Pseudomonas aeruginosa* (6,264,403 bp), *Escherichia coli*, 4,639,221 bp; *Bacillus subtilis*, 4,214,814 bp; *Mycobacterium tubercolosis*, 4,411,529 bp; *Synechocystis* sp., 3,573,470 bp; *Deinococcus radiodurans*, 2,648,638 bp; and *Neisseria meningiditis* 2,272,351 bp.

Figure 1.2 shows the distribution of genome sizes for Bacteria and Archaea completely sequenced so far, whose averages are 2.9 and 2.3 Mbp, respectively. Quite interestingly, it can be noticed that two peaks can be observed for eubacterial genome sizes, with the first one at about 1 Mbp, corresponding to *specialist species* (very small genomes, very specific niches), and the second one, around 2–4 Mbp, much more flattened out, for *generalist species* (larger genomes, a wide range of places to live). It can be also noted that Archaea show a narrower genome size distribution than that of Bacteria.

The large variation in chromosome size of bacterial genomes, which is much more evident when comparing higher taxonomic positions, may be due to rapid gene loss when a species chooses to live in a very specific ecological niche or when a gene is

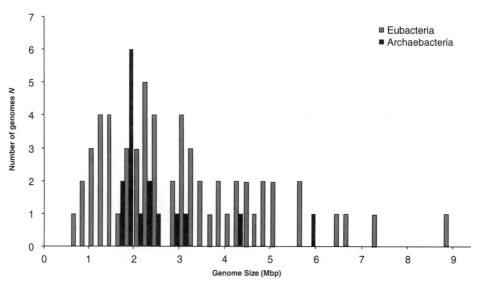

Figure 1.2. Distribution of genome sizes for the 76 bacterial and archeal completely sequenced genomes listed in Table 1.1.

gained via gene duplication or horizontal gene transfer (Casjens 1998). Gene duplication and the presence of repeated elements is undoubtedly at the basis of an increase in genome size; the genomes of *E. coli* and *B. subtilis* contain repetitive elements and cryptic prophage and phage remnants; *M. tubercolosis* contains repetitive DNA, insertion sequences (ISs), and duplicated housekeeping genes. Repeats in bacterial genomes seem to be important for genome plasticity, since they are involved in several processes, such as recombination, inversion, deletion, translocation, and transposition (Romero, Martinez-Salazar et al. 1999). Lateral gene transfer, also from distantly related organisms, is a fundamental process in bacterial genome evolution, since it is responsible for the acquisition and deletion of extensive amounts of DNA in the chromosome, thus producing extremely dynamic genomes (Ochman, Lawrence et al. 2000).

Gene content can be distributed in more than just a single chromosome, and extrachromosomal elements contribute greatly to gene equipment. *Deinococcus radiodurans* contains four genetic elements: two main circular chromosomes (total length: 3,060,986 bp), the megaplasmid, and the plasmid (total extrachromosomal length: 223,170 bp), which together amount to more than 3.2 Mbp. *Borrelia burgdoferi* harbors 17 linear and circular plasmids, whose combined size is about 533,000 bp, more than half the length of the main chromosome (910,725 bp).

An interesting question one might ask regards the minimum size compatible with independent life for a prokaryotic genome. It appears that the smallest genomes, like *Mycoplasma genitalium* (the smallest), *M. pneumoniae*, and *M. pulmonis*, belong to organisms that have a parasitic lifestyle and thus are adapted to specific niches and host–pathogen interactions. Indeed, among factors preventing evolution to a smaller size are the need for an adequate genome and the limited efficiency of the transcriptional and translational machinery (Koch 1996; see also Sec. 7.2).

A remarkable property clearly emerging from prokaryotic genome data is the striking correlation between the fraction of genes coding for regulatory proteins (e.g., 5.8% in *E. coli* and 9.4% in *P. aeruginosa*) and the genome size. This trend is particularly prominent for generalist species that can survive in diverse environments (Stover, Pham et al. 2000).

An interesting aspect of genome size is its content in coding versus noncoding regions. It should be recalled that most protein genes have only been predicted by the methods described in Sec. 6.8.7, and thus they should be regarded as hypothetical, especially those with no homologs in other species. In Bacteria, on average 85 ± 4.7% of the genome encodes for proteins (see Table 1.2), although a conspicuous part (on average 37.2 ± 10.6%) of the coding portion consists of unidentified open-reading frames. Although the functions performed by these unidentified proteins are unknown to date, they are most likely to represent specific and necessary products for the organism.

In some bacterial species, a process of reductive evolution of the number of genes in the genome has been documented. In obligate intracellular parasites such as *Rickettsia* and *Chlamydia* and some endosymbionts (Andersson and Andersson 1999), genes become inactive when their function is no longer required by the organism. To date, the most extensive genome degradation has been reported for the leprosy bacillus *Mycobacterium leprae*, in which only 49.5% of the genome codes for proteins, with over 1000 pseudogenes (27% of the potential coding capacity; Cole, Eiglmeier et al. 2001).

TABLE 1.2. Content of Protein Coding Genes in Completely Sequenced Microbial Genomes

Species	ORFs	Percent Functionally Identified	Protein Coding Regions (%)	Operons[a]	Lateral Transfer[a]
Archaea					
Aeropyrum pernix	2694	23.5	88.8		
Archaeoglobus fulgidus	2436	46.0	92.2	+	−
Halobacterium sp. NRC-1 (3 replicons)					
Methanobacterium thermoautotrophicum	1855	46.0	92.0	+	+
Methanococcus jannaschii	1743	36.0		+	+
Pyrococcus abyssi	1765				
P. horikoshii	2061	35.0	90.7		
Sulfolobus solfataricus	2977	32.2	83.9		
S. tokodaii	2826			+	−
Thermoplasma acidophilum	1509	55.0	87.0	+	+
T. volcanium					
Bacteria					
Aquifex aeolicus	1512		93.0	+	+
Bacillus halodurans	4066	52.7	85.0	+	+
B. subtilis	4100	58.0	87.0	+	+
Borrelia burgdorferi	853	59.0	93.0	+	−
Buchnera sp. APS	583	85.8	88.0	+	
Campylobacter jejuni	1654	77.8	94.3	− (few)	
Caulobacter crescentus	3767	53.9	90.6		
Chlamydia pneumoniae AR39	1052	60.0		+	−
C. pneumoniae CWL029	1052	60.0		+	−
C. trachomatis MoPn	924				+
C. trachomatis serovar D	894	68.0			+
Chlamydophila pneumoniae J138	1072				
Clostridium acetobutylicum	3740				
Deinococcus radiodurans R1 (2 chromosomes)	2633 / 369	69.0		+	+
Escherichia coli K-12	4288	62.0	87.8	+	
E. coli O157:H7 EDL933	5349			+	+
E. coli O157:H7 Sakai	5361		88.1	+	+

Haemophilus influenzae Rd	1743	57.7	85.0		+
Helicobacter pylori 26695	1590	68.6	91.0	+	+
H. pylori J99	1495	59.9	90.8	+	+
Lactococcus lactis subsp. lactis	2310	64.2	90.8	+	+
Listeria innocua	2973	54.0	86.0	+	+
L. monocytogenes EGD-e	2853		90.3	+	+
Mesorhizobium loti	6752		90.3		+
Mycobacterium leprae	1604	42.0	49.5		+
M. tuberculosis	3924	40.0	91.0	+	–
Mycoplasma genitalium	470	79.6	88.0	– (few)	?
M. pneumoniae	677	75.9	88.7	+	
M. pulmonis	782	62.0	91.4		+
Neisseria meningitidis MC58	2158	53.0	83.0	+	
N. meningitidis Z2491	2121		82.9		
Nostoc sp. PCC7120	5366				
Pasteurella multocida	2014				
Porphyromonas gingivalis W83					
Pseudomonas aeruginosa PA01	5570	54.2	89.4		+
Rickettsia conorii	1374		81.0	+	+
R. prowazekii	834	62.7	75.4	+	+
Salmonella enterica serovar typhimurium LT2	4330				
S. enterica subsp. enterica serovar typhi	4599		87.6	+	+
Sinorhizobium melitoti	6204	59.7	85.8	+	+
Staphylococcus aureus	2714		89.0		
Streptococcus pneumoniae R6	2043			+	+
S. pneumoniae TIGR4	2236	64.0		+	+
S. pyogenes	1752				+
Synechocystis PCC6803	3168	44.2	87.0	+	+
Thermotoga maritima	1877	54.0	95.0	+	
Treponema pallidum	1041	55.0	92.9	+	–
Ureaplasma urealyticum	613	53.0	88.6		
Vibrio cholerae (2 chromosomes)	2770, 1115	58.0	86.3	+	+
Xylella fastidiosa	2782	42.0	88.0		+
Yersinia pestis	4012	47.0	83.8	+	+

[a] +, Occurrence; –, absence.

A comparison of the two completely sequenced *E. coli* strains [i.e., the laboratory strain K-12 and the enterohemorrhagic strain O157:H7 (EDL933 and Sakai)] shows their complex relationships. They differ by about 860 kbp in length and share a clearly homologous backbone of 4.1 Mbp (with 75,168 scattered single nucleotide polymorphisms). Of the remaining genome, 1.34 Mbp represent islands specific to the pathogenic strain, nine of which encode putative virulence factors; 0.53 Mbp represent islands specific to the laboratory strain. Their atypical base composition suggests that most differences in overall gene content are attributable to horizontal transfer of relatively recent origin (Perna, Plunkett et al. 2001).

Noncoding regions represent the most variable part of bacterial genomes and they contain mostly repeated elements whose origin may be tracked in some cases. Prokaryotes, though to a lesser extent than eukaryotes, contain repeated sequences organized both in tandem and interspersed in the genome. The function of such repeated sequences is still largely unknown, yet data seem to suggest that they may represent multiple regulatory signals or serve other biological purposes mostly related with pathogenicity (van Belkum, Scherer et al. 1998; van Belkum 1999).

Tandem repeated elements found in microbial genomes might be prone to the same genetic variability encountered in the eukaryotic genome and could be used profitably as markers for identification and genotyping of bacterial strains. A simple way to detect tandem repeats in genomic sequences is the application of linguistic methodologies, in particular those that measure the linguistic complexity (LC) of DNA sequences (see Sec. 6.8.2). The LC profile along a genome, which can be computed by several methods, clearly highlights genome regions containing tandem repeats (see also Fig. 6.25), which correspond to local LC minima. Other bioinformatic approaches can be used, such as the dot-matrix plot (see Sec. 6.2) or algorithms specifically devoted to the search for tandem repeats (see Sec. 6.8.3).

Interspersed repeat elements have been identified in numerous bacterial species (Lupski and Weinstock 1992). Most of these elements are shorter than 200 bp, evenly distributed in the genome, and located primarily in noncoding regions. For example, REP (repetitive extragenic palindrome) and ERIC (enterobacterial repetitive intergenic consensus) sequence motifs have been found to be widespread in numerous enterobacterial genomes. The identification of interspersed repeats in genomic sequences can be accomplished either by similarity searches against collections of known repeat elements or by the application of bioinformatic tools specifically designed for such a task (see Sec. 6.8.3).

1.4 GENE CONTENT AND ORGANIZATION

Gene identification is one of the first steps in the analysis of microbial genomes and is usually performed with computer-aided methods that employ statistical gene prediction models. Microbial genomes tend to be gene-rich, typically containing about 85% coding sequences; thus gene discovery is much easier than in eukaryotic genomes, especially in higher eukaryotes such as humans whose genome has less than 2% coding sequences. In addition, unlike eukaryotic genes, microbial genes are not interrupted by introns. A survey of all entirely sequenced genomes has revealed the complete absence of spliceosomal introns and genes coding for components of the spliceosomal machinery from a wide range of bacterial and archaeal species (Logsdon 1998).

In microbial genomes, the identification of a significantly long ORF gives quite reliable proof of protein-coding regions. However, the most reliable way to identify a gene in a new genome is to find a close homolog from another organism. This can be done very effectively using database searching programs such as BLAST and FASTA (see Sec. 6.4 for details on these programs). If significant similarity is not found with known proteins, the search for protein signatures specific for one or more protein families can be carried out by comparing the unknown protein against databases of protein signatures such as ProSite (Hofmann, Bucher et al. 1999), ProDom (Corpet, Servant et al. 2000), and Pfam (Bateman, Birney et al. 2000) (see also Secs. 5.6 and 6.9). Unfortunately, many of the genes in newly sequenced genomes have no significant similarities to known genes or domains. For these genes we must rely on computational methods for their identification. Several algorithms have been devised that provide a very high prediction accuracy in microbial genomes. They are generally based on the observation of base compositional heterogeneity at the three codon positions of the predicted gene or on the recognition of a specific codon use strategy. These methods, whose detailed description can be found in Sec. 6.8.7, may reach over 97% prediction accuracy.

In discriminating between coding and noncoding regions, the analysis of the pattern of linguistic complexity (LC; see Sec. 6.8.2) may also reveal useful, since noncoding regions are, overall, less complex than coding ones. Furthermore, LC pattern analysis can help detect potential regulatory sites when typical LC patterns are displayed. Figure 1.3 shows the number of genes in the available microbial genomes that hit any combination of NCBI protein families (cluster of orthologous

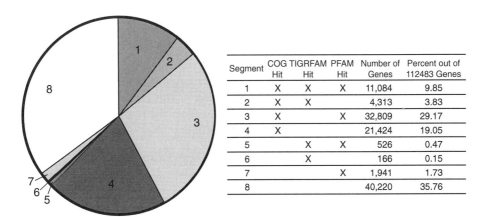

Segment	COG Hit	TIGRFAM Hit	PFAM Hit	Number of Genes	Percent out of 112483 Genes
1	X	X	X	11,084	9.85
2	X	X		4,313	3.83
3	X		X	32,809	29.17
4	X			21,424	19.05
5		X	X	526	0.47
6		X		166	0.15
7			X	1,941	1.73
8				40,220	35.76

Figure 1.3. Number of protein-coding genes in the complete microbial genomes (a total of 112,483 based on the TIGR-assigned annotation) showing hits (marked with ×) with any combination of the NCBI protein families (cluster of orthologous genes, COGs), the TIGR protein families (TIGRfam), and the Pfam domain database. The original color code is rendered here as a gray scale. (Data from the TIGR Comprehensive Microbial Resource, July 2001; see the URL in the Appendix.)

groups of proteins, COGs), the TIGR protein families (TIGRfam), or Pfam. It is striking to note that more than 35% of proteins predicted are still completely unknown, as no hit has been found to known proteins or protein domains. Microbial genes identified so far in completely sequenced genomes range from a minimum of 470 genes in *Mycoplasma genitalium* to a maximum of 6752 genes in *Mesorhizobium loti*. Figure 1.4 shows the distribution of genes in the diverse functional families.

Gene number seems to reflect bacterial lifestyle. Specialized parasites (e.g., *Mycoplasma*) have about 500 to 600 genes, while *Myxococcus xanthus*, which has a complex life cycle (with a sporulation phase, etc.) encodes about 10,000 proteins (Casjens 1998; Shimkets 1998). In *M. loti, P. aeruginosa, E. coli,* and *B. subtilis*, the number of genes exceed 4000, to cover more than 87% of the genome. In the latter cases the size of the genome reflects evolutionary events and certainly cannot be considered as the minimum size for independent life.

Another unexpected property is the extreme variability not only in gene content but also in gene order, even between closely related species and among independent isolated genomes of the same species (e.g., *Salmonella typhi*; Liu and Sanderson 1995).

Archaeal genomes have a coding capacity of about 90%, and unidentified proteins account for about 40% of the ORFs predicted, whose gene number ranges between 1700 and 2900 (see Table 1.2). A gene number around 2000 already exceeds what is necessary for a lithoautotrophic lifestyle.

In Archaea, duplicated regions are generally present even in the smallest organisms, such as *Methanobacterium thermoautotrophicum*, and it has been suggested that gene duplication could provide metabolic flexibility. Lateral transfer has been detected in the two methanogenic species (*M. thermoautotrophicum* and *Methanococcus jannaschi*) and in *Thermoplasma acidophilum*. Table 1.2 reports the number of open-reading frames (ORFs) for each completely sequenced genome, the percentage of proteins identified and of protein-coding regions, the presence or absence of lateral transfer events, and the degree of conservation of operonal gene organization (the latter topic is discussed below).

Comparative analysis based on data available for completely sequenced prokaryotic genomes has revealed a picture of generally well conserved protein sequences as opposed to the very scarce conservation of gene organization. Indeed, the structure of bacterial genomes has a high level of plasticity. This can be appreciated from the great variation in shape and size of genomes, from the number of replicons in the various organisms, and to a greater extent, from the different genome structures found even between closely related species.

Jacob and Monod (1961) were the first scientists to study a transcriptionally regulated system, the lactose metabolism system in *E. coli*. They discovered that to maximize gene regulation efficiency in prokaryotes, the enzymes for a particular metabolic pathway are often grouped in a cluster that is transcribed into a polycistronic mRNA from a single promoter sequence. This unit of bacterial gene expression and regulation, which includes structural genes and their control elements, is called an *operon*. An operon is made up of several elements: an *operator*, which is the binding site for repressor molecules; a *promoter*, which contains the binding site for RNA polymerase; and a *repressor*, a gene encoding a protein that binds to DNA at the operator and blocks the binding of RNA polymerase at the promoter.

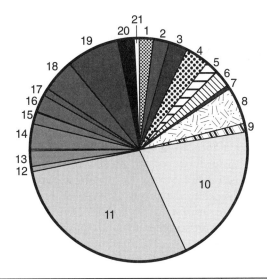

Segment	Gene Role	Number of Genes	Percent out of 112,483 Genes
1	Amino acid biosynthesis	2,398	2.13
2	Biosynthesis of cofactors, prosthetic groups, and carriers	2,444	2.17
3	Cell envelope	3,107	2.76
4	Cellular processes	3,707	3.29
5	Central intermediary metabolism	2,100	1.86
6	DNA metabolism	2,769	2.46
7	Disrupted reading frame	107	0.09
8	Energy metabolism	6,909	6.14
9	Fatty acid and phospholipid metabolism	1,614	1.43
10	Hypothetical proteins, conserved	23,544	20.9
11	Hypothetical proteins, not conserved	31,941	28.3
12	Other categories	999	0.88
13	Protein fate	2,763	2.45
14	Protein synthesis	4,670	4.15
15	Purines, pyrimidines, nucleosides, and nucleotides	1,744	1.55
16	Regulatory functions	3,301	2.93
17	Transcription	1,191	1.05
18	Transport and binding proteins	5,502	4.89
19	Unclassified	8,365	7.43
20	Unknown function	2,986	2.65
21	Viral functions	322	0.28

Figure 1.4. Distribution of genes from complete microbial genomes in the various functional categories. The original color code is rendered here as a gray scale. (Data from the TIGR Comprehensive Microbial Resource, July 2001; see the URL in the Appendix.)

At present, gene order appears not conserved over long evolutionary time, except for some essential operons, such as ribosomal protein operons (Mushegian and Koonin 1996; Koonin and Galperin 1997). There are several examples of differences in genome structure between closely related species. Rearrangement of gene order is frequently observed and can be due to deletions and insertions. Clearcut examples are *Mycoplasma genitalium* and *M. pneumoniae*, where seven blocks with the same gene order can be recognized, but they are shuffled in the two genomes. More intriguing still, differences occur in comparing two isolates of the same species; the genome of *Helicobacter pylori* strain J99 is about 24,000 bp smaller than that of strain 26695, and 6 to 7% of the genes are specific to each strain (Alm, Ling et al. 1999). These "unique" genes are mostly located in a *plasticity zone* (probably a pathogenicity island). The same occurs in *Salmonella typhi* isolates, which show extensive chromosomal rearrangements. This lack of syntheny can be due to homologous recombination events or to other dynamic processes whose nature remains to be understood.

Because of this fluidity, it is very difficult to establish general rules in bacterial genome organization. In many cases, genes belonging to operons in some genomes are found scattered in other genomes. This observation, although needing experimental validation, may suggest that the operonlike structure is not as diffuse as it was believed to be. However, the general finding of clusters of genes that perform similar or related functions, presumably operons, demonstrates that this is an important advantageous property which also applies in the case of horizontal transfers.

Another common feature is the presence of genes in a higher copy number near the origin of replication, and gene orientation in the same direction as replication (but *E. coli* is an exception), to minimize a head-on collision between transcription and replication (Casjens 1998).

1.5   BASE COMPOSITION

The base composition of prokaryotic genomes is extremely variable. In Bacteria the mean guanine and cytosine (GC) content of the genome ranges from 67% in *Deinococcus radiodurans* to 25.5% in *Ureaplasma urealyticum*. In Archaea such a range is closer, from 56.2% in *Aeropyrum pernix* to 31.3% in *Methanococcus jannaschi* (Table 1.3). Figure 1.5 shows the %GC distribution of completely sequenced genomes listed in Table 1.3.

Average base composition in protein coding genes is not very interesting per se, except in extremely biased cases, since owing to the peculiar properties of the genetic code (i.e., degeneracy) it should not affect expression patterns. In this respect, Dayhoff (1978) has shown that, in general, average proteins can be made by RNAs having the most variable base compositions. In this respect, Lobry (1997) has shown that the average amino acid composition of bacterial proteins is greatly influenced by genomic G + C content. However, this influence was often found to be lower than expected, assuming neutrality in the evolutionary pattern.

The average value of base composition as well as its variation along the genome is an extremely interesting feature that can shed light on the evolutionary history of a given organism and also be used in applications of genetic engineering. It is generally believed that the mean GC content is related to bacterial phylogeny

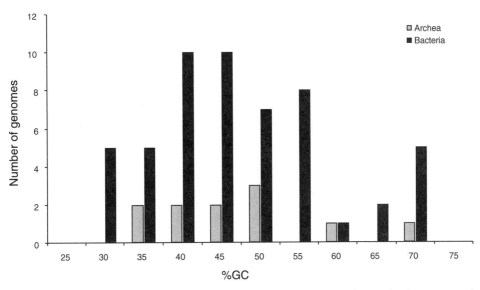

Figure 1.5. Distribution of %GC for the 64 bacterial and archeal completely sequenced genomes listed in Table 1.1.

(Muto and Osawa 1987) as supported by the phylogenetic tree based on 5S rRNA sequences (Hori and Osawa 1986). Among gram-positive bacteria, those with high or low genomic GC content cluster together. Other authors have also suggested the existence of a biased mutation pressure, called *AT/GC pressure* by Muto and Osawa (1987), which causes differences in the genomic GC content among different lineages. This AT/GC pressure could have played a major role in diversification of bacterial genomic sequences and codon use during evolution.

Completely sequenced genomes reveal that there is very often a nonhomogeneous pattern of base composition along the genome which may reflect important aspects of genomic organization: (1) formation of compartments (i.e., AT- or GC-rich traits of the genome, due to neutral or selective events; see below), (2) acquisition of new material from outside (e.g., horizontal genes transfer), and (3) insertion/deletion events. In all cases, this behavior interferes with the normal divergent pattern of vertical evolution and complicates the phylogenetic analysis of organisms and, even more so, the measurements of their genetic distances (see Sec. 7.3).

The presence of distinct regions having different %GC values has been described in several microbial genomes, such as the archaeon *Archaeoglobus fulgidus* (five regions) and in the bacterial species *Helicobacter pylori* (strain 26695: five regions; strain J99: nine regions; see Table 1.3). In some species, a region with GC content higher than in the remaining genome has been found: for example, in the rRNA operons of *Aquifex aeolicus* and *Mycoplasma genitalium*. Other organisms have a lower GC value in some regions: for example, in genes for polysaccharide production of the archaeon *Methanobacterium jannaschi* [probably a mark of a lateral transfer event (Bult, White et al. 1996)], and in *Bacillus subtilis*, where many (A + T)-rich islands are present, probably a remnant of bacteriophage lysogens or other inserted elements (Kunst, Ogasawara et al. 1997). Figure 1.6 shows the %GC

TABLE 1.3. Compositional Features of Completely Sequenced Microbial Genomes

Species	Nucleotide Composition						
	%A	%T	%C	%G	GC Skew	AT Skew	%G + C
Archaea							
Aeropyrum pernix	21.50	22.10	28.30	27.90	-0.0071	-0.0138	56.2
Archaeoglobus fulgidus	25.80	25.60	24.20	24.30	0.0021	0.0039	48.5
Halobacterium sp. NRC-1 (3 replicons)	16.00	16.00	34.00	33.90	-0.0015	0.0000	67.9
Methanobacterium thermoautotrophicum	25.00	25.30	24.70	24.80	0.0020	-0.0060	49.5
Methanococcus jannaschii	34.40	34.10	15.50	15.80	0.0096	0.0044	31.3
Pyrococcus abyssi	27.50	27.70	22.40	22.20	-0.0045	-0.0036	44.6
P. horikoshii	28.90	29.10	21.20	20.60	-0.0144	-0.0034	41.8
Sulfolobus solfataricus	31.90	32.20	17.80	17.90	0.0028	-0.0047	35.7
Thermoplasma acidophilum	27.10	26.80	22.90	23.00	0.0022	0.0056	45.9
T. volcanium	30.10	29.90	19.90	20.00	0.0025	0.0033	39.9
Sulfolobus tokodaii	33.43	33.78	16.29	16.50	0.0066	-0.0052	32.8
Bacteria							
Aquifex aeolicus	28.40	28.10	21.60	21.70	0.0023	0.0053	43.3
Bacillus halodurans	28.20	28.00	21.60	22.00	0.0092	0.0036	43.6
B. subtilis	28.10	28.30	21.80	21.70	-0.0023	-0.0035	43.5
Borrelia burgdorferi	35.40	35.90	14.32	14.20	-0.0042	-0.0070	28.5
Buchnera sp. APS	37.00	36.60	13.00	13.20	0.0076	0.0054	26.2
Campylobacter jejuni	34.80	34.60	15.30	15.20	-0.0033	0.0029	30.5
Caulobacter crescentus	16.40	16.30	33.60	33.50	-0.0015	0.0031	67.1
Chlamydia pneumoniae AR39	29.50	29.80	20.20	20.30	0.0025	-0.0051	40.5
C. pneumoniae CWL029	29.80	29.50	20.30	20.20	-0.0025	0.0051	40.5
C. trachomatis MoPn	29.80	29.80	20.10	20.20	0.0000	0.0000	40.2
C. trachomatis serovar D	29.40	29.20	20.60	20.60	0.0000	0.0034	41.2
Clamydophila pneumoniae J138	29.86	29.56	20.33	20.26	-0.0017	0.0051	40.6
Clostridium acetobutilicum	34.57	34.51	15.42	15.50	0.0026	0.0009	30.9
Deinococcus radiodurans R1 (chromosome I)	16.90	16.30	33.30	33.30	0.0000	0.0181	66.6
D. radiodurans R1 (chromosome II)	16.40	16.50	33.50	33.40	-0.0015	-0.0030	66.9
Escherichia coli K-12	24.60	24.50	25.40	25.30	-0.0020	0.0020	50.7
E. coli O157:H7 EDL933	24.50	24.40	24.90	24.90	0.0000	0.0020	49.8

E. coli O157:H7 Sakai	24.70	24.70	25.20	25.30	0.0020	0.0000	50.5
Haemophilus influenzae Rd	31.00	30.80	19.10	18.90	-0.0053	0.0032	38.0
Helicobacter pylori 26695	30.30	30.80	19.60	19.20	-0.0103	-0.0082	38.8
H. pylori 26696	16.80	16.50	33.50	32.90	-0.0090	0.0090	66.4
H. pylori J99	30.30	30.40	19.60	19.40	-0.0051	-0.0016	39.0
Lactococcus lactis subsp. lactis	32.30	32.20	17.50	17.70	0.0057	0.0016	35.2
Listeria innocua	31.27	31.29	18.87	18.57	-0.0080	-0.0003	37.4
L. monocytogenes	31.05	30.97	19.13	18.85	-0.0074	0.0012	38.0
Mesorhizobium loti	18.60	18.60	31.60	31.10	-0.0080	0.0000	62.7
Mycobacterium leprae	21.00	21.10	28.70	29.00	0.0052	0.0024	57.7
M. tuberculosis	17.10	17.10	32.80	32.70	-0.0015	0.0000	65.5
Mycoplasma genitalium	34.50	33.70	15.70	15.90	0.0063	0.0117	31.6
M. pneumoniae	29.40	30.50	20.00	19.90	-0.0025	-0.0184	39.9
M. pulmonis	36.90	36.30	13.30	13.30	0.0000	0.0082	26.6
Neisseria meningitidis MC58	24.20	24.20	25.50	25.90	0.0078	0.0000	51.4
N. meningitidis Z2491	23.90	24.10	25.80	25.90	0.0019	-0.0042	51.7
Nostoc sp. PCC7120	29.29	29.36	20.64	20.71	0.0016	-0.0013	41.3
Pasteurella multocida	29.80	29.70	19.90	20.40	0.0124	0.0017	40.3
Porphyromonas gingivalis W83	25.80	25.80	24.00	24.20	0.0041	0.0000	48.2
Rickettsia conorii	33.68	33.88	16.12	16.32	0.0063	-0.0030	32.4
R. prowazekii	35.30	35.60	14.30	14.60	0.0104	-0.0042	28.9
Salmonella enterica thyphi	23.92	23.99	26.01	26.09	0.0015	-0.0015	52.1
S. typhirium LT2	23.90	23.88	26.11	26.11	-0.0001	0.0004	52.2
Sinorhizobium meliloti	18.64	18.63	31.49	31.24	-0.0039	0.0001	62.7
Staphylococcus aureus	33.40	33.60	16.30	16.40	0.0031	-0.0030	32.7
Streptococcus pneumoniae	30.18	30.10	19.80	19.92	0.0029	0.0013	39.7
S. pneumoniae TIGR4	30.26	30.04	19.75	19.95	0.0049	0.0036	39.7
S. pyogenes	30.80	30.50	19.00	19.40	0.0104	0.0049	38.4
Synechocystis PCC6803	26.00	26.10	23.80	23.80	0.0000	-0.0019	47.6
Thermotoga maritima	26.90	26.70	22.70	23.40	0.0152	0.0037	46.1
Treponema pallidum	23.50	23.60	26.20	26.50	0.0057	-0.0021	52.7
Ureaplasma urealyticum	37.20	37.20	12.50	12.90	0.0157	0.0000	25.4
Vibrio cholerae (chromosome I)	25.90	26.30	23.70	23.90	0.0042	-0.0077	47.6
V. cholerae (chromosome II)	26.40	26.60	23.20	23.60	0.0085	-0.0038	46.8
Xylella fastidiosa	22.50	24.70	24.90	27.70	0.0532	-0.0466	52.6
Yersinia pestis	26.21	26.16	23.69	23.94	0.0052	0.0009	47.6

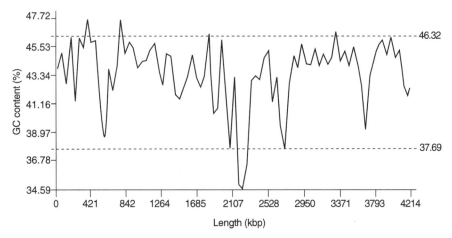

Figure 1.6. %GC content display for the *Bacillus subtilis* 168 DNA molecules generated by the TIGR facility. The genome is represented on the *X*-axis from left to right. Along the *Y*-axis of the plot is the average %GC for a "window" of 50 kbp. The two dashed lines represent the 5% lower limit and the 95% upper limit average GC content for the DNA molecule shown; 90% of the GC content of this genome falls within these two dashed lines.

variation along the *B. subtilis* genome where GC-low and GC-rich regions can be distinguished.

GC variations along the genome could be explained in many different ways, and it should be assumed that the concept of isochores in prokaryotes is probably different from that proposed for eukaryotes; this issue is discussed in Part III. Beside nucleotide frequency variation along the genome, we can have an asymmetric distribution of the two complementary base pairs in DNA strands, called *AT* or *GC skews*. These can be measured in terms of two parameters, defined as AT and GC skew:

$$AT_{skew} = \frac{A-T}{A+T} \qquad GC_{skew} = \frac{G-C}{G+C}$$

where A, C, G, and T are the occurrences of the relevant bases in the sequence under investigation. In the case of intrastrand parity of complementary bases (e.g., A ≈ T and C ≈ G), the skew index is close to zero and indicates the absence of strand compositional asymmetry. On the contrary, if one of the bases significantly outnumbers its complementary base (e.g., A ≫ T), a remarkable positive or negative skew index is observed. In general, the skew index may range between −1 (A = 0 or G = 0) and +1 (T = 0 or C = 0).

Figure 1.7 shows the distribution of global AT and GC skew calculated on microbially complete genomes, listed in Table 1.3. It is evident that both AT and GC skew do not significantly deviate from zero for archaeal and bacterial genomes, thus generally obeying the *second parity rule*, first put forward by Erwin Chargaff (1950), whereby there is intrastrand equivalence for complementary bases (i.e., A = T and G = C). The genome of *Xylella fastidiosa* represents the only notable exception, as a deviating GC and AT skew is observed (Fig. 1.7).

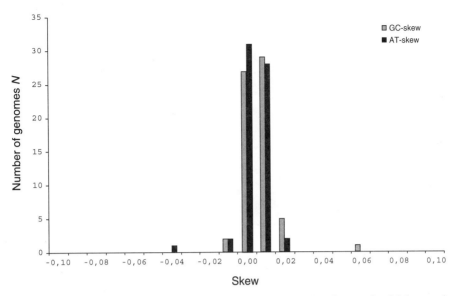

Figure 1.7. Distribution of global AT and GC skew calculated on microbial complete genomes listed in Table 1.3.

On the contrary, traits with positive or negative AT or GC skews have been found along the prokaryotic genomes. This has been correlated with regions involved in processes that generate single-stranded tracts of the genome, such as DNA replication or transcription. Lobry (1996) first demonstrated the existence of an asymmetric substitution patterns in the leading and lagging strands of three bacterial genomes (*E. coli, Bacillus subtilis*, and *Haemophilus influenzae*). This asymmetry divides the bacterial chromosome into two segments (chirocores) whose boundaries coincide with the origins of replication. The author suggests that this asymmetry could be due to a mutational bias (different mutation rates in the two strands as a consequence of asymmetries in replication or repair) rather than to a selective bias. Since then, asymmetries in base composition between the two strands have been observed in many prokaryotic genomes (Blattner, Plunkett et al. 1997; McLean, Wolfe et al. 1998; Mrazek and Karlin 1998), as in Fig. 1.8, where the base skew transition around the replication origin is shown for the *Borrelia burgdoferi* genome. The existence of mutational differences between the leading and lagging strands has been confirmed by Tillier and Collins (2000) to be the cause of base composition skew in bacterial genomes. The authors have demonstrated that the sequence of a gene product could be influenced by whether it is encoded on the leading or lagging strand.

The genome linguistic analysis (see Sec. 6.8) can also represent a powerful tool to investigate general genome properties. The analysis of *w*-mer use in complete bacterial genomes has revealed several peculiar properties. In *E. coli* the most frequent oligomers in the leading strand of replication form a family containing the trimer CTG; the palindromic tetramer CTAG is underrepresented (Karlin, Mrazek et al. 1997), and this could be explained by the hypothesis by some authors that

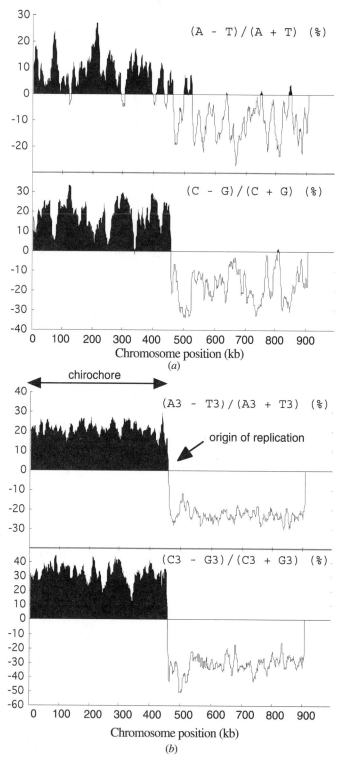

Figure 1.8. Base skew transition around the experimentally mapped replication origin of *Borrelia burgdoferi*: (*a*) genome sequence; (*b*) third codon positions where the chirochore structure is enhanced. Adapted with permission from CNRS UMR 5558, *Biométrie-Génétique et Biologie des Populations: Rapport d'activités, 1994–1998*, Universtité Claude Bernard, Lyon, France, June 1998. *http://biomserv.univ-lyonl.fr/umrfranc/sommaire.html.*

CTAG may "kink" DNA and thereby interfere with function (Medigue, Viari et al. 1991).

In *B. subtilis*, analysis of the abundance of oligonucleotides has revealed the existence of a dinucleotide bias: The most overrepresented ones are AA, TT, and GC, and those less represented are TA, AC, and GT. The distribution of words of 4, 5, and 6 nt seems to be significantly correlated to replication, as several of these words are very significantly overrepresented in one strand and underrepresented in the other (Kunst, Ogasawara et al. 1997).

1.6 CODON USE

Taking into account the genome compositional constraints reported above and the properties of the genetic code, it is conceivable that synonymous codons are not used equally in Bacteria. Until recently it was generally accepted that the most important factors affecting codon use variation in prokaryotic genomes were GC-base compositional bias and selection at the translational level; indeed, these were the criteria allowing discrimination between high- and low-expressed genes (Sharp, Stenico et al. 1993). According to Shpaer (1986) and Gouy (1987) in Bacteria, synonymous codon use may also be influenced by base composition at neighboring sites; these context-dependent codon biases are widespread but not conserved among the different bacterial species (McVean and Hurst 2000).

A compilation of codon use tables for several prokaryotic genomes is given in the CUTG database (Nakamura, Gojobori et al. 2000) and at the TIGR Comprehensive Microbial Resource (CMR; see the URL in the Appendix). Codon use tables can also be obtained by considering subsets of genes belonging to specific functional categories or falling within user-selected coordinates for a chosen chromosome. Figure 1.9 shows the codon use chart for all protein-coding genes of the *E. coli* O157 : H7 genome.

In comparative genome studies, some authors have defined several indexes to investigate codon use strategy (see Sec. 6.8.9). The preference of a specific codon within its family can be expressed by the ratio between the number of its occurrences and the total occurrences of the corresponding amino acid. If no codon preference is exhibited within a family of synonymous codons, the codon use percentage is simply equal to $100/n$, where n is the number of synonymous codons within the family.

The *effective number of codons* used in a gene (ENC index) and the *Codon Adaptation Index* (CAI) are the most used indexes to measure bias in codon use strategy (i.e., how far the codon use of a gene departs from equal use of synonymous codons). The effective number of codons (ENC), proposed by Wright (1990), is calculated through a method used in population genetics to determine the effective number of alleles segregating in a given population. The larger the variety of synonymous codons used by a gene, the larger the ENC, with a minimum expected value of 20 and a maximum of 61; if synonymous codon use in a given gene is random, the ENC value approaches 61 (see also Sec. 6.8.9).

Sharp and Li (1987) proposed the Codon Adaptation Index (CAI) to measure how closely the codon use of a specific gene matches the optimum codon use strategy in a set of reference genes from the same organism. To determine whether

T		C		A		G	
TTT Phe (F)	2.26% 36982	TCT	0.86% 14047	TAT Try (Y)	1.65% 26974	TGT Cys (C)	0.53% 8749
TTC Phe (F)	1.62% 26545	TCC Ser (S)	0.89% 14614	TAC Try (Y)	1.23% 20114	TGC Cys (C)	0.66% 10906
TTA Leu (L)	1.39% 22692	TCA	0.81% 13232	TAA STOP	0.20% 3374	TGA STOP	0.11% 1950
TTG Leu (L)	1.33% 21866	TCG	0.88% 14361	TAG STOP	0.02% 480	TGG Trp (W)	1.51% 24795
CTT	1.14% 18666	CCT	0.72% 11764	CAT His (H)	1.30% 21323	CGT	2.01% 32840
CTC Leu (L)	1.05% 17222	CCC Pro (P)	0.55% 9109	CAC His (H)	0.94% 15500	CGC Arg (R)	2.09% 34167
CTA	0.38% 6316	CCA	0.84% 13709	CAA Gln (Q)	1.46% 23970	CGA	0.38% 6290
CTG	5.07% 82751	CCG	2.23% 36422	CAG Gln (Q)	2.94% 48113	CGG	0.63% 10392
ATT Ile (I)	2.95% 48303	ACT	0.90% 14737	AAT Asn (N)	1.91% 31267	AGT Ser (S)	0.93% 15252
ATC Ile (I)	2.39% 39062	ACC Thr (T)	2.28% 37266	AAC Asn (N)	2.17% 35456	TGC Ser (S)	1.61% 26341
ATA Ile (I)	0.55% 9054	ACA	0.80% 13075	AAA Lys (K)	3.39% 55430	AGA Arg (R)	0.29% 4801
ATG Met (M)	2.69% 43913	ACG	1.50% 24511	AAG Lys (K)	1.09% 17841	AGG Arg (R)	0.19% 3115
GTT	1.82% 29762	GCT	1.53% 25093	GAT Asp (D)	3.26% 53271	GGT	2.40% 39261
GTC Val (V)	1.48% 24309	GCC Ala (A)	2.51% 41065	GAC Asp (D)	1.89% 30952	GGC Gly (G)	2.79% 45685
GTA	1.09% 17794	GCA	2.04% 33444	GAA Glu (E)	3.89% 63546	GGA	0.87% 14284
GTG	2.59% 42375	GCG	3.21% 52503	GAG Glu (E)	1.83% 29982	CGG	1.15% 18885

Figure 1.9. Codon use chart for all protein-coding genes of the *E. coli* O157:H7 genome. For each codon triplet the percent codon frequency and number of codon occurrences are shown. (Data from the TIGR Comprehensive Microbial Resource; see the URL in the Appendix.)

codon use was correlated with gene expression, they took a set of highly expressed genes in the bacterium *E. coli* as a reference set and showed that this was in fact correlated with the gene expression level in *E. coli*.

The availability of a higher number of completely sequenced genomes has contributed greatly to this scenario with new and greater details. Pan, Dutta et al. (1998) have shown that codon use in highly expressed genes of the eubacterial species *Haemophilus influenzae* and *Mycobacterium tuberculosis* is biased. In particular, they demonstrated the existence of a preference for G-starting codons by highly expressed genes. This could be a general feature of bacteria, irrespective of their overall GC content. Kanaya, Yamada et al. (1999), by correlating codon use and tRNA abundance for several unicellular organisms, observed that codons preferred in highly expressed genes were related to the codons that are optimal for the translation process, as predicted by the composition of isoaccepting tRNA genes.

McInerney (1998) introduced the concept of the influence of replicational/transcriptional selection on the codon use of a given gene. He has analyzed the complete genome sequence of *Borrelia burgdorferi*, demonstrating that in this species the genes may adopt two different codon use strategies, depending on whether the gene is located on the leading or lagging strand of replication. The GC skew, which is due to an asymmetric replication mechanism, causes codon use variations. In other words, the mutational bias between the two strands, together with transcriptional selection, would be responsible for the location of the most highly expressed genes in the leading strand of replication. This, in turn, correlates to GC skew (see the previous paragraph).

Also, in *Chlamidia trachomatis* the choice of synonymous codons is the result of several factors, the most important being gene location, hydropathy, and the degree of conservation of the protein under examination (Romero, Zavala et al. 2000). In *E. coli* the codon bias is influenced by both replication (Bulmer 1990) and translation selection; the latter mechanism is responsible for two codon biases, both of them context-dependent: the avoidance of AGG motifs and the avoidance of out-of-frame stop codons (Smith and Smith 1996). Recently, McVean and Hurst (2000) have suggested that the underrepresentation of AGG motifs in the *E. coli* genome is the result of selective forces, while the avoidance of out-of-frame stop codons is the consequence of the mutational bias.

In *B. subtilis* three classes of genes have been recognized, according to different codon use (Kunst, Ogasawara et al. 1997): class 1, including the majority of the genes (3375 genes), among which are most of the genes for sporulation; class 2 (188 genes), including genes expressed under exponential growth conditions; and class 3 (537 genes), with codons enriched in AT residues.

In *M. tuberculosis*, a comparative analysis has revealed a statistically significant preference for the amino acids Ala, Gly, Pro, Arg, and Trp, which are all encoded by GC-rich codons, and a comparative reduction in the use of the amino acid encoded by AT-rich codons (Asn, Ile, Lys, Phe and Tyr; Cole, Brosch et al. 1998). Himmelreich, Hilbert et al. (1996), in their analysis of the codon use of *M. pneumoniae*, were able to distinguish gene subsets with low (below 35%) and high (50 to 56%) GC content. The codon use of the low- and high-(G + C)-content subfractions is influenced by the base composition, favoring codons with either G–C or A–T at the third position; and it is also related to genes that are frequently expressed like those coding for ribosomal proteins.

Campylobacter jejuni has a genome with a very low GC content (30.4%); this seems to affect the codon use, since codons ending in A or T are strongly preferred (Gray and Konkel 1999). In *Helicobacter pylori* the low GC (39%) partially reflects the synonymous codon use, but there is no evidence for translational selection or biased mutation patterns among synonymous codons (Lafay, Atherton et al. 2000).

1.7 REPLICATION AND EXPRESSION

In Bacteria, most studies have been aimed at the recognition of enzymes involved in replication, transcription, and in protein synthesis. These studies have been based mainly on present knowledge; that is, they refer to the most studied prokaryotes, primarily *E. coli*. It is well known in this bacterium that the replication of circular double-stranded DNA is initiated by the *initiator proteins* (e.g., dnaA protein) binding to the origin of replication (a specific point on the DNA, which in *E. coli* is oriC) and bending the double-stranded DNA around it. The following step is the binding of helicases and primases to this complex. Helicases are able to unravel and separate the double-stranded DNA helix, while primases (primosomes), synthesize the RNA primer. At this stage, the DNA polymerase enters the replication forks and starts synthesizing a DNA strand complementary to the template strand. The speed of replication is about 1000 bp per second and thus the *E. coli* genome is fully replicated in about 40 min.

E. coli contains three distinct enzymes capable of catalyzing the replication of DNA: DNA polymerases (pol) I, II, and III. DNA polymerases I and II appear to perform *proofreading/editing functions*, moving along the DNA and correcting mismatched base pairings. Indeed, normal bacteria show errors in replication at a rate of approximately 10^{-9} per base pair, while mutants with defective pol I and pol II show errors at a rate of approximately 10^{-5} per base pair. DNA polymerase III is the main polymerase for replication. This enzyme is much less abundant than pol I; however, its activity is nearly 100-fold that of pol I. DNA pol I was the first replication enzyme to be characterized in *E. coli*. It is made of a single polypeptide of 103 kDa coded by the locus polA. The chain can be cleaved by proteolysis in two fragments, a 68-kDa Klenow fragment, which possesses the 3'-5' exonuclease and the polymerase activities, and a 35-kDa small fragment containing the 5'-3' exonuclease activity.

The complete sequencing of bacterial genomes has allowed us to discover the features distinguishing each species from the general pattern of *E. coli*. In all the complete bacterial genomes sequenced so far, the primary DNA polymerase corresponds to the DNA polymerase III (pol III) holoenzyme in *E. coli*, although some variation may occur, particularly in the number of subunits of the core structure. The *E. coli* pol III is made up of several subunits, and the assemblage of the holoenzyme is a multistep process. The core enzyme contains three subunits: α (130 kDa), which synthesizes DNA; ε (25 kDa), which has a 3'-5' proofreading exonuclease activity; and θ (10 kDa), which probably is required for assembly. The addition of the τ subunit (71 kDa), leads the core enzyme to dimerize, generating the pol III* complex; then the addition of the complex γ made up of several subunits, of which the best characterized are γ (52 kDa, alternative product of the gene coding for τ subunit) and δ (32 kDa) creates the pol III' complex, which finally generates the holoenzyme after the addition of the β subunit (40 kDa).

In *Aquifex aeolicus*, an additional member of the γ-τ/δ' family is present among the pol III subunits; in *Borrelia burgdorferi* and *Mycoplasma genitalium*, only 4 of the 10 polypeptides (α, β, γ, τ) of the *E. coli* pol III have been identified. *Mycoplasma pneumoniae* codes for two potential α subunits; and only the subunits β (dnaN), δ' (holB), γ, and τ (dnaX) are present. This may indicate a simplified replication complex compared to those of other gram-negative bacteria, such as *E. coli* and *H. influenzae*. Similar to what happens in other minimal genomes (e.g., *M. pneumoniae* or *B. burgdorferi*), the genome of *Treponema pallidum* encodes for the α, β, ε, γ, and τ subunits of *E. coli* DNA polymerase III. *Rickettsia prowazekii*, an endocellular parasite, has a smaller set of genes involved in DNA replication; only four genes for the core structure of DNA polymerase III have been identified: α (dnaE), ε (dnaQ), β (dnaN), θ, and γ (dnaX) subunits. Further details on the replication mechanism in the genomes sequenced remain to be elucidated.

Bacterial transcription [i.e., the transfer of DNA genetic information to a complementary sequence of RNA nucleotides by the DNA-dependent RNA polymerase (RNA pol)] is also well described in *E. coli* and in a few other bacteria. According to these notions, the basic critical components for transcription are the subunits of RNA pol, the promoters (i.e., the DNA sequence of the operon recognized by the DNA-dependent RNA polymerase), and the σ factors, which stabilize the polymerase in order to start polymerization at specific sites.

In *E. coli*, RNA pol is a complex enzyme, made up of four kinds of subunits: the α subunit (with a mass of 40 kDa, encoded by the rpoA gene), which binds to the promoter; the β subunit (155 kDa, encoded by the rpoB gene), whose function is the binding of ribose–triphosphate–organic base; the β′ subunit (160 kDa, encoded by the rpoC gene), which binds to DNA; and the σ subunit (70 kDa, encoded by the rpoD gene), which is responsible for the initiation of transcription. In bacterial cells, the RNA pol exists in two forms: the core enzyme ($\alpha_2\beta\beta'$), which elongates RNA, and the holoenzyme ($\alpha_2\beta\beta'\sigma$), which initiates RNA synthesis.

Recognition of individual promoters is determined by the kind of σ factor present. In *E. coli* there are several σ factors, each named after the molecular weight (e.g., the 70-kDa σ in *E. coli* is called σ70). Sigma size varies widely, from 32 to 92 kDa.

Comparative genomics has shown that the transcription mechanism is mostly similar to that of *E. coli*, as also proved experimentally in some cases. The number and type of σ factors are extremely variable and appear to be a species-specific feature; for example, in *Bacillus halodurans*, 11 σ factors have been identified, of which 10 are unique to this species; they are probably needed for its hyperalkaline lifestyle; in *B. burgdorferi*, beside σ70, two alternative σ factors (σ54 and rpoS) have been found; *R. prowazekii* encodes an alternative σ32 factor. In *B. subtilis*, 19 σ factors have been identified, which might be due to the sporulation phase present during its life cycle. Alternative σ factors play a key role in directing differentiation of the spore cell. During sporulation, many genes are expressed which are not expressed during vegetative growth, due to several alternative σ factors involved, each at a certain time, to initiate transcription of the gene subset needed at that time.

The consensus sequence recognized by the σ subunit of bacterial RNA polymerase is composed of two conserved boxes located at positions −35 and −10 with respect to the transcription start site. The distance between these two conserved boxes, whose consensuses are TTGACA and TATAAT for the −35 and the −10 box, respectively, may range between 15 and 19 bp. Figure 1.10 shows the base frequency matrices (see Sec. 6.8.4) for the rpoD recognition elements calculated from the known *E. coli* binding sites as well as the corresponding logo (see Sec. 6.8.6). Using pattern discovery methods (see Sec. 6.8.6), Thieffry, Salgado et al. (1998) predicted a large fraction of *E. coli* genome transcriptional regulatory sites. A comprehensive library of DNA binding sites for *E. coli* promoters can be found in the DPInteract Database (Robison, McGuire et al. 1998; see the URL in the Appendix).

In prokaryotes, transcription is coupled to translation—the synthesis of proteins by amino acid polymerization—because mRNAs are rapidly degraded, usually within a minute after transcription. Once a length of newly made RNAs has dissociated from the DNA, a ribosome can bind to it and protein synthesis can be initiated. So translation begins right after transcription and well before transcription termination. These processes have been studied extensively and their various steps are well known.

Genes encoding the 5S, 16S, and 23S rRNA are generally organized into an operon (called *rrn*) in the domain Bacteria. However, even if their clustering permits transcription of equimolar quantities of each rRNA gene, alternative organizations are observed in some prokaryotes, thus indicating that organization into operons is not a strict requirement. The number of rRNA operons per bacterial genome varies from 1 to as many as 15 copies, and this features seems to be required in species

Position	A	C	G	T	Consensus
01	16	19	13	202	T
02	18	10	12	210	T
03	21	24	161	44	G
04	137	46	20	47	A
05	55	126	20	49	C
06	143	24	41	42	A

(*a*)

Position	A	C	G	T	Consensus
01	12	22	17	199	T
02	210	12	9	19	A
03	46	32	36	136	T
04	159	33	29	29	A
05	158	37	31	24	A
06	11	21	9	209	T

(*b*)

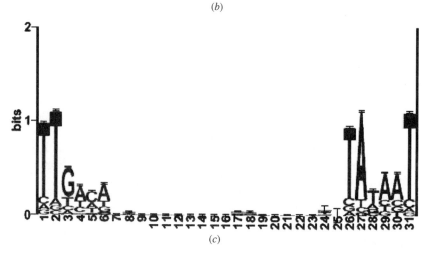

(*c*)

Figure 1.10. Position weight matrices for (*a*) the −35 and (*b*) the −10 boxes from 250 *E. coli* promoters collected in the DPInteract Database. (*c*) Corresponding sequence logo. [(*a*, *b*) From Robison, McGuire et al. (1998).]

with a high growth rate. For example, the pathogenic bacteria *Mycoplasma pneumoniae* has one rRNA operon, while the enteric bacterium *E. coli* possesses seven copies per genome. The spore-forming bacteria *Bacillus subtilis* (10 copies per genome) and *Clostridium paradoxum* (15 copies) show the highest number of rRNA operons (Klappenbach, Dunbar et al. 2000).

Ribosomal RNA operons can be also involved in recombination events; indeed, inversions can occur between rRNA operons, but they are generally unstable. The *rrn* operons contain genes for both rRNA and tRNA, each functional sequence being separated from the next by transcribed spacer regions. The genes encoding transfer RNAs (tRNAs) are often located in the internally transcribed spacer (ITS) region and distal to the 5S rRNA gene. The ITS region is variable in length and is responsible for most sequence diversity between multiple rRNA operons in many species of bacteria. The rRNA operon is transcribed as a single RNA molecule and then is cleaved by ribonucleases into mature rRNA and tRNA molecules. Very often the rRNA operons are localized around the origin of replication. This could provide a gene dosage effect during rapid growth; indeed, the effective number of rRNA operons in *E. coli* can be as high as 36 copies during exponential growth with multiple replications forks.

It is well known that tRNA number is species-specific. The most divergent pattern in protein synthesis in the various bacterial species regards tRNA synthetase genes. *Bacillus halodurans* lacks the glutaminyl-tRNA synthetase gene (glnS), one of two threonyl-tRNA synthetase genes (thrZ), and one of two tyrosyl-tRNA synthetase genes (tyrS), while a member of the same genus, *B. subtilis*, lacks only the glnS gene; the same occurs in *Thermotoga maritima, Treponema pallidum, Mycoplasma genitalium*, and *M. pneumoniae*. Both strains of *Helicobacter pylori* sequenced up to now lack the asparaginyl-tRNA synthase gene (asnS); in *Rickettsia prowazekii*, no asnS and glnS have been identified. In some cases the mechanisms used to overcome deficiencies are known, whereas in others they are still to be unraveled. In organisms lacking glnS gene, a single glutamyl tRNA synthase aminoacylates both tRNA^{Glu} and tRNA^{Gln} with glutamate. Subsequently, the amidation of glutamate to glutamine leads to the formation of glutaminyl tRNA synthetase (Freist, Gauss et al. 1997). Probably the lack of asnS gene may be overcome through a transamidation process forming Asn-tRNA^{Asn} from Asp-tRNA^{Asn} (Tomb, White et al. 1997).

In Archaea, many components of the replication and expression mechanisms are similar to those of eukaryotes. Indeed, this is the main reason why the Archaea were defined as a separate branch of prokaryotes in the tree of life, more closely related to eukaryotes (Woese, Kandler et al. 1990), even if from complete genome sequencing it appears that the resemblance of Archaea and Bacteria is more pronounced than expected (see below and Chapter 8).

The first studies of archaeal DNA replication in vivo showed that these species are sensitive to aphidicolin, a specific inhibitor of eukaryal but not bacterial replication polymerase [see Forterre and Elie (1993) for a review]. Since then, several eukaryotic-specific replication proteins have been identified in Archaea, including the origin recognition complex (ORC), the B DNA polymerase family, the Dna2 (3′-5′ helicase), and the ATP-dependent DNA ligase (Edgell and Doolittle 1997). In some cases, bacterial and eukaryal proteins are also homologous, but the archaeal/eukaryal counterparts still show a higher degree of similarity (e.g., replication factor C, which is the sliding-clamp protein, or ribonuclease H).

Data reported for the completely sequenced Archaea have revealed that the archaeal replication apparatus does not have eukaryotic features only. In *Archaeoglobus fulgidus*, a homolog of the proofreading ε subunit of the *E. coli* pol III not previously recorded in Archaea has been observed; the *Halobacterium* NRC-1 genome encodes for three DNA polymerase types: two family B polymerases, a bacteriophage-like family A polymerase, and the heterodimeric family D polymerase. In *Methanobacterium thermoautotrophicum*, the presence of two Cdc6 homologs and three histones would support a DNA replication initiation and chromosome packaging with eukaryal features; however, the presence of an ftsZ gene indicates a bacterial type of cell division initiation. Very recently, comparative genomic studies on the hyperthermophilic archaeon *Pyrococcus abyssi* showed that this species has a bacterial mode of replication but uses eukaryotic-like machinery.

The eukaryotic-like replication genes lie around the origin of replication, which is strongly conserved, while (as in bacterial genomes) the replication terminus is a hot spot of genome shuffling (Myllykallio, Lopez et al. 2000).

Similarly, archaeal transcription and translation mechanisms seem to be a mosaic of eukaryal and bacterial features (Bell and Jackson 1998). The first studies on transcription in Archaea showed the existence of striking parallels with the eukaryal transcriptional apparatus: like Bacteria, Archaea have a single RNA polymerase that transcribes all genes, but it is very similar to that of eukaryotes in that it contains three or four large and many small subunits; moreover, archaeal RNA polymerases are similar in sequence and antigenicity to eukaryal RNA polymerase II. Promoters in Archaea are a −30 TATA-like sequence that is a binding site for a transcription factor (TFB), not σ-like subunits. These findings also contributed to the prevailing view that Archaea and Eukarya share a peculiar machinery that differs from that of their bacterial counterpart. However, further investigations have highlighted that transcription in Archaea has a mixed character. Indeed, the search for transcription-associated proteins in four completely sequenced genomes (*Methanococcus jannaschi*, *Archaeoglobus fulgidus*, *Methanobacterium thermoautotrophicum*, and *Pyrococcus horikoshii*) has revealed that out of 280 transcription-associated proteins predicted, only 51 have homologs in Eukarya only, while 168 have bacterial homologs as well. The remaining proteins have homologs in both phylogenetic domains, while some elements (such as GvpE, a transcriptional activator involved in the regulation of gas vesicle synthesis in halophilic archaea) are unique to Archaea (Kyrpides and Ouzounis 1999).

Deeper in detail, the homology with Eukarya is confined essentially to the initiation factors TFB (similar to the eukaryotic TFIIB, even if the two proteins bind to the DNA in inverse orientation in the two structures), the components of the RNA polymerase core enzyme (up to 15 components versus 4 in *E. coli*), and the TATA-box-binding protein (TBP, which displays about 40% identity with eukaryal TBP).

In Archaea translation also appears again to be a mixture of prokaryotic and eukaryotic features. There is evidence for a bacterialike initiation process (i.e., leader sequence binding to ribosome), where several involved factors are related to bacterial analogs whereas others are of eukaryotic origin. The Archaea have 70S ribosomes (bacterial-sized), and the way they recognize the start codon resembles the bacterial process (Bell and Jackson 1998). However, archaeal translation is inhibited by diphtheria toxin, in a manner similar to what occurs for eukaryal ribosomes, but it is not inhibited by most bacterial-translation-inhibiting antibiotics. Archaeal pre-RNA processing resembles the bacterial pathway, but the Archaea possess a protein homologous to eukaryal fibrillarin (Dennis 1997). Translation is initiated with methionine (like Eukarya), not formylmethionine (like Bacteria); archaeal tRNAs contain introns like the eukaryotic introns. Finally, the Archaea have homologs of the eukaryotic elongation factors eEF-1a and eEF-2.

As to archaeal comparative genomics, as in Bacteria, the most divergent feature is the number of tRNA synthases: *Archaeoglobus fulgidus* lacks both glutamine and asparagine synthetases; *Methanobacterium thermoautotrophicum* and *Methanococcus jannaschii* lack glutamine, cysteine, lysine, and asparagine synthetases. In *Halobacterium* sp., no asparagine and glutamine tRNA synthetases have been detected. *A. fulgidus* and *M. thermoautotrophicum* lack selenocysteine-tRNA gene; the former has only one rRNA operon and lacks the 5S rRNA gene.

From the data reported above, which are based on a limited set of genomes, several notions have already been confirmed and new ones have emerged, thanks to a comparison of the complete genomes sequenced until now. One of the most limiting factors in the exploitation of genomic data is that a large portion of the genome protein coding region remains to be elucidated. Thus functional genomics is becoming a rapidly growing field exploiting new methods of both dry and wet biology. Among the latter, RNA interference (RNAi) is giving good results.

The identification of gene function based on comparative analysis is often difficult since the similarity of homologous genes could be so low as to be hardly detectable and the setup of the genome may vary dramatically even between closely related species. Events of lateral gene transfer and/or genome fusion may be responsible for very important divergences in some tracts of related genomes. These events are more frequent than expected and have already created great arguments regarding the classification and evolution of prokaryotes.

CHAPTER 2

EUKARYOTES

2.1 INTRODUCTION

This chapter is structured in two parts. In the first part we report data regarding a rough classification of eukaryotes, base compositional features, and some properties of genome organization, replication, and expression. As in Chapter 1, our aim is to provide the reader with some an outline on current knowledge in the field. For a deeper insight, we recommend that the reader consult the available literature. The second section is dedicated to more detailed descriptions of seven eukaryotic genomes completely sequenced and now available: *S. cerevisiae*, *S. pombe*, *A. thaliana*, *O. sativa*, *D. melanogaster*, *C. elegans*, and *H. sapiens*.

2.2 CLASSIFICATION AND TIME SCALE

The name *eukaryote* comes from the Greek *eu-karyon*, meaning "true nut" or "true kernel"; the "nut" is indeed the nucleus of the eukaryotic cell, which contains the genetic material. Indeed, unlike Bacteria and Archaea, Eukaryotes have their DNA in a separate cell compartment surrounded by a double membrane. A peculiar feature of the eukaryotic cell is the presence of complex membrane systems subdividing the cellular environment into compartments. Other features unique to eukaryotes are the presence of a cytoskeleton (a network of protein filaments, mostly actin and tubulin, anchored to the cell membrane, which constitutes a structural framework); the presence of organelles (mitochondria and plastids; see Chapter 3 for further details); and the presence of the endoplasmic reticulum (ER), a complex network of membranes that provides a transport system within the cell and is classified as rough or smooth. *Smooth endoplasmic reticulum* is the site for lipid formation; the *rough ER* (so-named because its surface contains ribosomes) is the site of protein production, from where proteins are transferred to the Golgi apparatus via transport vesicles and then to their final destinations.

Eukaryota, Archaea, and Bacteria make up the three major branches of the tree of living organisms. It is no easy task to draw a comprehensive taxonomy of Eukaryota; indeed, to date, about 60 lineages have been identified on the basis of their cellular organization (Patterson 1999), and the relationships among them are not clear. Table 2.1 reports the classification for the main eukaryotic taxa. Eukaryotes have been broken down into four kingdoms: animals (metazoa), plants (viridiplantae), fungi, and protozoa. The latter include very diverse life forms, and they should probably be classified in several kingdom-level taxa, as they are a paraphyletic group consisting of monocellular eukaryotes—neither animals nor true fungi nor green plants.

Eukaryotes probably emerged from prokaryotic ancestry about 2 billion years ago (Knoll 1992). The most realistic theories regarding the origin of eukaryotes are those based on symbiotic relationships, which have been and still are an important driving force in the evolution of eukaryotes. Apart from the old serial endosymbiotic theory (Margulis 1970), more recently other models have been developed, among which are the *hydrogen hypothesis* (Martin and Muller 1998) and the *syntrophic hypothesis* (Moreira and Lopez-Garcia 1998). These are discussed in detail in Sec. 7.5.

Whatever the origin, the first eukaryotic fossil record is that of the microfossils found in the 1-billion-year-old Bitter Springs Formation of northern Australia, which seem to have preserved nuclei. However, many authors doubt this interpretation (these fossils could easily be cyanobacteria, and the "nucleus" could be the product of cell content shrinkage). More likely, eukaryote fossils are acritarchs, spherical (probably algal) protists that began to appear about 1.8 billion years ago.

Table 2.2 reports the geological time scale with details on the appearance of life forms throughout time. Earth life can be divided arbitrarily into several periods, each corresponding to a particular vertical stratigraphic position in Earth's crust. The name commonly used for this classification, *chronostratic*, usually refers to the type and variety of fossils formed in each layer (e.g., the Cenozoic or "recent life" era is characterized by fossils of mammals and recent plants). The absolute timing of each era or period can be established with fair accuracy using radiometric dating methods, which measure the amount of radioactivity that occurs in appropriate rock types. In this way an integrated geochronological time scale can be obtained when all major events of life evolution on Earth can be placed. According to widely accepted estimates, Earth was born 4.5 billion years ago. Earth's history is usually subdivided into two main ages, the Precambrian [4500 to 570 million years ago (Mya)] and the Phanerozoic (>570 Mya to present). As the Precambrian makes up roughly seven-eighths of Earth's history, it cannot be considered as a single unified time period. During the Precambrian, the most important events in biological history took place. Eukaryotic cells evolved in the Palaeoproterozoic, and just before the end of the Precambrian, complex multicellular organisms, including the first animals, evolved. It is well known that most of the major invertebrate phyla are reported in the fossil record over a relatively short time interval, not exceeding 20 million years, 540 to 520 Mya ago. Indeed, in the early Cambrian period, intense diversification resulted in more than 35 new animal phyla: called the *Cambrian explosion*, a sharp and sudden increase in the rate of evolution.

The date for the base of the Cambrian period has been revised a number of times in the past decade. New discoveries show that the explosion started about

TABLE 2.1. Classification of the Main Eukaryote Taxa

Taxon	Representatives (Common Name)
Metazoa	
Placozoa	Trichoplax
Rhomobozoa	
Porifera	Sponges
Cnidaria	Jellyfish, sea anemones, sea fans, corals
Ctenophora	Comb jellies
Platyhelminthes	Flatworms
Nemertea	Ribbon worms
Mollusca	Snails, clams, squids
Annelida	Segmented worms, including pogonophorans, vestimentiferans, earthworms, leeches
Echiura	Spoon worms
Sipuncula	Sipunculids, peanut worms
Brachiopoda	Lampshells
Bryozoa	Bryozoans
Chaetognatha	Arrow worms
Rotifera	Rotifers, thorny-headed worms
Gastrotricha	Gastrotrichs
Nematomorpha	Horsehair worms
Nematoda	Roundworms
Arthropoda	Insects, spiders, crabs, scorpions, crustaceans, millipedes
Onychophora	Velvet worms
Tardigrada	Water bears
Echinodermata	Starfish, urchins, sea cucumbers
Hemichordata	Hemichordates, acorn worms
Chordata	Vertebrates and relatives
Fungi	
Chytridiomycota	Water molds, Allomyces
Zygomycota	Bread molds, *Rhizopus, Mucor*
Basidiomycota	Mushrooms, rusts, smuts
Ascomycota	Sac fungi, yeast, *Penicillium*
Viridiplantae	
Prasinophyta	
Chlorophyceae	
Trebouxiophyceae	
Ulvophyceae	
Chlorokybales	
Klebsormidiales	
Zygnematales	
Charales	
Coleochaetales	
Embryophytes	Land plants
Protozoa	
	Cryptomonads, euglenids, glaucophytes, ciliates

TABLE 2.2. Geological Time Scale with Details on the Appearance of Life Forms

Era	Period	Epoch	Age Began (My)	Life Forms
Cenozoic	Quaternary	Holocene	0.01	Modern life forms
		Pleistocene	1.64	Last ice age, large terrestrial mammals, mammoths, mastodons, first modern human, cave paintings
	Tertiary	Pliocene	5.2	First Australopithecines, toolmaking, Neanderthals
		Miocene	23.3	Large sharks, whales, first hominids
		Oligocene	35.4	First grasses, anthropoids
		Eocene	56.5	First marine and large terrestrial mammals
		Paleocene	65.0	Many types of mammals, first primates
Mesozoic	Cretaceous	Senonian Gallic Neocomian	144.2	Age of dinosaurs, mollusks, flowering plants
	Jurassic	Malm Dogger Lias	205.7	First belemnites, squids, frogs, birds, salamanders
	Triassic	Late Middle Early (Scythian)	248.2	First turtles, cycads, lizards, dinosaurs, mammals
Paleozoic	Permian		290.0	First mammal-like reptiles
	Carboniferous		362.5	Coal age, first conifers
	Devonian		408.5	Age of fish, insects, ammonites, first jawless fish, placoderms, amphibians
	Silurian		439	First land plants, ferns, lycopods, sharks, bony fish
	Ordovician		510	First corals, starfish, sea urchins, blastoids, eurypterids, bryozoans, scaphopods, vertebrates First trilobites, conodonts, forams, sponges, worms, brachiopods
	Cambrian		570	Nautiloids, chitons, clams, snails, monoplacophorans, crustaceans, crinoids, cystoids, carpoids
Archaeozoic	Precambrian		>570	First simple plants and invertebrate animals: algae, bacteria, jellyfish

Source: Data based on Jeff Poling's page on geological ages of earth history, http://www.palaeos.com/Geochronology/default.htm.

570 Mya, in the late Paleozoic era. Several independent molecular data sets suggest that invertebrates diverged from chordates about 1 billion years ago, about twice as long ago as the Cambrian period began. Paleontological data (Seilacher, Bose et al. 1998) indicate triploblastic metazoans as early as 1.1 billion years ago, in agreement with molecular data. This shows that animal body plans changed not very much before the explosive emergence of new designs in the Precambrian–Cambrian transition and before the onset of a more competitive style of Darwinian evolution. Another major step in the history of life on Earth is the invasion of land by plants, followed by the vertebrate invasion of land, which occurred in the late Devonian. Two mass extinctions seem to have occurred: a Permo-Triassic extinction, which affected mainly the marine biosphere, and the Cretaceous–Tertiary extinction, affecting 15% of marine families and 25% of terrestrial families, dinosaurs among them. Finally, in the Cenozoic era, mammals underwent extensive radiation as a result of the opening of new opportunities caused by the dinosaur extinction. The Pleistocene epoch saw the evolution and expansion of our own species, *Homo sapiens*.

2.3   GENOME SHAPE AND SIZE

As mentioned above, the nucleus contains the vast majority of the genetic material of the eukaryotic cell; however, a small but indispensable fraction of cellular genetic information is contained in mitochondria/chloroplast DNAs. In only a few cases, such as in Archaeozoa, which do not contain organellar DNA, is all cellular genetic information nuclear coded. A peculiarity of eukaryotes is that the genetic information contained in the nucleus is redundant, at least diploid during one or more phases of cell life. The DNA is packaged into chromosomes which are visible by optic microscopy only when condensed during mitosis or meiosis.

Eukaryotic chromosomes consist of DNA associated with a set of proteins forming a complex called *chromatin*. According to the condensation status, two types of chromatin can be recognized: *heterochromatin*, which is the condensed form of chromatin organization and is considered to be transcriptionally inactive; and *uncoiled euchromatin*, most abundant in active, transcribing cells. Interphase cells contain two classes of heterochromatin: *constitutive*, never transcribed, as in satellite DNAs, and *facultative*, which can be inactive in one cell lineage but expressed in others.

Seventy percent of chromatin mass is made of two classes of proteins: histones and nonhistone proteins. *Histones* are highly basic proteins forming the core structure of chromatin, called the *nucleosome*, in which double-stranded DNA is wrapped around an octamer of histones, a tripartite assembly of a $(H3/H4)_2$ tetramer flanked by two (H2A/H2B) dimers. These sets of eight histones are separated by spacer regions of free DNA and histone H1, whose function might be to cap the ends of the DNA surrounding the octamer core. Nucleosome structure causes the linear length of the DNA to decrease by about 10-fold. The structure of chromosomes is also largely determined by *nonhistone* chromosome-associated proteins, many of which belong to the SMC (structural maintenance of chromosomes) family. SMC proteins are involved in chromosome condensation, genetic recombination, and DNA repair [see (Strunnikov 1998) for a review] and are usually subdivided into five structural subgroups. Four SMC genes have been found in *S. cerevisiae*, *C. elegans*, and in mammals.

In eukaryotic chromosomes, the DNA is condensed further (>250-fold) through coiling and supercoiling processes. The passage from the 10-nm unit thread to the 30-nm chromatin fibers of the interphase is considered to be facilitated by histones, but the mechanisms for the additional 250-fold compaction, needed to reach the metaphase level, remain to be elucidated. Woodcock and Dimitrov (2001) have proposed a hierarchical classification for higher-order chromatin structure. The primary level is the linear arrangement of nucleosomes on DNA. The secondary level is that formed by interactions of nucleosomes and regulatory proteins. The tertiary level is represented by structures formed by long-distance interactions between secondary structures. A stable, large-scale chromatine structure also exists in interphase nuclei; this structure could be involved in regulating gene expression (Belmont, Dietzel et al. 1999).

In eukaryotic chromosomes the minimum features required for maintenance are the centromere, involved in the segregation process of sister chromatids at mitosis and meiosis; the telomeres, which stabilize the ends of chromosomes; and the origins of replication. The *centromere* is a constricted region of a chromosome, consisting of highly condensed heterochromatin; it contains the *kinetochore*, the attachment site for mitotic spindle microtubules. The centromere formation is a complex process that presumably occurs through modifications at the chromatin level [see Choo (2000) for a review]. Analysis of different organisms, from fungi to mammals, reveals that no universal centromere sequence exists, although their common feature is the presence of tandem repeats reiterated over hundreds of kilobase pairs that are found in organisms as diverse as *Arabidopsis thaliana*, *Drosophila melanogaster*, and *Homo sapiens*. A remarkable exception is *Saccharomyces cerevisiae*, whose chromosome centromeres, which have precisely been mapped, lack satellite sequences. A striking feature of satellite repeats is that despite the lack of a universal consensus sequence, they show a remarkably uniform repeat unit length. For example, the length of the basic alpha-satellite unit is 171 bp in primates, 186 bp in the fish *Sparus aurata*, 155 bp in the insect *Chironomus pallidivittanus*, and 180 bp in both *Arabidopsis thaliana* and *Zea mais* (Henikoff, Ahmad et al. 2001).

Telomeres are nonnucleosomal DNA–protein complexes at the ends of eukaryotic chromosomes that serve as protective caps (Lingner and Cech 1998). They perform at least two critical functions: they allow for the complete replication of chromosome ends, and they mask chromosome ends, which would otherwise resemble accidental DNA breaks and trigger DNA-damage response mechanisms that promote cell-cycle arrest and DNA repair. Perhaps as a consequence of these unique properties, the chromatin adjacent to telomeres is itself unusual, and in yeast, for example, it exists in a dynamic heterochromatic state. In human somatic tissues, telomere shortening with successive cell division is thought to serve as a counting mechanism that prevents unlimited proliferation, thus correlating with cellular aging and carcinogenesis.

Telomeric DNA usually consists of simple repetitive sequences that are T_2AG_3 in mammals and most other eukaryotes and TG_{1-3} in *Saccharomyces cerevisiae*. These lengths of telomeric DNA can vary considerably between species (e.g., 50 kbp in *Mus musculus*, 10 kbp in humans, and about 300 bp in *S. cerevisiae*), but each organism maintains a fixed average telomere length in its germline. Telomere

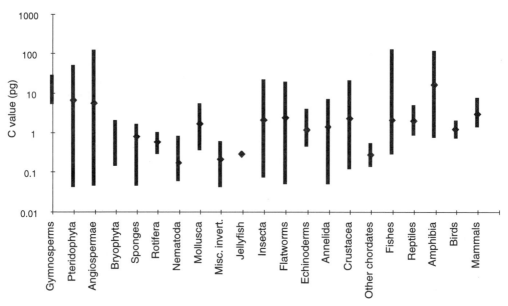

Figure 2.1. *C* values (pg) for different eukaryotic taxa. (Data from the animal genome size database, *www.genomesize.com.*)

repeats are generated by a specialized reverse transcriptase, called *telomerase*, which is a ribonucleoprotein that uses its stable associated RNA moiety as the template for synthesis of telomeric repeats.

In higher eukaryotes, chromosomes are studied most frequently at mitotic metaphase, in which they appear as identifiable shapes characteristic of the *karyotype* (nuclear type) of the species being studied, defined by chomosome number and other landmarks visible under the microscope during metaphase. A metaphase chromosome is identified morphologically both in its total length and from the position of the centromere. A chromosome is said to be *metacentric* if the centromere is near the middle, *acrocentric* if the chromosome arms are unequal, and *telocentric* if the centromere is at the very end. Another way to identify chromosomes is by means of banding techniques using stains (such as quinacrione HCl or Giemsa) that differentiate the chromosomes by patterns of transverse bands of different length. Obviously, the human is one of the most studied karyotypes. In human chromosomes, the short arm is named *p* (petite) and the long arm *q* (the next letter in the alphabet). In the 23 human chromosomes, six are acrocentric (13, 14, 15, 21, 22, and Y) and five are metacentric (1, 3, 16, 19, and 20), the remaining being submetacentric (intermediate between the former two).

The DNA content in the haploid genome, called the *C value*, is characteristic of each living species. Data reported in Fig. 2.1 show an enormous variation in the range of *C* values, from 0.05 pg in *Paraturbanella teissieri* (miscellaneous invertebrate group in Fig. 2.1) to 132 pg in the fish *Protopterus aethiopicus*. The *C* value varies over 600-fold among angiosperm species (Bennett and Leitch 1995). The smallest known plant genome belongs to *Arabidopsis thaliana*, and one of the largest is a member of the lily family, *Fritillaria assyriaca*. Nevertheless, the genomes of

TABLE 2.3. Approximate Genome Size of Several Eukaryotic Species

Higher Taxon	Species	Genome Size (Mbp)	Number of Chromosomes[a]
Amphibia	*Xenopus laevis*	3100	18
	Xenopus tropicalis	1700	10
Aves	*Gallus gallus*	1200	39
Fish	*Fugu rubripes*[b]	400	22
	Danio rerio	1700	25
Fungi	*Saccharomyces cerevisiae*[b]	13	16
Insecta	*Drosophila melanogaster*[b]	180	4
Mammalia	*Homo sapiens*[b]	2910	23
	Bos taurus	3651	30
	Canis familiaris	3355	39
	Mus musculus[b]	3454	20
	Rattus norvegicus	3135	21
Nematoda	*Caenorhabditis elegans*[b]	97	6
Plant	*Arabidopsis thaliana*[b]	125	5
	Oryza sativa[b]	400–430	12
	Zea mais	5000	10
Protozoa	*Leishmania major*	34	36
	Plasmodium falciparum[b]	26	14

[a] Refers to the haploid genome.
[b] Complete genome available.

Arabidopsis and *Fritillaria* code for about the same number of genes. At the beginning, quantitative estimates of DNA *C* values were made using slow chemical extraction methods, but the subsequent development of new techniques, including Feulgen microdensitometry and more recently, flow cytometry, have made estimating DNA amount both easier and faster.

Genome size is only roughly correlated with complexity; for example, humans have larger genomes than most insects, which have in turn larger genomes than fungi. This correlation breaks down further among chordates. For example, some amphibians have genomes almost 50 times larger than those of humans. The genome sizes of several eukaryotic species, including those of completely sequenced organisms, are shown in Table 2.3 (see also Sec. 7.2).

Not only DNA content, but also the number and size of chromosomes, are highly variable within eukaryotes. When only completely sequenced genomes are considered, *Arabidopsis* has a haploid set of five chromosomes; *Caenorhabditis* and *Drosophila* have five and six, respectively; *Saccharomyces*, 16; and humans, 23 (Fig. 2.2). There is no correlation between genome size and the number of chromosomes, or among the latter and organismal complexity.

The kinetics of reassociation between two strands in solution, defined by the product of nucleic acid concentration C_0 and time, provides a measure of genome complexity. Put simply, the higher the concentration of two complementary strands in solution, the shorter the time needed for their reassociation. C_0t curves plotting the percentage of DNA reassociation ($1–C/C_0$, with C being the single-strand DNA concentration at time t) versus the logarithm of C_0t are used to measure the complexity of a genome. The value of C_0t at 50% DNA reassociation is called $C_0t_{1/2}$ and

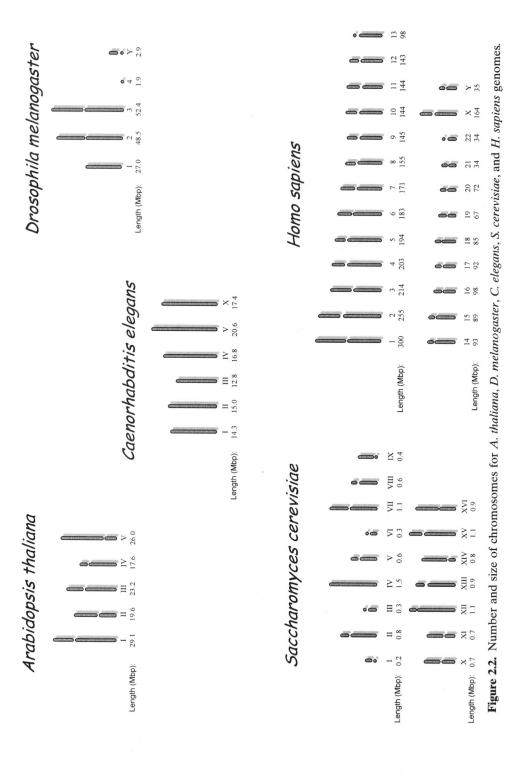

Figure 2.2. Number and size of chromosomes for *A. thaliana*, *D. melanogaster*, *C. elegans*, *S. cerevisiae*, and *H. sapiens* genomes.

is directly correlated to the nonrepetitive DNA content, also defined as *genome complexity*. Small genomes have lower $C_0t_{1/2}$ values than those of large genomes. A composite curve is generally obtained when analyzing complex eukaryotic genomes, such as that of human, where the repetitive fractions reassociate at much lower $C_0t_{1/2}$ values than does the single-copy fraction. In this way it is possible to determine the types and relative amount of repetitive DNA in a genome as well as the size of its unique fraction, which defines its "complexity" (see Sec. 6.8.2 for alternative definitions of DNA complexity).

The lack of correlation between total genome size and organism complexity, also defined as the *C-value paradox* [see Gregory (2001) for a review], can be explained primarily by the presence of a variable amount of repetitive DNA in different genomes. It has been found and now amply confirmed by complete genome sequencing that the eukaryotic genome consists mostly of noncoding DNA with repeated elements as the major component.

Repeated sequences in eukaryotic genomes can be classified into two major groups: tandem and interspersed repeats. *Tandem repeats* are a major component of specific chromosomal regions (telomeric repeats and centromeric satellites) or can be found interspersed in the genome (mini- and microsatellites). *Interspersed repeats* include mostly inactive copies of a wide variety of contemporarily and historically active transposable elements, such as retroelements and DNA transposons, which can each be subdivided further into distinct classes. The emerging picture is that transposable elements are not mere parasites; rather, they are integral players in genomic evolution, showing either a selfish or an altruistic nature, depending on different evolutionary circumstances. These repeated elements represent the bulk of the reverse flow of information from RNA to DNA in the genome. They could be fossil records of an evolving mechanism still going on; hence understanding their presence in the genome could help solve a broad range of biological problems both basic and applied in nature, from population and phylogenetic studies to genome engineering and genetic diseases (fragile chromosome, shortening of telomeres).

To date, over 1500 different repetitive sequence families have been compiled from all available eukaryotic sequence data and collected and organized into the RepBase specialized database (Jurka 2000). The number of annotated repeat families is expected to increase rapidly in the next years, concurrently with the progress of eukaryotic genome projects. The automatic annotation of repeat elements in genomic sequences can be produced by the RepeatMasker resource (see Sec. 6.8.3), where repeats are classified in five different classes: SINEs (short interspersed elements), LINEs (long interspersed elements), LTR (long terminal repeat) elements, DNA elements, and simple repeats (microsatellites). For a more detailed description of species-specific repeats, see Secs. 2.7.1 through 2.7.5, dedicated to the completely sequenced eukaryotic genomes.

Another factor contributing to the lack of compactness of the higher eukaryotic genome is the structure of the genes, which present striking differences from prokaryotic genes. Genes of higher eukaryotes are usually discontinuous, being split into exons and introns, with the latter removed by splicing from precursor RNA to produce functional transcripts; alternative splicing mechanisms may generate multiple products from the same gene (see Sec. 2.6.1).

Eukaryotic gene transcripts are generally monocystronic. However, operonlike structures such as those observed in bacteria are found in some eukaryotic genomes, such as *C. elegans* (Blumenthal 1998). To understand the origin, evolution, and genetic impact of eukaryotic genome structure in its uniqueness is one of the most challenging aspects of modern biology.

2.4 BASE COMPOSITION

At the level of the entire genome, base composition, usually expressed as GC content, varies greatly within and among major groups of organisms. Among eukaryotes the intragenomic variation in base composition is more dramatic than among prokaryotes. Regional differences in GC content within a single genome have been described in various taxa, including yeast, plants, trypanosomes, and vertebrates, and their meaning has been interpreted in various ways.

Long (>300 kbp) tracts of rather homogeneous base composition have been called *isochores*, a term used first by Bernardi in 1985 to describe the genomes of warm-blooded vertebrates (Bernardi, Olofsson et al. 1985). According to the *isochore model*, the nuclear genome of warm-blooded vertebrates can be described as a mosaic of isochores (= equal regions), very long DNA tracts in which the base composition can be regarded as homogeneous (see Sec. 7.3). Figure 2.3 shows the general scheme of isochores in warm-blooded vertebrate genomes and the resulting CsCl profile of major human DNA components.

The existence of isochores has been demonstrated experimentally by fractionating randomly sheared genomic DNA on cesium sulfate gradients. Each fraction from the gradient contains DNA banding at a specific density that is a direct reflection of its base composition. The different fractions have been grouped into five isochore classes: the GC-poor L1 and L2 classes and the increasingly GC-rich H1, H2, and H3 classes (Bernardi 2000).

It is well known that these GC-rich and GC-poor regions may have different biological properties. The G + C content of all three codon positions, introns, and flanking sequences varies according to the isochore class to which the gene belongs. In the same way, the amino acid content of proteins is constrained by isochore class; amino acids encoded by (G + C)-rich codons are more frequent in H isochores.

The density of genes is higher in H than in L isochores, and the distribution of repeated elements seems to be influenced by the compositional properties of the genome: LINEs are preferentially located in L isochores, while SINEs (particularly Alu in the human genome) are found preferentially in H isochores [see (Gautier 2000) for a review]. In mammals there is a correspondence between cytogenetic bands and isochores. Giemsa bands are composed primarily of L isochores; while reverse bands are more heterogeneous (H and L isochores). T bands found at the telomeres of chromosomes are usually made up of H isochores (Tenzen, Yamagata et al. 1997).

Isochore patterns differ remarkably between cold- and warm-blooded vertebrates. This finding has been considered a proof of the existence of selective constraints in genomic GC content. Uniform GC content has been reported for the nuclear genomes of fishes and *Xenopus*; in contrast, there is a shared presence of GC-rich isochores in homeothermic birds and mammals. GC-rich DNA produces a more heat-stable helix and more stable mRNA transcripts; thus it is postulated to

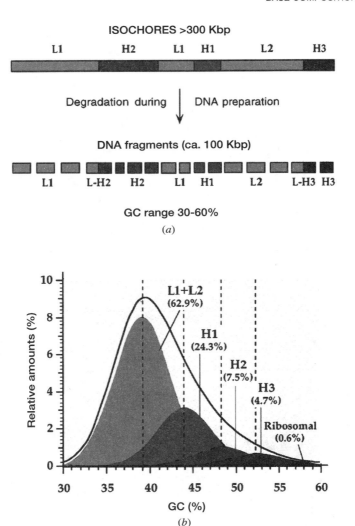

Figure 2.3. (*a*) The genome of warm-blooded vertebrates is a mosaic of large (≫300 kbp on average) segments, the isochores of which are compositionally homogeneous (above a size of 3 kbp) and can be partitioned into a number of families. Isochores are degraded during routine DNA preparations in fragments approximately 100 kbp in size. The GC range of the isochores from the human genome is 30 to 60%. (*b*) The CsCl profile of human DNA is resolved into its major DNA components: namely, the DNA fragments derived from isochore families L (i.e., L1 and L2), H1, H2, and H3. Modal GC levels of isochore families are indicated on the abscissa (dashed vertical lines). The relative amounts of major DNA components are indicated. [Courtesy of G. Bernardi; modified from Bernardi (2000).]

be selectively advantageous in animals with high body temperatures. Within the homeothermic taxa, isochore structure is variable; for example, myomorph rodents, fruit bats, pangolins, and some insectivores seem to be lacking most GC-rich isochores. This topic, in particular the evolutionary processes involved in isochore formation, is discussed in detail in Sec. 7.3.

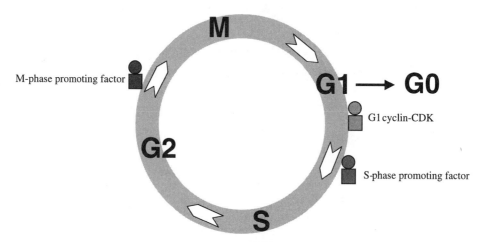

Figure 2.4. General scheme of the mitotic cell cycle in eukaryotes.

2.5   REPLICATION, REPAIR, AND RECOMBINATION

The replication machinery has evolved with the change from prokaryotic to eukaryotic chromosome structure, from a replication pattern of one origin to many origins and to a segregation mechanism (mitosis). The mitotic cell cycle in eukaryotes involves the duplication and separation of chromosomes and is coupled with the process of dividing one cell into two. The cell cycle consists of the alternating phases of DNA replication (S phase) and chromosome separation (mitosis, or M phase) interrupted by gaps known as G1 (interval between the M and S phases) and G2 (interval between the S and M phases), during which protein synthesis takes place. Important controls operate at the transition points as cells move from G1 into the S phase, and from G2 into the M phase, primarily through the regulated kinase activity of cyclin-dependent kinases (CDKs) (Fig. 2.4). The levels of cyclins in the cell rise and fall with the stage of the cell cycle.

The basic steps in DNA replication include (1) helicase-mediated unwinding of a portion of the double helix, (2) binding of a molecule of DNA polymerase to one strand of the DNA, which begins moving along it (in the 3′- to -5′ direction) for synthesis of the leading strand (in the 5′- to -3′ direction), and (3) binding of a different DNA polymerase to the other template strand for the discontinuous synthesis of the lagging strand (Fig. 2.5).

The multiple chromosomes of eukaryotic genome are coordinately replicated within a defined period, the S phase of the cell cycle (Zannis-Hadjopoulos and Price 1999). DNA synthesis is initiated at multiple replication origins in each chromosome and follows a defined temporal order within the cell cycle. The speed of replication is about 50 bp per second, which in the case of a single replication origin

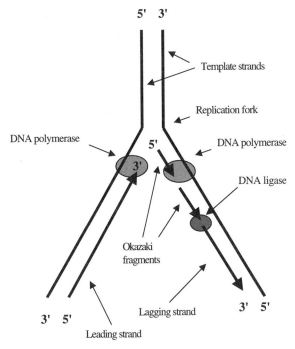

Figure 2.5. Model of DNA replication.

would add up to about one month for complete replication of an average human chromosome. Indeed, the multiple replication origins allow completion of chromosome replication in about 1 hour in vertebrates. Two parameters that appear to establish the order of replication are the proximity of the replication sites to centromers and telomers (Lingner and Cech 1998), and transcriptional activity. Recent studies also suggest a direct role for transcription factors in activating eukaryotic origins of DNA replication.

The activation of replication origins also seems to depend on structural determinants, which either facilitate the local unwinding of DNA or serve as recognition signals for the initiation of replication. Certain types of sequences and structures are common to most replication origins of prokaryotes and lower and higher eukaryotes. These feature include bent DNA, DNA unwinding elements (DUEs), matrix attachment regions (MARs), and inverted repeats that are capable of forming cruciform DNA (see Sec. 6.8.5 for a description of computational methods for MAR identification).

In addition to DNA sequences, epigenetic effects, such as DNA methylation and chromatin structure, may play important roles. In particular, DNA methylation of CpG dinucleotide clusters appears to mark specific origins for replication through changes in chromatin structure. Chromatin structure can inhibit both replication origin and transcriptional promoter activities (see Sec. 2.6.1) by blocking access of initiation factors to specific sites. Finally, the chromatin structure, although dispensable for DNA replication in vitro, appears to play a fundamental role in vivo, by regulating the concentration of proteins required to initiate DNA replication, by facilitating the assembly or activity of DNA replication forks, and by determining where initiation of DNA replication occurs in the genome. The importance of the nuclear structure is witnessed by the fact that assembly of prereplication complexes on cellular chromatin is delayed until mitosis is completed and a nuclear structure has been formed (DePamphilis 2000).

Multicellular eukaryotes face an additional problem with respect to single-cell eukaryotes in that the former must carry out a complex developmental program during which cell division cycle may vary in length from a few minutes to several hours. Shorter cell cycle times are found in rapidly cleaving embryos, where the G1 phase is skipped and the S phase is shorter, due to an increased number of initiation sites per genome. The same initiation sites are never activated twice during the same S phase. Multicellular eukaryotes have developed positive regulation, necessary for signaling the initiation, and stringent negative feedback controls to prevent reinitiation and to ensure that DNA is replicated once and only once at each cell division. At the same time, they retain the flexibility to change the number and locations of replication origins as cells undergo differentiation during development.

In *E. coli* and in other bacterial organisms, the *replicon model*, defined as a genetic element that replicates as a unit, with a unique origin of replication (replicator) serving as a target of a positive-acting initiator protein, was described thoroughly. This model was later confirmed in viruses and phages and partially in yeast. To what extent origins of DNA replication in multicellular eukaryotes are defined according to the replicon rules is still being debated.

Studies of replicators from fungi to metazoans suggest that they have diverged in sequence. The best characterized eukaryotic origin of replication is the autonomously replicating sequence (ARS) of the yeast *S. cerevisiae* (Gilbert 1998). These contain one conserved A and two or three not-conserved B elements, which are targets for transcription factors, plus DNA unwinding elements (DUEs) and the origin recognition complex (ORC), which are made of proteins forming the initiation complex. This structure cannot be generalized even to other unicellular eukaryotes (e.g., *S. pombe*), let alone to multicellular organisms, where attempts to identify specific consensus sequences in the origin of replication have failed to date.

At present, some 20 origins have been identified in complex eukaryotes, including preferential sites (e.g., ori β and ori γ), DUE elements, and binding sites for basic helix-loop-helix (bHLH) proteins (see Sec. 6.10.5 for prediction tools for HLH protein domains). In early embryos of *Xenopus laevis*, however, replication appears to occur randomly, with no sequence specificity. In mammalian cells, there is some evidence that replication initiation begins from preferred chromosomal locations, some more active than others. However, a detailed structure is still lacking and present data are far from sufficient to define any DNA element as an origin of replication.

Most of the proteins identified in bacterial replication complexes have their functional counterparts in eukaryotes, although they do not match in the number of homologous factors or in their levels of similarity. Indeed, some proteins have the same function in bacteria and eukaryotes, but their primary structure and their folding as well appear to be completely different (e.g., SSB in *E. coli* and RPA in eukaryotes); others show low amino acid similarity (not statistically significant), yet they show very similar folding [e.g., pol β (dnaN) in *E. coli* and proliferating cell nuclear antigen (PCNA) in eukaryotes].

Early steps in the initiation of DNA replication appear to be conserved from yeast to human, despite the low level of amino acidic similarity (20 to 30%) of the proteins involved (Leatherwood 1998). Proteins were isolated and characterized at first in yeast and then in *Xenopus*, where in vitro systems proved to be a useful tool to analyze the homologous proteins in vertebrates. The basic enzymatic activities required for the initiation of DNA synthesis are helicase, topoisomerases, homologs of the yeast ORC complex, the cdc6/cdc18 protein, the minichromosome maintenance (MCM) protein family, and the MCM-associated protein cdc45/sna41. Dna2, a helicase functionally homologous to DnaB and PriA in bacteria, unwinds the double-stranded template and defines the growing point of the replication fork. Topoisomerases relieve torsional strains that build up in the helix ahead of the growing fork.

ORC proteins (DnaA in bacteria) are the initiator proteins specifically binding to the ARS sequence in *S. cerevisiae*. In metazoans, they are required to initiate DNA replication, but their mechanism remains to be elucidated. The ORC complex is essential for the association of other initiation proteins, cdc6, and MCM, indicating that the role of ORC may be to mark the site where the pre-replication complex (pre-RC) assembles. Cdc6 is a loading factor whose role is to bind to ORC and recruit MCM proteins onto chromatin. MCM proteins comprise six related proteins which appear to have an important role in the regulation of origin firing. Different combinations of subcomplexes have positive or negative effects on initiation of DNA replication through variations in their phosphorylation state. In particular, in yeast and multicellular organisms, activation of S- and M-phase cyclin-dependent kinases (CDKs) are required for fine tuning each step of the assembly of the initiation complex and for the temporal program of origin activation. In addition, early and late replicons are affected by gene expression. As reported above, there is evidence of links between transcription and replication through factors involved in chromatin decondensation that is important for loading both transcription and replication proteins.

The following factors and enzymes have been identified in the replication mechanism of eukaryotes:

1. RPA (SSB in bacteria) was one of the first eukaryotic proteins shown to be involved in the initiation and elongation step of DNA replication. It is a hetero-trimeric complex whose role at the replication fork may be in unwinding and stabilizing single-stranded DNA assisted by MCM proteins.

2. PCNA (proliferating cell nuclear antigen) (pol β or dnaN in bacteria) is the *processivity factor*, also called the *sliding clamp*, responsible for loading the DNA polymerase onto the active template and ensuring that replisomes remain firmly attached to the template.

3. RFC (replication factor C) is a complex of five subunits, also called the *clamp-loading complex* (*gamma complex* in bacteria), which loads PCNA onto DNA.

In yeast at least eight DNA polymerases have been identified: α, β, δ, ε, ζ, η, and Rev1 protein and the mitochondrial (γ-like) polymerase (Budd and Campbell 2000). In mammals the number may be even greater. Pol-α, -δ, -ε, and -γ are essential replicative DNA polymerases; the others are nonreplicative, involved in DNA repair (see below).

DNA synthesis requires initial RNA priming. Priming in eukaryotes is carried out by the primase activity of the RNA primase pol-α complex (DnaG in bacteria) and is essential for lagging-strand synthesis as well as for the initiation of leading-strand synthesis. Once the RNA primer is synthesized, it is extended by pol-α generating a short DNA primer onto which the pol-δ system is assembled. The switch to replication by pol-δ requires RFC and PCNA.

In lagging-strand synthesis, the loaded pol-δ (pol III core in bacteria) extends the primer until it reaches the 5′ end of the next Okazaki fragment. RNA primers are removed by the concerted action of RNase H and FEN1 (pol I in bacteria) and the gap created is filled in by pol-δ. Finally, the resulting nick is then sealed by DNA ligase. The role of pol-ε in DNA replication is still obscure. One study suggests that pol-δ copies one strand and pol-ε copies the other. Pol-ε and -β are likely to be involved in DNA proofreading by their 3′-5′ exonuclease activity.

To ensure the accuracy of DNA sequencing and to maintain the genomic integrity required for cell survival and offspring, replication and repair mechanisms are tightly linked. Because repair is required during DNA replication, it makes sense, in terms of efficiency and cell economy, that it has not evolved separately from replication. Several DNA replication proteins also participate directly in repair: for example, DNA polymerases, DNA primase, RPA, PCNA, and RFC interact with repair proteins to repair breaks, fill gaps, and correct base alterations caused by DNA damaging agents.

In *S. cerevisiae* there are at least five pathways of repair: incision followed by excision [also called nucleotide excision (NER) to repair ultraviolet-induced pyrimidine dimers], repair of lesions occurring during replication, repair by a recombinatorial mechanism (often referred to as *double-strand break repair*), repair of double-strand breaks by direct ligation of broken ends (Ku pathway), and mismatch repair pathways, which excise damage in the form of noncomplementary pairs of normal bases arising during replication and recombination (Budd and Campbell 1997). These repair systems have widely divergent mechanisms, all requiring the synthesis of new DNA. The length of the DNA gaps to be repaired following incision and/or degradation and resynthesis range from a few nucleotides to several thousand. All polymerases participate in recognizing the damage and in the repair, although the DNA polymerases involved and the proteins that recruit them to the damage in the various pathways are still not defined completely.

Similarly, repair enzymes not required for replication per se interact with the DNA replication enzymes to resolve damage occurring during replication. Thus, DNA repair enzymes are targeted to sites where DNA replication creates mistakes (or encounters unrepaired damage), and DNA replication enzymes are targeted to sites of environmentally or metabolically induced DNA damage. The repair pathways in eukaryotes were initially characterized in *S. cerevisiae* because temperature-conditional mutants of most replicative genes can be created. More recently, studies have been carried out in metazoans in connection with human genetic repair diseases involving repair deficiency (e.g., *Xeroderma pigmentosus*).

Another fundamental biological process is recombination and its generation of genetic diversity. During the prophase of the first meiotic division, homologous chromosomes condense, pair, and synapse; in this way they exchange genetic material prior to segregation at metaphase. Recombination requires chromosomes to be broken [double-strand breaks (DSBs)] and rejoined, with some of the breakage–reunion events leading to recombinant products (visualized microscopically as chiasmata). A meiosis-specific structure, the synaptonemal complex (SC), facilitates synapsis and recombination between homologs. DSBs determine recombination in meiotic cells and are a major factor in recombination of mitotic cells. The processes of genetic recombination and repair of damaged DNA are intimately related. DSB repair events are classified as *homologous recombination* if the interaction involves nearly perfected matched base pairs, and as *illegitimate* or *nonhomologous repair* if DNA ends are joined with no complementary base pairs at the junction (Paques and Haber 1999).

Understanding the relationships between DNA repair, recombination, gene expression, and DNA replication, and the control of their interplay, is required for a more complete picture of cell division and the most fundamental process affecting biological life (Flores-Rozas and Kolodner 2000). It is also essential for clarifying the processes connected to normal and abnormal mammalian DNA replication and for developing new therapeutic strategies in cancer and other dysfunctions of replication.

2.6 GENE EXPRESSION

2.6.1 Transcription and Posttranscriptional Regulation

The uncoupling between transcription and translation is a major difference between eukaryotic and prokaryotic gene expression. In addition, regulation of gene expression in eukaryotes involves chromatin structures and DNA methylation because these can affect the ability of the transcriptional regulatory proteins and of RNA polymerases to gain access to specific genes. The expression of an eukaryotic gene is accomplished in five major stages: transcription, RNA processing, translation, protein modification, and subcellular localization. Usually, these processes have been studied separately, although they are intimately interconnected. Figure 2.6 shows the general structure of a eukaryotic gene and the processes of transcription and posttranscriptional mRNA maturation.

In eukaryotes, the transcription of nuclear genes is shared among RNA polymerase (pol) I, pol II, and pol III (Hampsey and Reinberg 1999; Paule and White 2000). Pol I synthesizes ribosomal (r) RNAs, pol II messenger (m) RNAs, and pol III 5S rRNA, transfer (t) RNA, 7SL RNA, U6 small nuclear (sn) RNA, and a few other small stable RNAs, many involved in RNA processing. Although pol II has received the most attention, due to the huge variety of pol II–transcribed genes that encode proteins, the activities of pol I and pol III together are responsible for the bulk of cellular transcription (about 80% of total RNA synthesis). Pol I is localized in nucleoli and synthesizes rRNA in the fibrillar centers where rRNA is then processed and assembled into ribosomes. Pol II and pol III transcription occurs at about 8000 and 2000 sites, respectively, spatially separate within the nucleoplasm.

The eukaryotic nuclear RNA polymerases are complex enzymes, made up of 12 or more subunits. Five of these subunits (ABC 10 α, ABC 10 β, ABC 14.5, ABC 23, and ABC 1027 in *S. cerevisiae*) are shared by all three enzymes. Moreover, pol I and pol III share two subunits (AC19 and AC40 in *S. cerevisiae*) which are not used in pol II. The presence of common subunits would seem to offer the opportu-

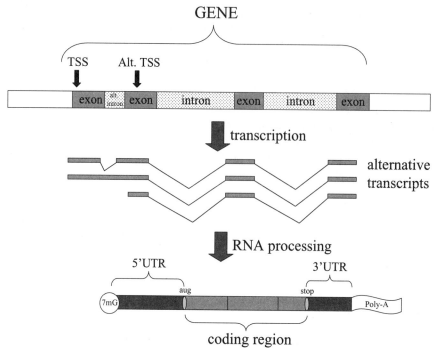

Figure 2.6. General structure of a eukaryotic gene where three alternative transcripts can be generated by two alternative transcription start sites (TSSs) and one facultative intron. The tripartite structure of one mature mRNA, composed of the 5′ and 3′ untranslated region (UTR) and the protein-coding sequence, is also shown.

nity for coordinated regulation. Although the transcription factors regulating pol I and pol III are different, there are fundamental similarities in their enzymatic activities. The polymerase is recruited through protein–protein interaction by a single protein binding upstream of the transcription start site, which positions the polymerase precisely to start transcription at a single site, called the *transcription start site* (TSS) or *+1 site*. Whereas the transcription machinery of pol II and pol III is often compatible with genes from widely different species, pol I transcription exhibits stringent species specificity, in the interaction of the transcription factors with the promoter and, to a lesser extent, in the protein–protein interactions between factors.

There is very little sequence similarity between rRNA promoters from different species, but the general organization of functional promoter elements is highly conserved from yeast to humans. Multiple copies of rRNA genes are found as repeated clusters, usually arranged tail to tail. The entire promoter is located in the intergenic spacer (IGS) between the units transcribed. It is composed of the core region (50 bp upstream of the initiation site), containing the only conserved sequence element, the ribosomal initiator (rINR). This AT-rich element serves to bind the core factor that recruits pol I. In most species, protein factors and additional DNA elements, including the upstream promoter element (UPE) (150 to 200 bp upstream of +1), the proximal terminator (PT), and the repeated enhancer elements (Henderson and Sollner-Webb 1990; Smith, O'Mahony et al. 1993), help to assemble and stabilize the complex formed on the core promoter. Spacing and helical side positioning between elements is critical for the interaction between proteins bound to the elements.

In many metazoans, the IGS also contains additional functional promoters called *spacer promoters* (SPs) from which transcripts with unknown function start. Phylogenetic comparisons and experimental data suggest that enhancers evolved from the spacer promoters by repetitive duplication and truncation. The most striking and unusual feature of pol III is that it can use three different promoter types. Most require sequence elements downstream of the initiation site (i.e., within the region transcribed). These regions are in general discontinuous structures composed of essential blocks separated by nonessential regions. A classic example is the promoter of 5S rRNA (type I promoters) found in vertebrates and in many lower organisms, including *Drosophila melanogaster* and *S. cerevisiae*.

The most common promoter arrangement used by pol III is found in the tRNA genes (type II). It contains two highly conserved sequence blocks, A and B, within the region transcribed. Unlike type I promoters, location and distances between blocks are extremely variable. Type III promoters have an extragenic organization. The best characterized type III promoter is for a human U6 snRNA. It contains a TATA box, a proximal sequence element (PSE), and a distal sequence element (DSE).

As gene expression is thought to be regulated predominantly at the level of transcription of RNA pol II, there has been a focused and sustained effort to dissect the mechanisms of transcriptional control (Greenblatt 1997). Eukaryotic organisms have evolved an impressive repertoire of tools and refined mechanisms to fine-tune gene expression and modulate the interplay among thousands of genes in essential events such as development and differentiation (Sauer and Tjian 1997).

mRNA synthesis by pol II requires the regulated assembly of multiple protein complexes and two distinct types of DNA elements: core promoter sequences and arrayed enhancers. The *core promoter*, whose canonical consensus is missing in several known promoters, specifies the accurate initiation site of transcription and is composed of a TATA box located closely upstream of the transcription start site (TSS) and an initiator element (INR), overlapping the TSS. Several additional sites, defined as *enhancers* and located upstream and downstream of the TSS, are used to modulate gene transcription in both a temporal (e.g., developmental specific) and a spatial (e.g., tissue-specific) framework.

The core and enhancer promoter elements are recognized by two classes of transcription factors: *general* (or *basal*) *transcription factors* (GTFs), which bind to the core promoter close to the site of transcription initiation, and *gene-specific activators*, which bind to the enhancer elements at varying distances upstream of the transcription start site (Muller 2001). Both GTFs and sequence-specific transcription factors contact RNA polymerase, either directly or via mediators in a multicomponent initiation complex that can contain more than 50 different polypeptides. These complexes have a dynamic nature, changing in composition and conformation during different stages of transcription. The basal transcription factors include pol II itself and at least six GTFs: TFIIA, TFIIB, TFIID, TFIIE, TFIIF, and TFIIH. All these factors contribute to assembly of the preinitiation complex (PIC) that initiates transcription. Table 2.4 summarizes general transcription factors associated with pol II.

TABLE 2.4. General Transcription Factors Associated with RNA Pol II in Humans

Factor	Number of Subunits	MW (kDa)	Function
TFIID-TBP (TATA-box-binding protein)	1	38	Recognize the core promoter and recruit TFIIB
TFIID-TAFs (TBP-associated proteins)	12	15–250	Assist transcription activation and promoter recognition
TFIIA	3	12, 19, 35	Stabilize binding between and promoter
TFIIB	1	35	Recruit RNA pol II and TFIIF
TFIIF	2	30, 74	Assist the binding of complex assembly at the promoter
TFIIE	2	34, 57	Recruit TFIIH; modulate helicase, ATPase, and kinase activity
TFIIH	9	32–89	Promote melting through helicase activity; DNA repair

Source: Modified from Roeder (1996).

In the absence of transcriptional activators, eukaryotic genes are transcribed weakly at the basal level. Thus, the principal mechanism for differential gene activity is via the action of sequence-specific enhancer-binding proteins that can direct activated levels of transcription. Unfortunately, very little is known about pol II promoter structure and organization, and consequently, prediction about its activity is still largely inefficient (see Sec. 6.8.5). The expression of a gene is dictated in part by integration of the cellular and environmental signals controlling the activity of transcription regulatory proteins. The response of a gene to incoming signals will depend on the physical state of the gene and the machinery that transcribes it. The physical state of the gene can fall in one of three broad categories: repressed, basal, or induced. A *repressed gene* is essentially "off" and might be encased within chromatin such that the transcription machinery or other factors cannot access the underlying promoter DNA. A *basely expressed gene* might reside within "open" chromatin and thus be accessible to the transcription machinery. However, in the absence of a functional activator to recruit the transcription machinery, the gene is expressed at low levels. *Induced genes* are likely to reside in open chromatin and be bound by transcriptional activators, which efficiently assemble the transcription machinery. Induced genes are typically expressed at high levels, although the level of induction will depend on the behavior of transcriptional regulators present in the transcription complex.

The production of mature mRNA following gene transcription requires at least three RNA-processing mechanisms: capping, splicing, and polyadenylation, all of which are interconnected with each other and with transcription (Siomi and Dreyfuss 1997), in particular with the elongation polymerase complex of pol II. Moreover, there is evidence that one process influences the others (Proudfoot 2000).

A surprising structural feature of pol II itself has provided a molecular connection between transcription and mRNA processing. The largest subunit of pol II contains an extraordinary C-terminal domain (CTD), in mammals comprising 52 seven amino acid repeats. Phosphorylation of its serine residues results in a huge change in the structure of this domain, which can switch transcription from initiation to elongation, probably releasing part of the initiation complex and at the same time creating space for interaction with RNA-processing factors. The first RNA-processing event, occurring soon after transcription has started, is *cap modification*. It consists of the addition of a cap structure at the 5′ end of mRNA by the unique 5′-5′ linkage of a 7-methylguanosine mediated by a 5′ cap guanine-N-7 methyltransferase. The capped 5′ end of the mRNA is then protected from endonucleases, and more important, it is recognized by specific proteins of the translation machinery. Therefore, capping is likely to mark the complete switch from transcription initiation to elongation.

The next RNA-processing reaction to take place on the nascent transcript is exon splicing. The complexity of splicing increases enormously from yeast to mammals. In yeast, few genes have introns, usually small and close to the start of the transcript (a complete list of yeast introns is collected in the Yeast Intron Database (YID; see the URL in the Appendix). However, most mammalian genes contain many introns, comprising well over 90% of the gene. Relatively small exons are hard to find in a sea of "intronic rubbish." Some splicing factors (SR proteins, CASP, SCAF) associate with CTD through its specific binding domain. A recent and surprising observation is that different types of promoters appear to dictate the splicing pattern of transcripts initiated on that promoter. In a context of alternatively spliced exons (e.g., fibronectin gene), it has been demonstrated that one exon was included or excluded depending on the promoter used for the transcription. Moreover, slowing down transcription by inserting pause elements between splicing regulatory signals [negative regulatory elements, (NREs)] may influence splicing patterns. Finally, the 3′ end of the mRNA is modified by the addition of a poly(A) tail (from 20 to 250 nt).

As with capping and splicing, 3′-cleavage and polyadenylation factors have also been clearly associated with the CTD during transcription elongation. A specific hexanucleotide, typically AAUAAA defined as poly(A) signal, is recognized by the endonuclease activity of the polyadenylation polymerase, which cleaves the primary transcript approximately 10 to 30 bp 3′ to the poly(A) signal. It was shown that the factor, called *cleavage polyadenylation stimulatory factor* (CPSF), is also associated with the general transcription factor, TFIID. The fact that the pol II CTD has a direct effect on mRNA 3′-end processing is highly consistent with the observation that poly(A)-signal recognition is required to terminate transcription. According to these findings, a new scenario of posttranscriptional mechanisms regulating gene expression is substituting the view that transcriptional regulation is the predominant regulatory mechanism.

Posttranscriptional regulation of gene expression can involve the on–off regulation of particular gene products in a temporally and spatially regulated manner, allowing cells of different types or at different developmental stages to fine-tune their patterns of gene expression. On–off regulation of RNA expression, obtained by controlling transcript degradation or by modulating its translation efficiency, can allow a cell to respond to environmental clues more quickly than de novo transcription would permit. Many important events in development, such as pattern formation and terminal differentiation, are regulated by an array of posttranscriptional mechanisms, controlling mRNA stability, localization, and translation. Posttranscriptional regulation of gene expression can also generate an enormous range of protein products from a single gene. Three forms of intranuclear fine-tuning generating RNA sequence diversity are well established: alternative splicing of pre-mRNA, alternative polyadenylation site selection, and RNA editing. In the cytoplasm, utilization of alternative translation start sites can also produce functionally different proteins from a single mRNA. In some cases, these processes constitute a switch that reverses function dramatically (e.g., transcriptional activators and repressors can be generated from the same gene). Using these sophisticated strategies, vertebrate cells can produce a much larger variety of proteins than the number of genes in the genome (Wang and Manley 1997; Smith and Valcarcel 2000; Graveley 2001; Will and Luhrmann 2001).

Eukaryotic mRNAs possess a tripartite structure consisting of a 5′ untranslated region (5′ UTR), a coding region made up of amino acid coding triplet codons, and a 3′ untranslated region (3′ UTR) (see Fig. 2.6). A crucial role in posttranscriptional regulation of gene expression is played by 5′ and 3′ untranslated regions of mRNA as they modulate nucleocytoplasmic mRNA transport, translation efficiency, subcellular localization, and stability.

The average length of 5′ UTRs, roughly constant among eukaryotes, is about 200 nt, whereas the average length of 3′ UTRs varies in the different taxa, ranging from about 800 nt in vertebrates to 200 nt in plants and fungi. A striking intraspecific length heterogeneity is also observed with both 5′ and 3′ UTR ranging from few to some thousand nucleotides. The genome regions corresponding to the mRNA UTRs may contain introns, generally more frequently in 5′ than in 3′ UTRs, with about 30% of genes in metazoa having one or more fully untranslated exons.

Regulatory activity is mediated by cis-acting oligonucleotide patterns located in 5′ and 3′ UTRs interacting with specific RNA-binding proteins. RNA-binding proteins play central roles in posttranscriptional regulation of gene expression. They contain regions that function as RNA-binding domains (RNP, zinc-finger, S1, etc.) and auxiliary domains that mediate protein–protein interaction and subcellular targeting. In addition, repetitive elements, such as microsatellites, SINEs, and LINEs, are fairly abundant in mRNA UTRs and may in some cases play a functional role (Tomilin 1999). Unlike DNA-mediated regulatory signals, whose activity is essentially mediated by their sequence, the biological activity of regulatory patterns acting at the RNA level relies on a combination of primary and secondary structure elements assembled in consensus structures that are the target of trans-acting RNA-binding proteins.

The major role of 5′ UTR is in the control of mRNA translation, which can be modulated by secondary structural features. In particular, long 5′ UTRs, upstream AUG or upstream ORFs, and stable secondary structures may hamper translation efficiency. Also, specific consensus structures may modulate translation efficiency. The iron responsive element (IRE) located in the 5′ UTR may inhibit translation through the iron-level-mediated binding of iron regulatory proteins (IRPs), which impedes the normal mRNA scanning process of the small ribosomal subunit in translation initiation (Hentze and Kuhn 1996) (Fig. 2.7). Cis-acting elements located in the 5′ UTR, such as the internal ribosome entry site (IRES) (van der Velden and

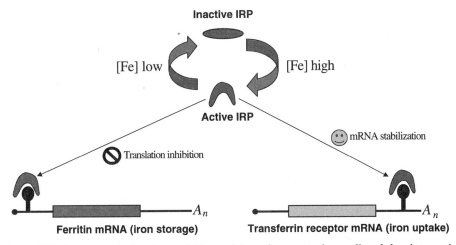

Figure 2.7. Posttranscriptional regulation of iron homeostasis mediated by interaction between the iron responsive element (IRE) and the iron regulatory proteins (IRPs). Low iron concentration activates IRPs that bind to IREs at the 5′ UTR of ferritin mRNA and to the 3′ UTR of transferrin receptor mRNA with opposite effects.

Thomas 1999) or short-sequence motifs complementary to 18S rRNA (Chappell, Edelman et al. 2000), may mediate cap-independent translation-promoting internal ribosome binding.

mRNA turnover is another crucial step in posttranscriptional regulation of gene expression. Induced variation in mRNA level may alter the expression of specific genes by altering the abundance of the corresponding protein. The poly(A) tail, which also contributes to translation efficiency by interacting with the 5′ UTR, protects the mRNA from degradation, as its shortening or removal initiates the mRNA degradation pathway. The turnover of mRNA is also regulated by cis-acting elements, such as the AU-rich elements (AREs) located in the 3′ UTR, which promote mRNA decay.

mRNA UTRs have a fundamental role in the spatial control of gene expression, particularly important during development. This control is obtained by specific sub-cellular localization of the mRNA mediated by cis-acting elements generally located in the 3′ UTR, also known as mRNA *zip codes*, which recognize zip-code-binding proteins. Examples include the *Drosophila* Stau protein, which is required for anchoring bicoid at the anterior pole of the oocyte. Zip codes can function in local control of translation or in active cytoplasmic transport.

2.6.2 Genetic Code and Codon Use

The universality of the genetic code has long been considered an essential feature of life. According to the *frozen accident theory*, the universal genetic code arose by accident at a certain time early in evolution and then become frozen because any possible code change, producing a completely altered protein inventory for a given organism, would have certainly been lethal. However, in recent years, a number of nonuniversal genetic codes have been reported not only in most nonplant

TABLE 2.5. Universal Genetic Code

First Base	Second Base U	C	A	G	Third Base
U	Phe	Ser	Tyr	Cys	U
	Phe	Ser	Tyr	Cys	C
	Leu	Ser	Stop	Stop	A
	Leu	Ser	Stop	Trp	G
C	Leu	Pro	His	Arg	U
	Leu	Pro	His	Arg	C
	Leu	Pro	Gln	Arg	A
	Leu	Pro	Gln	Arg	G
A	Ile	Thr	Asn	Ser	U
	Ile	Thr	Asn	Ser	C
	Ile	Thr	Lys	Arg	A
	Met	Thr	Lys	Arg	G
G	Val	Ala	Asp	Gly	U
	Val	Ala	Asp	Gly	C
	Val	Ala	Glu	Gly	A
	Val	Ala	Glu	Gly	G

TABLE 2.6. Known Deviations from the Universal Genetic Code in the Nuclear Genome

Codon	Deviation from the Universal Code	Taxon
CUG	Leu → Ser	*Candida*
AUA	Ile → unassigned	*Micrococcus*
UAA	Stop → Gln	Ciliated protozoan and green algae
UAG	Stop → Gln	Ciliated protozoan and green algae
UGA	Stop → Trp	Mycoplasma
UGA	Stop → Cys	Cyliate euplotes
UGA	Stop → selenocysteine	Bacteria, eukaryotes
CGG	Arg → unassigned	Spiroplasma
AGA	Arg → unassigned	Spiroplasma

mitochondrial genomes, but also in several nuclear systems (Tables 2.5 and 2.6), thus contradicting the frozen accident theory.

To allow for genetic code changes to occur during evolution, it is likely that some ambiguity has to be tolerated during code transition, with more than one meaning to be attributed to the same codon. This is possible if mutations take place in the tRNA or in its cognate synthetase, which make it possible for the same tRNA to be charged with different amino acids or that different tRNAs, charging different amino acids, come to possess the same anticodon. Indeed, it has been reported that the serine tRNA[CAG] responsible for the nonuniversal decoding of CUG into serine may, to some extent also charge leucine (Santos, Cheesman et al. 1999). Data reported in Table 2.6 show strikingly that with the exception of CUG reassignment in *Candida*, all the changes involve unassignment of sense codons or sense reassignment of termination codons. It has been shown that most of these codon changes involve mutations in the releasing factor protein eRF1 which render a particular stop codon ineffective for peptide chain termination. Subsequently, tRNA-mediated mutations can gradually produce a sense meaning for that codon without causing lethality. Mutations in the amino terminal domain of eRF1, whose tridimensional structure resembles the tRNA cloverleaf structure, might change the recognition specificity of a stop codon, allowing its subsequent recruitment as a sense codon. Comparative analysis of eRF1 from several species confirmed this hypothesis because the Ile32 found in all species with the canonical code was replaced by Val32 in all species that recruited UAR codons for glutamine (Lehman 2001).

In unicellular eukaryotes (e.g., *S. cerevisiae*), as in prokaryotes, the synonymous codon choice may regulate translation. In other words, highly expressed genes that show a codon use bias stronger than other genes are under strong constraints for codon choice in that they preferentially use synonymous codons that correspond to the most abundant tRNA species. In contrast, in multicellular eukaryotes, codon use bias is not corrected to the level of expression but to the base composition of the relevant genomic region. Quite interestingly, a relationship between codon use and protein structure has been observed in mammals but not in *E. coli* (Xie, Ding et al. 1998; Zeeberg 2002).

2.6.3 Translation and Posttranslation Modifications

Translation takes place in the cytoplasm. The translation of eukaryotic mRNAs is a highly competitive and tightly regulated step mainly in the initiation phase. Robust translation of cap-dependent mRNAs occurs in G1-to-S transition, whereas IRES-mediated translation (see Sec. 2.6.1 and below) is probably involved in the synthesis of key proteins and in G2-to-M transition. Global translation rates are reduced in mitosis; only translation of a subset of mRNAs is induced specifically (Pyronnet and Sonenberg 2001). These changes correlate with the activities of several translation eukaryotic initiation factors (eIFs), which are modulated during cell cycle. More than nine eIFs have been identified, and several of these factors are composed of multiple polypeptide chains. This large number of polypeptides suggests that protein–protein interactions are likely to play an important role in translation initiation. Translation initiation is divided into three phases: (1) binding of the 40S small ribosomal subunit with initiator tRNA and eukaryotic initiation factors (eIF) near the 5' end, (2) scanning along the 5' UTR, and (3) starting codon recognition and 60S subunit joining. The resulting 80S ribosome then begins translation.

According to the cap-dependent model of translation initiation, the 5' cap structure attracts the eukaryotic initiation factor 4F (eIF4F) complex to the mRNA. eIF4F is a heteromultimeric complex composed of the cap-binding protein eIF4E, the RNA-dependent ATPase eIF4A (an RNA helicase), and the modular factor eIF4G. The small (40S) ribosomal subunit binds to the 5' end of mRNAs as a 43S complex, including eIF3, a multisubunit initiation factor, and the ternary complex eIF2 with GTP and Met-tRNA. Movement of the 43S complex along mRNA, termed *scanning*, determines recognition of the AUG triplet as the start codon. Then the scanning mechanism of translation initiation identifies the first AUG from which to initiate translation. Indeed, several mRNAs have other AUGs or even ORFs upstream the functional AUG start codon. In these cases, leaky scanning, reinitiation, or ribosome shunting may take place to synthesize the correct product even if

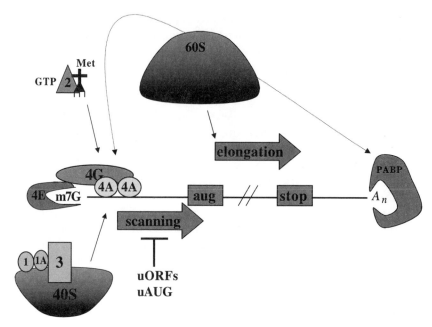

Figure 2.8. General model for translation initiation of eukaryotic mRNAs.

at a lower efficiency (Kozak 2001) (Fig. 2.8). Sequences flanking the AUG may modulate the efficiency with which this is recognized as the initiator codon to start translation. In particular, Kozak (1989) has first observed that in vertebrate mRNA, the context of the initiator site should conform to the sequence GCCRCCaugG, where the key nucleotides are two purines (R), usually A, in position −3 and guanine in position +4 with respect to the AUG start codon. The strong preference for A^{-3} and G^{+4} has also been observed in the initiator context of mRNAs from invertebrates, plant, and fungi. Although experimental results suggest that the contribution of other bases of the context to the start site selection is only marginal, base composition at these positions, particularly at −2 and −1, is strongly biased, thus they are likely to exert some kind of modulation of translation initiation efficiency (Pesole, Gissi et al. 2000b). Codon–anticodon base pairing between the initiator AUG and Met-tRNA triggers eIF2-bound GTP hydrolysis, catalyzed by eIF5, and this causes dissociation of initiation factors and a large (60S) subunit joining to form the 80S ribosome.

The poly(A) tail of mRNAs also participates in translation initiation and acts synergistically with the cap structure. RNA circularization is thus emerging as a central feature of eukaryotic translation initiation. The enhancement of cap-mediated initiation by the 3′ poly(A) tail involves direct interaction between poly(A)-binding protein (PABP), which is bound to the tail, and the eIF4G subunit of cap-bound eIF4F, which in effect circularizes mRNA. This interaction could enhance recruitment of terminating ribosome to the 5′ end of capped mRNAs; moreover, it appears to increase translational fidelity by ensuring that only properly processed and exported mRNA molecules are translated. This interaction has been observed in yeast, mammals, and plants.

Cells adapt to stress and changing growth conditions by altering the translational capability. Translation can be regulated globally by affecting initiation factor function (e.g., through phosphorylation of eIF4E, and eIF2, or by the disruption of the eIF4G adapter function). In particular, cleavage of eIF4G has been correlated with the shutdown of translation in cells infected by certain picornaviruses. It is proposed that the viral cleavage of eIF4G inhibits translation by dissociating the cap-binding function from the RNA-unwinding and ribosomal-binding activities of the cap-binding-protein complex. Cleavage of eIF4G is not restricted to virus-infected cells; indeed, cleavage of eIF4G by caspase 3 has been observed during apoptosis. IRES-mediated cap-independent translation is independent of eIF4F integrity and is likely to function during a cell cycle as a general mechanism to maintain the synthesis of a subset of proteins when cap-dependent translation is compromised.

An additional important role in the regulation of expression is played by post-translational modifications that make proteins functionally active. Mechanisms such as glycosylation, acetylation, phosphorylation, processing, folding, and oligomerization may determine the stabilization and maturation of the translation products and the targeting to specific cellular or extracellular compartments. Several of these modifications are accomplished by cellular protein trafficking, which through fusion of endoplasmic reticulum (ER) and/or Golgi exocytic vesicles delivers proteins to the specific destinations or secretes them from the cells. The various fates of proteins may be linked to the localization of their mRNAs. The formation of ER to Golgi transport vesicles have been identified in yeast, mammalian cells, and plants, determined by the COPII (coat protein II) complex. The targeting systems comprise three components: signals on the proteins (e.g., nuclear localization signals, nuclear export signals, matrix-targetin signals), chaperonin factors that ensure transport-competent folding, and membrane translocation apparatuses (receptors and channels) that represent active devices for protein-specific recognition and unidirectional translocation across or insertion into lipid bilayers.

Approximately one-third of the newly synthesized proteins are transported to the lumen of the ER. Alterations in homeostasis after various cellular stresses, which prevent protein folding and cause the accumulation of misfolding or malfolding proteins in the ER, have the potential to induce cellular damage. Eukaryotic cells have adapted for survival to deal with an accumulation of unfolded proteins in the ER. Their response includes transcriptional induction, translational attenuation, and degradation of unfolded proteins. In contrast, excessive and/or long-term stress in the ER results in apoptosis. Recent evidence indicates that ER stresses and accumulation of abnormal structures in cytoplasm or nucleus are associated with genetic or neuronal degenerative diseases (e.g., Huntington's, Parkinson's, or Alzheimer's disease).

Another intriguing posttranslational mechanism is *protein splicing*, when precursor proteins undergo a maturation process in which intervening protein sequences called *inteins* are removed to form a mature *extein* protein and the free inteins. Protein splicing involves an autoproteolytic reaction mediated by inteins themselves. Inteins can be found in eukaryotes, bacteria, and archea, thus suggesting their early origin (Pietrokovski 2001). Replication, transcription, and translation are basic processes that have been studied and proven in the few most-studied organisms. No doubt genomics will contribute to our knowledge by scaling up present data and showing how they fit entire genomes and different organisms.

Beside the common features, the structures and organization of eukaryotic genes show species-specific features illustrated in the following paragraphs, providing a summary description of completely sequenced eukaryotic genomes.

2.7 COMPLETELY SEQUENCED EUKARYOTIC GENOMES

The number of genomes in eukaryotes whose sequencing is in progress or which are completely sequenced is obviously much smaller than in prokaryotes. The major features of seven completely sequenced genomes are described here. They include two fungi (*Saccharomyces cerevisiae* and *Schizosaccharomyces pombe*), two plants (*Arabidopsis thaliana* and *Oryza sativa*), two invertebrates (*Drosophila melanogaster* and *Caenorhabditis elegans*) and one mammal (*Homo sapiens*). In addition, the draft genomes of an invertebrate, the mosquito *Anopheles gambiae*, and two vertebrates, mouse (*Mus musculus*) and the bony fish *Fugu rubripes*, have also been recently released. About 15,000 genes have been predicted in the 280 Mbp of mosquito (see the URL of the Mosquito Genome Draft in the Appendix).

The mouse sequence derives from whole genome shotgun of around 7× coverage. The current assembly (April 2002) comprises about 2.46 Gbp from 96% of total euchromatic DNA (see Mouse Genome Draft in the Appendix). Of the 22,444 genes so far predicted, about 75% have a firm counterpart in the human genome.

About 31,000 genes have been predicted so far in the Fugu genome (see the URL of the Fugu Genomics Project in the Appendix). The analysis of this genome, much smaller than other vertebrate genomes (around 400 Mbp) containing the same repertory of genes as in humans, should accelerate characterization of the complete inventory of human genes and of their controlling sequences. Other genome projects are in progress and will soon be completed (Table 2.7).

Some comments on the organisms selected follow. Apart from *Homo sapiens*, which always plays a major role in biology research, the choice for entire genome sequencing has been directed to species already studied in laboratory experiments whose genetics is already well known. This is true of the invertebrates *C. elegans* and *D. melanogaster* genomes as well as for the almost ready mouse genome, among mammals. For plants, the sequencing of *A. thaliana*, beside than of maize and oryza, is justified by the small size and simplicity of its genome. For unicellular organisms, *S. cerevisiae* has a priority for its genetic simplicity and because it is widely used in both laboratory experiments and biotechnological applications, whereas *S. pombe* represents a model organism for study of the cell cycle.

Also recently completed has been the tiny eukaryote genome (2.9 Mbp) of *Encephalitozoon cuniculi*, a microsporidian intracellular parasite that infects a wide range of eukaryotes, ranging from protozoans to humans (Katinka, Duprat et al. 2001). It was identified as the agent of opportunistic infections in humans after the emergence of AIDS and the use of immunosuppressor therapies for organ transplants. The *E. cuniculi* genome is organized in 11 chromosomes, ranging in size from 217 to 315 kbp. A total of 1997 genes have been predicted that do not include genes for some biosynthetic pathways (e.g., the tricarboxilic acid cycle), thus indicating the strong host dependence of this organism. Quite interestingly, *E. cuniculi* lacks mithocondria, but the identification in its genome of genes related

TABLE 2.7. Eukaryotic Ongoing Genome Projects

Organism	Information	Size (kp)
Cryptosporidium parvum	Protozoa (8 chromosomes)	10,400
Dictyostelium discoideum	Protozoa (6 chromosomes)	34,000
Entamoeba histolytica	Protozoa	20,000
Giardia lamblia	Protozoa (5 chromosomes)	12,000
Leishmania major	Protozoa (36 chromosomes)	33,600
Plasmodium chabaudi	Protozoa	30,000
P. falciparum	Protozoa (14 chromosomes)	30,000
P. vivax	Protozoa	
P. yoelii	Protozoa	
Tetrahymena sp.	Protozoa	
Theileria annulata	Protozoa (4 chromosomes)	10,000
T. parva	Protozoa	11,500
Trypanosoma brucei GuTat10.1	Protozoa	53,000
T. brucei TREU 927/4	Protozoa (11 Chromosomes)	35,000
T. cruzi	Protozoa	
Encephalitozoon cuniculi	Microsporidia	2,900
Spraguea lophi	Microsporidia	6,200
Aspergillus fumigatus Af 293	Fungi	35,000
A. nidulans	Fungi	31,000
A. niger	Fungi	38,000
Botryotinia fuckeliana	Fungi	
Candida albicans 1161	Fungi	15,000
C. albicans SC5314	Fungi	
Cryptococcus neoformans JEC 21	Fungi	
Neurospora crassa	Fungi	47,000
N. crassa 74-OR23-IVA	Fungi	43,000
Phanerochaete chrysosporium	Fungi	30,000
Pichia angusta	Fungi	
Pneumocystis carinii f.sp. *carinii*	Fungi	7,700
P. carinii f.sp. *hominis*	Fungi	7,500
Schistosoma mansoni	Fungi (8 chromosomes)	270,000
Schizosaccharomyces pombe	Fungi	14,000
Ustilago maydis	Fungi	20,000
Caenorhabditis briggsae	Nematodes	
Brassica napus	Plants	
Chlamydonas reinhardtii	Plants	100,000
Glycine max (soybean)	Plants	
Gossypium hirsutum (cotton)	Plants	
Medicago sativa	Plants	
M. truncatula	Plants	
Phaseolus vulgaris	Plants	
Phytophthora infestans	Plants	250,000
P. sojae	Plants	65,000
Sorghum bicolor	Plants	
Thalassiosira weissflogii	Plants	25,000
Zea mays (maize)	Plants	
Strongylocentrotus purpuratus (sea urchin)	Echinozoa	900,000
Danio rerio (zebrafish)	Fish	
Ciona intestinalis (sea squirt)	Phlebobranchia	
Takifugu rubripes (fugu)	Fish	400,000
Oreochromis niloticus (tilapia)	Fish	
Xenopus laevis	Amphibia	
Gallus gallus (chicken)	Birds	
Bos taurus (cattle)	Mammals	
Canis familiaris (dog)	Mammals	
Mus musculus (mouse)	Mammals	3,000,000
Ovis aries (sheep)	Mammals	
Pan troglodytes (chimpanzee)	Mammals	
Rattus norvegicus (rat)	Mammals	
Sus scrofa (pig)	Mammals	

Source: Data from GOLD (Genome OnLine Database), *wit.integratedgenomics.com/GOLD*.

to mitochondrial function suggests a loss of the organelle during evolution and supports single-fusion events for the origin of the eukaryotic cell (see Sec. 7.4).

Other eukaryotic sequencing projects completed to date are chromosome 2 (947 kbp, 205 ORFs) and 3 (1060 kbp, 220 ORFs) of *Plasmodium falciparum* (Gardner, Tettelin et al. 1998; Bowman, Lawson et al. 1999) and chromosome 1 (257 kbp, 79 ORFs) of *Leishmania major* (Myler, Audleman et al. 1999). The two latter protist species are particularly interesting because they are responsible for leishmaniosis and malaria, very common diseases that affect a large portion of the human population.

2.7.1 *Saccharomyces cerevisiae* Genome

The baker's yeast *S. cerevisiae* is considered an ideal eukaryotic microorganism for biological studies. Due to its importance as a model system, it is the first eukaryotic genome to have been sequenced completely, through an international effort involving some 600 scientists in Europe, North America, and Japan.

The yeast genome, released in April 1996, contains a haploid set of 16 well-characterized chromosomes, ranging in size from 200 to 2200 kbp, for a total of 12,068 kbp (Goffeau, Barrell et al. 1996). Other nucleic acid entities can also be considered to be part of the yeast genome: namely, plasmids (a 2-μm circle plasmid is present in most strains); dsRNA viruses, present in almost all strains and accounting for approximately 0.1% of total nucleic acid; and a 20S circular single-stranded RNA which acts as an independent replicon.

Base compositional analysis of the first yeast chromosome to be sequenced, chromosome III, showed striking variations along the chromosome; specifically, in chromosome III each arm displays one large peak in G + C content, evident particularly at the third positions of codons (GC3s).

These findings were confirmed by observations on the longer chromosomes, XI, II, and VIII, of similar multiple periodicity of G + C peaks (Dujon, Alexandraki et al. 1994; Feldmann, Aigle et al. 1994; Johnston, Andrews et al. 1994). GC-rich peaks, where ORFs of high GC content at third silent codon positions are preferentially located, have been found mostly concentrated in shorter chromosomes, thus resulting in a negative relationship between GC content and chromosome length (Bradnam, Seoighe et al. 1999). GC3s variation, however, seems not to be related to gene expression levels, because the set of "preferred" codons in highly expressed yeast genes has a G + C content similar to the mean content of the entire genome (Sharp, Stenico et al. 1993).

Similarly to the human and, in general, for mammalian genomes, it has been suggested that peaks in GC content correspond to an increase in gene density (Feldmann, Aigle et al. 1994; Johnston, Andrews et al. 1994; Galibert, Alexandraki et al. 1996; Bowman, Churcher et al. 1997; Churcher, Bowman et al. 1997; Dujon, Albermann et al. 1997; Jacq, Alt-Morbe et al. 1997). However, these apparently clear patterns of regional GC content variation along chromosomes and of GC content/gene density correlation are still a matter of discussion. Bradnam, Seoighe et al. (1999) found that variation in GC3s is not completely random because the clusters of ORFs of similar GC3s values observed can be accounted for by considering the very short range correlation between neighboring ORFs. High-GC3s ORFs are located preferentially on shorter chromosomes, and the distribution of all GC3s values is not normally distributed. In this context, the behavior of chromosome III (with the most pronounced regional variation in GC3s content) appears to be unique among yeast chromosomes, probably because it contains the mating-type loci, which are preserved by selective pressure.

Zenvirth, Arbel et al. (1992) demonstrated that the periodicity in base composition (the alternance of GC peaks) seen on chromosome III corresponds to a variation in recombination frequency along the chromosome arms, with the AT-rich regions coinciding with the recombination-poor centromeric and telomeric sequences and GC-rich peaks coinciding with high recombination regions in the middle of each arm of the chromosome. In particular, Zenvirth and co-workers have demonstrated that the relative incidence of double-strand breaks, which are thought to initiate genetic recombination in yeast (de Massy, Rocco et al. 1995), correlates directly with the GC-rich regions of this chromosome. The four smallest chromosomes (I, III, VI, and IX) seem to exhibit average recombination frequencies some 1.3 to 1.8 times greater than the average for the genome as a whole. Kaback (1989) suggested that high levels of recombination have been selected for these very small chromosomes to ensure at least one crossover per meiosis, and thus permit correct segregation.

Gene prediction in *S. cerevisiae* is still in progress (Johnston 2000), with several databases reporting slightly differing data: the Saccharomyces Genome Database (SGD) lists 6281 ORFs, the Yeast Protein Database (YPD) lists 6142 ORFs, and MIPS lists 6495. Goffeau, Barrell et al. (1996) report a total of 6275 open-reading frames (ORFs) of over 100 amino acids long, but 390 of them are unlikely to be translated; thus approximately 5885 ORFs were predicted to specify protein products in the yeast cell.

The yeast genome is highly compact, with almost 70% of the total sequence saturated with ORFs, and on average one protein-encoding gene is found every 2 kbp of the yeast genome. The average size of yeast genes is 1.45 kbp, or 483 codons, a ranging from 40 to 4910 codons; only 3.8% of the ORFs contain introns. Approximately 30% of the genes have already been characterized experimentally, and about one-half of the remaining 70% with unknown function contain either a motif of a characterized class of proteins or correspond to genes encoding proteins that are structurally related to functionally characterized gene products from the same or other organisms.

In addition to gene prediction, the most intriguing challenge at present is to define the functions of each of these products. Computer analysis of the yeast proteome allows classification of about 50% of the proteins based on simple and conservative homology criteria with other proteins of known functions (see Sec. 6.6) (Goffeau, Barrell et al. 1996). Table 2.8 reports the functional classification of protein according to the MIPS yeast genome resource. A similar classification is also reported by YPD (Costanzo, Hogan et al. 2000; Costanzo, Crawford et al. 2001). The still high number of proteins of unknown function (see Table 2.8) demonstrates that computational approaches are largely inadequate to elucidate the functions of the entire yeast proteome, and often provide only a general description of the biochemical function of the protein products predicted (such as "cell division" or "transcription factor") with no indication of their precise biological role. However, computational gene prediction represents a valuable guide to experimentation.

TABLE 2.8. Functional Classification of Yeast Proteins[a]

Cellular Role	N	%
Metabolism	1062	16.4
Energy	252	3.9
Cell growth, division, and DNA synthesis	833	12.8
Transcription	787	12.1
Protein synthesis	357	5.5
Protein destination	357	5.5
Transport facilitation	310	4.8
Cellular transport and transport mechanism	495	7.6
Cellular biogenesis (excluding organelle-located proteins)	205	3.2
Cellular communication/signal transduction	133	2.0
Cell rescue, defense cell death, and aging	363	5.6
Ionic homeostasis	123	1.9
Cellular organization	2254	34.7
Transposable elements, viral and plasmid proteins	116	1.8
Classification not yet clear-cut	150	2.3
Unclassified proteins	2437	37.5

Source: Data from the MIPS Yeast Genome Database, *mips.gsf.de/proj/yeast*.

[a] Of the total 6495 annotated ORFs, 3908 match at least one category. Protein function has been characterized by genetics, biochemistry, or similarity to proteins characterized.

In vivo and in vitro experimental studies proved much additional data. Velculescu, Zhang et al. (1997), using the Serial Analysis of Gene Expression (SAGE; see Sec. 4.2.3) have analyzed the transcription profile of the set of genes expressed from the yeast genome, called the *transcriptome*. In particular, based on the analysis of over 60,000 transcripts, they report the presence of about 4665 genes, among which only 1981 have known function. Because in many cases there is no correlation between mRNA concentration and protein abundance (Gygi, Rochon et al. 1999), a more accurate analysis of gene expression must be carried out at the protein level: for example, through two-dimensional gel electrophoresis followed by protein spot characterization (Washburn and Yates 2000). Apart from protein-coding genes, the yeast genome contains approximately 140 rRNA genes in a large tandem array on chromosome XII and 40 genes encoding snRNAs scattered throughout the 16 chromosomes. Yeast contains 275 tRNA genes belonging to 43 families, with 80 genes containing introns. In addition, *S. cerevisiae* chromosomes contain movable DNA elements, *retrotransposons*, that may vary in number and position in different strains.

In protein coding and noncoding regions of the yeast genome, microsatellites or simple sequence repeats (SSRs), tandemly repeated tracts of DNA 1 to 6 bp long, and ubiquitous units in prokaryotes and eukaryotes have been found and are inherently unstable. Two models have been proposed to explain microsatellite generation and instability: DNA polymerase slippage and unequal recombination. Toth and co-workers (Toth, Gaspari et al. 2000) investigated the abundance of microsatellites with repeated unit lengths of 1 to 6 bp in several eukaryotic taxonomic groups, among which there is the yeast. Intergenic *S. cerevisiae* regions and introns contain fewer hexanucleotide repeats than exons do, in contrast with what is observed in the other genomes completely sequenced. Similarly, Miret, Pessoa-Brandao et al. (1998) suggest that CAG and CTG trinucleotide repeats frequently undergo contractions in yeast.

There are five families of yeast retrotransposons, Ty1 to Ty5, with little variation within families in terms of both size and sequence (Jordan and McDonald 1999). The closely related Ty1 and Ty2 are the most numerous elements in the genome. Comparisons with intraelement 5′-3′-long terminal repeat (LTR) sequence comparisons indicate that almost all Ty elements have recently transposed, thus suggesting a rapid genomic turnover of these elements. This pattern of variation indicates that the majority of full-length Ty1, Ty2, and Ty1/2 insertions represent active or recently active elements and is consistent with a high level of genomic turnover.

Other than repeated regions, the genome of S. cerevisiae contains approximately 55 large duplicated chromosomal regions plus several smaller duplicated chromosomal regions (Seoighe and Wolfe 1998). These duplicated regions are traces of ancient tetraploidy, due to a whole-genome duplication that occurred approximately 10^8 years ago (Wolfe and Shields 1997; Seoighe and Wolfe 1998). This event was followed by the widespread deletion of superfluous duplicate genes, sequence divergence of the remaining duplicates, and successive genomic rearrangements. Seoighe and Wolfe (1998) have suggested in yeast that gene order evolution after the duplication could be due to a prevalence of successive random small inversions (Seoighe, Federspiel et al. 2000).

Fisher, Shi et al. (2000) tested the chromosomal speciation model whereby chromosomal rearrangements are the primary cause of reproductive isolation. They suggest that chromosomal rearrangements are not a prerequisite for speciation in yeast, and the rate of formation of translocations is not constant. These rearrangements appear to result from ectopic recombination between full-length transposons (the five families of Ty elements) or LTR elements or other repeated sequences.

2.7.2 *Schizosaccharomyces pombe* Genome

The fission yeast S. pombe is a single-cell free-living archiascomycete fungus sharing many features with cells of higher eukaryotes that diverged from budding yeast S. cerevisiae some 330 to 420 Mya. It has long served as a model organism for the study of cell-cycle control, mitosis, and meiosis.

The fission yeast genome, whose sequencing has recently been completed (Wood, Gwilliam et al. 2002), contains a haploid set of three chromosome of 5.7, 4.6, and 3.5 Mbp, for chromosome I, II, and III, respectively, totaling 13.8 Mbp. The unique fraction of the genome, about 12.5 Mbp, is comparable to that of S. cerevisiae. The overall G + C content is 36.2%, compared to 38.3% GC observed for the S. cerevisiae genome.

A total of 4940 protein-coding genes have been predicted, which account for about 57% of the total genone. The total number of genes is substantially less than the 5885 genes predicted for S. cerevisiae. It is interesting to note that both these eukaryotic genomes have fewer genes than the *Mesorhizobium loti* genome, the largest eubacterial genome sequence available to date, containing 6752 genes. This finding implies that a free-living eukaryotic cell can be constructed with fewer than

5000 genes and that the distinction between bacterial and eukaryotic cell organization is not simply linked to the total number of genes but mostly to the gene types and the way they interact with each other and with the environment.

Introns are much more frequent in *S. pombe*, where they are found in 43% of genes, than in *S. cerevisiae*, where only 5% of genes are found to contain introns. The average intron length is 80 nucleotides. The large number of introns could increase the protein inventory, through alternative splicing mechanism, thus balancing the relatively small number of genes predicted.

A total of 174 tRNAs, six spliceosomal RNAs (U1 to U6), 16 small nuclear RNAs (snRNAs), and 33 small nucleolar RNAs (snoRNAs) have been identified, mostly dispersed throughout the genome. The ribosomal genes (5.8S, 18S, and 26S rRNAs) are grouped as 100 to 200 tandem repeats in chromosome III, whereas the thirty 5S rRNAs are distributed throughout the genome.

2.7.3 *Caenorhabditis elegans* Genome

Caenorhabditis elegans is a small (about 1 mm in length), free-living soil nematode found in temperate regions and used widely as a model organism to investigate the genetics of development and neurology. Scientists have followed the fate of each of the cell divisions that produce the mature 959-cell adult worm from a single fertilized egg, and using electron microscopy and serial sectioning they have reconstructed the entire nervous system (Chalfie and Jorgensen 1998).

Although the nematode genome project is still in progress, the paper in *Science* (T.C.e.S. Consortium 1998) describes a number of features of this worm sequence, considering it as "essentially" complete. The nuclear genome of *C. elegans* consists of 97 Mbp organized into six chromosomes (see Fig. 2.2). At first sight, the *C. elegans* genome looks uniform, as its GC content (36% average value) is essentially unchanged across all the chromosomes. There are no localized centromeres as seen in most other metazoans. Instead, the extensive, highly repetitive sequences that are characteristic of centromeres in other organisms may be represented here by some of the many tandem repeats found scattered among the genes, particularly on the chromosome arms.

Gene density seems to be fairly constant across the chromosomes, although some differences can be found between the centers of the autosomes, the autosome arms, and the X chromosome. Autosome arms show higher density than the central region; in the X chromosome genes are present at a lower percentage than in autosomes. The DNA on the arms might be evolving more rapidly than in the central regions of the autosomes. This is confirmed by several findings. Both inverted and tandem repetitive sequences are more frequent on the autosome arms than in the central regions of the chromosomes or on the X chromosome; the fraction of genes with similarities to organisms other than nematodes tends to be lower on the arms. Finally, meiotic recombination is much higher on the autosome arms.

19,099 protein-coding genes have been predicted, for an average density of one predicted gene per 5 kbp. The number of genes is about three times that found in yeast; the coding sequence accounts for 27% of the genome, with an average of five introns per gene (total introns account for 26% of the total sequence).

The problem of gene prediction is more complex in *C. elegans*, due to the presence of transplicing and to the organization of as many as 25% of the genes into operons. Blumenthal (1995) demonstrated that in *C. elegans*, two or more genes can be transcribed from the same promoter, with one gene separated by no more than a few hundred nucleotides from the other. In genes undergoing transplicing, the 5′ exon begins with a splice acceptor sequence, making this 5′ exon more difficult to distinguish from internal exons. This combination of factors may result in two genes being merged into one.

The Genefinder algorithm (P. Green, unpublished) has been used by the *C. elegans* Sequencing Consortium (T.C.e.S. Consortium 1998) to identify putative coding regions and to provide an initial overview of gene structure. In a later phase, the computer-generated gene structure prediction has been refined by expert annotators using comparisons with the available EST, protein similarities, and cDNA sequence data from the related nematode *Caenorhabditis briggsae*. The results are stored in the *C. elegans* database (ACEdb) available via the Web (see the URL in Table 5.1). About 42% of proteins predicted have matches with proteins belonging to distantly related species, while 34% only match other nematode proteins.

The *C. elegans* Sequencing Consortium (T.C.e.S. Consortium 1998) used the Pfam protein family database (Bateman, Birney et al. 2000; see Sec. 5.9) to classify the 20 common protein domains found in the worm genome. The most frequent domains are those involved in transcriptional regulation and intercellular signaling. In addition to protein-coding genes, the nematode genome contains several hundred genes for noncoding RNAs. There are 659 widely dispersed transfer RNA genes and at least 29 tRNA-derived pseudogenes. It is striking to note that 44% of the tRNA genes are found on the X chromosome, even though it contains only 20% of the total sequence. Several other noncoding RNA genes (e.g., 5S RNAs, spliceosomal RNAs) occur in dispersed multigene families; some are organized in long tandem arrays, while others are found in the introns of protein-coding genes, which may indicate an active transposition process.

A significant fraction of the *C. elegans* sequence is repetitive, as in other multicellular organisms. Tandem repeats account for 2.7% of the genome and are found, on average, once per 3.6 kbp. Inverted repeats account for 3.6% of the genome and are found, on average, once per 4.9 kbp. Some repeats come in families within the genome; for example, the repeat sequence CeRep26, the tandemly occurring hexamer repeat TTAGGC, is seen at multiple sites internal to the chromosomes as well as in telomeres (Wicky, Villeneuve et al. 1996). Many repeat families are distributed nonuniformly with respect to genes, and in particular, they are more likely to be found within introns than between genes.

In the nematode genome, 38 dispersed repeat families have been identified. Most of them seem to be associated with transposition events and, indeed, they include *C. elegans* transposon, although these repeat elements may not explicitly encode an active transposon (Ketting, Fischer et al. 1997). For example, four new families of the Tc1/mariner type have been recognized in the worm, but their high divergence from one another and from other family members suggests that they are probably no longer active in the genome. Beside multiple-copy repeat families, a significant number of simple duplications have been observed in the *C. elegans* genome. These duplications involve segments ranging from hundreds of bases to tens of kilobases in length; in one case, a segment of 108 kbp containing six genes was duplicated tandemly with only 10 nucleotide differences observed between the two copies.

2.7.4 *Drosophila melanogaster* Genome

Genetic and physical mapping, whole-genome mutational screens, and functional alterations of the genome by gene transfer were pioneered in metazoans through use of the fruit fly *Drosophila melanogaster* (Rubin and Lewis 2000). Because of the large research community working on this organism and because of its modest genome size, *Drosophila* was rightly chosen as a test system to explore the applicability of whole-genome shotgun sequencing. Indeed, the public consortium Berkeley Drosophila Genome Project (BDGP) and Celera Genomics collaborated in 1999–2000 to sequence the *D. melanogaster* genome. Celera produced a whole-genome shotgun approach, whereas the BDGP produced a BAC-based physical map and defined a tiling path of overlapping BAC clones that span the euchromatic portion of the genome (see Sec. 4.1 for genome sequencing techniques).

Drosophila contains five chromosomes, for a total genome size of about 180 Mbp, a third of which is centric heterochromatin. In *Drosophila*, the euchromatic region is located mainly on two large autosomes and the X chromosome; the small fourth chromosome contains only about 1 Mbp of euchromatin. In March 2000 about 120 Mbp of the euchromatic portion of the genome had been published (Adams, Celniker et al. 2000).

Genomic data of *D. melanogaster* are stored in two principal databases. FlyBase (see the URL in Table 5.1), an electronic publication copyrighted 1993 by the Genetics Society of America, is a database of genetic and molecular information on all species from the family Drosophilidae, *D. melanogaster* being the most represented. GadFly (the Genome Annotation Database of *Drosophila*; see the URL in the Appendix) includes functional classifications for the sequence annotated at Celera. Transcripts and protein sequences, as well as protein function, have been predicted using computational approaches under the supervision of an "Annotation Jamboree" of more than 40 experts, each responsible for a protein family. To define the gene structure, the Genscan (Burge and Karlin 1997) and Genie (Reese, Kulp et al. 2000) programs were used, producing slightly different results (see Sec. 6.8.7 for a description of gene-finding programs). A total of 17,464 gene products were predicted using Genscan, as against 13,189 using Genie; the latter estimate was judged more reliable by the authors.

The annotated *D. melanogaster* sequence was released on March 24, 2000, and constituted Release 1 of the genomic sequence. In October 2000, Release 2 was made available via the Web: approximately 330 of the gaps in that sequence have now been filled. Some annotations have been corrected or added by Celera/BDGP. The BDGP is currently finishing the genomic sequence to a high level of quality (phase 3), and FlyBase/BDGP is reannotating this finished sequence to create Release 3, which will gradually be deposited in GenBank. At present, 13,601 genes have been annotated for *Drosophila*, which encode 14,113 transcripts, some of them generated through alternative splicing. About 3000 genes predicted by Genie have no database matches.

An attempt to classify the function of these gene products was performed through a controlled vocabulary using a scheme called *Gene Ontology* (Ashburner, Ball et al. 2000; see also Sec. 5.5). Based on this classification, the most represented gene products are 40.7% "function unknown/unclassified," 12% "enzyme," and 4.4% "nucleic acid–binding proteins." There are 56,673 predicted exons, an average of four per gene; the size of the average predicted transcript is 3058 nucleotides. There are at least 41,000 introns, occupying 20 Mbp of sequence. The average gene density in *Drosophila* is one gene per 9 kbp, and regions with high gene density seem to be correlated with high GC content.

5536 of the 13,601 *Drosophila*-predicted genes are duplicated, and approximately 70% of duplicated gene pairs are on the same strand. Local gene duplications result in gene clusters that are distributed randomly along the chromosome arms; the largest cluster in the fly contains 17 genes and encodes for proteins of unknown function.

Comparison of the core proteome of flies with that of *S. cerevisiae* and *C. elegans* showed that it shares 16% of its genes with yeast (mainly proteins involved in replication, transcription, and cell division) and 35% of its genes with the worm. Moreover, homologs of 289 human disease genes are found in the fly, including genes implicated in cancer, innate immunity, and metabolic diseases (Rubin, Yandell et al. 2000).

Transposable elements, ubiquitous among contemporary organisms, have been studied extensively in *Drosophila* [see Lozovskaya, Hartl et al. (1995) for a review] and are a major source of genetic change, including the creation of novel genes, the alteration of gene expression in development, and the genesis of major genomic rearrangements. Transposons are regulated in their activity. Early works suggested that they may serve as controlling elements mobilized under stress conditions; afterward, the activity of transposable elements was assumed to be autonomous and independent. New findings on the regulation of transposable elements indicate that both circumstances may occur.

In the fruit fly, over half of the mutations observed are transposon related. The natural transposons in the *Drosophila* genome were not sequenced. BDGP states that for many of the transposons, a consensus sequence was pasted into the genomic sequence; hence the transposon sequence is not the actual one at that location. They intend to replace the pasted consensus sequences with true sequences as soon as sequencing to high quality is over.

The distribution of dinucleotide microsatellites (one type of microsatellite, a special class of repetitive DNA often used as genetic markers) in the genome of the fruit fly shows that their average length is 6.7 repeats. Microsatellite genesis seems to be a random process; indeed, a significant positive correlation between AT content and (AT/TA)n microsatellite density is observed, thus indicating a neutral origin of microsatellites. However, there is evidence for a nonrandom distribution of microsatellites; the average microsatellite density seems to be higher on the X chromosome (Bachtrog, Weiss et al. 1999). Studies of nucleotide composition in *Drosophila* have revealed differences in nucleotide composition among lineages. Compositional differences are associated with an accelerated rate of nucleotide substitution in functionally less contrained regions (Rodriguez-Trelles, Tarrio et al. 1999; 2000).

Chesnokov, Gossen et al. (1999) report that in the fruit fly, genes encoding the basic DNA replication machinery are conserved, and in particular, all the proteins known to be involved in start-site recognition are encoded by single-copy genes. Among these genes there are members of the six-subunit heteromeric origin recognition complex (ORC), the MCM helicase complex, and the regulatory factors CDC6 and CDC45 (which determine processing of preinitiation complexes; see Sec.

2.5). In *Drosophila,* all four members of the protein family SMC which are involved in DNA repair and chromosome inheritance, particularly in sister chromatid condensation, have been identified. They are conserved from yeast to mammals.

The transcription machinery is not conserved completely among eukaryotes. *Drosophila* core RNA polymerase II and some transcription factors (TFIIA-H, TFIIIA, and TFIIIB) are similar in composition to those of both mammals and yeast, but core RNA polymerases I and III, and other factors, such as TBP, vary greatly in sequence composition in *Drosophila* and mammals compared to yeast. The RNA polymerase I transcription factors of flies and mammals have clear amino acid conservation. Yeast RNA polymerase I factors do not appear to be related to them. *Drosophila* encodes about 700 transcription factors, about half of which are zinc-finger proteins.

Most genes encoding translation factors are present in only one copy in the *Drosophila* genome; however, six genes encoding proteins highly similar to the RNA cap-binding protein eIF4E are present. In *C. elegans* three eIF4E isoforms have been recognized, which are involved in transplicing, a process which, however, does not exist in *Drosophila.*

2.7.5   *Arabidopsis thaliana* Genome

The flowering plant *Arabidopsis thaliana*, due to its characteristics—a short generation time, small size, large number of offspring, and a relatively small nuclear genome—is a widely used model system for genetic and molecular biology studies. In December 2000, 115.4 Mbp of the 125-Mbp genome, organized in five chromosomes, was published (Arabidopsis Genome Initiative 2000). The *Arabidopsis* genome is predicted to encode 25,498 genes. A combination of algorithms, all optimized with parameters based on known *Arabidopsis* gene structures, was used by the authors of the paper on the complete genome sequence to define gene structure.

Gene finding has involved several steps: (1) analysis of BAC sequences using computational gene finders (Genscan, GeneMark.HMM, Xgrail, Genefinder, and GlimmerA; see Sec. 6.8.7); (2) alignment of the sequence to the protein and EST databases; and (3) assignment of functions to each gene. Sixty-nine percent of the genes were classified based on sequence similarity to proteins of known function in all organisms. Only 9% of the genes have been characterized experimentally, and 31% could not be assigned to any functional category. *Arabidopsis, C. elegans*, and *Drosophila* have a similar range of 11,000 to 15,000 different types of proteins, suggesting that this is the minimal set required by extremely diverse multicellular eukaryotes. Among the non-protein-coding genes identified in the *Arabidopsis* genome, there are 589 cytoplasmic tRNAs, 27 organelle-derived tRNAs, and 13 pseudogenes. These account for 46 tRNA families needed to decode all possible codons. Other non-protein-coding genes are spliceosomal RNAs (U1, U2, U4, U5, U6) and snoRNAs.

Arabidopsis genes are rather compact, containing several exons (average length: 250 bp), punctuated by short introns. Exons in *Arabidopsis* are richer in guanosine and cytosine bases (44%) than are introns (32%), a typical feature of plant genes. The genes are closely spaced, about 4.6 kbp apart. The small genes of *Arabidopsis* may have helped it to tolerate extensive genomic reorganization (see below)

(Walbot 2000). Only 35% of the proteins predicted are unique in the genome. Pronounced redundancy results in segmental duplications and tandem arrays. As in other eukaryotic genomes, in *Arabidopsis* some gene families (17% of the total) are organized in tandem arrays of two or more units; in particular, 1528 tandem arrays containing 4140 individual genes have been identified. The proportion of proteins belonging to families of more than five members is substantially higher in *Arabidopsis* (37.4%) than in *Drosophila* (12.1%) or *C. elegans* (24.0%).

In the plant cell, three genomes (nuclear, mitochondrial, and chloroplast) live together and work in an integrated way, despite the strong differences in gene number, organization, and rate of evolution. The search for proteins of organellar origin in the *Arabidopsis* genome revealed that 806 proteins predicted are more similar to proteins from the cyanobacteria *Synechocystis* and thus are possibly of plastid descent. A continuous transfer of genes between the three genomes still seems to be going on; for example, plastid DNA insertions in the nucleus (17 insertions totalling 11 kbp) contain full-length genes encoding proteins or tRNAs, fragments of genes and an intron as well as intergenic regions. Another example is the set of DNA repair and recombination (RAR) genes: *Arabidopsis*, different from other eukaryotes, has several homologs of RAR genes. Evolutionary analysis indicates that the extra copies of RAR genes could be originated through relatively recent gene duplication or be due to gene transfer from chloroplast and mitochondrial genomes to the nucleus.

The duplicated regions encompass about 60% of the *Arabidopsis* genome and consist mostly of 24 large duplicated segments 100 kbp or larger. Many duplications appear to have undergone further shuffling (e.g., as local inversions) after the duplication event. The larger number of gene copies in *Arabidopsis* compared to other sequenced eukaryotes can be explained in two different ways: First, independent amplification of individual genes (probably by unequal crossing over) could have generated tandem and dispersed gene families to a greater extent; and second, segmental duplications could be due to ancestral duplication of the entire genome and subsequent rearrangements. These features of *Arabidopsis*, and presumably of other plant genomes, may indicate relaxed constraints on genome size in plants.

Indeed, polyploidy occurs widely in plants and is proposed to be a key factor in plant evolution. In *Arabidopsis*, segments are duplicated but not triplicated, therefore, like maize, it could have had a tetraploid ancestor. A comparative sequence analysis of *Arabidopsis* and tomato estimated that a duplication occurred 112 Myr ago to form a tetraploid (Ku, Vision et al. 2000). The diploid genetics of *Arabidopsis* and the extensive divergence of the segments duplicated could have masked its evolutionary history.

Transposons in *Arabidopsis* account for at least 10% of the genome, or about one-fifth of the intergenic DNA; they include 2109 class I transposons and 2203 class II transposons. Class I elements are much less abundant in *Arabidopsis* than in other plants, such as maize and occupy primarily the centromere. In contrast, class II transposons predominate in the pericentromeric domains (Le, Wright et al. 2000). Transposon-rich regions are relatively gene-poor and have lower rates of recombination. Hence, there is a correlation between low gene expression, high transposon density, and low recombination. In the *Arabidopsis* genome also, long terminal repeat retrotransposons (LTR), short interspersed nuclear elements (SINEs), and long interspersed nuclear elements (LINEs) have been identified.

More than 3000 proteins encoded by the *Arabidopsis* genome seem to be involved in gene expression. These products represent many nuclear proteins that modulate chromatin structure, contribute to the basal transcription machinery, or mediate gene regulation. In *Arabidopsis,* gene regulation can occur via DNA methylation as in other eukaryotes. Indeed, the genome encodes eight methyltransferases grouped into three types; two of the three types are orthologous to mammalian DMT69, whereas one, chromomethyltransferase, is unique to plants.

The *Arabidopsis* transcription machinery includes the three nuclear DNA-dependent RNA polymerase systems typical of eukaryotes; the RNA pol II and III transcription factors seem to be very similar to those of other eukaryotes, while most transcription factors for RNA pol I have not been identified. A similarity search on the entire genome revealed 1709 proteins with significant similarity to known classes of plant transcription factors. Of the 29 classes of *Arabidopsis* transcription factors, 16 appear to be unique to plants.

Plant cells share some exclusive features distinguishing them from animal cells: polysaccharide-rich cell walls, plastids, vacuoles, cytoskeletal arrays, and plasmodesmata linking cytoplasms of neighboring cells. Indeed, *Arabidopsis* lacks the typical animal cytoskeletal array component, such as vimentin or cytokeratin, but it has several variants of actin and α- and β-tubulin.

The regulation of development in *Arabidopsis*, as in animals, involves cell–cell communication, hierarchies of transcription factors, and the regulation of chromatin state. Plants and animals have converged on similar processes of pattern formation, but as we mentioned earlier, they seem to have used and expanded different transcription factor families as regulators. Moreover, several *Arabidopsis* genes needed for development appear to be derived from a cyanobacteria-like genome, as is the case of phytochromes, specialized light receptors involved in many developmental decisions.

Plants need to respond to environmental stimuli in a different way from animals, due to their sessile conditions. Comparative genomic studies revealed that plants have evolved their own pathways of signal transduction, and none of the components of the metazoan signaling pathways are found in *Arabidopsis*. A particular plant response is that to the light stimulus. The signal transduction cascade begins with the activation of photoreceptors; then the light signal is transduced, causing activation or derepression of nuclear and chloroplast-encoded photosynthetic genes, enabling the plant to respond adequately. In the *Arabidopsis* genome, about 100 candidate genes have been identified, involved in light perception and signaling, and 139 nuclear-encoded genes that function potentially in photosynthesis. The functional role has, however, been described for only 35 genes.

Another important response for a sessile organism is what follows the recognition of a pathogen. The *Arabidopsis* genome contains numerous resistance genes distributed at many loci, and none of them is similar to their metazoan counterparts. Some authors indicated that resistance gene evolution may involve duplication and divergence of linked gene families (Ellis, Dodds et al. 2000).

A great many of the *Arabidopsis* genes are involved in the metabolic pathways that ensure autotrophy, such as mineral acquisition, photosynthesis, respiration, intermediary metabolism, and the synthesis of nucleotides, cofactors, lipids, fatty acids, and amino acids. The *Arabidopsis* metabolic gene complement is very similar to that of the photoautotrophic cyanobacterium *Synechocystis* (see above), but it shows a certain degree of structural redundancy, as many single-copy genes in *Synechocystis* are present in multiple copies in *Arabidopsis* (as is the case of the enzyme pyruvate kinase). Furthermore, it also differs by the presence of many genes encoding enzymes for pathways that are unique to vascular plants; for example, more than 420 genes are probably involved in cell wall biosynthesis and modification. The high degree of apparent redundancy in such genes might reflect differences in substrate specificity by some of the enzymes.

2.7.6 *Oryza sativa* Genome

Rice is one of the most important crops for human consumption providing a staple diet for over half of the world population. A draft sequence of two subspecies of *Oryza sativa* has been independently reported by two research groups from Syngenta (*Oryza sativa* subspecies *japonica*; Goff, Ricke et al. 2002) and the Bejing Research Institute in China (*Oryza sativa* subspecies *indica*; Yu, Hu et al. 2002). This latter species is the most widely cultivated.

The genome size of *japonica* subspecies (466 Mb) resulted about 10% larger than that of the *indica* subspecies (420 Mb), both genomes being relatively smaller with respect to other grass genomes. It is not clear if the genome size difference is due to a bias in repeat abundance or to technical differences in sequencing and assembling. Both drafts were obtained by whole-genome shotgun sequencing and were claimed of a similar coverage (92–93%).

The average GC content—43.3% (*indica*) and 44% (*japonica*)—are higher than that of *Arabidopsis thaliana* genome (36%). Indeed, it has been reported that *Graminae* (grass) has a higher GC content with respect to other plants. Interestingly, a GC gradient was observed in rice (*indica*) coding genes with the 5' end up to 25% GC richer than the 3'end. A similar pattern has been observed for *A. thaliana* genes.

Repetitive DNA accounts for 42–45% of the rice genome, including simple sequence repeats (SSRs), short and long interspersed repeats, including gypsy- and copia-like transposons, and MITEs (miniature inverted-repeat transposable elements). The gene count ranged between 32,000 and 55,000, compared to about 25,000 estimated in *Arabidopsis*, the mean gene size being about 4.5 Kb. These figures need to be refined further by additional investigations and validated through transcript sequencing.

Most of the annotated *Arabidopsis* genes (80–85%) were shown to have a rice homolog with an average amino acid identity of 60%. About one-third of rice genes with *Arabidopsis* homologs were found to be specific of plants. The surprising observation that the number of genes in higher plants probably exceeds the number of human genes can be plausibly explained by the fact that plants are immobile and thus need a large complement of genes to scavenge nutrients, to respond to different environmental conditions, and to evade predators (Petsko 2002). However, more than 50% of rice genes code for proteins whose function is unknown, so the above explanation is just speculative.

Specific gene classes are overrepresented in rice and *Arabidopsis*, such as RING zinc-finger and F-box domain proteins mostly involved in intracellular protein turnover and degradation pathway. The rice and *Arabidopsis* genomes are extensively rearranged in respect to their gene order; however a detectable synteny for some genes can be observed.

Disease resistance genes (R-genes) responsible for specific recognition of pathogen attack and activation of defense mechanism appeared to have evolved since the divergence of rice and *Arabidopsis* occurred about 200 Mya. A remarkable conservation between rice and *Arabidopsis* is also observed for flowering time and flower development genes.

About 25% of rice genes is involved in metabolism accounting for all central processes (e.g., glycolysis, TCA cycle, photosynthesis and respiration, synthesis and degradation of amino acids, fatty acids, and nucleotides). Furthermore, extensive gene redundancy can be observed for major pathways, thus suggesting regulated expression in a tissue- or developmental-specific manner as well as in response to different environmental challenges. In particular, large gene families exist for genes encoding for enzymes involved in biosynthesis of secondary metabolites.

Other widely represented gene classes include genes for membrane transport, cell growth and maintenance, and cell communication/signal transduction. The two draft rice genome sequences will certainly provide an important knowledge basis for cereal genomics with the final goal of crop improvement.

2.7.7 *Homo sapiens* Genome

Genome Features In the year 2000 the complete sequencing of the human genome was announced. Due to the joint efforts of a public consortium and the Celera company, this enterprise is now described in two milestone papers published in the journals *Science* (Venter, Adams et al. 2001) and *Nature* (Lander, Linton et al. 2001), where the main steps and strategies followed and the achievements so far obtained are reported. The idea of sequencing the human genome arose in the early 1980s, but the Human Genome Project had been launched only by late 1990, with the creation of genome centers in several countries. Indeed, the human genome draft sequence currently available is the result of a collaboration involving 20 groups from the United States, the United Kingdom, Japan, France, Germany, and China.

The draft genome sequence was generated from a physical map covering more than 96% of the euchromatic part of the human genome and about 94% of the human genome, accounting for 2.91 Gbp distributed on 22 autosomes and X and Y chromosomes. Data on the average GC content of the human genome published by the International Human Genome Sequencing Consortium (IHGSC) and by Celera Genomics differ slightly (41% versus 38%, respectively). The draft genome sequence showed that the local GC content has substantial long-range excursions from its average. Indeed, there are huge regions (>10 Mbp) with GC content far from the average. For example, the average GC content on chromosome 17 q is 50% for the distal 10.3 Mbp but drops to 38% for the adjacent 3.9 Mbp. Long-range variation in GC content is also evident throughout the genome, as the distribution of average GC content across the draft genome sequence is 15-fold larger than predicted (by a uniform process). The distribution is also notably biased, with 58% below the average and 42% above the average, with a long tail of GC-rich regions. It is well known that the human genome, as representative of warm-blooded vertebrates, contains base composition compartments defined by Bernardi (2000) as isochores. This matter is hotly debated, and for a more exhaustive discussion, refer to Sec. 7.3.

The relation between GC content and gene density also has been investigated for this genome. Chromosomes 17, 19, and 22 have a high number of H3 bands (see Secs. 2.4 and 7.3) and show the highest gene density; conversely, chromosomes X, 4, 18, and 13 have the fewest H3 bands and the lowest gene density. There are some exceptions: the chromosome 15 shows few H3 bands but normal gene density, and the chromosome 8 shows low gene density but normal H3 band distribution. The correlation between GC-content domains and cytological properties has been confirmed in the human genome: GC-poor regions are strongly correlated with dark G-bands in karyotypes, and the density of genes is greater in GC-rich regions.

It is well known that the dinucleotide CpG, independent of the GC content, is greatly underrepresented in human DNA, occurring at only about one-fifth of the roughly 4% expected frequency. This occurs because most CpG dinucleotides, methylated on the cytosine base, give rise to T residues by spontaneous deamination of methyl-C residues. As a result, methyl-CpG dinucleotides steadily mutate to TpG dinucleotides. However, the genome contains many *CpG islands*, which are commonly defined as regions of DNA at least 200 bp in length that have a G + C content above 50% and a ratio of observed versus expected CpGs close to or above 0.6 (see Sec. 6.8.8 for CpG island identification tools; Gardiner-Garden and Frommer 1987). The CpG islands that are not methylated occur at a frequency closer to that predicted by the local GC content. CpG islands are of particular interest because many of them are associated with the 5′ ends of genes. The draft genome sequence contains 28,890 CpG islands, and their number varies substantially among some of the chromosomes. Most chromosomes have 5 to 15 islands per Mbp, with a mean of 10.5 islands per Mbp. Chromosome Y has an unusually low 2.9 islands per Mbp. Most of the islands, having 60 to 70% GC content, are short, with more than 95% less than 1800 bp long.

In human, coding sequences comprise less than 5% of the genome (see below), whereas repeated sequences account for at least 50% and probably much more. Broadly, the repeats fall into five classes (see Sec. 2.3 for an alternative repeat classification): (1) transposon-derived repeats (interspersed repeats); (2) inactive (partially) retroposed copies of cellular genes, including protein-coding genes and small structural RNAs (processed pseudogenes); (3) simple sequence repeats, [e.g., (A)n, (CA)n, (CGG)n, microsatellites (repeat units up to 5 bp) and minisatellites (more than 5 bp)]; (4) segmental duplications, blocks of around 10 to 300 kbp in length; and (5) blocks of tandemly repeated sequences [e.g., at centromeres and telomeres (not represented in the draft genome sequence)].

In mammals, almost all interspersed transposable elements can be distinguished into four types: long interspersed elements (LINEs), short interspersed elements (SINEs), LTR retrotransposons, and DNA transposons. The former three transpose through RNA intermediates and the latter transposes directly as DNA.

Human LINEs, about 6 kbp long, but often found in truncated form, harbor an internal pol II promoter and encode two open-reading frames. The LINE machinery is believed to be responsible for most reverse transcription in the genome, including the retrotransposition of the nonautonomous SINEs. Three distantly related LINE families are found in the human genome: LINE1, LINE2, and LINE3, accounting for 20% of the sequence. Only LINE1 is still active. SINEs are short (about 100 to 400 bp) nonautonomous transposons; they harbor an internal pol III promoter and encode no proteins. The human genome contains three distinct monophyletic families of SINEs (13% of the sequence): the active Alu and the inactive MIR and Ther2/MIR3.

LTR retroposons account for 8% of the draft human sequence; they are flanked by long terminal direct repeats that contain all the necessary transcriptional regulatory elements, and their transposition occurs through the retroviral mechanism, with reverse transcription occurring in a cytoplasmic virus-like particle, primed by a tRNA. Finally, the human genome contains seven major classes of DNA transposons, which can be subdivided into many families with independent origins. They account for 3% of the sequence and are similar to bacterial transposons, with terminal inverted repeats.

The transposable elements are not equally distributed in the human genome: LINE sequences occur at much higher density in AT-rich regions, while SINEs (MIR, Alu) show the opposite trend; LTR retroposons and DNA transposons show a more uniform distribution. The highest density is recorded in a 525-kbp region on chromosome Xp11, with an overall transposable element density of 89%, while the lowest density of interspersed repeats is in the four homeobox gene clusters, *hoxA*, *hoxB*, *hoxC*, and *hoxD*, containing less than 2% interspersed repeats.

The age of the repeated elements can be used as a "fossil record." The analysis of the human genome revealed that most interspersed repeats predate the eutherian radiation, with two major peaks of DNA transposon activity. At present, there is no evidence for DNA transposon activity in the past 50 Myr in the human genome, and LTR retroposons appear to be on the way to extinction. This seems to be a peculiar feature of the hominid lineage; indeed, comparison with the mouse showed that this genome has not undergone the decline seen in humans, and in contrast to their possible extinction in humans, LTR retroposons are alive in the mouse. Among the other repeated elements, simple sequence repeats (microsatellites and minisatellites, SSRs) are commonly found in the human genome. They comprise about 3% of the human genome; dinucleotide repeats AC and AT are the most frequent. Repeat identification and annotation in genomic sequences are described in Sec. 6.8.3.

The human genome harbors many segmental duplications of portions of genomic sequence. The finished sequence consists of at least 3.3% segmental duplications, which can be divided into interchromosomal duplications, defined as segments that are duplicated among nonhomologous chromosomes, and intrachromosomal duplications, which occur within a particular chromosome or chromosomal arm. Exhaustive analyses have been conducted on the two finished chromosomes, 21 and 22. These studies revealed that chromosome 22 contains a region of 1.5 Mbp adjacent to the centromere consisting 90% of interchromosomal duplication; and chromosome 21 presents a similar scenario. Segmental duplications seem to have a trend toward clustering. This finding could be explained considering that the genome could have a damage-control mechanism whereby chromosomal breakage products are preferentially inserted into pericentromeric regions, which are structurally very complex.

The availability of the entire draft genome sequence makes it possible to explore variation in the rate of recombination across human chromosomes. The analysis reported by IHGSC shows that the average recombination rate tends to be much higher in the distal region of chromosomes and on shorter chromosome arms in general. However, the recombination rate tends to be suppressed near the centromeres and higher in the distal portions of most chromosomes. The increase is most pronounced in the male meiotic map (Lynn, Kashuk et al. 2000). The authors suggest two possible explanations. The increased rate of recombination on shorter chromosome arms could be explained if once an initial recombination event occurs, additional nearby events are blocked by positive crossover interference on each arm. An alternative possibility is that a checkpoint mechanism scans for and enforces the presence of at least one crossover on each chromosome arm. Further studies on the draft genome sequence are needed to discriminate among the alternative hypotheses.

Prediction of the Total Number of Genes Gene finding is one of the major tasks of a genome annotation effort. In summary, gene prediction includes three main approaches: (1) detection of transcribed gene regions based on similarity to known ESTs or RNAs, (2) detection of likely coding regions based on similarity between six-frame translation of genomic regions or peptides predicted from the approach described below and known proteins or protein modules (e.g., Pfam; see Sec. 5.9), and (3) de novo gene recognition using computational methods based prevalently on hidden Markov models (HMMs: for example, Genscan, Genie, HMMgene; see also Sec. 6.8.7).

The definition of gene content is not an easy task, particularly in large genomes like those of vertebrates, which possess small exons separated by large introns and large intergenic enhancers. Additionally, only the first approach can be used for prediction of genes that do not code for proteins (e.g., snoRNAs) because they do not contain ORFs. It should be remembered that probably some 1000 human genes fall in this category.

In the human genome the total number of genes predicted is currently much lower than the number proposed (80,000 to 100,000 genes) by preliminary data obtained essentially with indirect and not very reliable methods. If we take into account known RNA species, the total number approaches 30,000, which shows that the human genome has just twice the number of genes of invertebrates. However, at least for protein-coding genes, the calculation of the number of proteins for which an IPI (Integrated Protein Index) is in progress predicts a number of proteins greater than 30,000, which would mean that each gene produces more than just a single protein through alternative information-processing methods such as alternative splicing. In particular:

1. *Transfer RNA genes*. In the draft genome sequence the estimate is 497 genes and 324 putative pseudogenes, for a total of 821 species. This number is lower than the number calculated previously on an experimental basis (e.g., hybridization), which produced 1310 transfer RNA genes. Such a discrepancy may be explained, at least in part, by the fact that earlier estimates of the size of the human genome assumed too high a value. The draft sequence contains 37 of the 46 human tRNA species contained in the tRNA Database (see the URL in Table 5.1) with the inclusion of selenocysteine tRNA. The only tRNA gene not found in the draft genome is the one coding for a tRNA Glu species (DE9990). The human set appears to be complete according to the genetic code rules; it is, however, more restricted than expected. Indeed, the human genome contains fewer tRNA species than the worm but more than the fly. This means that in humans, where general wobble rules are

followed, there is an economic use of the 61 anticodons, and some are not even used. It is well known that in the majority of organisms there is a codon bias correlated in turn with a gene expression level and thus under selective pressure. In human and other vertebrate genomes, the codon bias is not so obviously correlated with gene expression level; rather, it is linked to the location of genes in different genome compartments. Very interestingly, tRNA genes are dispersed throughout the human genome but in a nonrandom fashion. They are indeed clustered: more than 25% of the tRNA genes on chromosome 6 in a region of 4 Mbp which contains 140 tRNA genes with a representation of 36 of the 49 anticodons found in the complete set. Of the 21 isoacceptors, only the species decoding Asn, Cys, Glu, and SelenoCys are missing in this chromosome region. The remaining tRNA species are clustered in other regions, mainly on chromosome 7 in a 0.5-Mbp stretch (30 Cys tRNA) and on chromosome 1 (Asn and Glu tRNAs). More than half of the tRNA genes (280 out of 497) are located on chromosomes 6 and 1. Chromosomes 3, 4, 8, 9, 10, 12, 18, 20, 21, and X have fewer than 10 tRNA genes each.

2. *Ribosomal RNA genes.* The *large subunit* (LSU) rRNA, containing 28S and 5.8S rRNA, and the *small subunit* (SSU) rRNA, containing 18S rRNA, occur as a 44 kbp tandem repeat unit which is present in 150 to 200 copies on the short arms of the acrocentric chromosomes 13, 14, 15, 21, and 22. The 5S rDNA genes also occur in tandem arrays, the largest being near the telomere on chromosome 1, containing 200 to 300 5S genes. In total, about 2000 5S-related sequences are assumed. There are no true complete copies of the rDNA tandem repeats in the draft genome sequence, owing to a bias in the initial phase of the sequencing effort against sequencing BAC clones.

3. *Small nucleolar RNA genes.* These species, called *snoRNAs*, belong to two families: C/D box snoRNAs, involved in guiding site-specific 2'-*O*-ribose methylations of other RNAs, and H/ACA snoRNAs, involved primarily in guiding site-specific pseudouridylations. A total of 84 snoRNA sequences have been mapped in the draft genome, representing the 87% of the total known, usually present as single-copy genes.

4. *Spliceosomal RNAs and other ncRNA genes.* Of the 22 species known so far, at least one copy of the 21 ncRNA genes has been found, including the spliceosomal snRNAs. For several genes, multiple copies have been found, as, for example, 44 genes for U6 snRNA and 16 for U1 snRNA.

5. *Protein-coding genes.* For the gene annotation of the human working draft, both the public consortium (Lander, Linton et al. 2001) and Celera (Venter, Adams et al. 2001) developed automated rule-based gene prediction systems trying to mimic the work of a human annotator using different approaches and evidence.

Celera used a system called Otto, a rule-based expert system giving different levels of confidence to different types of evidence but assigning the highest priority to sequence similarity with known transcribed sequences found in the RefSeq library (Pruitt, Katz et al. 2000) to the Unigene set of human ESTs (Wheeler, Church et al. 2001) or to known proteins. The Otto procedure was quite conservative; therefore, a de novo gene prediction strategy was also followed using gene-finding algorithms such as Genscan (Burge and Karlin 1997). Only predictions supported by at least one type of sequence-similarity evidence (i.e., human ESTs, mouse ESTs, etc.) were considered. In this way a total of 39,114 gene annotations were generated provided either by Otto (17,764) or by the de novo gene prediction not overlapping Otto predictions (21,350).

A different approach, based on the Ensembl gene annotation system (see the URL in Table 5.1), was adopted by the public sequencing consortium. It started with the de novo predictions provided by Genscan (Burge and Karlin 1997), followed by prediction refinement based on similarity with known transcripts (mRNAs, ESTs) and protein motifs contained in the Pfam database (Bateman, Birney et al. 2000). To increase stringency, another gene-finding program was used, Genie (Reese, Kulp et al. 2000), merging all overlapping results, supported by at least one EST similarity evidence, to generate a nonredundant set of protein-coding genes. The process resulted into an initial integrated gene index (IGI) associated with an integrated protein index (IPI). The first version of IGI reports the following values: IPI.1 protein set account for 31,778 proteins, of which 14,882 are from known genes, 4057 from prediction based on Ensembl plus Genie, and 12,839 based on Ensembl alone. Approximately 26% and 12% of the Celera and Ensembl gene set did not show similarity to the Ensembl and Celera gene set, respectively. Comparison with RIKEN mouse cDNAs (Kawai, Shinagawa et al. 2001) showed that 69% of the IGI have matches in this data set. The comparison with genes on the fully annotated chromosome 22 was used to assess the portions of gene prediction corresponding to pseudogenes and hence the rate of overprediction. Nine percent of IGI predictions were found to correspond to pseudogenes.

It should be pointed out that gene-prediction methodologies are subject to artifacts, such as:

1. ESTs may be contaminated by unspliced mRNAs, nongenic transcription, and/or genomic DNA contamination.
2. Alignment with previously known genes may detect pseudogenes. Obviously with this procedure new species-specific genes cannot be identified.
3. Lack of similarity with homologous genes from other organisms may be due to highly different evolutionary rates, gene loss, or acquisition of new functions during evolution.
4. De novo prediction suffers from problems linked to signal-to-noise ratio and depends on the level of knowledge used to establish rule-based procedures; it is unable to predict untranslated exons at both the 5′ and 3′ gene ends.

To sum up, more accurate methods are needed for reliable prediction. As genomic annotation proceeds, and data from other genome or cDNA projects (e.g., rat, mouse, pufferfish, etc.) will be made available, we can expect that a more detailed and reliable human gene inventory will be obtained. Indeed, it is likely that some genes have been missed because they do not exhibit similarity with known proteins, have low expression levels, or belong to rare tissues. The growing full-length cDNA set, and the development of new techniques such as the oligo-capping method (Suzuki and Sugano 2001), which provides extensive sets of 5′ mRNA ends, will add a remarkable piece of information, providing, in particular, evidence of transcription starts and ends and of alternative splice mRNA forms. In humans, EST and cDNA alignment to genomic sequences indicates that about 35% of mRNAs contain alternatively spliced exons [see Gaasterland and Oprea (2001) and the references therein]. Yet this figure is probably underestimated. At present, however, a probable total number of genes around 30,000 to 35,000 is suggested. Thus, with an average coding region of 1400 bp on a total genomic length of about 30 kbp, only one-third of the human genome would be transcribed in genes and the coding sequence would account for just 1.5% of the total sequence.

The full gene annotation may allow us to study a variety of features, such as GC content, exon/intron structure, UTR regions, and many other properties very useful in medical studies. Data so far available show that exons have a mean length of 145 bp, whereas the mean length of introns is 3365 bp. Average mRNAs were 2410 bp long, made up of 8.8 exons, with coding region, 5' UTRs, and 3' UTRs averaging 1340, 300, and 770 bp, respectively. Intronless genes were found to occur with appreciable frequency in the human genome and prevalently, to correspond to histones, G-protein receptors, olfactory receptors, and cytokines (Wright, Lemon et al. 2001). Splice junctions were found generally to be of canonical type, with a 5'-GT and 3'-AG pattern, with only 0.76% using 5'-GC/3'-AG and about 0.10% using the rare alternative 5'-AT/3'-AC pattern recognized by the splicing machinery U12.

Data on the draft sequence confirm that variation in gene size can be explained partially by the fact that GC-rich regions tend to contain a larger number of genes than do the AT-rich regions. However, the latter contain genes with larger introns. Conversely, whereas gene density increases more than tenfold when GC content rises from 30% to 50%, intron size tends to become smaller. About half of the genes were associated with upstream CpG islands. The classification of human proteins in functional categories has been accomplished by the public consortium using the InterPro annotation protocol (Apweiler, Attwood et al. 2001) based on databases of protein domains associated with specific biochemical and cellular functions (see also Sec. 5.9).

Analogously, the Celera classification of molecular functions of predicted proteins was carried out using the Pfam (Bateman, Birney et al. 2000) and SMART (Schultz, Copley et al. 2000) protein domain databases as well as with the Celera Panther Classification (CPC) [see ref. 116 in Venter, Adams et al. (2001)]. Overall, compared to the worm and fly genomes, humans appear to have many more proteins involved in cytoskeleton, defense, immunity, transcription, and translation, processes typically expanded in vertebrate physiology. However, a remarkable fraction of proteins predicted neither matched known functional domains nor showed a significant match to other known proteins. The 113 gene transfers from bacteria to vertebrate germ cell, predicted for the greater similarity between human and bacterial proteins than between human and yeast, worm, fly, or mustard weed, have later been revealed to be highly improbable (Salzberg, White et al. 2001; Stanhope, Lupas et al. 2001). Indeed, the fact that some genes have been found in very distantly related organisms, such as bacteria and human, and absent in more closely related taxa, may be attributed to the frequent gene loss taking place along some lineages.

To perform comparative proteome analysis, a more global perspective is needed which may provide information on the commonalities and differences among eukaryotes. The scenario emerging from a limited number of studies already indicates that new domains have been invented in vertebrates, together with expansion of families and evolution of new protein architectures.

CHAPTER 3

ORGANELLES

3.1 MITOCHONDRIA

3.1.1 General Structure and Function

Mitochondria are cytoplasmic organelles present in all eukaryotic organisms, but in a small group of phagothrophous or micropinocytotic nonphotosynthetic protists, such as *Giardia*, *Trichomonas*, and microsporidia. They were called *Archaezoa* (Cavalier-Smith 1993) to denote a primitive amitochondriate phase of eukaryotic evolution, but their phylogenetic position has recently been questioned. Their origin is discussed in Sec. 7.5.

Mitochondria are dynamic organelles, undergoing frequent changes in number and morphology, whose diameter ranges from 0.5 to 1 μm and whose length can reach up to 7 μm. They are usually described as bean- or oval-shaped, but their shape can change rapidly during their lifetime, as they can swell or contract in response to various stimuli, such as drugs or hormones, or during the process of ATP synthesis. These changes are believed to depend on the variation of cell need for energy. Indeed, the number of mitochondria per cell is strictly correlated with energy requirements; for example, in cardiac tissue, which requires large amounts of energy, mitochondria comprise about 50% of the volume of cardiac cells.

Recent studies have revealed that mitochondrion structure varies extensively during the cell cycle. In particular, the structure is determined by a balance of fission and fusion events required to maintain normal mitochondrial morphology and function. The mitochondrial division (fission) is a complex process requiring the intervention of multiple factors, the most characteristic of which are the dynamin-related proteins Dnmp1p and Mgm1p in yeast (Otsuga, Keegan et al. 1998; Shepard and Yaffe 1999). Fusion events have been demonstrated to occur in mating yeast, where the parental mitochondria fuse, with mixing of matrix and membranes but not of mtDNA (Nunnari, Marshall et al. 1997). Structure heterogeneity has been detected in populations of higher plant mitochondria, possibly associated with some functional specificity for the varying levels of respiratory capacity and the specific morphology of membrane folding (Dai, Lo et al. 1998).

Recent studies have demonstrated that in animal growing cells, mitochondria are found as dynamic reticular networks; each differentiation stage seems to have a peculiar shape and distribution of mitochondria, and alterations have been associated with such pathological situations as muscular dystrophy or cancer [see Garesse and Vallejo (2001) for review].

Mitochondria are formed by two membranes, an outer membrane that is highly permeable and an inner membrane where electron transport and oxidative phosphorylation occur (see below). The inner membrane shows numerous foldings (called *cristae*), whose function is to increase the surface area where adenosine triphosphate (ATP) synthesis takes place.

Recent evidence also suggests that both the mtDNA and the translational machinery are tethered to the inner membrane (Sanchirico, Fox et al. 1998; Hobbs, Srinivasan et al. 2001). The inner space is called the matrix: it contains the enzymes involved in the citric cycle, enzymes catalyzing fatty acid oxidation and certain amino acid reactions, as well as other enzymes which are specific for a given organism or tissue. The matrix space also contains the DNA and the genetic apparatus for the replication, transcription, and translation of the mitochondrial genome.

Mitochondria can be regarded as the powerhouses of the cell; indeed, they are responsible for converting nutrients into the energy-yielding molecule ATP through the process of oxidative phosphorylation. The oxidative phosphorylation couples a series of oxidation–reduction reactions to ATP formation. The electron flux through the respiratory complexes to the final O_2 acceptor to form water results in a drop of free energy, which is used to pump protons from the matrix to the intermembrane space. The energy conserved into this proton gradient is used to promote ATP synthesis by ATP synthase through the inversion of proton flux from intermembrane space to matrix. As mentioned above, the mitochondrial electron transport proteins are clustered into complexes known as complex I (also called NADH:CoQ oxidoreductase), II (succinate:CoQ oxidoreductase), III (CoQ:cytochrome c oxidoreductase), and IV (cytochrome c oxidase); ATP synthesis occurs in complex V, the ATP synthase. Figure 3.1 shows a schematic representation of mitochondrion and of the oxidative phosphorylation process occurring at the level of the inner membrane.

Mitochondrial efficiency has been reported to be close to 70%, markedly higher than that of internal combustion engines (about 10% efficient) or hydrogen–oxygen fuel cells used in spacecraft (approximately 40% efficient). Since mitochondria represent closed compartments in the cell, cytosolic–mitochondrial relationships are made possible by the existence of specialized transport systems (translocators, carriers), which allow the transport of metabolites from the cytosol to the mitochondria, and vice versa. The carrier family, common to all cells, includes only a few translocators whose existence is necessary for cell survival. This is exemplified by the ATP/ADP translocator, the phosphate (Pi) carrier, and the carriers for the translocation of citric acid intermediates. Mitochondrial permeability includes compounds that are not involved in energy metabolism. This is true for vitamins and coenzymes, which have been shown to enter mitochondria on the basis of specific translocators and other mechanisms. Active research on molecular characterization and isolation of translocators is still going on.

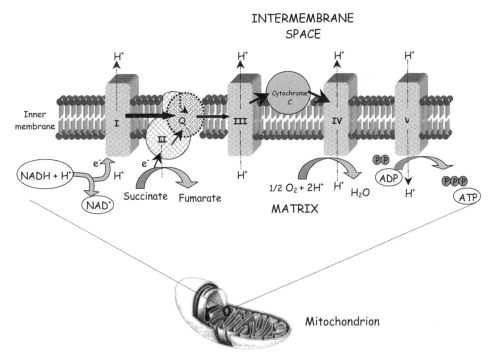

Figure 3.1. General scheme of the mitochondrion and of the oxidative phosphorylation process, which occurs in the inner mitochondrial membrane involving the four respiratory complexes (I–IV), the ubiquinone (Q), and the cytochrome c which mediate electron transport from NADH (or FADH$_2$) to the oxygen to form water and a protonic and potential gradient to be converted to energy. Energy conversion occurs at level of complex V (or ATP synthase) by the synthesis of ATP.

Hitherto, particular attention has been paid to the problem of translocation through the mitochondrial membrane of proteins which are synthesized in the cytosol, and are able to reach the mitochondrial compartments. Proteins destined to be imported into mitochondria are synthesized as precursors. Four protein translocation complexes have been characterized: TOM (translocase of the outer membrane), TIM 23 (translocase of the inner membrane), TIM 22, and Oxa1 (inner membrane protein). Their task is to mediate, in a concerted action, the targeting of preproteins into mitochondrial subcompartments. This multisubunit translocation machinery comprehends a number of import receptors and protein-conducting channels (Herrmann and Neupert 2000).

Recent discoveries have shown that mitochondria are involved in other central cellular processes, such as those responsible for the destiny of the eukaryotic cell (i.e., programmed cell death or apoptosis). Thus these organelles are involved in several metabolic disorders and play a pivotal role in aging. Indeed, the reactive oxygen species (ROS), generated by mitochondria themselves, appear to be the major source of the oxidative lesions that accumulate with age (Beckman and Ames 1998; Raha and Robinson 2000). The age-related oxidative injuries to mitochondrial macromolecules include damage to the structure and function of mitochondrial DNA as well as to mitochondrial proteins and lipids (Ames, Shigenaga et al. 1995; Gadaleta, Cormio et al. 1998).

3.1.2 DNA and Genetic System

Although derived from the same ancestral progenitor (see Chapter 7), the mtDNA has followed multiple and different evolutionary pathways in the various taxa; for this reason the mitochondrial genome shows a great variability in terms of structure, gene content, organization, and mode of expression in the various organisms. Several features are, however, common to the majority of mitochondrial genomes. The circular double-stranded structure appears to be almost a constant feature of mtDNA, but it is not the rule. Suyama and Miura (1968) first discovered the existence of linear mtDNA molecules in the ciliate protozoan *Tetrahymena pyriformis*. This finding, initially considered to be an artifact, was confirmed in 1975 when a linear mtDNA was reported in another ciliate, *Paramecium aurelia* (Goddard and Cummings 1975). Up to now, linear mitochondrial genomes have been reported in an unexpected high number of organisms from very distant taxa, including protozoans (Apicomplexa and Ciliata), algae, slime molds, oomycetous fungi, yeasts, and also in the metazoan Cnidaria [see Nosek, Tomaska et al. (1998) for a review]. Investigating linear genomes is important for human health, since they are found in a number of human pathogens and their peculiar replication mechanisms would allow the development of targeted drugs not interfering with human circular mtDNA.

mtDNA exhibits an extraordinary variability in length, particularly in the lower eukaryotes and in plants. The size ranges from only about 6 kbp in some protists (e.g., *Plasmodium* and *Theileria*) to 2000 kbp in plants (e.g., *Cucumis melo*) (see also Table 3.1 for the genome size of complete mitochondrial genome). In animal cells, the size of the mtDNA is rather constant, ranging between 15 and 17 kbp; whereas its genetic content, though constant, has a variable organization in the various phyla. Despite such a variety of structures and sizes, the information content of all mitochondrial genomes is not dramatically different in the various organisms. Indeed, mtDNA encodes essentially for two types of genes: (1) genes for several proteins which, together with products coded by the nuclear genome, form the oxidative phosphorylation complexes; and (2) genes involved in the protein synthesis machinery: namely, at least two species of rRNAs and a more or less complete set of genes for tRNAs.

The mt genome is a good example of genetic conservation versus structure diversity. It demonstrates the various strategies the eukaryotic cell can use to express the same information content. Among these, the most peculiar is certainly RNA editing (see below), which is particularly active in protozoans and in plants (Gray 1993).

As for the features of the mt genetic system, their nature can be either prokaryotic (e.g., naked DNA, absence of introns in metazoans) or eukaryotic (e.g., presence of introns in lower eukaryotes and plants, presence of 3'-end polyadenylated mRNAs in metazoans). In any case, the genetic information contained is always very limited. All other products of the mitochondrion are coded for by nuclear genes and translated in the cytoplasm. Beside the subunits of complexes forming the oxidative phosphorylation apparatus, the nuclear genomes also specify the enzymes of the mitochondrial matrix and hence the proteins involved in replication, transcription, RNA processing, translation, and protein (and RNA) import (see Sec. 5.15 for a description of databases collecting nuclear-coded genes coding for proteins imported in the mitochondrion or involved in mitochondrial biogenesis). Protein import into mitochondria depends on sorting signals generally corresponding to an

TABLE 3.1. Completely Sequenced Mitochondrial Genomes

Taxon	Accession Number	Length (bp)[a]
Protozoa		
Acanthamoeba castellanii	U12386	41,591
Cafeteria roenbergensis	AF193903	43,159
Chlamydomonas eugametos	AF008237	22,897
Chlamydomonas reinhardtii	U03843	15,758*
Chondrus crispus	Z47547	25,836
Chrysodidymus synuroideus	AF222718	34,119
Cyanidioschyzon merolae	D89861	32,211
Dictyostelium discoideum	AB000109	55,564
Laminaria digitata	AJ344328	38,007
Leishmania tarentolae	M10126	20,992
Malawimonas jakobiformis	AF295546	47,328
Naegleria gruberi	AF288092	49,843
Nephroselmis olivacea	AF110138	45,223
Ochromonas danica	AF287134	41,035*
Paramecium aurelia	X15917	40,469*
Pedinomonas minor	AF116775	25,137
Physarum polycephalum	AB027295	62,862
Plasmodium falciparum	AJ276844	5,967*
Plasmodium reichenowi	AJ251941	5,966*
Porphyra purpurea	AF114794	36,753
Prototheca wickerhamii	U02970	55,328
Pylaiella littoralis	AJ277126	58,507
Reclinomonas americana	AF007261	69,034
Rhodomonas salina	AF288090	48,063
Scenedesmus obliquus	AF204057	42,781
Tetrahymena pyriformis	AF160864	47,296*
Tetrahymena thermophila	AF396436	45,577
Theileria parva	Z23263	5,895*
Fungi		
Allomyces macrogynus	U41288	57,473
Candida albicans	AF285261	40,420
Hyaloraphidium curvatum	AF402142	29,593
Hypocrea jecorina	AF447590	42,130
Phytophthora infestans	U17009	37,957
Pichia canadensis	D31785	27,694
Podospora anserina	X55026	100,314
*Rhizophydium sp.*136	AF404306	68,834
Saccharomyces castellii	AF437291	25,753
Saccharomyces cerevisiae	AJ011856	85,779
Schizophyllum commune	AF402141	49,704
Schizosaccharomyces pombe	X54421	19,431
Spizellomyces punctatus	AF404303-5	58,830
Yarrowia lipolytica	AF338709	47,916
Plant		
Arabidopsis thaliana	Y08501/2	366,923
Beta vulgaris	AP000396/7	368,799
Marchantia polymorpha	M68929	186,609
		(*Continued*)

TABLE 3.1. *Continued*

Taxon	Accession Number	Length (bp)[a]
Metazoa		
Annelida		
Lumbricus terrestris	U24570	14,998
Platynereis dumerilii	A178678	15,619
Arthropoda		
Anopheles gambiae	L20934	15,363
Anopheles quadrimaculatus A	L04272	15,455
Apis mellifera ligustica	L06178	16,343
Artemia franciscana	X69067	15,822
Bombyx mandarina	AB070263	15,928
Bombyx mori	AF149768	15,643
Ceratitis capitata	AJ242872	15,980
Chrysomya chloropyga	AF352790	15,837
Cochliomyia hominivorax	AF260826	16,022
Crioceris duodecimpunctata	AF467886	15,880
Daphnia pulex	AF117817	15,333
Drosophila mauritiana	AF200830	14,964
Drosophila melanogaster	U37541	19,517
Drosophila sechellia	AF200832	14,950
Drosophila simulans	AF200833	14,972
Drosophila yakuba	X03240	16,019
Heterodoxus macropus	AF270939	14,670
Ixodes hexagonus	AF081828	14,539
Limulus polypehemus	AF00264	14,985
Lithobius forficatus	AF309492	15,695
Locusta migratoria	X80245	15,722
Narceus annularus	AY055727	14,868
Ostrinia furnacalis	AF467260	14,536
Ostrinia nubilalis	AF442957	14,535
Pagurus longicarpus	AF150756	15,630
Penaeus monodon	AF217843	15,984
Pyrocoelia rufa	AF452048	17,739
Rhipicephalus sanguineus	AF081829	14,710
Tetrodontophora bielanensis	AF272824	15,455
Triatoma dimidiata	AF301594	17,019
Brachiopoda		
Laqueus rubellus	AB035869	14,017
Terebratulina retusa	AJ245743	15,451
Cnidaria		
Metridium senile	AF000023	17,443
Acropora tenuis	AF338425	18,338
Echinodermata		
Arbacia lixula	X80396	15,719
Asterina pectinifera	D16387	16,260

TABLE 3.1. *Continued*

Taxon	Accession Number	Length (bp)[a]
Florometra serratissima	AF049132	16,005
Paracentrotus lividus	J04815	15,696
Strongylocentrotus purpuratus	X12631	15,650

Mollusca

Albinaria coerulea	X83390	14,130
Cepaea nemoralis	U23045	14,100
Crassostrea gigas	AF177226	18,224
Euhadra herkiotsi	Z71693-701	
Katharina tunicata	U09810	15,532
Loligo bleekeri	AB029616	17,211
Mytilus edulis	M83756-62	
Pupa strigosa	AB028237	14,189
Venerupis (Ruditapes) philippinarum	AB065375	22,676

Nematoda

Ancylostoma dudenale	AJ417718	13,721
Ascaris suum	X54253	14,284
Caenorhabditis elegans	X54252	13,794
Echinococcus granulosus	AF346403	13,598
Echinococcus multilocularis	AB018440	13,738
Fasciola hepatica	AF216697	14,462
Hymenolepis diminuta	AF314223	13,900
Necator americanus	AJ417719	13,604
Onchocerca volvulus	AF015193	13,747
Paragonimus westermani	AF219379	14,964

Platyhelminthes

Schistosoma japonicum	AF215860	14,085
Schistosoma mansoni	AF216698	14,415
Schistosoma mekongi	AF217449	14,072
Taenia crassiceps	AF216699	13,503
Taenia solium	AB086256	13,709
Trichinella spiralis	AF293969	16,706

Hemichordata

Balanoglossus carnosus	AF051097	15,708

Cephalochordata

Branchiostoma floridae	AF098298	15,083
Branchiostoma lanceolatum	Y16474	15,076

Urochordata

Halocynthia roretzi	AB024528	14,771

Myxiniformes

Myxine glutinosa	AJ404477	18,909
Eptatretus burgeri	AJ278504	17,168

(Continued)

TABLE 3.1. *Continued*

Taxon	Accession Number	Length (bp)[a]
Vertebrates: Petromyzontiformes		
Lampetra fluviatilis	Y18683	16,159
Petromyzon marinus	U11880	16,201
Vertebrates: Chondrichthyes		
Chimaera monstrosa	AJ310140	18,580
Heterodontus francisci	AJ310141	16,708
Mustelus manazo	AB015962	16,707
Raja radiata	AF106038	16,783
Scyliorhinus canicula	Y16067	16,697
Squalus acanthias	Y18134	16,738
Vertebrates: Osteichthyes		
Anguilla japonica	AB038556	16,685
Antigonia capros	AP002943	16,508
Arctoscopus japonicus	AP003090	16,577
Ateleopus japonicus	AP002916	16,650
Aulopus japonicus	AB047821	16,653
Beryx splendens	AP002939	16,529
Caelorinchus kishinouyei	AP002929	15,942
Carassius auratus	AB006953	16,578
Chauliodus sloani	AP002915	17,814
Chlorophthalmus agassizi	AP002918	16,221
Cololabis saira	AP002932	16,499
Conger myriaster	AB038381	18,705
Coregonus lavaretus	AB034824	16,737
Crenimugil crenilabis	AP002931	16,019
Crossostoma lacustre	M91245	16,558
Cyprinus carpio	X61010	16,575
Danio rerio	AC024175	16,890
Diaphus splendidus	AP002923	15,985
Diplophos taenia	AB034825	16,418
Elassoma evergladei	AP002950	15,780
Engraulis japonicus	AB040676	16,675
Exocoetus volitans	AP002933	16,527
Gadus morhua	X99772	16,696
Gasterosteus aculeatus	AP002944	15,742
Gonostoma gracile	AB016274	16,436
Harpadon microchir	AP002919	16,061
Helicolenus hilgendorfi	AP002948	16,728
Hoplostethus japonicus	AP002938	16,528
Ictalurus punctatus	AF482987	16,497
Ijimaia dofleini	AP002917	16,645
Lampris guttatus	AP002934	15,598
Latimeria chalumnae	U82228	16,407
Lepidosiren paradoxa	AF302934	16,403
Mastacembelus favus	AP002946	16,498
Monopterus albus	AP002945	16,622

TABLE 3.1. *Continued*

Taxon	Accession Number	Length (bp)[a]
Mugil cephalus	AP002930	16,685
Myctophum affine	AP002922	16,239
Myripristis berndti	AP002940	16,531
Neoceratodus forsteri	AF302933	16,572
Neoscopelus microchir	AP002921	16,686
Oncorhynchus mykiss	L29771	16,642
Oncorhynchus tshawytscha	AF392054	16,644
Osteoglossum bicirrhosum	AB043025	16,006
Pagrus major	AP002949	17,031
Pantodon buchholzi	AB043068	15,845
Paralichthys olivaceus	AB028664	17,090
Percopsis transmontana	AP002928	16,079
Platichthys bicoloratus	AP002951	15,973
Plecoglossus altivelis	AB047553	16,537
Polymixia japonica	AB034826	16,481
Polymixia lowei	AP002927	16,473
Polypterus ornatipinnis	L42813	16,624
Poromitra oscitans	AP002935	16,387
Protopterus dolloi	U62532	16,646
Rivulus marmoratus	AF283503	17,329
Rondeletia loricata	AP002937	16,530
Salmo salar	U12143	16,665
Salvelinus alpinus	AF154851	16,659
Salvelinus fontinalis	AF154850	16,624
Sardinops melanostictus	AB032554	16,881
Saurida undosquamis	AP002920	15,737
Scopelogadus mizolepis	AP002934	16,375
Stephanolepis cirrhifer	AP002952	16,306
Trachurus japonicus	AP003091	16,559
Zenopsis nebulosus	AP002942	16,065
Zeus faber	AP002941	16,715
Zu cristatus	AP002926	15,987

Vertebrates: Amphibia

Taxon	Accession Number	Length (bp)[a]
Mertensiella luschani	AF154053	16,650
Rana nigromaculata	AB043889	17,804
Typhlonectes natans	AF154051	17,005
Xenopus laevis	M10217	17,553

Vertebrates: Reptilia

Taxon	Accession Number	Length (bp)[a]
Alligator mississippiensis	Y13113	16,646
Caiman crocodilus	AJ404872	17,900
Chelonia mydas	AB012104	16,497
Chrysemys picta	AF069423	16,866
Dinodon semicarinatus	AB008539	17,191
Dogania subplana	AF366350	17,289
Eumeces egregius	AB016606	17,407
Iguana iguana	AJ278511	16,633
Pelomedusa subrufa	AF039066	16,787

(*Continued*)

TABLE 3.1. *Continued*

Taxon	Accession Number	Length (bp)[a]
Vertebrates: Aves		
Anomalopteryx didiformis	AF338714	16,716
Apteryx haastii	AF338708	16,816
Arenaria interpres	AY074885	16,710
Aythya americana	AF090337	16,616
Buteo buteo	AF380305	18,674
Casuarius casuarius	AF338713	16,757
Ciconia boyciana	AB026193	17,622
Ciconia ciconia	AB026818	17,347
Corvus frugilegus	Y18522	16,932
Coturnix japonica	AP003195	16,697
Dinornis giganteus	AY016013	17,070
Dromaius novaehollandiae	AF338711	16,711
Emeus crassus	AY016015	17,061
Eudromia elegans	AF338710	15,302
Falco peregrinus	AF090338	18,068
Gallus gallus	X52392	16,775
Haematopus ater	AY074886	16,589
Pterocnemia pennata	AF338709	16,747
Rhea americana	Y16884	16,714
Smithornis sharpei	AF090340	17,344
Struthio camelus	Y12025	16,591
Tinamus major	AF338707	16,702
Vidua chalybeata	AF090341	16,895
Vertebrates: Mammalia		
Arctocephalus forsteri	AF513820	15,413
Artibeus jamaicensis	AF061340	16,651
Balaenoptera musculus	X72204	16,402
Balaenoptera physalus	X61145	16,398
Bos taurus	V00654	16,338
Canis familiaris	U99639	16,728
Cavia porcellus	AJ222767	16,801
Cebus albifrons	AJ309866	16,554
Ceratotherium simum	Y07726	16,832
Chalinolobus tuberculatus	AF321051	16,818
Cynocephalus variegatus	AJ428849	16,748
Dasypus novemcinctus	Y11832	17,056
Didelphis virginiana	Z29573	17,084
Dugong dugon	AJ421723	16,850
Echinops telfairi	AJ400734	16,555
Echinosorex gymnura	AF348079	17,088
Equus asinus	X97337	16,670
Equus caballus	X79547	16,660
Erinaceus europaeus	X88898	17,447
Felis catus	U2075	17,009
Glis glis	AJ001562	16,602
Gorilla gorilla	D38114	16,364
Halichoerus grypus	X72004	16,797

TABLE 3.1. *Continued*

Taxon	Accession Number	Length (bp)[a]
Hippopotamus amphibius	AJ010957	16,407
Homo sapiens	D38112	16,569
Hylobates lar	X99256	16,472
Isoodon macrourus	AF358864	16,582
Lama pacos	Y19184	16,652
Lemur catta	AJ421451	17,036
Lepus europaeus	AJ421471	17,734
Loxodonta africana	AJ224821	16,866
Macaca sylvanus	AJ309865	16,586
Macropus robustus	Y10524	16,896
Mus musculus	V00711	16,295
Myoxus glis	AJ001562	16,602
Nycticebus coucang	AJ309867	16,764
Ochotona collaris	AF348080	16,968
Ornithorhynchus anatinus	X83427	17,019
Orycteropus afer	Y18475	16,816
Oryctolagus cuniculus	AJ001588	17,245
Ovis aries	AF010406	16,616
Pan paniscus	D38116	16,563
Pan troglodytes	D38113	16,554
Papio hamadryas	Y18001	16,521
Phoca vitulina	X63726	16,826
Physeter catodon	AJ277029	16,428
Pongo pygmaeus	D38115	16,389
Pongo pygmaeus abelii	X97707	16,499
Pteropus dasymallus	AB042770	16,705
Pteropus scapulatus	AF321050	16,741
Rattus norvegicus	X14848	16,300
Rhinoceros unicornis	X97336	16,829
Sciurus vulgaris	AJ238588	16,507
Soriculus fumidus	AF348081	17,488
Sus scrofa	AJ002189	16,613
Tachyglossus aculeatus	AJ303116	16,360
Talpa europaea	Y19192	16,884
Tarsius bancanus	AF348159	16,927
Thryonomys swinderianus	AJ301644	16,626
Trichosurus vulpecula	AF357238	17,191
Tupaia belangeri	AF217811	16,754
Ursus americanus	AF303109	16,841
Ursus arctos	AF303110	17,020
Ursus maritimus	AF303111	17,017
Volemys kikuchii	AF348082	16,312
Vombatus ursinus	AJ304826	16,996

[a] All genomes are circular except for those with an asterisk.

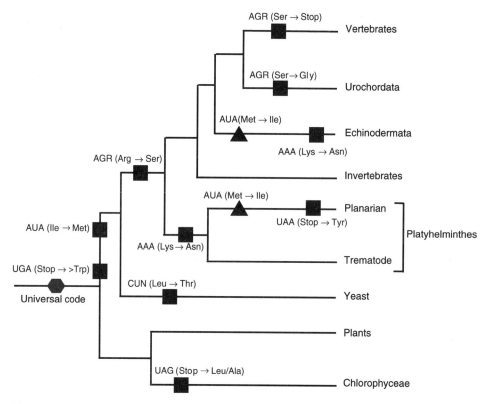

Figure 3.2. Evolution of the genetic code in mitochondrial genomes. Deviations from the universal code are indicated by squares and reversals by triangles.

N-terminal extension of the targeted proteins. The mitochondrial-targeting peptides (mTPs) contain all the relevant information to target the cytosolically synthetized peptides to the correct submitochondrial compartments (Emanuelsson, von Heijne et al. 2001) (see Sec. 6.10.3 for mTP prediction methods). Indeed, the biogenesis of mitochondria requires a well-regulated and integrated mechanism of crosstalk between the two—three when plastid DNA is present—genomes, which is still almost unknown. The availability of the complete sequence of nuclear genomes will speed enormously the understanding of these networks in physiological and pathological conditions.

The mitochondrial genetic code presents several deviations from the universal genetic code. In different species the code is slightly modified with respect to the universal code and it also varies from phylum to phylum. Figure 3.2 shows the deviations from the universal genetic code that have occurred so far in mitochondria evolution. In mtDNA of most phylogenetic groups, UGA is used as a tryptophan codon rather than as a termination codon. On the other hand, AGR (R = A or G), coding for arginine in the universal code, is a stop codon in the mtDNA of vertebrates, but it codes for serine in the mtDNA of echinoderms and for glycine in the urochordate *Ciona intestinalis*. CTN codons (N = A, C, G, U) code for threonine in yeast, and AUA is an additional codon for methionine in most metazoans and in yeast. The mitochondrial termination codons are UAA and UAG; in vertebrates, they also are AGA and AGG. In metazoans, some gene transcripts end with the incomplete termination codons U or UA and the complete termination codons are created by posttranscriptional polyadenylation, which is not present in yeast mitochondria.

Another surprising feature of the mitochondrial genetic system is the use of an oversimplified decoding mechanism allowing translation with a reduced set of tRNA species. In vertebrates, this reduction is achieved with a wider wobbling between the third base of the codon and the first base of the anticodon, where a U is able to recognize all bases in the third codon position of four-codon families. This implies that only 22 tRNA species are required to translate all sense codons for the 20 amino acids. In general, mitochondrial tRNAs are shorter and may present unusual structures with respect to tRNAs involved in cytoplasmic translation.

3.1.3 Genome Features

Due to high taxon-specific variability, we describe the mt genomes features considering the most studied taxa separately: namely, fungi, protists, land plants, and metazoans, with particular reference to those species whose genomes have been sequenced completely (Table 3.1).

Fungi Facultatively anaerobic species of the kingdom Fungi have been very useful as a model system for research in mitochondrial genetics. In the early 1970s, mitochondrial biogenesis focused greatly on the yeast mitochondrial genome; then since the complete sequence of nuclear genome became available, interest has been focused on nuclear genes coding for mitochondrial products.

A great advantage in studying mitochondrial biogenesis in fungi depends on the fact that some fungal species, exemplified by *Saccharomyces cerevisiae*, readily form respiratory-deficient mutants (petite), which can be due to mutations in nuclear as well as in mt genomes. On this basis, several classes of petite mutants can be identified: nuclear (usually recessive: pet-); mtDNA point mutation (mit-); mtDNA deletion mutation (rho-); and total absence of mtDNA (rho 0). Interesting to note, rho 0 mutants still have mitochondria, although they are poorly developed, lacking all of the respiratory complexes.

Fungal mtDNAs have been studied extensively in the "higher fungi," Ascomycota and Basidiomycota, while the "lower fungi," Chytridiomycota and Zygomycota, are less well known, although they probably represent most of the genetic diversity of fungi. To provide a wider data set, especially for phylogenetic studies, the Fungal Mitochondrial Genome Project (FMGP; see the URL in the Appendix) is sequencing the complete mitochondrial genomes for a representative sample of the major fungal lineages (Paquin, Laforest et al. 1997). The FMGP focuses on the lower fungus early-diverging Chytridiomycota (traditionally referred to as the group of *zoosporic fungi* or *aquatic fungi*), whose mitochondrial genome shows peculiar molecular features.

Up to now, 14 fungal mt genomes have been sequenced completely (Table 3.1). The informational content of the mt genome is quite constant in fungi, although genome size (ranging from 19,431 to 100,314nt) and gene order and organization can vary among the species. Furthermore, gene variability (e.g., gene length, intron number, and distribution), has been observed even between strains of the same species).

The circular shape seems to be the rule for fungal mitochondria, with some exceptions. *Saccharomyces cerevisiae* mtDNA (Foury, Roganti et al. 1998) has been shown to exist in both a heterogeneous linear and a circular form in vivo. Among ascomycetes, in species of the genera *Candida*, *Pichia*, and *Williopsis*, linear mtDNAs have been identified.

As referred to *S. cerevisiae*, which is one of the most studied yeast genomes, mtDNA contains the genes for cytochrome *c* oxidase subunits I, II, and III, ATP synthase subunits 6, 8, and 9, apocytochrome *b*, the ribosomal protein var1, and several ORFs. Some introns of the split genes (see below) code for proteins involved in RNA processing or intron transposition (maturases, reverse transcriptases, and site-specific endonucleases). In addition, the genes for 21S and 15S rRNAs, 24 tRNAs, and the 9S RNA component of RNase P, as well as seven to eight replication origin-like elements, have been identified. The lysil tRNA is imported from the cytoplasm (Kolesnikova, Entelis et al. 2000). For the NADH-dehydrogenase complex, in both *S. cerevisiae* and *Schizosaccharomyces pombe*, all genes for the subunits of this complex are present in the nuclear genome. In *Neurospora crassa* and *Allomyces*, as in animals, some subunits are encoded by the mtDNA.

Differently from metazoans (see below), many fungi and plants have the *atp9* gene. *N. crassa* (van den Boogaart, Samallo et al. 1982) and *Aspergillus nidulans* have two distinct copies of this gene, one in the nucleus and another in mitochondria. In *Neurospora* both copies are functional at certain times during its life cycle (Bittner-Eddy, Monroy et al. 1994).

Allomyces macrogynus has a pseudogene coding for a ribosomal protein (*rps3*) and an insert in the *atp6* gene that may have been acquired by interspecific transfer. Moreover, *A. macrogynus* rRNA structures resemble more closely those of bacteria than to their counterparts in other fungal mitochondria, thus confirming the ancestral character of this fungal mitochondrial genome.

The chytridiomycete *Spizellomyces punctatus* mtDNA (61.3 kbp long) has a peculiar genome organization, since it has three (one big and two small 1.5-kbp) circular chromosomes. One of the two small circles codes for the *atp9* gene, whereas the other one carries only a conserved promoter-like element but no identified gene.

In *S. cerevisiae*, gene distribution is asymmetric because all genes, except one tRNA, are coded by the same strand, and 16 of the 24 tRNAs are clustered. Its gene organization is very loose, and several genes, in particular the apocytochrome *b* (*cob*), the cytochrome oxidase subunit I (*cox1*), and the 21S rRNA, contain introns, as in the majority of fungal mtDNAs. In *S. cerevisiae* the number of introns and GC- and AT-rich mini-inserts within the genes are strain dependent.

The *S. cerevisiae* mtDNA has an average GC content of 17.1%; the genome contains long AT-rich stretches interrupted by many GC-rich clusters. The latter represent the major source of yeast polymorphism. Many studies strongly suggest that these clusters are recombinogenic mobile elements (Seraphin, Boulet et al. 1987; Weiller, Bruckner et al. 1991; Foury, Roganti et al. 1998). Indeed, it is well established that yeast mtDNA undergoes genetic exchange and recombination events; this property has also been demonstrated in other fungi, such as the Ascomycota *Candida albicans* (Anderson, Wickens et al. 2001). Similar clusters are also present in other yeast mtDNAs, while in others they are lacking or structurally different

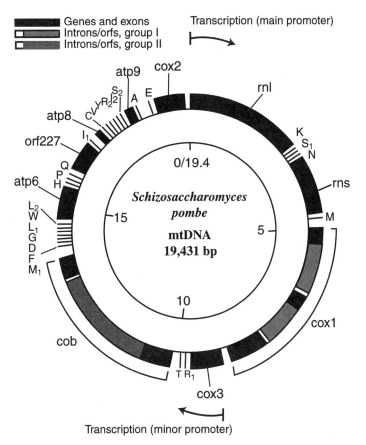

Figure 3.3. Organization of the *Schizosaccharomyces pombe* mitochondrial genome. *Rns* and *rnl* represent small and large rRNA genes. tRNA genes are represented by the single-letter code of the charged amino acid.

with taxonomic significance, as in the case of the yeast *Kluyveromyces marxianus* var. *lactis*, which contains a large number of GC-rich sequence clusters (Ragnini and Fukuhara 1988).

The two rRNA genes (15S and 21S) are not contiguous, as in the majority of other mitochondrial genomes, but separated by about 50 kbp. In *Saccharomyces douglasii*, the gene order is different from that of *S. cerevisiae* mitochondrial DNA. In particular, a segment of approximately 15,000 bp, including the gene coding for small rRNA, has been translocated to a position between the genes coding for varl and large rRNA (Tian, Macadre et al. 1991).

Figure 3.3 shows the organization of the *S. pombe* (fission yeast) mtDNA. Concomitant with the small genome size (about 19,000 bp), this genome is tightly packed, with genes separated by short noncoding sequences, and three introns, belonging to groups I and II, are present (Fig. 3.4). In this species most of the tRNAs are scattered along the molecule and make up sort of punctuation signals, as in metazoan mitochondrial genomes.

Ascomycete mitochondria often contain a large number of introns: In *Podospora anserina*, 36 introns occupy about 60% of the mitochondrial genome (Cummings,

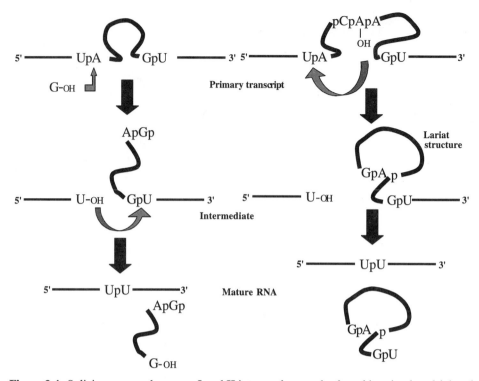

Figure 3.4. Splicing process by group I and II introns that can be found in mitochondrial and chloroplast genes. Many of the group I and II introns are self-splicing (i.e., no additional protein factors are needed for the intron splicing out). Group I introns (left) require an external guanosine nucleotide as a cofactor whose 3′ OH attacks the 5′ phospate of the 5′ nucleotide of the intron. A similar mechanism occurs in group II introns (right), which, instead, involve the 2′ OH attack of an adenine residue within the intron. In this case the intron removed forms a lariat structure. A specialized database of organellar introns has been established which allows several types of retrieval. (From the FUGOID database, *wnt.cc.utexas.edu/~ifmr530/introndata/introduction.htm*.)

McNally et al. 1990); however, two Monoblepharidales (*Harpochytrium* and *Schizophyllum commune*; Paquin, Laforest et al. 1997) lack introns completely. Organellar introns have been classified into groups I and II according to their conserved secondary structures and to the occurrence of short, primary sequence motifs (Michel, Jacquier et al. 1982; Jacquier 1996) (Fig. 3.4). The majority of fungal mitochondrial introns belong to group I, and they are located preferentially in protein-coding genes. Other than mobile intron group I elements, lower fungi mitochondrial genome harbor two novel types of putative mobile DNA elements: one encoding a site-specific endonuclease that confers mobility on the element, and the other constituting a class of highly compact, structured elements (Paquin, Laforest et al. 1997). In particular, the chytridiomycete fungus *A. macrogynus* contains 81 (G + C)-rich sequence elements, 26 to 79 bp long and can be folded into double-hairpin elements (DHEs), whose overall distribution pattern in the genome suggests that they are mobile elements (Paquin, Laforest et al. 2000).

Rearrangements (translocations or inversions) seem to be the major source of divergence in yeast mtDNA evolution. The important role of gene rearrangements is indicated by the fact that hot spots of point mutations, identified in *cox1*, *cox3*, and *atp6* genes of *S. cerevisiae*, located specifically in the proximity of insertion sites of optional mobile group I intron-related elements, inhibits genome shuffling (Foury, Roganti et al. 1998).

Replication of the mitochondrial genome of yeast starts from multiple origins (*ori* sequences) which are scattered along the yeast genome (four on one strand and three on the other). The peculiar feature of these regions is that they are homologous and are formed by three GC-rich clusters separated by four AT-rich stretches. These regions can be folded in a hairpin structure, thanks to the presence of inverted repeats, and contain little open-reading frames and a sequence *r*, from which events of transcription and replication start (de Zamaroczy and Bernardi 1985). The process is bidirectional and RNA primed. RNA polymerase and a primase synthetize the primers starting from the *r* sequences, and the initiation of DNA synthesis on one strand is activated by transcription and/or replication on the other strand.

In yeast mitochondria, replication and transcription are related mechanisms. The enzymes known to be involved in yeast mtDNA replication are the RNA polimerase Rpo41p, which synthesizes the RNA primers, the RNase MRP, involved in primer processing, single-stranded binding proteins (SSBs), Mip1p with DNA polymerase activity, and Abf2p, showing similarity with animal mtTFA, which is involved in recombination intermediate stabilization.

Several transcription units (at least 13) are present in the loose gene organization of yeast mtDNA. The transcripts are synthesized as polycistronic precursors, starting from several initiation sites within highly conserved nonanucleotide motifs (TTATAAGTA) that constitute promoter elements. An RNase P, containing an essential 9S RNA component, coded for by the mtDNA, cuts the polycistronic RNAs at the level of a conserved dodecamer motif (AATAATATTCTT) present in nearly all protein-coding genes. Polygenic transcripts carrying at least five to seven tRNA sequences have been identified, and the high number of processing intermediates has suggested that several alternative processing pathways, involving cleavage at the 3′ and 5′ ends of the tRNA sequences and in the long intergenic regions, may occur (Palleschi, Francisci et al. 1984). The processing of the split genes appears to be very complex, as both mitochondrial products, called *maturases*, coded by their own introns, and nuclear-coded components are involved. Yeast mRNAs are not polyadenylated, and usually have flanking untranslated regions. It has been reported that translocation events do not appear to alter the transcription process.

The yeast mitochondrial RNA polymerase is composed of a core element showing similarity to bacteriophage T7 RNA polymerase and has a promoter recognition factor similar to bacterial σ factors. The catalytic subunit Rpo41p has an additional transcription factor termed sc-mtTFB (also called MTF1), required for the specificity of the promoter binding.

In *Spizellomyces punctatus* the existence of tRNA editing (see below) has been demonstrated (Laforest, Roewer et al. 1997). This fungus codes only for eight tRNAs, in contrast with the minimal set of 24 or 25 tRNAs, which is normally found in fungi. It has an editing pattern virtually identical to that described in the amoeboid protozoan *Acanthamoeba castellanii*. The same type of tRNA editing also occurs in another lower fungal lineage, the Monoblepharidales (*Harpochytrium* and *Monoblepharella* spp.; see Table 3.2).

Fungal mitochondria have their own (nonuniversal) translation code (see Fig. 3.2), with some major exceptions: in *S. pombe*, the universal translation code is used for all ubiquitous mitochondrial genes. The universal translation code is also used in mitochondria of several lower fungi, such as the zygomycetes and certain lineages of the chytridiomycetes (e.g., *Allomyces*, *Monoblepharella*, and *Harpochytrium*). The universal translation code in *S. pombe* can be explained as a primitive character that was inherited from its lower fungal ancestors.

The crosstalk nucleus-mitochondrion has been studied extensively in yeast. Parikh, Morgan et al. (1987) demonstrate that *S. cerevisiae* cells can respond to changes in the state of their mitochondria via an intracellular nucleus-mitochondrion signaling system called the *retrograde response*. This response is induced by mitochondrial dysfunction and is mediated by products of the *RTG1*, *RTG2*, and *RTG3* genes (Epstein, Waddle et al. 2001). Recently, Kirchman, Kim et al. (1999) demonstrated that induction of the retrograde response in yeast cells results in extension of the life span. The retrograde signaling pathway acts continuously as a stress response mechanism, compensating mitochondrial state alterations by adjusting biosynthetic and metabolic activities, with an enhancement of longevity (Liu and Butow 1999; Jazwinski 2000, 2001).

The mitochondrial genome of fungal pathogens seems to be involved in pathogeneticity. Olson and Stenlid (2001) have demonstrated that there is mitochondrial control of virulence in hybrid species of fungal plant pathogens. The virulence of hybrid species of the basidiomycete fungus *Heterobasidion annosum* (a casual agent of root and butt rot in conifers) is controlled by the mitochondrial genome, probably via mitochondrial factors that determine the virulence of hybrids, or through interplay between the organism's mitochondrial and nuclear genomes which gives rise to a mitochondrially determined phenotype in hybrids.

Yeast as a model for aging was first proposed in 1950 by Barton (Sinclair, Mills et al. 1998), who showed that individual yeast cells are mortal. Since then, numerous models for yeast aging have been proposed, including bud scar excesses, membrane defects, cell wall changes, surface-to-volume limits, and increases in membrane fragility. However, recent studies seem to give support to an aging model in which the activation of a molecular aging clock results in the replication and accumulation of a senescence factor that eventually overwhelms old cells. One of the best studied fungus is undoubtedly the filamentous fungus *Podospora anserina*. Its vegetative growth is not unlimited, stops definitively at a given point, and is accompanied by death of the apical cells. This phenomenon has been called *senescence*. At the cellular level, senescence is accompanied by a drastic reorganization of the internal membrane network (mitochondria, endoplasmic reticulum, nuclear membrane). At the molecular level, it is correlated with mtDNA modifications; indeed, senescence is coincidental with the accumulation of extrachromosomal circular DNAs (senDNAs; Jamet-Vierny, Begel et al. 1980). Three distinct groups of senDNAs (α, β, and γ) originate from separate regions of the mitochondrial genome. Of these, α-senDNA is observed most frequently in senescing cultures, and this sequence is absent from the genomes of many long-lived strains (Sinclair, Mills et al. 1998). Thus, for *P. anserina* (and it is valid also for *S. cerevisiae* aging), genome stability appears to be an important factor in age-dependent senescence (Jamet-Vierny, Begel et al. 1980).

Recent studies have shown that yeast can be used as a model system in human mitochondrial neurodegenerative pathology investigations (Rohou, Francisci et al. 2001). In humans, deletion mutations of mtDNA accumulates exponentially with age in nerve and muscle tissue. Biochemical studies also show an increase in mtDNA point mutations with age and a concurrent decline of the overall bioenergetic capacity of mitochondrion (Cortopassi and Wong 1999).

Protists   Despite the fact that protists comprise the bulk of biological diversity of the eukaryotic lineage, only 22 (11%) of the complete mitochondrial genome sequences available in public databases belong to this group of organisms. To stimulate activities in this field, the Organelle Genome Megasequencing Program (OGMP; see the URL in the Appendix) was established in 1992 with the specific aim of determining complete protist mtDNA sequences systematically and comprehensively. The size of protist mtDNA ranges from 6 kbp in the apicomplexan species (the smallest known mtDNAs) to 77 kbp in the choanoflagellate *Monosiga brevicollis*. The majority of protist mtDNAs are compact, gene-rich genomes, with few or no large noncoding regions (Gray, Lang et al. 1998).

Protist mtDNA shows different shapes. In Ciliata (e.g., *Tetrahymena pyryformis* and *Paramecium aurelia*), the genome is linear, 46 and 40 kbp long, respectively. In the algae *Chlamydomonas reinhardtii*, besides the linear (major species) mitochondrial genome (16 kpb long), a circular (minor species) mtDNA might also be present. In the kinetoplasts of the Trypanosomatidae, *Crithidia fasciculata*, *Leishmania tarentolae*, and *Trypanosoma brucei*, the mitochondrial genome has a very peculiar structure, being composed of an intricate network containing two types of molecules: many thousands of minicircles (1 to 3 kbp, depending on the species), accounting for 90 to 95% of the DNA, and 25 to 50 maxicircles (20 to 40 kbp). Minicircles lack long amino acid–coding frames and maxicircles are the counterparts of mtDNAs in other organisms (Benne 1985). It should be assumed, however, that more than half of the genes are cryptic and require extensive editing at the RNA level (see below). Minicircle-encoded RNAs are necessary to guide the editing process required to generate functional mRNAs from the maxicircle transcripts.

As far as gene content is concerned, protist mtDNAs generally resemble plant rather than animal or fungal mtDNAs. The mitochondrial genome of the heterotrophic flagellate *Reclinomonas americana* contains the largest gene number so far identified in any mtDNA (97 genes). Because the genes in the other sequenced mtDNAs are all subsets of the *R. americana* set, some authors have suggested the *R. americana* pattern is closest to the ancestral pattern of genes carried by the protomitochondrial genome (Lang, Burger et al. 1997). During evolution, gene loss has probably occurred through gene transfer to the nucleus, with many respiratory chain genes and almost all ribosomal protein genes having already been eliminated in the common ancestor of animal and fungal mtDNAs.

Most protist mtDNAs contain a number of conserved but unidentified ORFs, and most of them are unique (Gray, Lang et al. 1998). As to genes not encoding proteins, protist mtDNAs, with only few exceptions, encode large (LSU) and small (SSU) rRNAs whose potential secondary structures deviate minimally from their bacterial counterparts. A minority of protist mtDNAs encode rRNA genes whose structure and/or the structure of their products is very unusual. It is the case of try-panosomatid protozoans (e.g., *Leishmania tarentolae* and *Trypanosoma brucei*) 9S (SSU) and 12S (LSU) mitochondrial rRNAs, which are among the smallest and structurally most divergent of known rRNAs (Gutell 1994). Similar to animals and fungi, most protist mtDNAs lack the 5S rRNA gene, exceptions being *Prototheca wickerhamii, Nephroselmis olivacea, Chondrus crispus*, and *R. americana*.

In *Tetrahymena pyriformis* the mtDNA shows some atypical features of mito-chondrial gene organization and expression (e.g., split and rearranged large subunit rRNA genes; split *nad1* gene, whose two segments are located on and transcribed on opposite strands). The latter feature has also been found in *Paramecium aurelia* (Burger, Zhu et al. 2000).

Several protist mtDNAs appear to encode the minimal tRNA set required for translation. However, in most cases, the mitochondrial genome lacks some tRNAs, and import from the cytosol is needed. That is the case with *Acanthamoeba castel-lanii, Dictyostelium discoideum, Paramecium aurelia, Tetrahymena pyriformis, Chlamydomonas* spp., and *Pedinomonas minor* (Gray, Lang et al. 1998). An extreme situation is found in Apicomplexa and Trypanosomatidae, whose mitochondrial genomes have no tRNA genes and import a full set of tRNAs from the cytoplasm (Simpson, Suyama et al. 1989; Hancock and Hajduk 1990; Dietrich, Weil et al. 1992).

The generation of RNA molecules having nucleotide sequences differing from those encoded by genes, called *RNA editing*, was first discovered by Benne, Van den Burg et al. (1986) in trypanosome mitochondria. In these organisms the posttran-scriptional RNA maturation process involves the addition or removal of uridine residues at precise sites in encoded transcripts. Editing may generate start and less frequently termination codons, corrects frameshifts, and even build entire open-reading frames from nonsense sequences thus giving rise to the functional RNA [see Estevez and Simpson (1999) for a review]. The information for the correct pattern of insertions/deletions is usually provided by small guide (g)RNAs, which are antisense with respect to the edited RNA regions to be edited and possess 3′-U tails of 5 to 24 residues. These gRNAs are mostly encoded by the minicircle com-ponent of the mtDNA.

In recent years, different types of RNA editing have been described (Table 3.2). In *A. castellanii*, sequencing of the mtDNA has demonstrated a novel type of tRNA editing that affects most mtDNA-encoded tRNAs. This editing modifies the accep-tor stem in 13 of 16 tRNAs (Lonergan and Gray 1993; Borner, Morl et al. 1996). In the case of tRNAs encoded by *D. discoideum*, mtDNA secondary structure model-ing strongly suggests that several of them undergo a similar type of editing.

In *Physarum polycephalum*, extensive RNA editing is necessary to create func-tional mitochondrial transcripts, since transcripts exhibit a combination of C-, U-, and dinucleotide insertions as well as base conversion of C to U. Table 3.2 summa-rize known different forms of RNA editing in mitochondria.

The protists' mitochondrial genomes also harbor noncoding regions. In particu-lar, introns are not always present. Only small numbers of group I introns have been found in the amoeboid protozoa *A. castellanii* and *D. discoideum*, the green algae *Prototheca wickerhamii, Nephroselmis olivacea*, and *Chlamydomonas eugametos*, and the choanoflagellate *Monosiga brevicollis*. Very few group II introns have been found in protist mtDNAs (a total of seven such introns in five of the 22 completely sequenced protist mtDNAs).

TABLE 3.2. Forms of RNA Editing in Mitochondria

Type	Organism (Genome)	Transcripts	Cis-Acting Elements	Trans-Acting Factors	Mechanism
		Insertion/Deletion Editing			
U insertion/deletion	Kinetoplastids	Many mRNAs	Anchoring sequence	gRNAs, terminal U transferase (TUT), RNA ligase, endonuclease, U exonuclease, other factors	Cleavage, TUTase or U exonuclease action, ligation
C insertion (also U, UA, AA, CU, GU, GC)	*Physarum polycephalum*	Many mRNAs, also tRNAs, rRNAs	?	?	Linked to transcription
C to A, A to G, U to G, and U to A	*Acanthamoeba castellanii*	tRNA	?	?	Replacement of the first three 5' nt
3'-terminal A addition	Vertebrates	mRNAs	Flanking tRNA structure	Endonuclease, terminal A transferase (TATase)	Cleavage and TATase action
3'-terminal A or C addition	Animals	tRNAs	?	?	3' CCA addition
		Modification Editing			
C to U	Marsupials	tRNA (Gly → Asp anticodon) cox1 mRNA	?	?	?
	Physarum polycephalum		?	?	?
	Land plants	Many mRNAs, also itRNAs, rRNAs	?	?	C deamination

105

Some group I introns seem to have been inherited vertically from a mitochondrial ancestor of fungi, green algae, and plants, much as in *cox1* group I introns in *P. wickerhamii* (Wolff, Burger et al. 1993). Conversely, evidence suggests acquisition of some of these introns by horizontal transfer, as appears to be the case for the group II introns found in the *rnl* gene of the brown alga *Pylaiella littoralis* (Fontaine, Rousvoal et al. 1995), and in *P. purpurea*, whose group II introns have been acquired through a very recent lateral transfer from cyanobacteria (Burger, Saint-Louis et al. 1999).

Plants   Mitochondrial genomes of plants differ from those of other higher eukaryotes for their impressive diversity, displayed in the large size, structure complexity, gene complement, and specific modes of expression and evolution. Among the score of land plant species investigated to date, the mitochondrial genome ranges from 200 to 2500 kbp. This remarkable variation is in sharp contrast with the rather constant size of other mitochondrial genomes and also of chloroplast genomes of plants (120 to 180 kbp; Levings and Brown 1989). Size in higher plant genomes can vary even eightfold within the same family (e.g., Cucurbitaceae; Lilly and Havey 2001).

To date, the precise structural organization of plant mitochondrial genomes is still poorly understood, although extensive analysis with a variety of different approaches has led to some substantial advance (Scissum-Gunn, Gandhi et al. 1998). The entire genetic complexity of the extant plant mitochondrial genome is currently represented as a single large circular molecule termed the *master chromosome*. Inter- and intramolecular recombination at active repeated sequences interspersed throughout the genome as well as amplification by replication is believed to produce a heterogeneous population of linear, circular, and more complex molecules, including catenate- and nucleoid-like rosette structures (Fig. 3.5) (Backert, Nielsen et al. 1997). The available data suggest that linear forms and subgenomic circles are the most predominant in vivo, while the master chromosome exists only in a very low copy number per mitochondrion or only in certain rare cell types. The only known exceptions to this model are found in *Brassica hirta* and *Marchantia polymorpha*.

Recombination at large and small repeats creates a very dynamic genome structure in plant mitochondria and plays an important role in the rapid and extensive rearrangements that characterize the evolution of plant mtDNA. Plant mitochondrial genomes encode necessary but insufficient information for the organelle biogenesis and function. Many nuclear gene products are required to express organelle genes and to assemble the organelle enzyme complexes. This results in an active protein trafficking from the cytosol to the mitochondria. The mitochondrial protein import process is a multistep process and in plants is made even more complicated by the presence of a chloroplast. Indeed in plants, coordinated control over the three genomes of the cell is indispensable for proper cell function (Macasev, Newbigin et al. 2000).

The complete DNA sequence has been determined for the liverwort *Marchantia polymorpha* (Oda, Yamato et al. 1992), the flowering plant *Arabidopsis thaliana* (Unseld, Marienfeld et al. 1997), and *Beta vulgaris* (Kubo, Nishizawa et al. 2000) mitochondrial genomes, demonstrating that the plant mitochondrial genome contains many more genes than its animal, fungal, and some protist counterparts.

Land plant mtDNA encodes SSU (18S) and LSU (26S) rRNAs as in all eukaryotes, a 5S rRNA, and a partial number of tRNAs ("native") and proteins essential

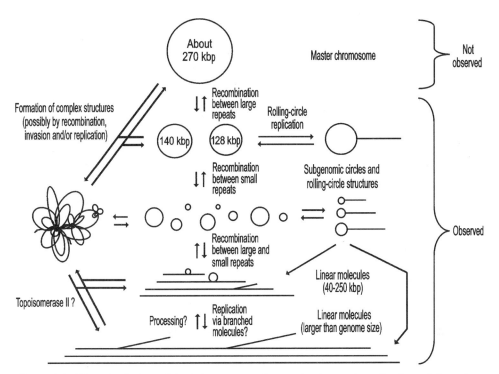

Figure 3.5. Hypothetical model of the structural organization and replication of the mitochondrial DNA in *Chenopodium album*. Circular molecules (including circles with tails, which comprise 13 to 26% of the molecules, about one-third of which are represented by plasmid *mp1*), linear molecules (56.5 to 81.5%), and more complex molecules (4.2 to 17.5%) are expected to exist in dymanic equilibrium, as observed in electron microscopic studies. The organization of DNA may be more simple or even more complicated in the mitochondria of other species. The enormous diversity in the size of molecules (including molecules larger than the putative master chromosome, such as oligomers and concatenamers), and in their shape, may be generated by various mechanisms: inter- and intramolecular recombination events between large and small repeats; the formation of higher-ordered (nucleoid-like rosette) structures, and specific types of replication. [Modified from Backert, Nielsen et al. (1997) and Oldenburg and Bendich (2001)].

for the formation of a functional mitochondrion. Mitochondrial proteins are components of the respiratory chain, including the three large subunits of complex IV, subunits 3 and 4 of complex II, cytochrome *b* of complex III, nine subunits of complex I, and four subunits of complex V (Gray, Burger et al. 1999). In addition, in plants but not in animals and most fungi, some proteins essential for the biogenesis of mitochondrial ribosome and for heme *c* cytochrome are also encoded in the mtDNA (Kubo, Nishizawa et al. 2000).

Other open-reading frames (ORFs) have been found in plant mitochondria which may be functional genes, but their functions are as yet unclear (Kubo, Nishizawa et al. 2000). In other cases, frequent intra- or intermolecular recombination events result in the creation of novel, chimeric genes whose products lead to cytoplasmic–nuclear incompatibilities known as *cytoplasmic male sterility* (CMS; Schnable and Wise 1998).

Moreover, interorganellar DNA exchange involving both noncoding and coding sequences of chloroplast, nuclear, and viral origin result in the formation of mosaic genomes (Gray, Cedergren et al. 1989; Levings and Brown 1989; Palmer 1990) and contribute to the size expansion of the mitochondrial genomes in plants. For instance, a variable number of plastid sequences, including some transcribed and functional tRNAs ("chloroplast-like") (Dietrich, Small et al. 1996), have been transferred to the mitochondrion of several plant species (Oda, Yamato et al. 1992; Unseld, Marienfeld et al. 1997; Kubo, Nishizawa et al. 2000). In land plants, mitochondria tRNAs have different genetic origin. The missing tRNAs are replaced by nuclear-encoded tRNAs (Dietrich, Small et al. 1996), imported from the cytosol sometimes in association with precursor aminoacyl-tRNA synthetases and/or replaced by mitochondrially encoded chloroplast-like tRNAs (Ramamonjisoa, Kauffmann et al. 1998; Mireau, Cosset et al. 2000; Glover, Spencer et al. 2001). Furthermore, most of the nuclear sequences identified to date in mitochondria are related to transposable elements, particularly retrotransposons (Knoop, Unseld et al. 1996). Numerous duplicated coding regions, noncoding large intergenic sequences, duplications, introns, and intron-ORFs also contribute to the large size of land plant mitochondrial genomes.

In flowering plants, several protein-coding genes are known to contain introns, mostly belonging to group II (Michel, Jacquier et al. 1982; Jacquier 1996). The only group I intron known from vascular plants is located in the mitochondrial *cox1* gene (Adams, Clements et al. 1998; Cho and Palmer 1999) and was acquired independently by horizontal transfer, probably from a fungus, many times separately among the angiosperms (Cho, Qiu et al. 1998).

Some group II introns, such as those in the genes coding for *nad1*, *nad2*, and *nad5*, have been split by recombination, dispersing exons to different genomic locations. In all instances, the splits are located in domain IV, which is the most variable region of group II introns and often contains insertion sequences with or without reading frames, even in continuous introns. The most distal intron of the *nad1* gene in *Arabidopsis* and in most flowering plants mtDNAs encodes an ORF, termed *mat-r*, for an intron maturase-like protein.

There are well-documented cases of loss and gain of introns in higher plants. Loss of these intervening sequences probably occurs via reverse transcription of mature transcripts followed by insertion of the cDNAs (Oda, Yamato et al. 1992). Introns are probably gained through duplicative transpositions; this can be deduced from the similarity between introns of different genes (Laroche and Bousquet 1999), which implies a common evolutionary origin and suggests that introns may have moved from gene to gene in a common land plant ancestor. Recent evidences has, however, also identified noncoding and coding sequence movements from the plant mitochondrial genome into the nuclear genome.

Differences in gene content of present-day plant mitochondrial genomes suggests that the gene flux to the nucleus is an ongoing evolutionary process, involving both respiratory chain and ribosomal protein genes and has occurred in independent and separate events during angiosperm evolution (Adams, Daley et al. 2000). The best studied examples of recent mitochondrial gene transfer during the evolution of flowering plants are the cytochrome oxidase subunit 2 gene (*cox2*) in legumes (Adams, Song et al. 1999) and the ribosomal protein genes (*rps7*, *rps10*, *rps12*, *rps11*, *rps14*, *rps19*) in different plant species (Figueroa, Gomez et al. 1999; Kubo, Harada et al.

1999). Pseudogene remnants of ribosomal protein genes have also been identified in plant mtDNA and are thought to reflect an evolutionary stage in which degeneration of the mitochondrial copy has occurred after successful transfer of a functional copy to the nucleus (Sanchez, Fester et al. 1996).

The mitochondria-to-nucleus transfer of essential genes appears to be RNA-mediated, with the nuclear copy representing an edited form (or a spliced form, in the case of intron-containing mitochondrial genes such as *rps10*). The functional activation of such transferred mitochondrial genes requires acquisition of sequences conferring proper expression and also targeting of the new cytoplasmically synthesized protein to the mitochondrion, although such targeting does not necessarily involve an obvious presequence (Adams, Daley et al. 2000). Analysis of other recently transferred ribosomal protein genes shows the different mechanisms employed for their functional activation. In maize and rice, for instance, the mitochondrial-*rps14* gene has ended up within a nuclear gene encoding succinate dehydrogenase, and both proteins are produced through alternative splicing of the resulting mRNA and have the same mitochondrial-targeting signal (Figueroa, Gomez et al. 1999; Kubo, Harada et al. 1999). Rice *rps11* was duplicated in the nucleus after transfer and acquired its targeting sequences from two different nuclear genes for mitochondrial proteins (*atpB* and *coxVb*; Kadowaki, Kubo et al. 1996).

In addition, several mitochondrial intron sequences are frequently found in adjacent upstream or downstream regions of nuclear genes (Knoop and Brennicke 1991). Although the integration of transferred mitochondrial intron segments may be connected with tissue-specific expression of the nuclear genes, in other cases the functional role, if any, for these sequences are unclear (Knoop and Brennicke 1994).

The large size of plant mtDNAs, compared to their coding functions, suggests that their mode of expression might resemble more closely that of yeast than of mammalian mtDNA. As described previously, plant mitochondrial genomes sustain extensive reshuffling of genes by recombination through directly repeated sequences. In this context, how transcriptional regulatory elements, located upstream and downstream of the coding regions, are maintained is unknown.

In plant mitochondria the general transcriptional strategy involve multiple transcription units under the control of separate promoters. For example, more than 15 individual promoters were identified in the *Oenothera berteriana* mitochondrial genome (Schuster and Brennicke 1994). As a result, most protein-coding genes seem to be expressed as monocistronic transcripts. Nevertheless, multicistronic messages resulting from co-transcription of linked mitochondrial genes have also been found in all the higher plants investigated. However, co-transcribed genes in one plant (e.g., *cox3* and *orf25* in rice) may be unlinked in a related plant species (e.g., wheat).

Moreover, different genes in the same plant can have either simple or complex transcript patterns. The same holds for a given gene in different plants. Multiple transcripts may be the result of the presence of several copies of the same gene in the mitochondrial genome and alternatively, may be the result of either the use of multiple transcription start and termination sites or processing events at different sites within noncoding regions of the primary transcripts frames [see Gray (1992) and Schuster and Brennicke (1994) for reviews].

The promoter structure of protein-coding genes such as that of rRNA and tRNA genes in plant mitochondria have similar motifs but are different from mammalian and fungal mitochondrial promoters and also from nuclear eukaryotic and prokaryotic promoters. Considerable progress has been made in elucidating the structures of transcription initiation sites in monocot and dicot plant species, by a combination of in vitro capping and primer extension analyses (Binder, Marchfelder et al. 1996). In particular, in vitro linker scanning and point mutational analysis in dicots have revealed that the nonanucleotide motif 5′-CRTAAGAGA-3′ conserved forms at least part of the mitochondrial promoter in this group of plants (Binder, Hatzack et al. 1995). Part of this motif, the 5′-CRTA-3′ tetranucleotide is also present in promoters in monocot plants and is the only sequence element conserved between monocot and dicot promoters (Rapp, Lupold et al. 1993; Binder, Hatzack et al. 1995).

Additional sequences upstream of the conserved elements, mainly AT-rich, are also required for optimal promoter-specific transcription (Binder, Hatzack et al. 1995). Furthermore, transcription initiation sites have been identified in both plant lineages, which do not contain the conserved core motif, suggesting that different types of promoter might exist, perhaps requiring their own specific factors.

The recent identification of a phage-type mitochondrial RNA polymerase encoded by a nuclear gene in plants clarifies its potential to recognize one or more different promoter types in plant mitochondria (Hedtke, Borner et al. 1997). The isolation and functional analysis of the mtDNA-binding proteins necessary for promoter recognition have also been reported recently (Ikeda and Gray 1999).

Far less is known about the processing mechanisms of plant mitochondrial mRNAs. Various processes of RNA degradation have been identified in plant mitochondria. Plant mitochondrial transcripts have specific and often conserved 3′-inverted repeats that are predicted to fold into stem–loop structures. Functional analyses show that such structures are processing and stabilizing signals rather than transcription terminators (Dombrowski, Brennicke et al. 1997). The presence of nongenomically encoded nucleotides at the 3′ end (including mostly C and A residues) and polyadenylation of plant mitochondrial mRNAs have recently been reported, suggesting a role of these modifications in RNA metabolism regulation (Lupold, Caoile et al. 1999; Williams, Johzuka et al. 2000). In plant mitochondria, as discussed above, group II introns interrupt several protein-coding genes. Proper transcript processing for these genes requires cis- and trans-splicing reactions to allow the association of the split exons and thus produce mature transcripts (Malek, Brennicke et al. 1997).

Another feature, already observed in protists (see above), peculiar to plant mitochondrial genomes and required for mitochondrial gene expression, is RNA editing (see Table 3.2). In plant mitochondria, RNA editing is widely observed in all extant land plants except Marchantiidae liverworts and algae (Steinhauser, Beckert et al. 1999). In most cases, C-nucleotides are converted by enzymatic action (deamination) to U-nucleotides in RNA, with very few U-to-C transitions (Marchfelder, Binder et al. 1998). Nearly all plant mitochondrial protein gene transcripts are edited, although with different frequencies. A similar activity also occurs in plastids of higher plants, although it is less frequent and affects only some transcripts (Maier, Zeltz et al. 1996). RNA editing sites are mostly found in coding regions of mRNAs, although some C-to-U events involve intronic and other nontranslated sequences. Some of the editing sites affecting group II introns are predicted to improve the quality of the intron folding and thus are very likely to improve functional splicing.

While the rRNAs in plant organelles appear to undergo very little RNA editing, in tRNA molecules this process restores essential base pairings and is absolutely required to maturation and processing of the tRNA precursors (Table 3.2) (Marechal-Drouard, Ramamonjisoa et al. 1993).

The biochemical mechanisms of RNA editing are well described, indicating a deamination reaction being involved in the C-to-U conversion, while the reverse U-to-C reaction would require a transaminating mechanism (Marchfelder, Binder et al. 1998). A fundamental but unsolved question in plant mitochondria is the specificity of RNA editing, allowing the editing enzyme to choose only some of the multiple C residues in the mRNAs to be converted into U.

So far, neither shared specific sequences or potential secondary structures that might define editing sites could be identified from the systematic analysis of hundreds of editing sites in plant mitochondrial transcripts (Mulligan, Williams et al. 1999). Despite the current lack of evidence, it is very likely that RNA editing determinants acting either in trans, such as the guide RNAs in trypanosomatid editing, or in cis, as primary and secondary RNA structures directing A-to-I conversions in animals are, however, involved in editing site recognition in plant mitochondria (Brennicke, Marchfelder et al. 1999). It is also imaginable that other factors (proteins) are required to drive the site specificity of the editing enzyme (Bock 2001).

Evidence to date indicates that in both mitochondria and plastids, the upstream sequences are more critical than the downstream region in determining the editing-site specifications. In some of the mitochondrial editing sites, 25 to 30 nucleotides upstream appear to be sufficient, whereas for other sites, 200 nucleotides may not be enough to specify a C to be edited (Marchfelder, Binder et al. 1998). The frequent C-to-U replacement in plant mitochondrial protein coding genes usually results in a change in the amino acid specified by the genomic codon and typically changes an aberrant amino acid encoded by the unedited sequence to the evolutionarily conserved residue. RNA editing has also been found to occur at a reduced frequency in the third position of the codon, silent editing.

The creation of initiation or stop codons do occur, suggesting that mRNA editing may be used as a key control mechanism in gene expression in plant mitochondria. Plant mitochondrial transcripts are edited with various degrees of efficiency. Transcripts for many genes are essentially completely edited, but some plant mitochondrial transcripts exhibit pronounced heterogeneity as a result of incomplete RNA editing (Maier, Zeltz et al. 1996). Nuclear genotype as well as tissue type, developmental stage, and growth condition may influence the extent of transcript editing (Grosskopf and Mulligan 1996).

For a particular gene, both fully and partially edited transcripts may then be observed within the polysomal fraction of plant mitochondria, which are therefore apparently translated into a family of variant proteins. In the case of the ribosomal protein gene S12, immunological experiments in maize and petunia mitochondria gave different results on the integration of the "unedited" protein into the mature ribosomes; in maize this variant is not detectable in the ribosome (Phreaner, Williams et al. 1996), but it is in *Petunia hybrida* (Lu, Wilson et al. 1996).

In almost all cases, however, only one type of protein, synthesized from completely edited RNAs, appears to be selected accurately with the aid of chaperones and incorporated into the polypeptide complexes of the respiratory chain, as deduced from the *atp9* protein sequencing (Begu, Graves et al. 1990), all the other polypeptide variants presumably being rapidly degraded. RNA editing is thus required for the synthesis of functionally competent protein products. In tobacco, it has been demonstrated that the import of variant proteins into mitochondria leads fatally to the appearance of organellar dysfunction and generate the male sterile phenotypes (Hernould, Suharsono et al. 1993).

As in animal cells, also in plants, programmed cell death (PCD) is an indispensable aspect of many processes, such as development, defense responses, and cell architecture (McCabe and Leaver 2000). Regulation of PCD is just at the beginning of investigation by using tissue cultures. These studies, still at their infancy, have not yet indicated if plant mitochondria, as the animal counterparts, participate in this basic cellular process.

Metazoa

Genome Organization In contrast with the high variability displayed by the mitochondrial genomes of lower eukaryotes and plants, metazoan mtDNA is a molecule characterized by compact arrangement, constancy of gene content, and the presence of a single noncoding region about 1 kbp long. With the exception of the replication origin region(s), the genome is saturated with genes, lacking intronic sequences, and flanking untranslated regions. These genes are often contiguous and sometimes slightly overlapping or separated by only few nucleotides. In general, metazoan mtDNA consists of a single circular molecule. However, in the phylum Cnidaria, linear DNA molecules, present as a single 16-kbp molecule or two 8-kbp molecules, have been reported in species from the classes Hydrozoa (e.g., *Hydra fusca*, *H. attenuata*), Scyphozoa (e.g., *Cassiopea* sp.), and Cubozoa (e.g., *Cariodea marsupialis*).

Owing to the reduced size of the molecule, the sequencing of the metazoan mt genome has become very popular, providing a lot of information about its variations both between and within phyla. Table 3.1 lists the metazoan mitochondrial genomes that have been completely sequenced. The length of mtDNA has been reported in the range 13 to 19 kbp. The smallest size, 13,503 bp, is recorded for the platyhelminth Tenia crassiceps, and the largest, 19,507 bp, for *Drosophila melanogaster* (see Table 3.1). However, genome size can reach up to 20 kbp in *Meloidogyne javanica* and 42 kbp in *Placopecten magellanicus* due to the presence of repeated sequences. Length variation is more frequent in invertebrates and in poikilothermic vertebrates, and it appears to generate very rapidly and distribute both within (heteroplasmy) and between individuals (Wolstenholme 1992). Recently, changes in the copy number of tandem repeated sequences have been reported to be consistent with a mechanism of mtDNA recombination, although recombination in mtDNA remains undocumented (Lunt and Hyman 1997).

The uniformity of gene content is another remarkable feature of the majority of metazoan mtDNAs (see Sec. 3.5). The same set of genes is found in all metazoan mtDNAs (with only a few exceptions): namely, two ribosomal RNA species, a reduced but complete set of 22 tRNAs, and a set of 13 proteins. In addition, there is at least one region that does not encode any structural gene and which has been shown to include elements for the initiation and control of replication and transcription; this is called the *control region* and, in vertebrates, the *displacement* (D) *loop region*.

A rare exception to the rule of constant gene content are Cnidaria, where only two tRNA genes are found in *Metridium senile* and only one in *Renilla koellikeri* and *Sarchophyton glaucum*. In Nematoda the gene for ATP8 is missing, with the exception of *Trichinella spiralis*. ATP8 gene is also missing in the tapeworm *Hymenolepis diminuta*. The mollusk *Mytilus edulis* not only lacks the ATP8 gene but also hosts an additional tRNA expected to recognize the AUA codons for methionine. It is commonly accepted that gene order varies between lineages and that conservation of gene order is frequently observed among recent evolutionary neighbors. Indeed, the progressive acquisition of new sequences into mtDNA databases has shown an increasing number of mt genomes whose structure deviates from the previous congruent picture of this molecule. The gene organization of some representative animal mtDNAs is shown in Fig. 3.6.

In the phylum Cnidaria, generally considered to be one of the most primitive groups of metazoans, there are peculiarities not shared with other mitochondrial systems. In a member of the class Anthozoa subclass Hexacorallia, *Metridium senile*, two mt protein genes COI and ND5 contain a group I intron. Furthermore, COI intron encodes a putative endonuclease and the ND5 intron contains the ND1 and ND3 genes (Beagley, Okada et al. 1996). Other Anthozoa members of the subclass Octocorallia, *Renilla kolikeri* (Beagley, Macfarlane et al. 1995) and *Sarcophyton glaucum* (Pont-Kingdon, Okada et al. 1995, 1998), do not have introns but an extra gene coding for a protein similar to a bacterial mismatch repair protein.

In protostomes (Arthropoda, Mollusca, Annelida, and Nematoda), mt gene arrangement is not stable within major groups. Among nematodes, *Ascaris suum* differs from *Caenorhabditis elegans* only in the location of the AT-rich region, but extensive rearrangements have occurred in the mtDNA of *Meloidogyne javanica* (Okimoto, Chamberlin et al. 1991; Okimoto, Macfarlane et al. 1992). *Trichinella spiralis* mt genome arrangement can be related to that of coelomate metazoa but not to that of secernentean nematodes (pseudocoelomates). Particularly intriguing is the case of Mollusca, where different species-specific gene arrangement have been found in different genomes. Interestingly, the two species of pulmonate gastropods, *Euhadra kerklotsi* and *Cepaea nemoralis*, although classified in the same superfamily, Helicoidea, have quite a different arrangement of tRNA and protein-coding genes. On the other hand, the mtDNA of *Albinaria cerulea*, another pulmonate gastropod only distantly related to the previous ones, shows in a portion of the genome the same gene order as in *Euhadra* mtDNA and in another portion the gene order of *Cepaea* mtDNA. This demonstrates that a gene rearrangement has occurred in members of the same superfamily and suggests an accelerated rate of gene rearrangement in some pulmonate lineages (Yamazaki, Ueshima et al. 1997). The *Pupa strigosa* mitochondrial gene arrangement is almost identical to those of pulmonate land snails, but it is radically divergent from those of the prosobranch gastropod *Littorina saxatilis* and other mollusks (Kurabayashi and Ueshima 2000). The mtDNA of another mollusk, *Katharina tunicata*, has highly dissimilar gene arrangement with respect to all other mollusk mtDNAs available so far, but notably similar to that of *Drosophila* (Arthropoda). Furthermore, this mollusk contains two additional sequences which can be folded into tRNA-like structures, but it is not clear whether these are functional genes (Boore and Brown 1994). Finally, the mollusk

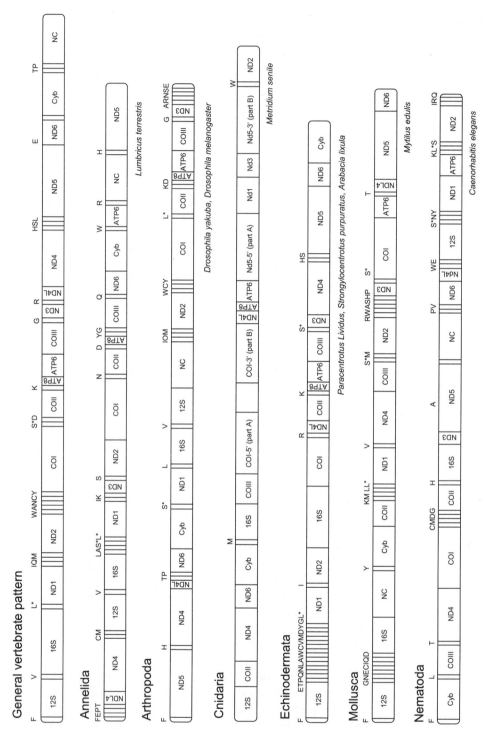

Figure 3.6. Comparative gene organization of representative metazoan linearized mtDNAs. The tRNA genes are indicated according to the amino acid transported: L, Leu(CUN); L*, Leu(UUR); S, Ser(AGY); S*, Ser(UCN). Ψ indicates a pseudogene. NC, noncoding region.

114

Mytilus edulis has a mitochondrial gene arrangement radically different from those reported above (Hoffmann, Boore et al. 1992). The Brachiopoda *Laqueus rubellus* shows an extreme compact gene organization with no long noncoding region; its genome has an overall gene arrangement drastically different from that of any other organism reported so far (Noguchi, Endo et al. 2000).

The only two representatives of Annelida, *Lumbricus terrestris* and *Platynerei dumerii*, have a gene organization different from one another and from that of the other major protostome groups for the location of the tRNA genes. Interestingly, the ATP8 gene is not immediately upstream ATP6, a condition found only in pulmonate gastropods (Boore and Brown 1995).

The tapeworm *Hymenolepis diminuta* (Platyhelminthes, Cestoda) mitochondrial gene arrangement is similar to that of other platyhelmints and supports a phylogenetic position of flatworms as members of the Eutrochozoa (von Nickisch-Rosenegk, Brown et al. 2001). Extensive gene rearrangements can also be observed in Arthropoda. In Insecta, the comparison of the gene order between *Apis mellifera* (Crozier and Crozier 1993) and *Drosophila* (Clary and Wolstenholme 1985; Lewis, Farr et al. 1995) showed a rearrangement of 11 tRNA genes, whereas the organization of the genome of *Artemia franciscana* (Crustacea) is very similar to that of *Drosophila*, showing a single rearrangement merely affecting two tRNAs (Valverde, Batuecas et al. 1994). Recently, the entire mt genome from two different ticks, *Ixodes hexagonus* and *Rhipicephalus sanguineus*, representative of the subgroup prostriate and metastriate of the subclass Acari, have been reported (Black and Roehrdanz 1998). The sequences show an extensive rearrangement which also implies duplication of the control region in the metastriate tick. The basal hexapod *Tetrodontophora bielanensis* mitochondrial genome shows numerous tRNA translocations (Nardi, Carapelli et al. 2001). The arrangement of genes in the wallaby louse, *Heterodoxus macropus*, mtDNA is different from that of other metazoans studied so far. All tRNA genes have moved and/or inverted relative to the ancestral gene arrangement of insects (e.g., that of the fruit fly *Drosophila yakuba*), and at least nine protein-coding genes are moved (Shao, Campbell et al. 2001). All these data, which show extensive gene arrangements even between closely related lineages suggest that phylogenetic studies based on gene arrangement comparisons may be very misleading in many cases, although some attempts have been made (Boore and Brown 1998).

A survey of deuterostome mtDNA sequences would suggest a less frequent gene rearrangement. In three different classes of Echinodermata (i.e., Asteroidea, Echinoidea, and Crinoidea), *Asterina pectinifera* (Asteroidea), and *Florometra serratissima* (Crinoidea) show similar gene organization with few tRNA transpositions, although this is slightly different from that of Echinoidea (*Arbacia lixula*, *Paracentrotus lividus*, and *Strongylocentrotus purpuratus*), owing to a major inversion of a 4.6-kbp fragment spanning from the 16S rRNA gene to 13 of the 15 tRNA genes in the main tRNA region (Jacobs and Elliot 1988; Cantatore, Roberti et al. 1989; De Giorgi, Martiradonna et al. 1996a). Therefore, limited gene rearrangements have occurred in echinoderms, despite the fact that according to paleontological evidence, the different classes diverged approximately 550 to 450 Mya.

The overall gene order of the hemichordate *Balanoglossus carnosus* shows substantial differences from echinoderms, but it is quite similar to that of vertebrates. In contrast, it shares with echinoderms some aspects of the genetic code [i.e., sequence motifs in the control region, close similarity of the two Leu-tRNA genes, and the presence of an N-terminal extension of ND5 (Castresana, Feldmaier-Fuchs et al. 1998)].

Mitochondrial gene organization seems relatively stable only in Chordata, where few transpositions are tolerated. The gene arrangement described in eutherian mammals has been found to be identical to that of prototherian mammals, Amphibia, Testudines (*Pelomedusa subrufa*), Osteichthyes, and Chondrichthyes. Marsupialia show rearrangements of tRNAs only, whereas the tRNA^{Lys} has been lost and the corresponding sequence has become a pseudogene. Therefore, marsupial mitochondria must import a nuclear-coded tRNA^{Lys} to get the complete set of tRNAs (Janke, Xu et al. 1997). In Aves, tRNA^{Glu} and ND6 are translocated immediately adjacent to the control region (Desjardins and Morais 1990). Interestingly, a novel arrangement of bird mtDNA has been described in a recent study (Mindell, Sorenson et al. 1998), suggesting that the ancestral translocation involving ND6 + tRNA^{Glu} was accompanied by a duplication of the main noncoding region. The presence or absence of the duplicated control region would account for the two alternative arrangements observed in the mtDNA of extant birds investigated so far (Boore 1999). It is also remarkable to note that a nontranslated extranucleotide has been found in the ND3 gene of some birds and turtles, located at an internal codon, thus further supporting diapsid affinity of turtles (Mindell, Sorenson et al. 1998; Zardoya and Meyer 1998). In *Alligator mississipiensis* (Crocodilia) only three tRNAs moved to different locations (Janke and Arnason 1997), while in *Dinodon semicarinatus* (Serpentes) not only one tRNA is transposed but also the control region is duplicated and flanked by a pseudogene consisting of the 5′-half portion of the tRNA^{Pro}, which can be folded into a rather stable secondary structure (Kumazawa, Ota et al. 1998). In *Petromyzon marinus* (Agnatha) the control region is transposed, and it is interrupted by the presence of two tRNAs (Lee and Kocher 1995). In *Branchiostoma lanceolatum* (Cephalochordata) the control region is translocated and reduced to only about 80 nucleotides, and few tRNAs have been transposed (Spruyt, Delarbre et al. 1998). A notable exception is *Halocynthia roretzi* (Urochordata) whose mt genome has a highly divergent organization with respect to the other Chordata.

When comparing mt gene organization in vertebrates to that of other animals, one of the most remarkable feature is the different distribution of tRNA genes. In vertebrates, tRNA genes are scattered along the molecule, while in other metazoan species tRNAs are often clustered. In general, studies on tRNA gene localization in animal mt genomes clearly suggest their accelerated mobility relative to other mitochondrial genes (Blanchette, Kunisawa et al. 1999). On the basis of comparative analyses, events of duplication and remolding of tRNA genes during the evolutionary rearrangement of mt genomes in the sea urchin *Paracentrotus lividus* have been suggested (Cantatore, Gadaleta et al. 1987). In particular, it has been assumed that during evolution a tRNA gene lost its function and became part of a protein-coding gene. This loss was accompanied by the gain of a new tRNA through duplication and divergence from a tRNA gene specific to a different family of codons, namely tRNA$^{Leu(UUR)}$ gene. On the basis of these assumptions and of other observations indicating that tRNAs are present at the end of duplications and deletions, tRNAs should be considered as mobile elements involved in gene rearrangement (Moritz, Dowling et al. 1987; Saccone, Attimonelli et al. 1990). To explain this property, the hypothesis has been put forward that a gene flanked by two tRNAs could be considered to be very similar to a transposable element, with the two tRNAs corresponding to LTRs (long terminal repeats), each having short inverted sequences (amino acid stems) at its ends (Saccone, Attimonelli et al. 1990). In addition, as a consequence of the compactness of metazoan mt genome, tRNA genes have been forced to assume multiple roles and regulatory functions. It has been indicated that mt tRNAs might play a role in the origin of replication (Maizels and Weiner 1995). It is known that tRNAs act as recognition signals in vertebrates, where they are scattered along the molecule and make up a sort of punctuation signal for the processing of polycistronic transcripts (Ojala, Merkel et al. 1980). On account of the acquisition of such multiple roles by tRNAs, we have speculated that in the course of evolution, animal mt tRNAs, which had to cope with the dramatic size reduction of mtDNA, have adapted in order to fulfill new tasks. This has caused a less stable L-shaped structure and wider ambiguity in decoding of the genetic code (De Giorgi, Martiradonna et al. 1996b).

All metazoan mt genomes studied so far code for two ribosomal (r) RNA species, lacking 5S rRNA, which is present in plants only; the ribosomal proteins are nuclear coded. The two rRNA species have an unusually short length, the longer and smaller species having a sedimentation coefficient of 12S and 16S, respectively. Indeed, they possess a reduced structure missing several helices of the common core structure derived from prokaryotic and eukaryotic small rRNAs. Alignments and structures of small and large rRNAs are collected in specialized databases, such as the rRNA Web Server and the RDP Database (see the URLs in the Appendix). Nice secondary structure models for the mitochondrial rRNAs can be found at the Comparative RNA Web site (see the URL in the Appendix).

Base Composition   A striking heterogeneity in base composition can be observed between different metazoan lineages as well as within the same lineage. Figure 3.7 shows the %GC range and average content of several metazoan groups. Protozoal,

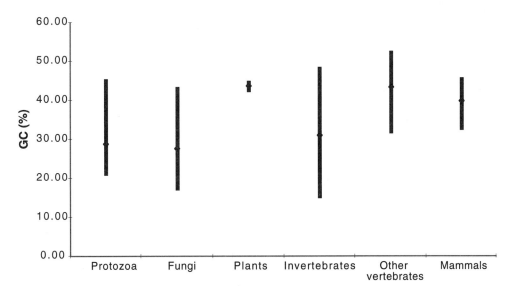

Figure 3.7. GC content range of mtDNA from representative metazoan groups. The average GC content for each taxon is also indicated.

fungal, and invertebrate mtDNAs show a lower GC content than those of plant and vertebrate mtDNAs. The highest GC content is observed in *Diplophos tenia* (Teleostei; %GC 52.3) and the lowest in *Apis mellifera* (Arthropoda; %GC 15.1), whose genome is extremely AT rich. The general compositional pattern is also reflected in the nucleotide composition percentage at the third codon positions. In vertebrates a conserved compositional pattern is observed where a remarkable asymmetric distribution of the complementary nucleotides between the two strands can be observed. Such an asymmetry is so strong as to cause differences in buoyant density in CsCl gradient between the (heavy) H strand, rich in G, and the (light) L strand, poor in G.

The compositional asymmetry can be quantified in terms of GC and AT skew (see Sec. 1.5 for a definition of compositional skew). In chordates the GC skew is always considerably negative, with the exception of cephalochordates and urochordates, whereas the AT skew is very low, but generally slightly positive. A taxon-specific pattern can be observed for other organisms both between different taxa and within the same taxon. Urochordates, nematodes, and platyhelminthes show a strong compositional asymmetry with positive GC skew and negative AT skew. The same pattern but to a lesser extent is shown by molluska and cnidarians, where as in all the remaining taxa, the compositional asymmetry is very low. Asymmetry is also evident in other properties of mtDNA: namely, gene distribution on the two strands and the replication mechanism. Hence there appears to be a correlation between compositional asymmetry and other forms of mtDNA asymmetry. The H-strand codes for most mitochondrial genes (29 of a total 37); among these there are 12 of the 13 protein genes (only the ND6 gene is coded by the L strand). Due to the high G content of the H strand, mRNAs coded on this strand are particularly poor in G. This is most evident at the third codon positions of protein-coding genes where G-ending codons are avoided.

Different models can be suggested to explain the peculiar compositional bias of the vertebrate genome. Metabolic discrimination between nucleotide bases and/or replication errors followed by biased repair could account for this property. Other explanations may be based on the mechanism of mtDNA replication, which, being asymmetric, leaves one DNA strand as a single, unprotected filament for two-thirds of the replication cycle (see below). The possibility of damage directly at the level of RNAs cannot be excluded (Saccone and Sbisà 1994). The need to protect mtDNA transcripts against the attack of free radicals, preferentially affecting G (Ames, Shigenaga et al. 1995), could have originated the dramatic reduction in G content of transcripts, hence an increase in G percentage on the template strand of transcription (the H strand) and the consequent compositional asymmetry (Saccone 1999). The strong compositional bias, in particular the trend to avoid G at the third codon position, might also explain several deviations of the vertebrate mt code with respect to the universal one. In vertebrate mitochondria, the most used initiation codon is AUA instead of AUG, because this letter is a G-ending codon. One of the three nonsense codons, UGA, becomes an additional tryptophan codon, because of the two contiguous G in the canonical UGG tryptophan codon, which is very rare in the sense strand of metazoan mtDNA. In other words, the compositional bias has been so strong as to influence the genetic code itself.

To gain insight in the cause of mtDNA compositional asymmetry and to identify possible differences in the substitution processes acting on the two DNA strands,

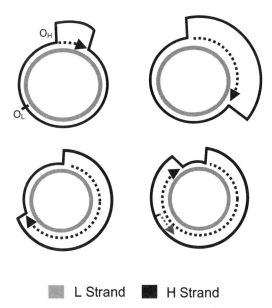

L Strand   ■ H Strand

Figure 3.8. General mechanism of mtDNA replication in vertebrates.

the compositional properties of mtDNA from several mammalian species have been studied, in particular their possible relationship with the asymmetry of the replication process. It is well known that the replication of vertebrate mtDNA is asymmetric and unidirectional (Clayton 1982). According to the D-loop replication mechanism shown in Fig. 3.8 (see also *Replication and Expression*) the parental L strand is never left as a single helix during replication, while parental H strand lives as a single helix until the corresponding complementary region of the L strand is newly synthesized. Since replication takes place rather slowly (Clayton 1982), the different regions of parental H strand remain as single helix for a rather long time, varying according to their distance from O_L (distance measured in the replication direction of the H strand). During this time, the single helix of parental H strand is protected only partially by mtSSB proteins, thus it could be more exposed to oxidative or hydrolytic damage and more prone to mutations than the L strand. Therefore, it has been suggested that differences in mutational pressure on the two H and L strands could cause the compositional asymmetry of mtDNA (Brown, Prager et al. 1982). It is striking to note that DNA polymerases γ used for mtDNA replication of both strands are the most accurate eukaryotic polimerases (Kunkel and Alexander 1986; Kunkel and Soni 1988; Pinz, Shibutani et al. 1995).

AT and GC skews, the base composition of the third positions of the quartet codons (P3Q), and the degree of gene conservation are correlated to the duration of the single-stranded state of the H-stranded genes during replication (D_{ssH}) calculated according to Reyes, Gissi et al. (1998). Figure 3.9*a* shows the correlation between the base composition on P3Q of the H-strand protein-coding genes and D_{ssH}, where a significant increase in A and C frequencies and a corresponding significant G and T frequency decrease with the D_{ssH} is observed. The hydrolytic

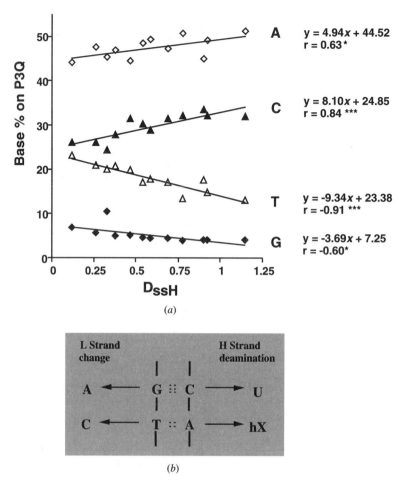

(a)

(b)

Figure 3.9. (a) Correlation between nucleotide percentages calculated on the third position of quartet codons (P3Q) and the single-stranded state duration (D_{ssH}) for each H-stranded gene, as mean values for 34 mammalian species. The linear regression equations are also reported (*, $p < 0.05$; ***, $p < 0.005$). (b) Suggested deamination processes that may take place in the single-stranded H strand: cytosine (C) into uracil (U) and adenine (A) into hypoxanthine (hX), which would imply a change of G into A and of T into C on the L strand.

deamination of both C and A, depending on the duration of the single-stranded state of the H (heavy) strand during replication, might be responsible for such a correlation (Reyes, Gissi et al. 1998). The mutation pattern hypothesized (see Fig. 3.9b) shows that spontaneous deamination of C on the H strand produces U, which base-pairs with A rather than G, and consequently, the percentage of G decreases and that of A increases on the L (light) strand according to D_{ssH}. In the same way, the deamination of A results in hypoxanthine (Lindahl 1993) which base-pairs with C rather than T, resulting in a reduction of T and increase of C correlated to D_{ssH}.

Replication and Expression A peculiar feature of metazoan mt genomes is the presence of a main noncoding region containing the regulatory elements for the replication and expression of the mtDNA. Regardless of the high degree of conservation of the coding genes, this region shows great variability in length and base composition. It ranges from only 121 bp in sea urchin to 4601 bp in *Drosophila* (Saccone 1999). Different from vertebrates, in invertebrates the structure and evolution of the regulatory region has not yet been fully characterized. In *Ascaris* and *Drosophila* this region is called an AT-rich region for its extremely high A + T content. In *Ascaris* the main AT region is 886 bp long, and a smaller noncoding sequence of 117 bp, which can be folded into a stem-and-loop structure, is also present. In *Drosophila* the AT region is extremely polymorphic; it varies in sequence and length both in different species (1077 bp in *Drosophila yakuba* and 4601 in *Drosophila melanogaster*) and within individuals of the same species, or in different mtDNA molecules of a single fly. The putative promoters and the replication origin are contained in two conserved regions, one of which can form a hairpin structure.

In echinodermates the main noncoding region (121 to 445 bp) is located in the tRNA cluster. It appears as a condensed version of the vertebrate replication origin, and the nascent strand coincides with a very stable stem–loop structure. In vertebrates the main noncoding region is called the D-loop-containing region, because starting from its replication origin (O_H), the nascent heavy strand displaces (D) the parental one, creating a triple-strand structure. This region, ranging from 879 bp (mouse) to 2134 bp (*Xenopus*), also contains the promoters for both the heavy (HSP) and light strands (LSP). The other noncoding region, only 30 bp long, contains the origin of the light strand replication (O_L) and is located at a distance of about 10 kbp from O_H within a cluster of five tRNA genes. This region can be folded in a stable stem-and-loop structure that is very conserved in all vertebrates except birds (Fig. 3.10). Indeed, the sequence equivalent to the O_L has not been found at the same position in the bird mt genomes sequenced so far. In rat and humans the two origins of replication have been shown to have an intrinsic DNA curvature, correlated with periodic distribution of dinucleotides in the sequence and involved in protein interactions.

The replication of mtDNA has been studied extensively in vertebrates. Figure 3.8 shows the general asymmetric mechanism of mtDNA replication in veretebrates. The replication starts with the elongation of the nascent H strand and the expansion of the D loop. When the displacement exposes the O_L as a single-stranded template, the synthesis of the L strand starts in the opposite direction. Both H- and L-strand replications require the presence of RNA primers which are produced by mtRNA polymerase at the level of O_H and by a mitochondrial primase at the level of O_L. In particular, in the D-loop region, the replication priming for O_H and the transcription of the L strand initiate from precisely the same site starting from the LSP. Thus, although the initiation of replication and transcription appears to be modulated in a coordinated fashion, the precise mechanism remains unclear. An interesting finding is that the transition from RNA to DNA synthesis occurs at the level of short conserved sequence blocks (CSBs), which were proposed to be regulatory signals for the start of the nascent strand in a region that can be folded in a termodynamically stable cloverleaf-like structure. An RNAse activity of a ribonucleoprotein capable of processing primer sites specifically has been discovered and

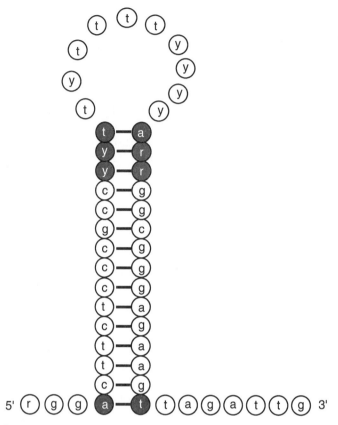

Figure 3.10. Consensus secondary structure of the OriL in mammalian mtDNAs. Shaded circles represent paired bases not present in all species.

termed MRP (mitochondrial RNA processing; Karwan, Bennett et al. 1991). An intriguing problem is the localization of MRP in the cellular compartments, since the amount of the MRP RNA component is very low in organelles (Kiss and Filipowicz 1992; Topper, Bennett et al. 1992). Mutations in the MRP RNA gene have recently been found associated with the human disease cartilage-hair hypoplasia (Ridanpaa, van Eenennaam et al. 2001).

The mtDNA primase can recognize the stem-and-loop structure of the O_L and initiate RNA priming within the T-rich loop of the template; the RNA/DNA transition occurs near the stem base (Clayton 1991). Interestingly, in the rat, stable complementary transcripts have been found at the level of both origins of replication (Sbisà, Nardelli et al. 1990; Sbisà, Tullo et al. 1992). Quite intriguingly, an alternative symmetric mode of replication has been observed in vertebrate mtDNAs involving a coupled leading- and lagging-strand DNA synthesis (Holt, Lorimer et al. 2000). The DNA polymerase γ (nuclear coded) is the only DNA polymerase activity isolated so far from mammalian mitochondria. It has associated $3' \rightarrow 5'$ exonuclease activity, which might confer proofreading capability (Insdorf and Bogenhagen 1989). Table 3.3 lists protein factors known to be involved in replication and transcription of the mitochondrial DNA in animals.

TABLE 3.3. Factors and Enzymes Involved in Animal Cells mtDNA Replication and Transcription

Enzyme/Factor	Function	References
γ-DNA Polymerase (α and β)	Catalyze mtDNA synthesis	Carrodeguas, Theis et al. (2001); Fan and Kaguni (2001)
mtTFA	Promotes transcription initiation and functions in maintenance of mtDNA	Moraes, Kenyon et al. (1999); Larsson, Wang et al. (1998)
mtMRP-RNA	Processes primer working on a DNA–RNA triple strand	Garesse and Vallejo (2001); Taanman (1999)
DNA-ligase	Acts in mtDNA repair and has a putative role in replication	Perez-Jannotti, Klein et al. (2001)
Primase	Catalyzes primer synthesis for L-strand replication	Taanman (1999); Lecrenier and Foury (2000)
DNA elicase	Unwinds double-stranded DNA	Clayton (1992, 2000)
TAS-binding protein	Controls D-loop extension	Lecrenier and Foury (2000)
DNA topoisomerase	Changes DNA topology during replication, eliminating or introducing supercoils	Taanman (1999); Lecrenier and Foury (2000)
RNA polymerase	Transcribes mtDNA in concert with mtTFA	Masters, Stohl et al. (1987)
mtTFB	Acts in promoter-directed transcription selectivity of the core mitochondrial RNA polymerase	McCulloch, Seidel-Rogol et al. (2002)
mTERF	Acts in termination of transcription binding the termination signal located downstream of the 16S rRNA gene and induces bending of the DNA helix	Fernandez-Silva, Martinez-Azorin et al. (1997) Shang and Clayton (1994)
mtSSB	Binds to single-stranded DNA during replication and stimulates the action of mitochondrial DNA polymerase	Maier, Farr et al. (2001); Ruiz De Mena, Lefai et al. (2000); Thommes, Farr et al. (1995)
Rnase P	Cleaves mitochondrial tRNA at the 5′ end	Rossmanith, Tullo et al. (1995)

The peculiar features of the gene structure and organization described previously are reflected in the mode of expression of the genomes. The pathways of mitochondrial transcription and posttranscriptional processing have been investigated extensively in animals.

As already mentioned, the vertebrate mtDNA is extremely compact and transcribed starting from two distinct promoters (HSP and LSP, one for each strand) in a polycistronic way (Cantatore and Saccone 1987; Clayton 1991). In contrast, the transcription mechanism in the sea urchin *Paracentrotus lividus* takes place via multiple and probably overlapping transcription units. The wide variation in the steady-state levels of the mature mRNAs indicates that mitochondrial expression is also regulated at the posttranscriptional level (Cantatore, Roberti et al. 1990).

The main peculiar feature of mitochondrial transcription in vertebrates is, indeed, complete symmetry, a phenomenon discovered in HeLa cells even before the sequence and the genetic content of the genome was determined. The two strands of mtDNA are transcribed, in opposite directions, starting from the two promoters at the 5' end of the main regulatory region of the mitochondrial genome. The two promoters share a 15-bp consensus (5'-CANAC(C/G)C(C/A)AAAGAYA, the initiation site is underlined) in humans (Hixson and Clayton 1985).

It is not yet known how starting from polycistronic transcripts of both strands, some RNAs are degraded completely while others have different levels of stability. The different concentrations of the mtRNA species in cells is probably the result of numerous events which, at different levels (transcriptional, posttranscriptional, and translational), regulate expression of the mitochondrial genome. Analyses of the kinetics of RNA labeling have suggested that L transcripts are synthesized more efficiently than H transcripts. Gene expression is also largely regulated posttranscriptionally. L transcripts are far less stable and degraded rapidly with the exception of those corresponding to structural genes (tRNAs and ND6) and to the D-loop region. Studies on the stability of the various RNA species have demonstrated that the half-life of tRNAs, mRNAs, and rRNAs are different even if they belong to a common precursor. Controls at both the initiation (more efficient promoter) and termination sites (presence of termination factors) can regulate the differential expression of RNA species.

Differential stability presumably explains the different steady-state amounts of individual mRNAs derived from the same transcript. Translational regulations must occur, since the different mRNAs appear to be translated with different efficiency. The other possibility to be taken into account is that the mitochondrial products of the respiratory complexes are controlled by the availability of the cytoplasmic subunits of the same complex (Attardi, Chomyn et al. 1990). The development of transcriptionally faithful in vitro systems has greatly streamlined the characterization of promoters and the purification of the components of the transcription mechanism in vertebrates and fungi. Transcription in both systems is directed by at least two protein factors: an RNA polymerase and a transcription factor (mtTFA), which is essential to support a correct and efficient transcription initiation. The mtTFA binds to the promoters and has the capability to unwind and bend DNA. Interesting that the relative binding affinity is higher for the LSP, which is the most active promoter. A protein similar to mtTFA has been isolated from yeast mitochondria (Parisi, Xu et al. 1993).

It is noteworthy that a transcription terminator factor (mTERF), binding the 16S rRNA end downstream, has been purified. This protein recognizes a sequence of 13 bases (consensus TGGCAGAGCCCGG) within the tRNA[Leu] and this activity could explain the higher abundance of rRNA species compared to mRNAs (all located downstream from the termination site) in mitochondria. Interest in this peculiar "regulatory region" is increased by the fact that a mutation in the target sequence has been linked to the mitochondrial MELAS myophaty. The mutation is an A → G transition (position 3243) and correlates with a reduced affinity of the termination factor for the MELAS template (Hess, Parisi et al. 1991).

The polycistronic mode of mitochondrial transcription implies the activity of one or more endoribonucleases for processing of the primers involved in the replication and for maturation of the structural genes. According to the tRNA excision processing model, an RNAse P-like activity recognizes the secondary structure of the tRNAs that punctuate the genome. tRNA removal from the primary transcripts generates mature forms of the RNA species, which are subsequently polyadenylated. Besides polyadenylation, which often generates stop codons in processed transcripts (UAA formed from incomplete stop codons U or UA), other forms of RNA editing are observed in animal mitochondria (see Table 3.2). These include C or A addition to form the 3′ terminus of tRNAs and C → U modification to form tRNA^{Asp} in marsupials.

Ribonuclease P is a ubiquitous ribozyme that has been isolated and extensively studied in bacteria, nuclei, and yeast mitochondria (Darr, Brown et al. 1992). In human, mouse, and rat, RNAse P activity co-purifies with the closely related MRP ribonuclease activity (see the earlier discussion of replication). These enzymes contain both protein and RNA components essential for the processing activity. Interestingly, the RNA components of these ribonucleases can be folded into similar, conserved structures, suggesting a possible common mechanistic conservation (Altman, Kirsebom et al. 1993; Karwan 1993; Schmitt, Bennett et al. 1993). In vivo mapping performed by Sbisà, Tullo et al. (1992) demonstrated the presence of novel processing intermediates (e.g., RNA precursor containing mRNA + tRNA species). This suggests that the mechanism of RNA processing in mitochondria proceeds stepwise, thereby producing several precursors of the mature forms. In other metazoans, however, the sites and the mechanisms of replication and transcription are still unkown.

A very interesting issue in the regulation and expression of the mitochondrial genome is how the nuclear-coded enzymes and mitochondrial substrates evolved in concert (Garesse and Vallejo 2001). This has important implications for the study of vertebrate mitochondrial genome evolution, which is discussed in Chapter 7.

3.2 CHLOROPLASTS AND OTHER PLASTIDS

Plastids are cellular organelles limited by a double membrane and characterized by a small circular DNA–plastid DNA. For the most part the plastid genome (plastome) encodes proteins involved in photosynthesis and gene expression (Sugiura 1992). However, like mitochondria, plastids are only semiautonomous and the genetic machinery required to synthesize proteins depends on both plastome and nuclear genome. For example, the regulation of gene expression in chloroplasts involves complex coordination between the nucleus and the organelle; data suggest that signaling between these two compartments occurs primarily to regulate photosynthetic and photomorphogenetic activities (Somanchi and Mayfield 1999). This is the case of the enzyme ribulose bisphosphate carboxylase oxygenase (Rubisco), which adds CO_2 to ribulose bisphosphate to start the Calvin cycle. This enzyme consists of multiple copies of two subunits, a large one encoded in the chloroplast genome and synthesized within the chloroplast, and a small subunit encoded in the nuclear genome and synthesized by ribosomes in the cytosol.

TABLE 3.4. Completely Sequenced Chloroplast Genomes

Species	Genbank AC	Length (bp)
Arabidopsis thaliana	NC_000932	154,478
Astasia longa	NC_002652	73,345
Chlorella vulgaris	NC_001865	150,613
Cyanidium caldarium	NC_001840	164,921
Cyanophora paradoxa	NC_001675	135,599
Epifagus virginiana	NC_001568	70,028
Euglena gracilis	NC_001603	143,172
Guillardia theta	NC_000926	121,524
Lotus japonicus	NC_002694	150,519
Marchantia polymorpha	NC_001319	121,024
Mesostigma viride	NC_002186	118,360
Nephroselmis olivacea	NC_000927	200,799
Nicotiana tabacum	NC_001879	155,939
Odontella sinensis	NC_001713	119,704
Oenothera elata subsp. *hookeri*	NC_002693	163,935
Oryza sativa	NC_001320	134,525
Pinus thunbergii	NC_001631	119,707
Porphyra purpurea	NC_000925	191,028
Toxoplasma gondii	NC_001799	34,996
Triticum aestivum	NC_002762	134,545
Zea mays	NC_001666	140,384

As for mitochondria, protein import in chloroplasts depends on sorting signals (chloroplast transit peptides, cpTPs) located at the N terminus of targeted proteins (Bruce 2000). The availability of many completely sequenced plastid genomes (listed in Table 3.4) allows comparative genomic analyses.

The functional unit of the plastome is the *nucleoid*, a large complex consisting of plastid DNA, various proteins, and RNAs. Whereas the cyanobacterial origin of plastid DNA is not disputed, the origin of the nucleoid proteins appears complex. Biochemical and comparative genomic analysis suggests that proteins of eukaryotic origin replaced most of the original prokaryotic proteins during the evolution of plastids (Sato 2001). Until recently, plastid inheritance was thought to be, with few exceptions, maternal. Now over 40 examples of paternal plastid inheritance are available. All gymnosperms thus far studied exhibit biparental inheritance, as do several outcrossing angiosperms.

Among plastids, the chloroplast is undoubtedly best characterized. This organelle, the central site of the photosynthetic process, is generally flattened and lens-shaped (Fig. 3.11) and characterized by three membrane systems. Plastids are surrounded by two membranes: a continuous smooth membrane, the external limiting membrane, below which extends the internal limiting membrane, which is differently folded. Internally there is a compartment filled with an amorphous material, called *stroma*, which resembles the mitochondrial matrix and has a nucleoid region housing the circular, naked DNA (cpDNA) and 70S ribosomes. The third membrane system is found dispersed in the stroma: It is represented by pho-

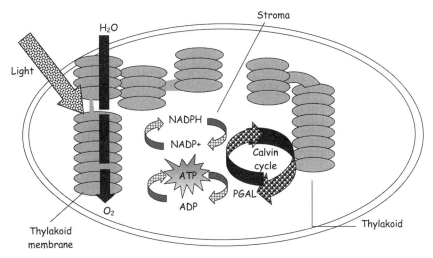

Figure 3.11. General scheme of choroplast structure and function.

tosynthetic membranes arranged in tiny membrane sacs, the thylakoids, which are stacked in groups called *grana*. All thylakoids of a chloroplast seem to be strictly interconnected, thus forming a single internal compartment that functions as a deposit for the H^+.

The stroma is the site of the light-independent reactions of photosynthesis (i.e., the Calvin cycle, a series of enzyme-catalyzed chemical reactions that produce carbohydrates and other compounds from carbon dioxide). The thylakoid membranes are the site of oxygenic photosynthesis (photosynthetic light reactions), an energy-transducing multistep process whereby light energy is trapped and converted into biochemical energy. In thylakoids, a series of electron transfer reactions from water molecules to $NADP^+$ (nicotinamide adenine dinucleotide phosphate), coupled to ATP synthesis, occur. There are three major groups of molecules embedded in these membranes: (1) the photosystems [photosystem I cores (CCI) and photosystem II cores (CCII)], containing photosynthetic pigments and the reaction centers; (2) the electron transport systems, such as cytochrome b6f complexes; and (3) the protein complexes involved in the chemiosmotic ATP synthesis [e.g., proton-ATP synthases (CF1 and CF0)]. These proteins function in a concerted manner to generate ATP and develop a reducing power in the presence of light.

The chloroplast genome has been a major focus in plant evolution and genetics studies (Golenberg, Clegg et al. 1993; Clegg, Gaut et al. 1994; Morton 1999; Stoebe and Kowallik 1999). The cpDNA was the first plant genome to be characterized, thanks to its small size and abundance in foliar tissues. It is circular and ranges in size from 30 to 200 kbp (Martin and Herrmann 1998). In angiosperms and most algae, its structure is highly conserved and arranged as two inverted repeats (IRs) separating a large single-copy (LSC) region from a small single-copy (SSC) region

(Turmel, Otis et al. 1999). The large (10 to 76 kbp) inverted repeat section is found in most species except the Leguminosae and, when present, it always contains the rRNA genes. The substitution rate in cpDNA is much lower than in nuclear DNA but higher than in mtDNA (Wolfe, Li et al. 1987).

cpDNA can exist in two orientations; this implies that the molecule can undergo an isomerization event; moreover, it has been shown that spinach, corn, tomato and pea can all exist as multimers. RNA editing involving C $\rightarrow$ U or U $\rightarrow$ C changes also occurs in chloroplasts, although at a much lower frequency than in land plant mitochondria (Brennicke, Marchfelder et al. 1999).

Chloroplast genes are generally close together, with spacers usually only a few hundred base pairs in length. Land plant cpDNA normally possesses two unique sequence regions; the known products of most cpDNA genes function either to promote chloroplast gene expression (four rRNAs, 30 or 31 tRNAs, 20 ribosomal proteins, four RNA polymerase subunits) or photosynthesis (28 thylakoid proteins plus one soluble protein, rubisco, large subunit).

Beside chloroplasts, other plastids have been identified. Protozoan parasites of the phylum Apicomplexa (e.g., *Plasmodium*, *Toxoplasma*, and *Cryptosporidium*) contain three genetic elements: the nuclear and mitochondrial genomes characteristic of virtually all eukaryotic cells and a circular extrachromosomal DNA. The latter is a maternally inherited genome contained in a vestigial nonphotosynthetic plastid, the apicoplast. In situ hybridization studies showed that the apicoplast is a distinct organelle surrounded by four membranes (Kohler, Delwiche et al. 1997).

Despite the paucity of data on apicoplast function, evidence of parasite delayed cell death when apicoplasts are impaired has been reported; it appears to be involved in the fatty acid and isoprenoid biosyntheses (Zuther, Johnson et al. 1999), and its tight association with mitochondria might be due to a functional cooperation (Marechal and Cesbron-Delauw 2001).

The size of this extrachromosomal genome ranges between 27 and 35 kbp; analysis of sequence data proved this remnant plastid genome has lost all genes encoding photosynthetic function. The complete genome of the 35-kbp DNA of the apicoplast in *P. falciparum* showed that it has low complexity and that its gene content and organization are clearly those of an algal plastid genome (Wilson, Denny et al. 1996), encoding almost exclusively components involved in gene expression. The gene map includes genes for duplicated large and small subunit rRNAs, 25 species of tRNAs, three subunits of a eubacterial RNA polymerase, 17 ribosomal proteins, and a translation elongation factor. In addition, it codes for an unusual member of the Clp family of chaperones, as well as for an open-reading frame of unknown function found in red algal plastids.

The apicoplast completely sequenced in *Toxoplasma gondii* is similar to chloroplast genomes. It contains an inverted repeat of ribosomal RNA genes and genes typically found in chloroplasts but not in mitochondria (rpoB/C, tufA, and clpC) and is also predicted to encode a complete set of tRNAs, numerous ribosomal proteins, and several unidentified ORFs (Kohler, Delwiche et al. 1997).

Recently, the complete sequence of the plastid genome of the nonphotosyntetic flagellate *Astasia longa* has been published (Gockel and Hachtel 2000). This 73,345-bp plastome has a different structure from plant cpDNAs but resembles *Euglena gracilis* chloroplast in its low GC content (22% on average), very small group III introns (group II-like), and rRNA genes organized in three tandem repeats. All chloroplast genes for photosynthesis-related proteins are absent except for *rbc*L, and as there is evidence that this plastid is functional, it is likely that the *A. longa* plastid genome has been maintained to allow the expression of proteins with no photosynthetic function when the flagellate lost this activity.

PART II

METHODOLOGIES

## CHAPTER 4

MOLECULAR BIOLOGY TECHNIQUES FOR GENOMICS

4.1 GENOME DNA SEQUENCING

4.1.1 DNA-Sequencing Techniques

Determination of the sequence of DNA is one of the most important aspects of modern molecular biology. Despite the increasing demand for high throughput in genome-sequencing efforts, the basic DNA-sequencing methods still remain those developed in the late 1970s: the chemical degradation method (Maxam and Gilbert 1977) and the enzymatic chain termination method (Sanger, Nicklen et al. 1977), which is the one currently used. Both methods generate, in different ways, a nested set of single-stranded DNA fragments, which can be separated according to their size by an electrophoresis procedure on high-resolution polyacrylamide gel. In the Maxam–Gilbert method, 5′ radioactively labeled DNA fragments are denatured and subjected to four specific chemical reactions: (1) dimethyl sulfate (DMS) + piperdine cleaves phosphodiester backbone at G; (2) DMS + piperdine + formic acid cleaves at A or G; (3) hydrazine + piperdine cleaves at C or T; (4) hydrazine in 1.5 M NaCl + piperdine cleaves at C. Reaction products are electrophoresed on a polyacrylamide denaturing gel and sequence read from bottom (5′) to top (3′) of the gel. This method can be used for DNA-sequence stretches, which due to a particular DNA sequence or structure, cannot easily be sequenced by the enzymatic method.

The Sanger method utilizes a DNA polymerase to synthesize a complementary copy of a single-stranded DNA template in the presence of chain-terminating dideoxy nucleoside triphosphates, which, when incorporated, determine termination of the chain elongation. In the Maxam–Gilbert method, a labeled DNA strand is subjected to a base-specific hazardous chemical reagent that randomly cleaves single-stranded DNA at one or two specific nucleotides. As compared to Maxam–Gilbert sequencing, the Sanger method is rapid and generates more easily interpretable raw data; therefore, it has become the most widely used method for

133

sequencing. Since 1977, significant improvements has been obtained in Sanger technology to allow DNA sequence to be determined faster and more efficiently. Modifications in the original technique consisted of the use of fluorescently labeled DNA molecules that allow automatic detection by laser-based technology, the development of capillary-based sequencing instruments, the engineering of more effective cloning vectors, and the use of more efficient sequencing enzymes. An important contribution to DNA sequence determination has derived also from the polymerase chain reaction (PCR) technology (Mullis, Faloona et al. 1986), which has revolutionized methods used to analyze DNA in many aspects related to genome analysis and has led to the development of linear amplification sequencing (cycle sequencing). More recent progress has derived from the development of automated sequencing instruments and robotic workstations for the automation of specific steps in the sequencing process, including clone library construction, picking and analysis of subclones, preparing sequencing reactions, and sample loading. All these innovations have led to a quick and dramatic increase in large-scale sequencing, and complete sequencing of large genomes has become possible.

Regardless of differences in size and complexity between genomes of different organisms, all genomes have to be divided into small pieces for sequence determination. The major approaches used for large-scale genome sequencing are the *random method*, also called *whole-genome shotgun sequencing*, and *clone-by-clone shotgun sequencing* based on physical maps. The two methods differ in the procedure for cloning, size of inserts, and sequencing strategy.

The whole-genome shotgun sequencing strategy (Fig. 4.1) consists in the fractionation of the whole genome in small fragments (2 to 10 kbp) to be cloned in M13 or plasmid vectors, so to construct high-quality libraries with uniform representation and a variety of insert sizes. A large number of these clones are then isolated and sequenced using standard vector-specific primers. Pairs of sequence reads, also called *mate pairs*, are then obtained for each insert, one read from both ends. Shotgun sequencing needs a large number of clones to be sequenced in order to cover the region of interest, but on the other hand, all sequencing reactions can be performed using the same set of primers in a very standard procedure. The high redundancy also ensures reliable sequence quality. The sequence data are then combined and assembled to form contiguous stretches of sequence (*contigs*) with sequences represented on both strands. Groups of sequence contigs are organized into scaffolds on the basis of linking information provided by mate pairs. To fill the gaps that remain after the shotgun phase, sequencing by primer walking is usually performed on subclones or PCR products bridging the gap. In the shotgun approach, the sequence coverage is usually five to eight, which means that every base is represented five to eight times (redundancy) by different sequence reads. The genome fragmentation and subcloning using several size classes of inserts is crucial in the assembly process to provide a short- and long-range genome organization framework and to avoid misassemblies due to the presence of differently sized repeated elements.

The clone-by-clone shotgun sequencing strategy (Fig. 4.1) is based on a fragmentation of the whole genome into pieces of about 100 to 200 kbp which are then cloned into large-fragment cloning vectors, such as BAC (bacterial artificial chromosomes) and PAC (P1-derived artificial chromosomes) vectors, to construct genomic libraries. The genomic DNA fragments represented in the library are then

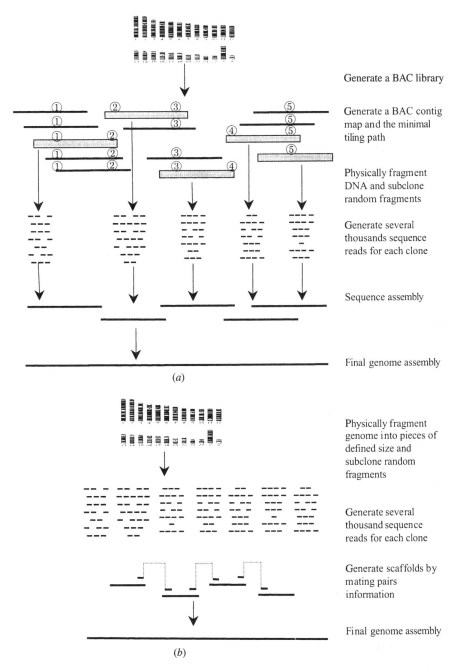

Generate a BAC library

Generate a BAC contig
map and the minimal
tiling path

Physically fragment
DNA and subclone
random fragments

Generate several
thousands sequence
reads for each clone

Sequence assembly

Final genome assembly

(*a*)

Physically fragment
genome into pieces of
defined size and
subclone random
fragments

Generate several
thousand sequence
reads for each clone

Generate scaffolds by
mating pairs
information

Final genome assembly

(*b*)

Figure 4.1. Strategies for the sequencing of complex genomes: (*a*) clone-by-clone shotgun sequencing [in this case a physical map, based on the information of genome landmarks (numbered circles) has been constructed previously]; (*b*) whole-genome shotgun sequencing strategy.

organized into a physical map, created by ordering the clones over chromosomal regions, using information derived from cross-hybridization results and from STS (sequence-tagged sites). These sequences are unique DNA landmarks of 100 to 1000 bp which occur at one position in the genome and can be used to order the genomic clones by PCR screening. Information about the sharing of common landmarks is used to allow detection of the overlapping series of clones, or *clone contigs* used to define the minimal ordered set of clones, also called the *minimal tiling path*, which provides complete coverage across a genomic region. The ultimate goal is a physical map covering the entire genome. For each BAC selected, the cloned DNA is purified and randomly fragmented, usually by physical shearing methodology. After broken-ends enzymatic repair and size fractioning, the DNA fragments in the size range selected are recovered and subcloned into a plasmid or phage vector. The sequence reads obtained by both insert ends (mate pairs) are then assembled to reconstruct the BAC clone sequence. Finally, the BAC clone sequences are assembled to reconstruct the genome sequence.

The preferable approach for the sequencing of whole genomes is still under discussion (Green 2001). The main advantage of the whole-genome shotgun approach is that it does not require earlier construction of a genome physical map. For many sequencing efforts on large eukaryotic genomes such as *S. cerevisiae* (Goffeau, Barrell et al. 1996) and *C. elegans* (T.C.e.S. Consortium 1998), overlapping arrays of large insert clones have been constructed, followed by complete sequencing of these clones. A similar strategy has been used with smaller bacterial genomes, such as *M. pneumoniae*, *E. coli*, and *B. subtilis*. In all these cases, genome sequencing has been initiated after the construction of a physical map. Several other genomes of bacteria and archaea have been sequenced using the whole-genome shotgun approach. Indeed, the two strategies are not mutually exclusive, but in fact, quite complementary. Thus, the advent of hybrid strategies that incorporate the advantages of both are now being used for the sequencing of mouse, rat, and zebrafish genomes (Green 2001).

4.1.2 The Human Genome Project

For sequencing of the human genome, the clone-by-clone and shotgun approaches have both been used. The clone-by-clone strategy was used by the public consortium of the Human Genome Project (HGP; Lander, Linton et al. 2001), whereas the shotgun strategy was used by the private company Celera Genomics (Venter, Adams et al. 2001). Both used automated high-throughput DNA sequencing and computational infrastructure to manage and assemble the enormous amounts of sequence information.

The Human Genome Project began formally in 1990 to produce in its first stage a genetic linkage map of the entire human genome. After the completion of high-resolution physical maps for all chromosomes, sequence production built up only in 1999–2000. It has led to a finished sequence covering more than 96% of the euchromatic part of the human genome and corresponding to about 94% of it, with an accuracy of 99%. Choice of the clone-by-clone strategy by the HGP was justified on several grounds. First, due to the presence of a great number of interspersed repeated sequences in the human genome, this approach would have lowered the probability of misassembly in the finished product compared to the shotgun strategy. Second, in the clone-by-clone strategy, each large insert clone derives from a single haplotype, whereas in the shotgun strategy, sequence derives from different copies of the genome, complicating the accuracy of the sequence assembly. Third, the clone-by-clone approach would have been able to deal with the inevitable cloning bias. Fourth, as the project involved 20 different centers, this method allowed better distribution of work among all members.

Before the phase of large-scale sequence production, a genome-wide physical map of overlapping clones was constructed through a systematic analysis of human BAC clones, obtained by digestion with restriction enzymes, to produce fingerprint clone contigs. These contigs have been then positioned along chromosomes by anchoring them with STS markers from existing genetic and physical maps. Fingerprint clone contigs were also tied to specific STSs by probe hybridization and by direct sequencing of the clones' ends. Then, from the fingerprint clone contigs, clones that presented minimum overlapping of a genomic region were selected for the next phase of sequencing using the random shotgun strategy.

In 1998, the private company Celera Genomics, launched the sequencing of the whole human genome by a shotgun strategy, aiming to completing the genome over a three-year period, with a 10-fold sequence coverage of the genome. The goal originally set was modified, and in less than a year Celera generated a sequence of 2.9 billion base pairs, with 5.1 coverage of the entire human genome. Access to Celera data is not free. They are not available to the scientific community in public databases such as GenBank or EMBL; their access requires an agreement between a scientist and the company.

At the moment, it is known that the two approaches have often provided conflicting results (Hogenesch, Ching et al. 2001), but it remains to be seen which of the two approaches has been more successful by estimating the accuracy of the assembly of the clones sequenced. Both sequences of the human genome are still rich in a considerable number of gaps, which include euchromatic sequences, recalcitrant to be cloned, heterochromatic regions, such as centromeres, and ribosomal gene clusters of the genome, which lack representation in standard libraries. For the HGP, the next efforts are targeted to the development of new techniques to close all remaining gaps as well as to screen additional libraries.

4.2 ANALYSIS OF THE TRANSCRIPTOME

4.2.1 Analysis of Gene Expression

The classical approaches for the analysis of gene expression, such as Northern-blot technique (Alwine, Kemp et al. 1977), RNase-protection assay (Berk and Sharp 1977), and quantitative PCR (Freeman, Walker et al. 1999), are able to analyze the expression pattern of a single or a few genes at a time and cannot easily be automated. For the large-scale analysis of gene expression at the transcript level, new methods have been developed allowing the profiling of thousands of genes simultaneously in different cells, tissues, and physiological conditions. Such methods allow the simple identification of differentially expressed genes or the more complex quantification of mRNA levels in a given condition. Some of them depend on the prior knowledge of gene sequences and on the availability of cDNA clones, whereas others are able of detecting new transcripts without previous information. Each

TABLE 4.1. Principal Methods Used for mRNA Qualitative and Quantitative Analysis

	Method Based on[a]:			Quantitative Method
	Hybr	PCR	Seq	
Northern blot	+			+
Quantitative PCR		+		+
Differential display		+		
Representational difference Analysis	+	+		
DNA microarray	+			+
Subtractive hybridization	+			
Suppression subtractive Hybridization	+	+		
SAGE			+	+
EST			+	+/−

[a] Hybr, hybridization; PCR, polymerase chain reaction; Seq, sequencing.

method shows a number of advantages and drawbacks; therefore, the choice of method depends on the goals of the study and on the possible association with data obtained from different sources.

Gene expression profiling methods can be classified into three categories (Table 4.1), based on their principles: (1) hybridization, (2) polymerase chain reaction (PCR), and (3) sequencing. A further category recently developed is represented by methods based on restriction enzymes, such as TOGA (total gene expression analysis; Sutcliffe, Foye et al. 2000) and TALEST (tandem arrayed ligation of expressed sequence tags; Spinella, Bernardino et al. 1999).

4.2.2   Expressed Sequence Tags

The Expressed Sequence Tags (ESTs) are short (about 300 to 500 nucleotides) single-pass sequence reads derived from cDNA clones selected randomly from cDNA libraries (Adams, Kelley et al. 1991; Adams, Kerlavage et al. 1995). Depending on cDNA synthesis, library construction, and sequencing strategies, ESTs can correspond to 3′ untranslated regions (UTR), 5′ UTR, or internal protein-coding regions of the original mRNAs. The first goal of the EST project was the acceleration of gene discovery in eukaryotic genome sequencing projects. EST sequences also contain information on genes expressed in spatially and/or temporally specific manner, since they derive from several cDNA libraries prepared from different tissues and cell lines, including normal and cancer cells, different developmental stages, different growth states, and so on. Therefore, ESTs provide at least qualitative expression data for a given source, although a great deal of sequencing is necessary because of the low sensitivity of the random cDNA strategy to the presence of less abundant mRNAs. In some cases, mRNA expression levels can also be inferred.

The sequenced cDNA libraries include unaltered, normalized, or subtracted libraries; PCR-amplified and unamplified libraries; and so on. As a general rule, the frequency of occurrence of a clone in a cDNA library is equivalent to that of the corresponding mRNA in the cell. Thus, in addition to the identification of expressed genes, a deep EST coverage of an unaltered cDNA library is potentially able to measure mRNA abundance in the library source. The normalization methods selectively reduce the level of representation of abundant transcripts so that the resulting mRNA preparations contain both abundant and rare transcripts at similar levels (Soares, Bonaldo et al. 1994; Bonaldo, Jelenc et al. 1996). Avoiding the redundant identification of commonly expressed genes, normalization allows the identification of new unknown and low-expressed genes but eliminates the quantitative information obtained by partial cDNA sequencing. The subtraction methods allow us to compare two mRNA populations and to remove from one population (tester) the genes that are present in the other (driver). This is a powerful method to identify and isolate cDNAs of differentially expressed genes, such as tissues-specific genes, but again it eliminates quantitative information. As regards PCR-amplified libraries, a biased amplification of some mRNAs compared to others can modify the relative abundance of the various mRNA species in the sample, allowing only a qualitative analysis of the cDNAs obtained. In conclusion, the EST strategy is more useful for the identification of new genes and transcripts than for determining their expression pattern in quantitative terms.

4.2.3 Serial Analysis of Gene Expression

Serial analysis of gene expression (SAGE) is a clever and efficient variation of the random cDNA sequencing approach (Fig. 4.2). Such sequence-based methods allow the identification and quantification of transcripts derived from new genes as well as from unknown genes and do not require a preexisting availability of clones or gene sequences (Velculescu, Zhang et al. 1995). Three principles underlie SAGE methodology: (1) a short sequence tag (10 to 14 bp) contains sufficient information to identify a transcript uniquely, provided that the tag is obtained from a unique position within each transcript; (2) tags can be linked together to form long serial molecules that can be cloned and sequenced; and (3) quantitation of the number of times a particular tag is observed provides the expression level of the corresponding transcript.

In the original SAGE protocol (Velculescu, Zhang et al. 1995), the mRNA sample is reverse transcribed using a biotinylated oligo(dT) primer and digested with a frequent-cutting restriction enzyme (the anchoring enzyme). The 3′ portion of the cDNAs is purified through binding to streptavidin-coated beads. Then the sample is separated into two aliquots, which are ligated each to a linker A or B containing an internal tagging enzyme site (the tagging enzyme is a type-IIS restriction enzyme which cleaves DNA at a defined distance from the recognition site). After digestion with the tagging enzyme, the unbound short DNA fragments are eluted, ligated, and amplified using the primers A′ and B′ complementary to A and B linkers, respectively. After PCR, the linkers A and B are removed by cleavage with the anchoring enzyme, and the sticky ends generated allow the generation of concatemers, which can be cloned and sequenced. The concatenamers are composed of a number of *ditags* (tags of two independent cDNAs) separated by an anchoring enzyme site. Based on sequence information, the presence and frequency of a transcript is inferred from the presence and frequency of the corresponding tag in the sample. This procedure is performed for every mRNA population to be analyzed.

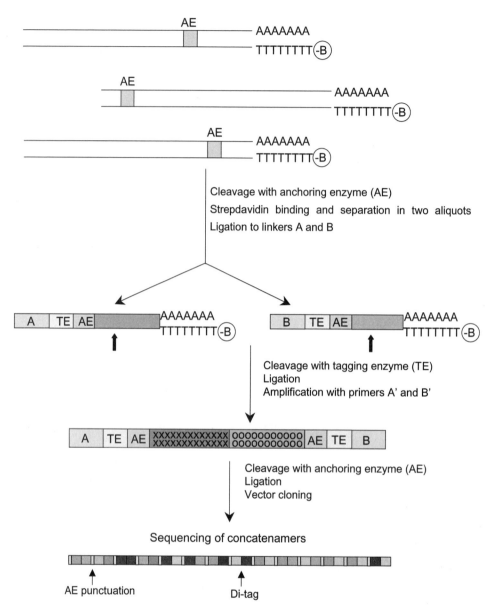

Figure 4.2. Experimental procedure for the serial analysis of gene expression (SAGE). cDNA obtained by poly-A+ mRNAs by a biotininylated oligo(dT) primer is digested with a frequently cutting anchoring enzyme (AE) and purified on streptavidin. After separation of purified cDNA into two aliquots, two linkers are ligated, A and B, that contain a site for a tagging enzyme (TE; the arrows indicate its downstream cleavage site). After ligation to form a ditag and amplification, a polytag is formed after cleavage with AE and ligation. The sequencing of the cloned concatenamers provides a measure of the abundance of a specific transcript as a function of the representation of its corresponding tag sequence (the AE punctuation between ditags allows tag identification).

The classical SAGE protocol has been used in the analysis of tissue samples, each usually consisting of millions of cells, but the modified microSAGE strategy allows us to perform gene expression analyses from as few as 5000 cells (Datson, van der Perk–de Jong et al. 1999). MicroSAGE strategy uses oligo(dT)-coated tubes or magnetic beads to capture poly(A) mRNAs, so that the SAGE steps occur all in the same vessel. This approach increases the efficiency of certain enzymatic steps and reduces the number of purification steps. Another modification of the SAGE to analyze limited amounts of RNA has been to use PCR to amplify cDNA mixture immediately following reverse transcription (Peters, Kassam et al. 1999; Neilson, Andalibi et al. 2000). However, a general concern in this approach regards a potential amplification bias during PCR, resulting in a distortion of the true transcript levels, especially for low-abundance transcripts. Finally, the use of different reverse transcriptases is very important for the generation of more robust cDNA levels (Velculescu, Vogelstein et al. 2000).

As a general observation, the SAGE approach requires a good-quality library, because the tag frequency data may be interpreted improperly in cases of inefficient ligation and/or amplification steps. Moreover, the transcripts lacking the anchoring enzyme site will be completely missed in the analysis, whereas rare transcripts will be detected only with high depth in the sequencing step, with its inherent cost and throughput limitations. The major strength of the SAGE method is that data represent absolute expression levels, calculated by dividing the observed abundance of any tag by the total number of tags analyzed. Thus, results from any new SAGE experiment are directly comparable to existing gene expression databases, and significant differences can be identified using a variety of standard statistical tests.

A limitation of SAGE is that the gene corresponding to a tag can be identified only if it is already recorded in a public database. Thus SAGE works best when the analysis is conducted in conjunction with EST and genome sequencing projects, so that each tag can be aligned readily and uniquely with a known gene or genome fragment (Stollberg, Urschitz et al. 2000). However, several techniques have recently been described that allow for more efficient conversion of a SAGE tag to cloning of unknown transcripts (Chen, Rowley et al. 2000). Several Web resources have been made available for online SAGE data analysis that allow differential expression analysis (e.g., colon cancer versus normal colon) as well as the mapping of a SAGE tag to a gene, or vice versa, to visualize SAGE data for a specific gene (see the SAGEmap URL in the Appendix).

4.2.4 Differential Display

The differential display (DD) was introduced by Liang and Pardee (1992). Its rationale is to sample a subset of transcripts as 3′-end cDNA fragments by means of PCR and then display as gel fingerprinting the cDNAs selected. The standard protocol can be illustrated as follows. Total RNA of high purity is reverse transcribed in cDNAs using a 3′-anchored oligo(dT) primer, usually of the form T_nX or T_nXY (X = A, C, G; Y = A, C, G, T), to fix the priming site to the beginning of the poly(A) tail of a subset of total mRNAs (about $\frac{1}{4}$ in the case of T_nX primers and $\frac{1}{16}$ in the case of T_nXY primers). Then a PCR step is performed on the cDNAs using the same anchor primers and arbitrary primers. Each of these primers will amplify a subset of all cDNAs, resulting in the generation of up to 100 cDNA fragments in one reaction tube. A portion of the amplified products is then resolved on a high-resolution denaturing polyacrylamide gel to obtain a fingerprint pattern. Radioactive nucleotides (Liang and Pardee 1992) or fluorescently labeled amplification primers (Ito and Sakaki 1997) allow for signal detection. After comparison of the cDNA band patterns derived from multiple RNA samples, cDNAs differentially amplified can be eluted from the gel, reamplified, cloned, and sequenced. Because of the high incidence of false positives in the original DD protocol, special care is necessary to confirm that only the target band, and no contaminating bands, have been cloned (Debouck 1995).

Optimization of the initial DD protocol has ensured high reproducibility and reliability. The improvements deal primarily with anchor and arbitrary primer design, reverse transcription and PCR conditions, cDNA size range, elimination of false positives, confirmation of differentially expressed candidate genes, postrun gel processing, signal detection, and so on (Liang and Pardee 1995; McClelland, Mathieu-Daude et al. 1995; Diachenko, Ledesma et al. 1996; Ito and Sakaki 1997; Jurecic, Nachtman et al. 1998). The DD compares transcribed genes systematically in a bidirectional way and can detect even low-abundance transcripts starting from a limited amount of RNAs. Moreover, the immediate availability of cDNA probes and the cloning of genes of interest is very useful. As drawbacks, the DD lacks quantitative information, because it does not quantify the relative amounts of differentially expressed genes. Moreover, the use of arbitrary primers increases the number of primer combinations to be tested for statistical coverage of complex transcript populations.

4.2.5 Representational Difference Analysis

Representational difference analysis (RDA), schematically described in Fig. 4.3, was originally developed to identify differences between two samples, the *tester* and the *driver*, without quantification (Lisitsyn and Wigler 1993). The method is based on both subtractive hybridization and PCR and consists in the enrichment of rarely expressed tester sequences followed by their specific amplification (O'Neill and Sinclair 1997). The two mRNA samples derived from two different populations, the tester and the driver (control), are reverse transcribed, digested with a frequently cutting restriction enzyme, ligated to adapters, and amplified with primers complementary to the adapters. The linkers of both cDNA pools are removed by digestion and a new linker is added only to the tester. After denaturation, the two samples are mixed so that the driver is in large excess compared to the tester (typically, the tester/driver ratio is 1:100) to promote hybridization between cDNA common to both the tester and the driver. Then a PCR exponentially amplifies only duplexes generated by the tester cDNA, using the priming sites on both ends of the double-stranded cDNA. Indeed, reamplification produces only a tester fragment which does not have a counterpart in the driver pool. To isolate low-abundance cDNA species expressed differentially, it is possible to modify the hydridization stringency and the tester/driver ratio. Moreover, the kinetic enrichment is usually repeated once or twice to achieve high efficiency. RDA allows unidirectional identification of differentially expressed genes and can also be used in the expression analysis of prokaryotes (Bowler, Hubank et al. 1999).

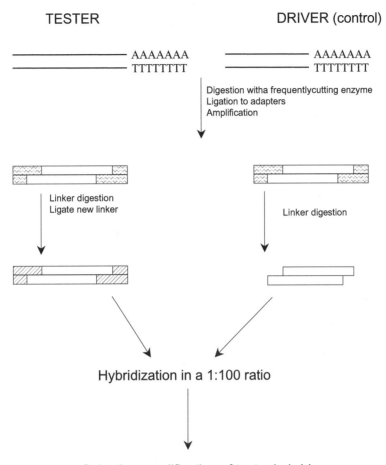

Figure 4.3. Schematic representation of the experimental procedure of representational difference analysis (RDA). The mRNA of different origin is reverse transcribed, digested with a frequently cutting endonuclease, and amplified using primers from the linkers ligated. The linkers of both the tester and driver cDNA pools are then removed and a new linker is added only to the tester cDNA. After hybridization of the tester and driver cDNAs, only those cDNAs present in the tester population will be selectively amplified.

4.2.6 DNA Microarrays

DNA microarray is a hybridization-based technology that depends on knowledge of gene sequences and/or the availability of cDNA clones. The microarray is a miniaturized high-density array of oligonucleotides or cDNAs, which can represent more than 250,000 different oligonucleotide probes or 10,000 different cDNAs per square centimeter (Bowtell 1999; Lipshutz, Fodor et al. 1999). The DNA fragments synthesized or deposited on the array are based on sequences of known genes or sequences of genes with unknown function and only partial sequence information, such as ESTs. The method works by hybridization of labeled cRNA or cDNA in solution to the DNA molecules attached at specific positions on a solid surface. The labeling and detection system is based on fluorescence.

Such a powerful method is able to monitor simultaneously the expression level (mRNA relative abundance) of thousands of genes in parallel in different tissues, cells, and developmental stages. Multiple experiments carried out in successive steps of a biological process (various stages of differentiation and disease process, at different times after application of a stimulus, and so on) allow us to build up a gene expression profile that characterizes the dynamic functioning of each gene during a specific biological process. Apart from discussions on technical issues and performance characteristics, the main questions now posed by the microarray technology concern the analysis, interpretation, storage, and access to the large amount of data produced.

Array Construction   Microarrays are usually made by synthesis or deposition of DNA spots on a solid support, such as coated glass slides, using two main procedures: photolithography (Lockhart, Dong et al. 1996) and mechanical gridding (Schena, Shalon et al. 1995). The photolithography, commonly used in the computer chip industry, utilizes an ultraviolet light passing through a photolithographic mask to direct a photochemical reaction of oligonucleotide synthesis on a siliconized glass surface. The in situ synthesis technology limits the probe length to 25 nucleotides; thus on such an array a given gene is represented by about 20 different 25-mers probe pairs. Each probe pair consists of a perfect match probe (PM), complementary to a specific mRNA, and a mismatch control probe (MM) that is identical to the perfect-match partners except for a single base difference in a central position. The analysis of PM/MM pairs allows low-intensity hybridization patterns from rare mRNAs to be recognized sensitively and accurately even in the presence of cross-hybridization and local background. This trick should eliminate the problem of the greater specificity and affinity expected for the hybridization of cDNA probes than for oligonucleotide probes.

Mechanical gridding methods are based on ink-jet or physical deposition of the material on solid support with high precision and can even be produced in-house. Because of the direct surface contact and the use of small amounts of liquid, these systems are susceptible to evaporation and spot contamination. Moreover, a large number of cDNAs or PCR products must be prepared, purified, quantitated, and cataloged before deposition on the solid support. As already mentioned, the probes arrayed can be oligonucleotides (photolithography and gridding) or cDNAs, PCR products, and cloned DNAs (gridding). Hybridization reaction conditions and sample preparations are optimized depending on the probes. Indeed, with short arrayed probes (20 to 50 nucleotides), the sample is fragmented to reduce suboptimal hybridization by possible interference from secondary structures and to minimize the effects of multiple interactions with closely spaced molecular probes (Lockhart, Dong et al. 1996). With amplified probes (300 to 2000 nucleotides) sample fragmentation is not required.

Sample Preparation, Hybridization Protocol, and Image Processing   Usually, labeled cDNA or cRNA target samples are prepared directly from cellular mRNA. The sample preparation presents the problems of mRNA extraction from very small quantities of tissues/cells and the necessity to preserve linearity in the mRNA labeling reaction. When a sufficient amount of mRNA is available, the single round of mRNA reverse transcription is used to incorporate fluorescently labeled ribonucleotides into the cDNA synthetized. The RNA can also be labeled directly using a psoralen–biotin derivative or by ligation to an RNA molecule carrying biotin (Wodicka, Dong et al. 1997). If only a small amount of mRNA is available, two different approaches are possible. The coupling of an amplification strategy to the reverse transcription reaction is efficient and reproducible (Schena, Shalon et al. 1995); however, the relative abundance of the cDNA products seems not well correlated with the original mRNA levels, because of the lack of linear amplification of all transcripts introduced by too many PCR cycles (Lockhart and Winzeler 2000). The second approach avoids PCR and uses multiple rounds of linear amplification based on double-stranded cDNA synthesis and subsequent in vitro transcription reaction (Lockhart, Dong et al. 1996). Indeed, the mRNA sample is reverse transcribed to cDNA using oligo(dT) primers, including the T7 promoter element at the 5 end. Once the double-stranded cDNA is made, a cRNA is synthesized by in vitro transcription using T7 RNA polymerase in the presence of labeled (fluorescent dye or biotin) ribonucleotide triphosphates. This cycle of linear amplification can be repeated until enough cRNA is produced. Importantly, this protocol can generate full-length products, whereas most oligo(dT) labeling protocols have a significant 3-end bias (Wang, Miller et al. 2000). Such a procedure can amplify the entire mRNA population by 20- to 250-fold in an apparently unbiased and reproducible fashion (Lockhart, Dong et al. 1996). In conclusion, the critical concern is whether the skewing introduced by PCR does not affect the relative concentration of the various mRNAs (Bashiardes and Lovett 2001). Commonly used labels include fluorophores or the nonfluorescent biotin, which is subsequently labeled by staining with a fluorescent streptavidine conjugate.

In the case of cDNA-spotted microarray, a two-color fluorescence competitive assay is used as hybridization strategy to minimize experimental variations when comparing independent hybridizations. The test and the control samples are labeled by reverse transcriptase incorporation of two different fluorescent dyes. The two samples are mixed together in equal proportions and co-hybridized to the same array. Then the array is scanned at two different wavelengths to detect emission of the two fluorescent dyes. The relative abundance of a particular transcript is determined as the ratio between the tester and the control signal intensities. In an oligonucleotide-spotted microarray, a single labeled sample is hybridized to the array.

For cDNA-spotted array, an mRNA at known concentration is used as the internal standard for calibration and added to the mRNA sample before the labeling reaction. The results of parallel Northern blot and microarray assays indicate that the expression levels correlate well within a factor of 5 but differ in terms of absolute values (Schena, Shalon et al. 1996). Because of the two-color fluorescence protocol, the transcript levels are provided as expression ratios relative to the control, and use of the same reference sample is required to compare different experiments.

In oligonucleotide-array experiments, cRNAs of known concentration are added to the hybridization mixture for direct quantitation of the mRNA levels: The hybridization intensity is demonstrated to be linearly related to the RNA target concentrations spiked into the mRNA sample for calibration (Lockhart, Dong et al. 1996). Thus, the hybridization intensities are an estimation of the absolute transcript level in individual samples and can be converted to mRNA copies per cells.

The raw data produced by a microarray are monochrome images derived from measuring the fluorescence for each spot and for each fluorescent dye. The processing of such an image into a gene expression matrix involves the solution of some not-trivial problems: The spots should be identified and their boundaries determined, and the fluorescent intensities should be measured for each spot and compared to the background intensity. Moreover, the fluorescent intensities must be rescaled to adjust to differences in labeling and detection efficiencies and differences in the quantity of initial RNA in the two samples examined. Thus appropriate normalization should be applied to enable data comparisons. Finally, for each gene, the data are reported as an *expression ratio*, that is, simply the normalized intensity values in the query sample divided by its normalized value in the control. Commonly, it is assumed that a twofold increase or decrease of expression ratio is indicative of a significant change in the expression level; however, there is no theoretical basis for selecting this level as significant. In general, some replicate analyses are required to get statistically meaningful results.

Profile Expression Comparisons and Interpretation   Expression levels of thousands of genes can be measured simultaneously by EST, SAGE, and microarray methodologies in different cells/tissues, under various conditions, and at different developmental stages. Such expression data can be organized in a table, called a *gene expression matrix*, where each row represents a gene, each column represents a sample (e.g., a tissue, developmental stage, or growth condition) and each cell contains a measure of the expression level of a given gene in a particular sample. In EST and SAGE studies, such a measure is the cDNA tag frequency, whereas in DNA microarray studies the measure is the expression ratio. A gene expression matrix can be analyzed by comparing the rows (i.e., the expression profiles of genes), or by comparing the columns (i.e., the expression profiles of samples). The expression data can also be visualized as points in an N-dimensional space or as N-dimensional vectors, where N is the number of genes or the number of samples, depending on the comparison. To look for similarities and differences in such complex data, we first need a way to measure the distance between objects and then a way to sort such objects, called *clustering algorithms* [reviewed in Brazma and Vilo (2000) and Quackenbush (2001)]. Finally, to interpret the results easily, it is helpful to have an intuitive visual representation of the data.

There are various methods for measuring distances (Claverie 1999). Metric distances, such as the Euclidean distance, and correlation coefficients, such as the Pearson correlation coefficient, are the most commonly used distance measurements. The correlation coefficient measure treats the N-dimensional vectors as random variables. Currently, there is no theory on how to choose the best distance measure; the choice probably depends on the questions we are asking.

Clustering methods are divided into supervised and unsupervised methods. The *unsupervised methods* analyze the internal structure of the expression data without using any external biological information. The *supervised methods* use biological information about probe genes, such as functional classification of genes or disease/normal state of the samples, to guide the clustering of expression data; thus a typical goal is to build up a classifier able to assign predefined classes to a given expression profile. The supervised methods use a positive and a negative training set to learn to distinguish between members and not-members of a class on the basis of expression data. After assessing the quality of the prediction, they can be applied to data of unknown classification. Brown, Grundy et al. (2000) applied various supervised methods, such as support vector machine (SVM), decision tree, Panzen windows, and Fisher's linear discriminant to yeast expression data and showed that SVM provides the best prediction accuracy.

The unsupervised methods can be classified into hierarchical and nonhierarchical (or partitioning) techniques. The *hierarchical methods* are "bottom-up" approaches which (usually) start with single-member clusters and gradually join them to form groups which are further joined, until formation of a single hierarchical tree. They differ in the rules governing the way distances are measured between clusters as they are constructed. The hierarchical methods are mostly derived from molecular phylogeny; however, there is no reason to believe that expression data are naturally organized in hierarchical descendents and bifurcating trees as in molecular evolution.

Nonhierarchical clustering simply partitions objects into different classes without trying to specify the relationships between individual elements. These are "top-down" clustering methods, which start with all elements in one cluster and gradually break it down into smaller clusters. Unlike hierarchical clustering, such methods require a priori assumptions on the number and structure of clusters. The most used among the latter are the self-organizing map (SOM), which partitions objects into a fixed number of groups by mapping onto nodes (Tamayo, Slonim et al. 1999; Toronen, Kolehmainen et al. 1999), and the K-means, which partitions objects by finding centroids (Tavazoie, Hughes et al. 1999). Additional unsupervised methods are filtering methods such as principal component analysis (PCA). PCA is a mathematical technique that reduces the number of variables in a data set without significant loss of information; thus it is used to reduce the effective dimensionality of gene expression space and to provide projection into a low-dimension subspace. Although the PCA does not delimitate clusters, it is useful for initial exploration and visualization of expression data and can help define the number of clusters to be specified in a nonhierarchical method.

The complexity of expression data implies that there is no "correct way" to analyze each data set, but the application of various distance metrics, clustering algorithms, and data filtering methods might stress different features in the data. Finally, the results have to be evaluated in the context of biological knowledge. The expression levels can be related to biological information such as protein function, structure and localization, protein–protein interactions, gene regulation, and so on. Relating expression profiles to protein function is currently hampered by intrinsic difficulties in functional classification (i.e., the multifunctional nature of many proteins, ambiguous or inappropriate naming of functions, and so on), whereas structure and localization of proteins have been more easily related to the expression profiles (Gerstein and Jansen 2000). Most important, genes having a similar expression profile probably share common upstream regulatory elements, already known or to be investigated (Bucher 1999).

Recently, a number of computational analysis tools and packages have been constructed from both companies and academic labs (some of them are listed in Table 4.2). The products range from software that can perform various types of

TABLE 4.2. Expression Data Analysis Tools

Program	URL	Description	Availability
Expression Profiler	ep.ebi.ac.uk	Set of tools for clustering, analysis, and visualization of gene expression and sequence data	Free for academic and nonprofit researchers
Cluster, TreeView	rana.lbl.gov/EisenSoftware.htm	Cluster analysis and visualization	Free for academic and nonprofit researchers
TIGR Multiexperiment Viewer, TIGR ArrayViewer	www.tigr.org/softlab	Analysis and presentation of microarray data	Free for academic and nonprofit researchers
SAM (significance analysis of microarrays)	www.stat.stanford.edu/~tibs/SAM	Supervised learning software for genomic expression data mining	Free for academic and nonprofit researchers
XCluster	genome-www.stanford.edu/~sherlock/cluster.html	Perform hierarchical clustering and self-organizing maps	Free for academic and nonprofit researchers
ArraySCOUT	www.lionbioscience.com/eng/index_c_1.htm	Enterprise-wide expression data analysis	License
GeneMaths	www.applied-maths.com/ge/ge.htm	Complete analysis of high-density microarrays and gene chips	License
GeneSpringer	www.sigenetics.com/cgi/SiG.cgi/Products/GeneSpring/index.smf	Visualization and manipulation of gene expression data derived from microarrays, Affymetrix chips, SAGE, etc.	License
GeneData Expressionist	www.genedata.com/products/expressionist	Computational system for analyzing large amounts of gene expression data from microarrays	License

clustering and dimensioning analyses to packages that attempt storage, handling, analysis, and presentation of data. The improvements made by commercial enterprise and academia in developing and testing algorithms and tools for expression data mining allow researchers to analyze the same raw data by a variety of methods in order to extract all the information present in them. Indeed, data analysis is the final rate-limiting step in performing expression profiling experiments.

Data Access   The nearly exponential growth of expression profiling data and the necessity for comparisons between data derived from different labs makes the standardization, organization, storage, and access of expression data a very important issue. A GenBank-like expression database would be the best solution to the problem. The MGED group (see the URL in the Appendix) proposed that "minimal information about a microarray experiment" must be reported for each microarray experiment to ensure the interpretability of the results and potential data verification from other labs. The minimum information should encompass array design, sample preparation, experimental design, hybridization, spot quantification, image analysis, and normalization controls. In addition, data should be submitted in a standardized format to facilitate data exchange for different array technologies. For example, some platforms, such as Affymetrix, yield absolute fluorescent intensities from a single sample, whereas others, such as the two-color hybridization protocols, yield intensity ratios between two samples. Many organizations, both private and public, are now compiling their own expression databases (Table 4.3), but in many cases availability of data is up to the authors, leading to a wide spectrum of presentations and formats. Among the expression databases, some are simply public data depositions, whereas others are available for public queries (Gardiner-Garden and Littlejohn 2001).

4.3 ANALYSIS OF THE PROTEOME

The identification and functional characterization of the protein inventory in an organism, tissue, or cell under different physiological or pathological conditions, which defines the field of *proteomics*, will certainly represent a major challenge in biology for many years to come. Integrated knowledge of the genetic information provided by large-scale sequencing efforts of the expression profile of transcripts and of their protein products as well as of the interaction network between different genes and proteins will ultimately result in a thorough understanding of the cellular function in all its aspects. This will certainly have profound implications in our daily life, making possible many unexpected applications in biotechnology and medicine.

Large-scale study of proteins is a much harder task than determining genome sequences or characterizing transcript expression. Indeed, considering alternatively spliced genes and posttranslation modifications, the total number of structurally and functionally different proteins may exceed by up to 200-fold the total number of genes of an organism (Watson 2001). This means that in the case of humans, where 30,000 to 40,000 genes have been predicted, more than 1 million structurally different proteins could be expressed. Full characterization of each protein should include determination of the (1) amino acid sequence; (2) posttranslation modifications;

TABLE 4.3. Partial List of Expression Databases

Database	URL	Description	Organization
Gene Expression Omnibus (GEO)	www.ncbi.nlm.nhi.gov/geo	Gene expression and hybridization array data repository, as well as an online resource for the retrieval of gene expression data from any organism or artificial source. Gene expression data from platform types such as spotted microarray, high-density oligonucleotide array, hybridization filter, and SAGE data, are accepted, accessioned, and archived as public data sets.	National Center of Biotechnology Information
GeneX	www.ncgr.org/genex	Freely Internet-available repository of gene expression data with an integrated tool set to analyze and compare data derived from different technologies.	NCGR and University of California (Irvine)
ExpressDB	twod.med.harvard.edu/ExpressDB	Relational database containing yeast and *E. coli* microarray-based expression data. It contains more than 20 million pieces of information loaded from numerous published and in-house expression studies (Aach, Rindone et al. 2000).	Harvard University
Stanford Microarray Databases (SMD)	genome-ww4.stanford.edu/MicroArray	SMD stores raw and normalized data from microarray experiments, as well as their corresponding image files. In addition, it provides interfaces for data retrieval, analysis, and visualization.	Stanford University
ArrayExpress	www.ebi.ac.uk/arrayexpress	Database of gene expression data derived from microarray experiments (Brazma, Robinson et al. 2000).	European Bioinformatics Institute

(3) three-dimensional structure; (4) abundance in different tissues, cells, or developmental stages; (5) subcellular localization; and (6) functional interactions with other proteins or DNA/RNA molecules. To afford this gigantic challenge, high-throughput techniques and suitable computational tools are absolutely required.

The poor correlation between mRNA and protein levels, due to the many diverse posttranscriptional regulation pathways, in many cases makes unreliable a measure of gene expression just in terms of transcript abundance, a task now quite easily accomplished by high-density microarrays or other high-throughput techniques (e.g., SAGE; see Sec. 4.2.3). Then a proteomic approach is a necessary complementary to genomics and transcriptomics. Ultimately, the proteomic approach is also necessary to confirm predicted genes, even if the corresponding transcript has been identified. It has been estimated that gene annotation could be affected by an error rate as high as 10% (Brenner 1999). Furthermore, a number of small proteins which could have escaped prediction analyses can be identified experimentally only by means of proteomic approaches. Large-scale characterization of the protein inventory of a specific cell type is the first fundamental step in proteome analysis.

4.3.1   Two-Dimensional Gel Electrophoresis

Two-dimensional gel electophoresis (2DE) is the most widely used tool to resolve complex protein mixtures. This technique separates sampled proteins in two dimensions, combining isoelectric focusing on immobilized pH-gradient strips (first dimension) and SDS-PAGE for molecular-weight separation on the second dimension. After the 2DE is complete, the resolved proteins can be visualized by a number of stainings. Figure 4.4 shows as an example the 2-DE map of control rat serum and of serum from an acutely infected rat (Haynes, Miller et al. 1998; Miller, Haynes et al. 1998). Depending on gel size, up to 10,000 different spots, corresponding to proteins in the size range 7 to 200 kDa and isoelectric point between 3.5 and 11.5, can be resolved to be characterized further. However, very low abundant proteins as well as very hydrophobic proteins (e.g., membrane proteins) or very acidic or basic proteins cannot be detected using this technique, whose resolution can somehow be improved by prefractionation of protein mixtures. The gel image pattern from different samples provides a fingerprint of the proteome of a particular biological system and can be compared using commercially available software to identify spots corresponding to quantitatively differentially expressed proteins.

4.3.2   Protein Identification

After protein separation, the next step is protein identification. The use of mass spectrometry (MS) for protein characterization has represented a major breakthrough in this field (Fig. 4.5). Since protein mass is insufficiently discriminating for its identification, protein is first converted to shorter peptides by proteolysis; then peptide masses are determined by MS. These masses can be compared with databases of masses of *in silico* digested proteins or six-frame translations of nucleic acid sequences (e.g., EST or genomic sequences) resulting from virtually the same proteolytic treatment. The process of spot excision, tryptic cleavage, and peptide extraction has been well defined but can be done manually for only a few spots. Therefore, robotic devices have been developed to automate this task. Ionization

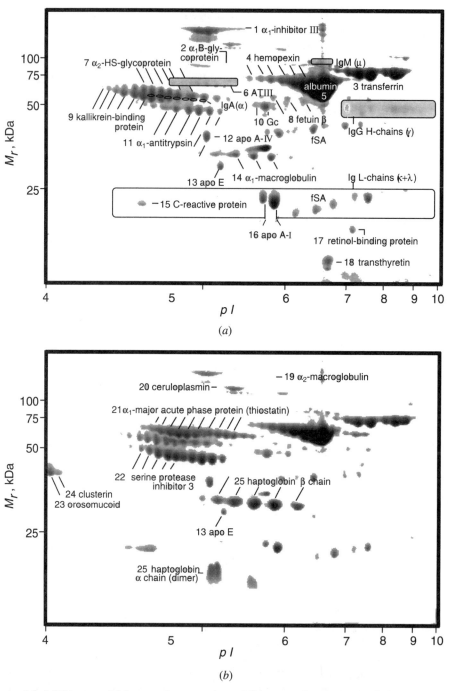

Figure 4.4. 2-DE map of (*a*) control rat serum and (*b*) serum from an acutely inflamed rat (day 2 after intremusculer injection of 5 mL/kg turpentine). Protein identifications are marked across the pattern. [Maps courtesy of E. Gianazza, modified from Haynes, Miller et al. (1998) and Miller, Haynes et al. (1998); regularly updated at linux.farma.unimi.it.]

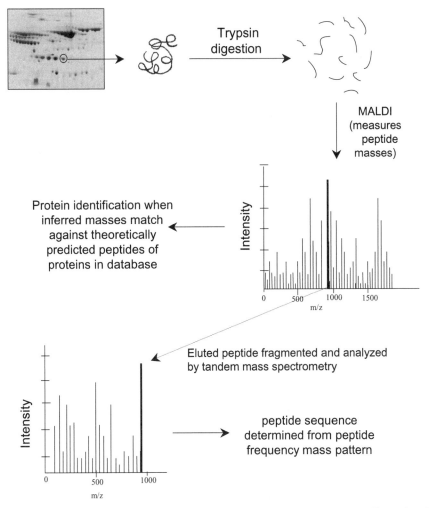

Protein identification when
inferred masses match
against theoretically
predicted peptides of
proteins in database

Eluted peptide fragmented and analyzed
by tandem mass spectrometry

peptide sequence
determined from peptide
frequency mass pattern

Figure 4.5. Schematic representation of standard proteome analyses by two-dimensional gel electrophoresis and mass spectrometry.

and volatilization of sampled proteins or peptides is needed for MS analysis. Several techniques have been developed for this task, such as matrix-assisted laser desorption ionization (MALDI), or electrospray or nanoelectrospray ionization (ESI or nESI). Different from MALDI, where laser-induced desorption volatilizes charged peptides, in ESI or nESI ion fragmentation occurs in the liquid phase formed by a spray of fine droplets. ESI can easily be coupled to separation devices based on high-performance liquid chromatography (HPLC), capillary isoelectric focusing (IEF), or capillary electrophoresis (CE), which different from 2DE, eliminates the steps required to transfer proteins from the separation device to the mass spectrometer, thus facilitating automated high-throughput analysis. These techniques combined with time-of-flight (TOF) or new hybrid mass spectrometers allow peptide mass fingerprinting.

In MALDI-TOF a peptide sample is mixed with a large excess of ultraviolet-absorbing matrix to make a dried spot that is hit in vacuum by a pulsed laser that induces ionization and volatilization of peptides without fragmentation. The peptide ions are then allowed to drift through a flight tube, where they separate according to mass and are then recorded by a suitable device. Although MALDI-TOF represents a milestone in proteomic analysis, it has some limitations. First, the ionization is not quantitative, with different equimolar peptides giving a different signal intensity. Second, if the number of different peptides generated from a single spot is proportional to the amount of proteins, low-abundant proteins could not be identified, due to the detection of an insufficient number of peptide masses. Finally, MALDI-TOF has limited resolving power with protein mixtures that can be found rather frequently in apparently clear spots. To further improve MALDI-TOF performance, the peptide sequence information should be added to the peptide mass information. This can be achieved using a tandem mass spectrometer (Fig. 4.5). In this case each of the peptides analyzed previously can undergo further fragmentation. The mass spectra obtained from the fragmentation pattern of each peptide may generate sequence information if they match virtual mass spectra obtained from known proteins or nucleotide sequences (after six-frame translation).

Tandem MS (MS/MS) can also allow the detection of posttranslational modifications (e.g., phosphorilation through the detection of phosphopeptides and phosphate fragmentation masses). Indeed, posttranslational modifications are extremely important for protein function and in no way can be inferred from genomic sequences or mRNA expression data. In searching databases with mass-spectrometric information, the fragment masses measured experimentally are compared with the fragment masses calculated by protein sequence databases, and a similarity score is calculated for the comparison. Several search tools are available for this task (Beavis and Fenyo 2000; Fenyo 2000), such as MOWSE (Pappin, Højrup et al. 1993) or MASCOT (Perkins, Pappin et al. 1999), (au URLs in the Appesolin) which also takes into account the relative abundance of peptides and compensates for protein size. The searching algorithm could also be used in characterizing proteins generated by comparing experimental MS and MS/MS results with hypothetic MS and MS/MS data predicted from proteins whose sequences have been deduced from corresponding genes using different exon–intron combinations.

The approaches described above are not effective for quantitative characterization of protein mixtures due to variable ionization efficiencies for different protein/peptides. Indeed, quantitation of resolved proteins is an important step in proteome analysis. This can be, in some cases, problematic also due to the very large protein abundance range, up to eight orders of magnitude. To perform a quantitative comparison of protein profilings from different samples, the isotope labeling technique can be used. The isotope-coded affinity tag (ICAT) strategy has proved to be very convenient because it allows protein labeling after protein extraction, avoiding expensive and time-consuming animal feeding or cell cultures with isotope-containing media. The ICAT reagent consists of an affinity tag (biotin) which allows the separation of ICAT-labeled peptides, a linker that incorporates heavy or light stable isotopes (hydrogen or deuterium) and a reactive group specific for cysteine groups. ICAT-labeled proteins are then digested with trypsin, purified by biotin-affinity chromatography and then analyzed by HPLC-MS. The intensity ratio for the pair of heavy and light ICAT-labeled ions provides a quantitative estimate of the relative abundance of each protein considered in the two samples under investigation. 2-DE data and analyses from different cellular systems and states are usually collected in comprehensive databases whose aim is the integration of protein, transcript, and genome information (Celis, Kruhoffer et al. 2000).

4.3.3 Study of Protein–DNA and Protein–Protein Interactions

The availability of fully sequenced genomes of both prokaryotic and eukaryotic organisms has led to large-scale studies on gene expression and function (functional genomics) and more recently on the proteome. In living organisms, proteins play some of the most important roles, acting on other macromolecules, such as nucleic acids, lipids, and other proteins. Inside cells, protein–DNA and protein–protein interactions are at the base of important mechanisms that govern signal transduction, cell division, and DNA replication and transcription. Outside cells, protein interactions allow cells to talk with each other. Thus it has been tempting to envisage large-scale studies for protein–protein interactions to provide exhaustive protein interaction maps (i.e., the "interactome").

One- and two-hybrid systems are techniques used to detect and study protein–DNA and protein–protein interactions respectively. They made use of yeast as a "test tube." In particular, the *one-hybrid system* provides a basic tool for identifying proteins that bind in vivo to a target sequence such as a cis-acting regulatory element or, alternatively, for identifying DNA fragments containing the binding sites for a protein of interest. The *two-hybrid system* is particularly useful to identify and investigate protein–protein interactions in vivo (Fields and Song 1989). At the present time, the yeast two-hybrid assay remains the only large-scale technology that is available to build protein interaction maps. The basic concept of one- and two hybrid systems is to detect the interaction between DNA and protein or between two proteins via transcriptional activation of one or several reporter genes.

In the basic version of the one-hybrid system, the DNA-binding sequence is cloned in the GAL promoter of a reporter gene (usually, HIS3) instead of the upstream activating sequences (UASs) (Fig. 4.6). The upstream activating sequence or operator is one type of cis-acting transcription element in yeast promoter, which is recognized by a specific transcriptional activator and enhances transcription from adjacent downstream TATA regions. The GAL promoter is activated by the GAL-4 transcription factor, which contains a domain that specifically binds to DNA sequences [binding domain (BD)] and a domain that recruits the transcription machinery [activation domain (AD)]. The cDNA library is cloned in vectors that express inserts as a fusion to a GAL-4 transcriptional activation domain (GAL4-AD).

Both recombinant vectors are used to transform yeast nutritional mutant strain. The two plasmids should bear different nutritional markers, thus could be selected independently. If any protein of the cDNA library does not interact specifically with the DNA sequence cloned upstream of the reporter gene, the GAL promoter is not transactivated. In a positive interaction, the library-encoded protein interacts with the DNA-binding sequence, and the fused GAL4 transcriptional activation domain (GAL4-AD) can activate the transcription of the reporter, causing growth on medium lacking HIS (Fig. 4.6).

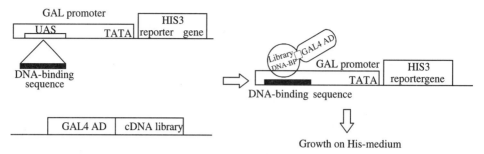

Figure 4.6. General scheme of the one-hybrid assay. If the cDNA library expressed protein specifically binds the DNA-binding sequence, the transformed yeast can grow on a medium lacking histidine.

In the basic version of the two-hybrid system, the cDNA library is cloned into a plasmid as a fusion to an activation protein. The *probe* or *bait protein* is cloned in an expression plasmid as a fusion to the heterologous DNA-binding protein LexA (Fig. 4.7). The major requirements for the bait protein are that it not be actively excluded from yeast nucleus and it not possess an intrinsic ability to strongly activate transcription. The plasmids expressing the LexA-fused bait protein and the activation domain-fused cDNA library are used to transform yeast possessing a dual reporter system responsive to transcriptional activation through the LexA operator.

In one of two reporter systems, binding sites for LexA protein are located upstream of the LEU2 gene, in the other reporter system, binding sites for LexA are located upstream of the LacZ gene. These two reporters allow selection for transcriptional activation by permitting selection for viability when cells are plated on medium lacking Leu, and discrimination based on color when the yeast is grown on medium containing X-gal. The basis of the assay is that transcription of the reporter gene will occur only if the bait protein and the library-encoded protein interact together, because only in this case are the AD and the BD domains of the transcriptional activator fused.

After yeast transformation, if any protein of the cDNA library does not interact specifically with the bait protein, the two reporter genes are not activated (Fig. 4.7a). In a positive interaction, the library-encoded protein interacts with the bait protein, resulting in activation of the two reporter genes, thus causing growth on medium lacking Leu and blue color on medium containing X-gal (Fig. 4.7b). Many variations of these assays have been described, altering the choice of BD and AD sequences, copy number of the plasmids encoding these sequences, strength of promoters, nature of selectable markers, and the nature of reporter genes.

It should be mentioned that the standard two-hybrid assay cannot take into account the specificity of all protein–protein interactions and will always give rise to a certain proportion of false positive and false negative results. Since its first description in the late 1980s, several versions of the two-hybrid system have been developed to overcome these limitations. Two strategies, the matrix approach and the library screening approach, are now being tested to find the most efficient way to explore proteomes for interactions (Finley and Brent 1994; Fromont-Racine, Rain et al. 1997; Ito, Tashiro et al. 2000; Uetz, Giot et al. 2000).

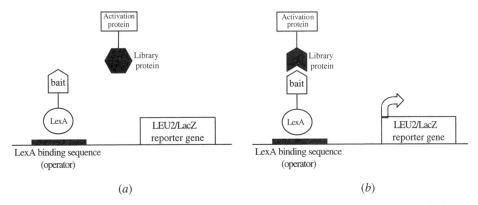

Figure 4.7. General scheme of the two-hybrid assay: (*a*) no binding; (*b*) binding. Binding of the cDNA library-expressed protein to the bait probe protein is tested through the expression of the reporter genes (LEU or LacZ).

Alternative approaches have been developed in organisms other than yeast. On the basis of throughput and cost of genome-wide studies, *E. coli* appears to be a more suitable host than *S. cerevisiae* because the generation time is much lower and molecular biology techniques are more adapted to bacteria than to yeast. Also, the high degree of competence of transformation of *E. coli* allows for the full coverage of highly complex genomic or cDNA libraries. Several bacterial two-hybrid systems have been described to date, and can be split into two categories: those based on transcriptional activation/repression of reporter genes and those based on reconstitution of an enzyme (Hu, O'Shea et al. 1990; Dmitrova, Younes-Cauet et al. 1998). In any case, the results obtained with these approaches should be confirmed by independent methods, such as co-immunoprecipitation or column affinity chromatography.

4.3.4 Proteome Analysis Using Biochips

Chip-based methods represent a very promising emerging technology for protein analysis (Fung, Thulasiraman et al. 2001; Lee 2001). In array-based methods small spots of proteins are immobilized on the surface of the chip, made up of silicon-based substrates of polyvinylidene difluoride (PVDF) membranes, such that they retain their native structure and functional activity. In this way it is possible to print on the array the entire complement of proteins produced by a cDNA library. These arrays can then be used to screen complex protein mixtures for particular binding affinities or other interactions. Several current arrays techniques are based on antibodies to identify antibody–antigen interactions by enzyme-linked immunosorbent assay (ELISA). A versatile strategy is offered by the ProteinChip system (Davies 2000), which uses a variety of affinity surfaces to capture the target proteins and then uses a technology based on surface-enhanced laser desorption ionization technology (SELDI) coupled to MS-TOF for molecular-weight determination. A ProteinChip array has been used for epitope mapping after protein binding to antibodies covalently immobilized on the array surface. The antigen proteins captured are digested enzymatically by on-chip incubation with endoproteases.

Following a wash to remove unbound fragments, the retained fragments defining the epitopes of bound proteins can be revealed by SELDI-TOF analysis.

A microarray containing the almost complete protein inventory for an organism has been produced for yeast where about 80% of the total known proteins are represented (Zhu, Bilgin et al. 2001). A high-quality yeast ORF library has been used to produce, under control of the inducible GAL1 promoter, yeast proteins fused at their NH_2 termini to glutathione-S-transferase-polyhistidine (GST-HisX6). The resulting protein collection has been printed onto nickel-coated slides to which the fusion proteins attach through their HisX6 tag. Anti-GST antibodies can be used to probe the level of protein attachment quantitatively. The yeast proteome chip has been used to test interactions with specific proteins or lipids previously biotinylated to allow interaction detection using fluorescent-labeled strepdavidin.

To make large-scale protein characterization really effective, the functional role of each protein should be determined carefully to derive protein therapeutics or protein-based drugs. This implies the availability of high-throughput strategies for deriving protein three-dimensional structural information as well as their dynamic flexibility (four-dimensional structure), which are crucial for drug discovery and optimization. New computational approaches need to be developed to provide a fundamental complement to any proteome initiative.

CHAPTER 5

BIOLOGICAL DATABASES IN THE GENOMIC ERA

5.1 INTRODUCTION

Starting in the 1980s, we have witnessed the tumultuous growth and evolution of biological databases. The evolution of sequencing technologies toward ever more efficient and sophisticated methodologies has resulted in the production of an ever-greater quantity of data—hence the need of computerized archives to store data in order to allow both management of such a huge quantity of information and interfacing to data analysis tools.

Biological databases is one of the most relevant topics in bioinformatics. The prerequisites to assure well-grounded computational studies based on the biological data available through databases are (1) good quality of annotated information, (2) up-to-date databases with respect to the daily huge production of data, (3) highly structured databases, (4) interoperability among databases, and (5) interface to good retrieval systems. In this chapter the most relevant biological databases, mainly those related to genomic and proteomic analysis, are described, in addition to some general aspects linked to the prerequisites noted above. A large number of the databases described and cited in this chapter are generally available through Web-based resources whose addresses are reported in Table 5.1. A wide overview of bioinformatics resources can be obtained at the Bioinformatics Resource Web site (TBR; see Table 5.1).

5.2 PRIMARY AND SPECIALIZED DATABASES

The core of biological databases are nucleic acid sequence databases, usually *primary databases* (PDs), as opposed to all the other databases, which are called *specialised databases* and can be further classified according to the type of data they store. A summary description of all available biological databases can be found in

159

TABLE 5.1. Major Bioinformatic Centers (a) and Internet Resources (b) Worldwide

(a) Bioinformatic Center (Acronym)	URL
Celera Genomics	www.celera.com
Center for Information Technology	www.ddbj.nig.ac.jp
EBI (European Bioinformatics Institute)	www.ebi.ac.uk
Jackson Laboratory	www.informatics.jax.org
National Human Genome Research Institute	www.nhgri.nih.gov
NCBI (National Centre of Biotechnology and Information)	www.ncbi.nlm.nih.gov
Sanger Centre	www.sanger.ac.uk
SIB (Swiss Institute of Bioinformatics)	www.expasy.ch
TIGR (The Institute for Genomic Research)	www.tigr.org/tdb

(b) Internet Resource (Acronym)	URL
ACEdb	www.sanger.ac.uk/Projects/C_elegans/
ACNUC	pbil.univ-lyon1.fr/software/acnuc.html
BankIt and Sequin	www.ncbi.nlm.nih.gov/BankIt/
Bioknowledge Library	www.incyte.com/sequence/proteome/ index.shtml
CluSTr	www.ebi.ac.uk/clustr
COG	ncbi.nlm.nih.gov:80/COG
CORBA (common object request broker architecture)	corba.ebi.ac.uk/intro.html
DBcat	www.infobiogen.fr/services/dbcat/
DbEST	www.ncbi.nlm.nih.gov/dbEST
DbSNP	www.ncbi.nlm.nih.gov/SNP
DBStats	www3.ebi.ac.uk/Services/DBStats
EBI_genomes	www.ebi.ac.uk/genomes
EcoCyc ontology	ecocyc.pangeasystems.com/ecocyc/ecocyc.html
EMBL Feature Table	www.ebi.ac.uk/embl/Documentation/ FT_definitions/feature_table.html
EMBL user manual	www.ebi.ac.uk/embl/Documentation/ User_manual/usrman.html
Ensembl	www.ensembl.org
Entrez	www.ncbi.nlm.nih.gov/Entrez
EuGENES	iubio.bio.indiana.edu:8089/
EXPASY	www.expasy.ch
FingerPRINTScan package	www.bioinf.man.ac.uk/fingerPRINTScan/
FlyBASE	www.flybase.org
FMGP (Fungal Mitochondrial Genome Project)	megasun.bch.umontreal.ca/People/lang/ FMGP/FMGP.html
GenProtEC	genprotec.mbl.edu
GMOD (Generic Model Organism Database)	www.gmod.org
GO	www.geneontology.org
GOBASE	megasun.bch.umontreal.ca/gobase
GOBrowser	www.informatics.jax.org/go/
GXD (Mouse Resource)	www.informatics.jax.org/mgihome/GXD/ gxdgen.shtml
HGBASE	hgbase.cgr.ki.se
HSSP	www.sander.ebi.ac.uk/hssp
HvrBase	db.eva.mpg.de/Hvrbase

TABLE 5.1. *Continued*

(b) Internet Resource (Acronym)	URL
InterPro	ebi.ac.uk/interpro
IPI (International Protein Index)	www.ebi.ac.uk/ibi/IPIhelp.html
KEGG (Kyoto encyclopedia of genes and genomes)	star.scl.genome.adjp/kegg
LION Bioscience	www.lionbio.co.uk/
LocusLink	ncbi.nlm.nih.gov/LocusLink/
MGD (Mouse Genome Database)	www.informatics.jax.org
MitBASE	www3.ebi.ac.uk/Research/Mitbase/mitbase.pl
MITOCHONDRIOME	bighost.area.ba.cnr.it/mitochondriome
MITODAT	www.lecb.ncifcrf.gov/mitoDat
MITOMAP (Human Mitochondrial Genome Database)	infinity.gen.emory.edu/mitomap.html
MITOP	www.mips.biochem.mpg.de/proj/medgen/mitop
MITSNP	www-genome.wi.mit.edu/SNP/human
MSD (Macromolecular Structure Database)	msd.ebi.ac.uk
NAR Database Issue	www3.oup.co.uk/nar/database
NCBI Human Genome Map Viewer	www.ncbi.nlm.nih.gov/cgi-bin/Entrez/ map_search
NCBI_Genomes	www.ncbi.nlm.nih.gov:80/entrez/ query.fcgi?db=Genome
OGMP (Organelle Genome Megasequencing Project)	megasun.bch.umontreal.ca/ogmpproj.html
OMIMALLELE	www3.ncbi.nlm.nih.gov/omim
P53 database	www.iarc.fr/p53/homepage.htm
PDB	www.rcsb.org/pdb
Pfam	sanger.ac.uk/Software/Pfam/
PID (Protist Image Database)	megasun.BCH.Umontreal.CA/protist/
PIR-PSD	pir.georgetown.edu
PLMitRNA	bio-www.ba.cnr.it:8000/BioWWW/#PLMItrna
PRINTS	www.bionf.man.ac.uk/dbbrowser/PRINTS/
ProDom	prodes.toulouse.inra.fr/prodom/doc/ prodom.html
ProSite	www.expasy.ch/prosite
Proteome Analysis Database	www.ebi.ac.uk/proteome/
RefSeq	ncbi.nlm.nih.gov/LocusLink/refseq.html
RHdb	www.ebi.ac.uk/RHdb
SGD (Saccharomyces Genome Database)	genome-www.stanford.edu/Saccharomyces
SMART	smart.embl-heidelberg.de
SRS (Sequence Retrieval System)	srs.ebi.ac.uk
SWISSPROT+TREMBL	www.expasy.ch/sprot
TBR (The Bioinformatics Resource)	www.hgmp.mrc.ac.uk/CCP11/
tRNA database	www.uni-bayreuth.de/departments/biochemie/ sprinzl/trna/index.html
UCSC Genome Browser	genome.ucsc.edu
UniGene	www.ncbi.nlm.nih.gov/UniGene/
UniSTS database	www.ncbi.nlm.nih.gov/genome/sts/index.html
UTRdb	bighost.area.ba.cnr.it/BIG/UTRHome
VAST (Vector Alignment Search Tool)	www.ncbi.nlm.nih.gov/Structure/VAST/ vasthelp.html
WEBIN	www.ebi.ac.uk/embl/Submission/webin.html
WormBase	www.wormbase.org

the first yearly issue of *Nucleic Acids Research* (NAR database issue; see Table 5.1) and through the Dbcat Database, the database of databases (Discala, Benigni et al. 2000).

The primary databases are the EMBL data library (Stoesser, Baker et al. 2001), managed at the European Bioinformatics Institute (U.K.); GenBank (Wheeler, Church et al. 2001), managed at the National Center for Biotechnology Information (NCBI, United States); and DDBJ (Tateno, Miyazaki et al. 2000), managed at the Center for Information Biology in Mishima (Japan). Primary databases store nucleic acid sequences produced worldwide, collected both from the literature and, prevalently, through the authors' direct submission. Moreover, all the data derived from genomic large-scale sequencing projects are stored in the primary databases through massive submission on behalf of the sequencing center. The three databases EMBL, GenBank, and DDBJ are managed by an International Collaboration aimed at exchanging data submitted by the scientific community to each center on a daily basis.

Other databases collecting biological data related to a well-defined class of functional or taxonomic groups can be classified as specialized databases. In these databases, which are frequently cared for by researchers expert in the field and hence competent in data validation, the data are much more accurate and validated than in primary databases. Among the specialized databases, we can include those produced by a *data warehousing* procedure, which implies selection from primary databases of a subset of rough data, homogeneous from the functional and/or taxonomic point of view, thoroughly revised and analyzed by experts to produce *curate databases* (Schonbach, Kowalski-Saunders et al. 2000; Strivens 2000). However, the classification of biological databases into primary and specialized databases is becoming somewhat obsolete with the availability of entire nuclear and organelle genomes. Approaches to the theoretical study of biological data are changing completely, and hence the organization of data in databases is rapidly evolving toward integrated databases worldwide distributed and interfaced to advanced analysis tools. These integrated resources should support a transition from the study of single genes to a study of high-throughput genomics, transcriptomics, proteomics, and any other "-omic" that could help in an understanding of organism development and evolution.

5.3   DATABASE STRUCTURES

The *database structure* or the *database scheme* is a description of how stored information is organized. The database scheme is called *ontology*, a written formal description of entities (or objects) and relationships based on an abstract conceptualization of the world (Karp 2000).

Before starting a description of database structure, some general concepts will be useful to the reader. A database is designed with the aim of collecting information related to a well-defined *class* of data; in the biological context classes of data are the worldwide available nucleic acid sequences, protein sequences, transcription factors, enzymes, and so on. The component elements of these classes are called *entities*, to which, in turn, specific *properties* can be attributed. The full set of entities and attributes representing the relevant information for a specific class of data is referred to as a database *entry*. A database entry is usually identified by the database name, the entry name, and the *accession number*, an alphanumeric code assigned to the entries progressively as they are stored in the database. Historically, biological databases have been managed in a very simple format, the *flat-file format*. This format is such that users can read the data understanding the database content, and computer programs can extract rigorously the information reported there. The classical example of the flat-file format is the one defined in the EMBL data library (Fig. 5.1). Here each entry in the database is available in a sequential file, and each attribute is reported in a line identified with a two-letter code (i.e., ID is for the entry identification line reporting entry name and sequence length, KW for the keywords lines, and so on). A complete description of the EMBL flat-file format is available in the EMBL User Manual at EBI (see the URL in Table 5.1).

The growth of biological databases both in number and in content, besides the advent of the genomic era, which is requiring more sophisticated tools for computational analysis of biodata, is strongly driving the adoption of more sophisticated database structuring technologies, such as those implementing hierarchical, relational, or object-oriented architecturs. These architectures are managed by database management systems (DBMSs). The *relational model* describes all the entities in the database through tables, as defined relations. Each row in a table is a record and each field in the row is an attribute. When defining the database structure, the type of values each attribute can assume must be declared. The relationships through the tables are based on Boolean algebra. The relational DBMS architecture has been the first to be adopted in the bioworld after the flat-file format but is not the one best fitting description of the biological world; the hierarchical and object-oriented architectures are better adapted to this purpose. In the *object-oriented architecture*, classes and subclasses of objects are defined to give a complete description of the biological data described in a database. Although this architecture is better suited for biological databases, the query language implemented through object-oriented DBMSs is less structured than in the relational or hierarchical databases. However, despite the tendency to adopt more and more complex architectures, the majority of distributed databases are still released in flat-file format. A justification of this is the fact that the flat file is the easiest to manage by the analysis tools and packages, such as the GCG package (Womble 2000), the EMBOSS package (Rice, Longden et al. 2000), or the SRS retrieval system (Etzold, Ulyanov et al. 1996), widely used by biologists.

5.4 LINKED DATABASES AND DATABASE INTEROPERABILITY

As stressed in Sec. 5.1, biologists pose much more complex queries to analyze a biological problem through biodatabases; this implies the need to develop programs allowing access to integrated resources distributed worldwide. This need existed as early as the beginning of the 1990s, when genomic data were a chimera and the quantity of data available was much smaller, although the available biological specialized databases were growing. At that time, the need to link databases was already strongly felt; a cross-referencing mechanism was invented and introduced in European databases (EMBL, SWISS-PROT, and related databases). The implementation of such a mechanism is usually made through the definition in each entry of the *cross-referencing lines*, allowing one to refer to an entry in another

```
ID   CPO9673      standard; RNA; ROD; 2180 BP.
XX
AC   AJ009673;
XX
SV   AJ009673.1
XX
DT   20-MAY-1999 (Rel. 59, Created)
DT   20-MAY-1999 (Rel. 59, Last updated, Version 1)
XX
DE   Cavia porcellus mRNA for p53 protein
XX
KW   p53 gene.
XX
OS   Cavia porcellus (domestic guinea pig)
OC   Eukaryota; Metazoa; Chordata; Craniata; Vertebrata; Euteleostomi; Mammalia;
OC   Eutheria; Rodentia; Hystricognathi; Caviidae; Cavia.
XX
RN   [1]
RP   1-2180
RA   D'erchia A.M.;
RT   ;
RL   Submitted (16-JUL-1998) to the EMBL/GenBank/DDBJ databases.
RL   D'erchia A.M., Department of Biochemistry and Molecular Biology, University
RL   of Bari, Via Orabona, 4 Bari, 70126 ITALY.
XX
RN   [2]
RA   D'erchia A.M., Pesole G., Tullo A., Saccone C., Sbisa E.;
RT   "Guinea Pig p53 mRNA: Identification of New Elements in Coding and
RT   Untranslated Regions and Their Functional and Evolutionary Implications";
RL   Genomics 58:50-64(1999).
XX
DR   SWISS-PROT; Q9WUR6; P53_CAVPO.
XX
FH   Key             Location/Qualifiers
FH
FT   source          1..2180
FT                   /db_xref="taxon:10141"
FT                   /organism="Cavia porcellus"
FT                   /tissue_type="spleen"
FT   CDS             234..1409
FT                   /db_xref="SWISS-PROT:Q9WUR6"
FT                   /gene="p53"
FT                   /product="p53 protein"
FT                   /protein_id="CAB43196.1"
FT                   /translation="MEEPHSDLSIEPPLSQETFSDLWKLLPENNVLSDSLSPPMDHLLL
FT                   SPEEVASWLGENPDGDGHVSAAPVSEAPTSAGPALVAPAPATSWPLSSSVPSHKPYRGS
FT                   YGFEVHFLKSGTAKSVTCTYSPGLNKLFCQLAKTCPVQVWVESPPPPGTRVRALAIYKK
FT                   SQHMTEVVRRCPHHERCSDSDGLAPPQHLIRVEGNLHAEYVDDRTTFRHSVVVPYEPPE
FT                   VGSDCTTIHYNYMCNSSCMGGMNRRPILTIITLEDSSGKLLGRDSFEVRVCACPGRDRR
FT                   TEEENFRKKGGLCPEPTPGNIKRALPTSTSSSPQPKKKPLDAEYFTLKIRGRKNFEILR
FT                   EINEALEFKDAQTEKEPGESRPHSSYPKSKKGQSTSCHKKLMFKREGLDSD"
FT   polyA_signal    2166..2171
FT                   /gene="p53"
XX
SQ   Sequence 2180 BP; 492 A; 654 C; 543 G; 491 T; 0 other;
     gtcatggcga ctgtccagct ttgtgccagg agcctcgcga gggctgctgg gatagggatt        60
     tccccctccc acgtgctcag actggcgcta aaagttttga gctgctctaa agtccagggc       120
     caccatcctg agcacaggta gctgctgggc tccggggggta cctggcgtcc gggcttggag       180
     tgtccctccc aagacacgct cccttgagaa ccagcagctg tcctgctgcc gctatggagg       240
     agccacactc ggacctcagc atcgagcctc cactgagtca agaaacattt tcagacttat       300
     ggaaactact tcctgaaaac aacgttctgt cggactcact gtcccctccc atggaccatc       360
     ttctgctgtc cccagaagag gttgcaagct ggctgggaga aaacccggat ggagatggcc       420
     atgtgtcagc agctcctgta tcagaggccc caacatcagc aggccctgcc ctggtggccc       480
     ctgcaccagc tacttcctgg cctctgtcat cctcagtccc ttcccataag ccataccgtg       540
     gcagctatgg tttcgaggtg cacttcctta agctgggac agccaagtct gtcacatgca       600
     cgtactcccc tggcctcaac aagctcttct gccagctggc aaagacctgt cctgtgcaag       660
     tatgggtgga atcaccaccc ccacctggca cccgagtccg tgccctggcc atctacaaga       720
     agtcacagca catgacggag gtggtgagac gctgcccca tcacgagcgc tgttccgata       780
     gtgatggcct ggcccctcct cagcatctca tccgggtgga aggaaatctg cacgcagaat       840
     acgtggatga cagaaccact ttccgccaca gcgtggtggt gccctatgaa ccgcctgagg       900
```

Figure 5.1. Sample Embl entry.

```
tcggctccga ctgcaccacc atccactaca actacatgtg caacagctcc tgcatgggag     960
gcatgaaccg gaggcccatc ctcaccatca tcacgctgga agactccagt gggaaactgt    1020
tgggtcgaga cagctttgag gtacgtgttt gtgcctgtcc tggaagagat cggcgcacag    1080
aggaggaaaa tttccgcaag aaaggaggcc tgtgccctga gccaacacca ggaaacatta    1140
agcgagccct gcccaccagc accagctcct ctccacagcc aaagaagaaa ccactggatg    1200
cagaatactt cacccttaag atccgtgggc gtaaaaactt cgagatactc cgagagataa    1260
acgaggcctt ggaattcaag gatgcccaaa ctgagaagga gccgggggag agcaggcctc    1320
actcgagcta cccgaaatcc aagaaggggc agtctacctc ctgtcacaaa aaactgatgt    1380
tcaagagaga aggactcgat tcagactgac gtcctcggcc tcctgtctgc catcctcagt    1440
gtaccctacc ccagcccgtc ccctccccgg gatttggaac cccgccgttg aatctttttcc    1500
cgctgtaggt gtgcctcaga aatgcccaag agttctgcca tttgcctttc ctggaccccа    1560
ctcaaggaat tggcctgtac tggtgtttgg ggaagggtgg tgggagccag gatacaccac    1620
ctgagttttt aaggtttttt accatgagca gagttaggac agagaagaga atgttcttgc    1680
atataaagat cagagtttac aatcagccat atgcccggct ggaagcccag tcctcagact    1740
gagctcactg ggaaagccag ggcaggtcat ttaatgtctc ctaacttgaa tattcacatc    1800
tatgaagtgc taatcatcac agctacctcc caggatttgt tgggggggct gatgaaagaa    1860
cctgtgtctg cccgtgaacc ccctcttgtt acttgggggc tggtgcgtat ttccagggtg    1920
aggcttcagc aggtgcctcc acctcttgct gacccttggc cctgaaaggc aacctcaccg    1980
ggtctcacac cctgcgggat cccatctcca gctcccagag ggtctgtctc ccccgagacc    2040
cgttttactg tcccgcttcc cctactcagg gtcagtttct tttgcactct tgaaggcaca    2100
tctgtatttg tcacccccac tcttcccttt ttgtatcgcc tttttatatc agtttcttat    2160
tttacaataa aactttacta            2180
//
```

Figure 5.1. *Continued*

database whose information is strictly correlated. The syntax used for the cross-referencing is usually "database_name:entry_name" and/or "database_name:accession_number." As an alternative, the databases implemented at the NCBI through Entrez (see Sec. 5.6) adopt the *neighbors mechanism*, which allows one to find database entries (nucleotide or protein sequences, scientific articles collected in the PubMed Database, or protein structures collected in the Structure Database) related to an entry already selected. Protein and nucleotide sequence neighbors are determined by performing BLAST similarity search (see Sec. 6.4.2), whereas PubMed neighbors are calculated by comparing the text and the MESH terms by using a very efficient algorithm. MESH terms constitute the NCBI-controlled vocabulary used for indexing texts available in the databases. Also, structure neighbors are calculated by comparing, through a suitable algorithm called VAST (Vector Alignment Search Tool; see the URL in Table 5.1), the three-dimensional architectures of other structures stored previously.

Cross-referencing and neighbors offer good results but are in some way restrictive in relation to the most sophisticated query that the "-omic" user poses to databases. The problem is that the remote resources are heterogeneous both in their structure (relational, object-oriented, flat-file) and in the way they are implemented (different DBMSs operating systems, programming languages). Hence, more advanced technologies need to be implemented, thus allowing interoperability among remote resources. Worth mentioning are CORBA (see the URL in Table 5.1), the Common Object Request Broker Architecture (Stevens and Miller 2000), and the Extensible Markup Language (XML). CORBA "takes an object-oriented view of the world," thus allowing the integration of data not managed as objects in the object-oriented paradigm. Through CORBA a user accessing distributed objects is not requested to know the object location or the language used for their implementation, thus guaranteeing interoperability between distributed resources. IDL, the Interface Definition Language, is the CORBA specification language, as HTML is for Web documents. CORBA consists of a client–server application where the server "serves" information to a client program. CORBA has been thoroughly developed at the EBI, where client and servers are available; in particular, CORBA clients for the retrieval of dbEST (see Sec. 5.12), Pfam (see Sec. 5.9), and RHdb databases (see Sec. 5.10) have been developed using JAVA applets, allowing more elaborate queries than simple Web searches.

XML is a standard protocol that describes the structure of data available in a database. The XML syntax is very similar to HTML (HyperText Markup Language), used to develop Web pages. Each database entity is defined through tags into a XML file. The description of both the tag function and the relationships between tags is defined through a DTD (document type definition) file. Recently, LabBook, Inc. and the European Bioinformatics Institute have announced that EBI's database output will be made available to researchers worldwide in a particular XML implementation, the LabBook's Bioinformatic Sequence Markup Language (BSML) format. LabBook's BSML is an open, Extensible Markup Language (XML) format to communicate genomic information. For the life science researcher, BSML enables the visualization, creation, and exchange of documents containing complex genetic sequence information and annotations. DDBJ also releases its database in XML format.

5.5 DATABASE ANNOTATION

Another important problem relates to the quality of annotations in databases. *Annotations* are the information reported in each database entry to classify the sequence(s) or any other type of data constituting the main object of the database (i.e., mutation in the mutation database, pattern in a pattern database, as well as nucleotide sequence in the EMBL, GenBank, or DDBJ databases, the protein sequence in the SWISS-PROT database, or the protein multialigned sequences in the HSSP database). Biological databases are the source for many relevant studies aimed at understanding gene and protein functions and all their interactions within an organism and between organisms. Hence data must be annotated with great care, avoiding duplication of information (redundancies), misspelling in typing, reporting of false or uncertain results, and uncontrolled vocabulary in gene and protein naming. At the beginning of the biodatabase era, these aspects had not been taken into account carefully, thus producing, especially in the primary databases, inconsistency in data, which is still present in the data released. Hence a careful bioinformatics researcher should avoid using data in primary databases whose entry date is before the 1990s without a careful revision. At present, data are also much more accurate because author submission tools have been developed guiding the data submitter or annotator through controlled vocabulary lists. The EBI has developed the WEBIN package (see the URL in Table 5.1), an Internet-based tool for submission of nucleotide and protein sequences, allowing to submit either one, several, or a large number of sequences guided by a controlled vocabulary, reducing typing mistakes and the generation of long lists of synonyms, thus improving the standardization of the information stored. Recently, the WEBIN–ALIGN submission tool has been developed to submit multialigned nucleotide or protein sequences. Similarly, the NCBI has developed BankIt, a Web-based submission tool and Sequin, stand-alone software that can be downloaded on Macintosh, PC/Windows, or Unix computers (see the URL in Table 5.1). The latter should be used for long or complex submissions and provides graphical viewing and editing options.

Primary databases contain a great quantity of redundant sequences. This is due to imprecise control mechanisms on behalf of the database annotator each time a new entry has been stored, thus causing the presence of the same sequence from the same authors and the same source twice, or more than twice, in the database (this problem has been bypassed in recent years thanks to more careful updating procedures); and biological redundancy due to the sequencing of the same DNA fragment from different sources. The EST data exacerbates redundancy due to the fact that the same EST fragment can be sequenced several times in relation to its expression level. Hence, before applying statistical analyses on nucleotide sequence databases, use of the CleanUP software (Grillo, Attimonelli et al. 1996) to produce nonredundant databases is surely of great help (see also Sec. 6.3.5).

The other big problem is *gene naming*. The commonest interrogation criteria for biological databases, both primary and specialized, are based on gene and protein names, but in the first two decades of the biodatabase era these have been annotated incorrectly, thus causing inconsistencies in data selected. As to protein names, the majority of problems have been due to the absence of controlled vocabularies guiding the data submitter in the introduction of new terms, thus causing naming of the same protein with different synonyms, sometimes differing simply in the presence or absence of a blank character or hyphen, sometimes due to mistyping. For gene names the problem is much more complex: some genes have several names (organism specific), and unrelated genes often share common names. Any attempt to obtain agreement among researchers to use the same name for the same function has failed (Pearson 2001); hence the solution is to generate controlled vocabularies where different names for the same function or equal names for different functions are in some way highlighted. KEYnet represented a first effort in this direction by providing a hierarchical classification of protein and gene names derived from the keywords (KW) line of the EMBL data library and properly linking possible synonyms (Catalano, Licciulli et al. 2000).

Recently, a new project has been launched, the *Gene Ontology* (GO) project (Ashburner, Ball et al. 2000; see the URL in Table 5.1), whose objective is to provide controlled vocabularies for description of the *molecular function*, *biological process*, and *cellular component of gene products*. These terms are to be used as attributes of gene products by collaborating databases, to facilitate uniform queries across them. The controlled vocabularies of terms are structured to allow both attribution and querying to be at different levels of granularity. In December 2001, GO collaborating databases were the Saccharomyces Genome Database (SGD), FlyBase, Mouse Genome Informatics (MGI), Gene Expression Database, The Arabidopsis Resource, Pombase, SWISS-PROT, TrEMBL, PIR, GenBank, EMBL, DDBJ, MEDLINE, PUBMED, and TIGR. Gene Ontology is not an integrated database but an independent database supporting interoperability among databases. Figure 5.2 shows the MGI entry for the p53 tumor suppressor gene (SWISS-PROT ID: P02340), also reporting the GO classification.

Mouse Genome Informatics
The Jackson Laboratory

Main Menu | MGI Home | User Support | Help Documents | Submissions | Chr Comm
Genes | Molecular | Homology | Mapping | Expression | Strain/Polymorphism | Refs | AccID

Query Forms ◄►

Your Input Welcome

Genes, Markers and Phenotypes

Query Results -- Details

Type: Gene
Symbol: Trp53
Name: transformation related protein 53
Chromosome: 11
Cytogenetic Offset: B2-C
cM Position: 39.0
MGI Accession ID: MGI:98834

Synonyms: p53

Additional Information:

- Mammalian Homology
- Marker Mapping Data (36)
- Phenotype (MLC)
- Phenotypic Alleles (7)
- RFLP/PCR Polymorphism (10)
- Gene Expression Data (115 results in 6 assays)
- GXD Index Data (27)
- Molecular Probes and Segments (53)
- References(202)

Chromosome map (cM):
```
0  ─── Lif
10 ─┬─ Egfr
   ├─ Hba
20 ─┤
   ├─ Gabra6
30 ─┤
   ├─ Csf2
40 ─┤
   ├─ Myhs
50 ─┤─ Trp53
   ├─ Nf1
60 ─┤
   ├─ Hoxb
70 ─┤─ Krt1
   ├─ Timp2
80 ─┴─ Tk1
```

Gene Classifications: (*You can* browse the Gene Ontology (GO) Classifications).

Category	Classification Term	Evidence	Reference
Biological Process	apoptosis	electronic annotation	J:60000
Biological Process	transcription regulation	electronic annotation	J:60000
Biological Process	cell growth and maintenance	electronic annotation	J:60000
Cellular Component	nucleus	electronic annotation	J:60000
Molecular Function	DNA binding	electronic annotation	J:60000

Figure 5.2. Mouse Genome Informatics (MGI) entry for p53 tumor suppressor gene, also reporting the relevant Gene Ontology classification.

5.6   RETRIEVAL SYSTEMS

A major prerequisite of biological databases is the availability of flexible, diversified, and user-friendly retrieval systems allowing navigation among databases, the retrieval of complete data sets with respect to the query issued, the minimization of spurious data selection, and suitable data extraction tools for use in further analyses. The retrieval systems most popular and used worldwide are SRS (Etzold and Argos 1993; Etzold, Ulyanov et al. 1996) and Entrez (see the URL in Table 5.1); noteworthy are ACNUC (Gouy, Gautier et al. 1985) and ACEdb (Kelley 2000) (see the URLs in Table 5.1). Moreover, any integrated resource where one or more databases are available has developed its own retrieval system, allowing either simple string searches or more sophisticated queries based on the Oracle SQL query language or the SRS object modules.

5.6.1   SRS

SRS, the sequence retrieval system designed by Thure Etzold, is a retrieval system allowing the user to apply, in the same session, one query to more databases, by restricting the query to entities common to the databases selected. SRS server tools can be downloaded from the LION Bioscience server (see the URL in Table 5.1) and implemented on mainframes equipped with central processors and storage capacity sufficient to manage big databases. The local SRS manager can install databases already structured at other SRS servers by downloading the scripts and indices or can structure new databases in SRS autonomously using modules available from the SRS server. Hence not all SRS servers have the same database set implemented. The format of the database to be stored in SRS is usually the flat file, but it can also be made of simple tables or much more elaborate Oracle relational tables. Such flexibility allows worldwide availability of SRS servers, whose list can be browsed clicking on "List of Public SRS Servers" on the SRS home page of the EBI (see the URL in Table 5.1). Recently, Celera Genomics has implemented SRS on its own Web server, where the genomic data can be queried. Databases in SRS are listed in the Top Page, in subsets defining different classes of databases. Each database is more or less described by its function, structure, and content. Once the databases to be queried are selected, it is possible to choose either the Simple or Extended Query form, which allows one to perform queries through different criteria, combining them with Boolean operators (AND, OR, BUT NOT). The results obtained can be managed further using the option View, Save, Link, or Launch. The View and Save options allow display and storage of the data retrieved in predefined formats or in formats defined by the user through the View Page. The Link option allows linking query results to data available in other databases to be selected; a prerequisite for the success of this option is that the databases involved be cross-referenced. The Launch option allows the use of computer programs such as FASTA, BLAST, or CLUSTAL with the set of data selected.

5.6.2   Entrez

Entrez (see the URL in Table 5.1) is the NCBI database retrieval system. Different from SRS, it is managed completely at the NCBI. Databases available in December 2001 through Entrez are the nucleotide sequence database (GenBank), protein sequence databases, the biomedical literature database (PubMed), genomic resources, the taxonomy database, the three-dimensional macromolecular structure database, OMIM (Online Mendelian Inheritance in Man), and PopSet (population study data sets). Retrieval can be done through a string search system where the words can be combined using Boolean logic operators (AND, OR, BUT NOT; e.g., "p53 AND apoptosis BUT NOT cancer"). The Limit option allows one to limit the retrieval to selected fields, such as the journal name and author name. The data selected can be saved using the Clipboard function for up to a limited time, also after network disconnection and reconnection. The History option offers a list of the set of data retrieved, thus allowing further refinement of the search. Each entry selected is displayed briefly, and by clicking on the related entities at the right of the list, it is possible to obtain further related data. This option is based on the neighbors definition described in Sec. 5.4.

5.6.3 Other Retrieval Systems

ACNUC is a relational-like database structure and retrieval system for querying nucleic acid or protein sequence databases. ACNUC is a package designed specifically for Unix-compatible mainframe computers, although a less flexible Web-based version has been developed (see the URL in Table 5.1). Although ACNUC can retrieve data from the EMBL, GenBank, and SWISS-PROT databases only and does not allow navigation among databases, it may still be of great help for the very flexible Select, Modify, and Extract commands, which, for example, allow a user to extract sequence regions defined by specific features or around such regions (e.g., a given sequence fragment upstream of the CDS feature).

ACEdb is a package originally developed (Stein, Sternberg et al. 2001) to manage *Caenorhabditis elegans* genome data related to the mapping and sequencing project. It has then been adopted for several other genomic projects. It is a stand-alone package that can be downloaded on both Unix- and Windows NT-equipped computers to create, query, and update any genomic database and give access to it via the Web (Durbin and Thierry-Mieg 1994; Kelley 2000).

5.7 NUCLEOTIDE DATABASES

Nucleotide databases are EMBL, GenBank, and DDBJ. Thanks to international collaboration, the content of these databases is the same, although the EMBL format is slightly different from the ones adopted by GenBank and DDBJ. The three databases group data in subsets defined on the basis of taxonomic divisions (bacteria, fungi, invertebrates, rodent, human, etc.) or based on sequence type, such as ESTs (expressed sequence tags), STSs (sequence tagged sites; see Sec. 6.7.4), HTGs (high-throughput genomic sequences), HTCs (high-throughput genomic cDNAs), PAT for patent sequences, and so on. The information reported in these databases is general: Beside sequence length, taxonomic information, and references, a description of the functional role of regions and sites located in the sequence is reported in the Feature Table. Each feature is identified by a Feature Key (i.e., CDS for fragment coding for a protein), a Feature Location, and a set of Feature Qualifiers describing the

Feature Key. The complete format of the EMBL Feature Table is available on the Web (see the URL in Table 5.1) and can be used as a guide when designing a new database. At present (release 7.1, June 2002) the EMBL data library reports 20,020,556,107 bases in 17,226,422 entries. A more detailed statistic of the nucleotide sequence databases is updated daily (see the DBStats URL in Table 5.1). Here it is possible to see that the nucleotide sequence databases report very biased data as concerning organism distribution: 61% of entries are from humans, and 85% of entries are from only 10 different species. This is a severe limitation to evolutionary studies.

5.8 PROTEIN DATABASES

Protein databases report protein sequences, produced directly through protein sequencing or, for the most part, derived from nucleotide databases after translation of the coding for protein genes. The major protein databases are SWISS-PROT + TrEMBL (Bairoch and Apweiler 2000; see the URL in Table 5.1) and PIR-PSD (Barker, Garavelli et al. 2001). SWISS-PROT is a highly curate and manually maintained protein database where functional and structural descriptions of the protein are annotated, also reporting information on polymorphisms, mutations, posttranscriptional modifications, and so on. Naming of genes and proteins has always been performed carefully, and recently, SWISS-PROT began interfacing to Gene Ontology (Ashburner, Ball et al. 2000). Redundancy is reduced to a minimum. Integration with the most relevant specialized databases, developed at the Swiss Institute of Bioinformatics (SIB) or by other groups, is available. TREMBL reports all the translated EMBL coding sequences not yet annotated in SWISS-PROT. Indeed, TREMBL is split into SPTREMBL (translated sequences to be moved into SWISS-PROT) and REMTREMBL (translated sequences that will never be introduced into SWISS-PROT; these are immunoglobulins and T-cell receptors, synthetic sequences, patent application sequences, small fragments, and CDSs not coding for real proteins).

PIR-PSD (see the URL in Table 5.1) is the result of a collaboration between the Protein Information Resource at Georgetown University, the MIPS (Munich Information Centre for Protein Sequences, summary), and the JIPID in Japan. PIR-PSD is a nonredundant, expertly annotated, fully classified, and extensively cross-referenced protein sequence database.

Based on SWISS-PROT, TREMBL, and other information available at the genome or transcript level, the EBI has now released the International Protein Index (IPI; see the URL in Table 5.1), reporting, in release 2, 33,013 human proteins. The index can be browsed through SRS at the EBI.

5.9 OTHER PROTEIN DATABASES

Besides the protein databases described above, a great number of specialized protein databases have been developed by the same groups that maintain SWISS-PROT and PIR-PSD and other expert groups. A list of the specialized protein databases can be obtained from the compilation in the *Nucleic Acids Research* database issue (see the URL in Table 5.1). Among the noteworthy specialized protein databases is InterPro.

InterPro (Apweiler, Attwood et al. 2001; see the URL in Table 5.1) is an integrated resource that has been created by pooling together different protein functional site and domain databases with the aim of supporting biologists in genome studies in assigning functions to new genes. The databases integrated in InterPro are ProSite (Hofmann, Bucher et al. 1999), Pfam (Bateman, Birney et al. 2000), PRINTS (Attwood, Croning et al. 2000), ProDom (Corpet, Servant et al. 2000), and SMART (Schultz, Copley et al. 2000) (see the URLs in Table 5.1). The advantage of using InterPro is that a single query provides results based on all integrated databases in a more homogeneous and easily interpretable way. InterPro has also been interfaced to Gene Ontology (Ashburner, Ball et al. 2000); hence, searching InterPro for a given function or a given cellular localization can be made through the GO-controlled vocabulary. InterPro can be queried through a Text search, through SRS, or through a sequence-based search tool. An example of an InterPro entry is shown in Fig. 5.3. Below is a description of the InterPro integrated databases.

PROSITE is a database reporting protein motifs characterizing protein families and domains. Other motifs obtained through *in silico* analysis but of unknown structure or function are also annotated. Data annotated in PROSITE can help users to reliably identify to which known protein family (if any) a new sequence belongs or which known domain(s) it contains. The entry annotates a consensus pattern or a profile detected through the use of pattern-searching methods on sequence data available in SWISS-PROT and TREMBL. Numerical results from the *in silico* analysis and the cross-referencing to SWISS-PROT entries where the pattern has been detected (hits) are also reported. The hits are classified as True, Potential, or False on the basis of experimental data obtained from literature or through a peer review by experts in the field; cross-referencing to a documentation form, where the biological function of the patterns (if any) is described and documented with bibliographic references, is also available. ProSite can be used in two different ways: by applying queries to the database using the search system at the ExPASy site (see the URL in Table 5.1) or through SRS "all text" searching; or by searching for a PROSITE pattern in a new protein sequence by applying pattern searching methods (a list of programs that make use of ProSite is available at the ExPASy site).

Pfam is a collection of protein families and domains. Pfam contains multiple protein alignments and profiles of these families derived from the application of HMMER (the Hidden Markov Model software; Eddy 1998). Proteins annotated in the Pfam family are cross-referenced to the ProSite and SMART domains. Hence users can perform a database search of a new protein against Pfam families. The result of this application reports the positive scoring domains in Pfam (if any) and their features as derived from the InterPRO integrated databases. Alternatively, it is possible to search for a SWISS-PROT + TREMBL ID or accession number against Pfam in order to learn if a given protein is part of a previously characterized protein family and/or if it contains functionally known domains.

PRINTS is a protein fingerprint database; a *fingerprint* is a group of conserved motifs characterizing a protein family. The motifs are usually distantly located in the proteins but may be near in three-dimensional structure. Each entry in PRINTS reports the description of the family and of the structural and functional roles (if

InterPro Home
Text Search
Databases
Documentation
FTP Site
Sequence Search

Project Outline
Collaborators
Example Entry
Dataflow Scheme
Release Notes
User Manual
Deleted IPRs
References

InterPro Entry IPR001304

C-type lectin domain

Database	InterPro
Accession	IPR001304; lectin_c (matches 928 proteins)
Name	C-type lectin domain
Type	Domain ⓘ
Dates	08-OCT-1999 (created)
	28-JUN-2000 (last modified)
Signatures	PS00615; C_TYPE_LECTIN_1 (401 proteins)
	PS50041; C_TYPE_LECTIN_2 (672 proteins)
	PF00059; lectin_c (483 proteins)
Found in ⓘ	IPR001491; Thrombomodulin (13 proteins)
Children ⓘ [tree]	IPR002352; Eosinophil major basic protein (9 proteins)
	IPR002353; Type II antifreeze protein (44 proteins)

Abstract ⓘ

A number of different families of proteins share a conserved domain which was first characterized in some animal lectins and which seems to function as a calcium-dependent carbohydrate-recognition domain [1, 2, 3]. This domain is known as the C-type lectin domain (CTL) or as the carbohydrate-recognition domain (CRD). The categories of proteins in which the CTL domain has been found include,

1. type-II membrane proteins where the CTL domain is located at the C-terminal extremity of the proteins,

2. proteins, sometimes called 'collectins', that consist of an N-terminal collagenous domain followed by a CTL-domain [4],

3. selectins (or LEC-CAM), cell adhesion molecules implicated in the interaction of leukocytes with platelets or vascular endothelium [5, 6]. Structurally, selectins consist of a long extracellular domain, followed by a transmembrane region and a short cytoplasmic domain. The extracellular domain is itself composed of a CTL-domain, followed by an EGF-like domain and a variable number of SCR/Sushi repeats,

4. large proteoglycans that contain a CTL-domain followed by one copy of a SCR/Sushi repeat at the C-terminus. In addition they may also contain, in their N-terminal domain, an Ig-like V-type region, two or four link domains and up to two EGF-like repeats,

5. type-I membrane proteins, and

6. various other proteins that uniquely consist of a CTL domain.

References

1. Drickamer K.
   Two distinct classes of carbohydrate-recognition domains in animal lectins.
   J. Biol. Chem. 263: 9557-9560(1988). [MEDLINE:88257070] [PUB00002490]

2. Drickamer K.
   Evolution of Ca(2+)-dependent animal lectins.
   Prog. Nucleic Acid Res. Mol. Biol. 45: 207-232(1993). [MEDLINE:93342185] [PUB00004941]

Database links   PROSITE doc; PDOC00537
                  Blocks; IPB001304

Matches ⓘ   Table all   Graphical all

Figure 5.3. InterPro entry for the C-type lectine domain.

any) of the motifs composing the fingerprint, cross-referencing to related databases, literature references, and a list of the sequences where the fingerprint, or part of it, is matched. As in the databases above, fingerprints in PRINTS can be obtained by searching for citations, number of motifs, free text, or by scanning PRINTS with one protein sequence (using the FPscan function) or a bulk of protein sequences (using the MULscan function); both these functions are components of the Finger-

PRINTScan package (see the URL in Table 5.1). The resulting entries can be viewed in the InterPRO format where the domains are described or through use of the CINEMA software (Parry-Smith, Payne et al. 1998), allowing the display and management of protein multialignments.

ProDom reports protein domain families generated automatically by recursively applying PSI-BLAST (Altschul, Madden et al. 1997). Each entry in ProDom is formed by a set of homologous proteins. Estimation of the distances among homologous proteins, based on PAM values, allows display the family as a phylogenetic tree. Cartoons, with a graphic representation of all proteins containing selected domains are also provided. Recently, the ProDomCG database has been released: It is a subset of ProDom for domains of proteins from organisms whose genomes have been sequenced completely.

SMART is a database reporting over 500 domain families found in signaling, extracellular, and chromatin-associated proteins detected by applying the Simple Modular Architecture Research Tool based on a hidden Markov model. The domains are annotated extensively with respect to phylogenetic distribution, functional class, tertiary structures, and functionally important residues.

A detailed review by Attwood (2000b) is recommended further reading to clarify the pros and cons of the above-described and other pattern databases and the principles of the algorithms used to classify patterns, profiles, and domains stored in protein pattern databases.

Based on the InterPRO resource and other protein specialized databases (CluSTr and HSSP), the PROTEOME Analysis Database (Apweiler, Biswas et al. 2001) has been set up to provide comprehensive statistical and comparative analyses of the proteomes predicted for fully sequenced organisms. The analysis is performed on the nonredundant complete proteome sets of SWISS-PROT and TrEMBL entries (Anderson, Matheson et al. 2000). CluSTr (Kriventseva, Fleischmann et al. 2001; see the URL in Table 5.1) is a database of clusters of homologous proteins derived from pairwise comparison and is cross-referenced to the InterPro database and to the HSSP and PDB databases. HSSP (Dodge, Schneider et al. 1998; see the URL in Table 5.1) is a database of multialigned homologous proteins based on their secondary structural features. PDB (Berman, Westbrook et al. 2000; see the URL in Table 5.1) is a database collecting the atomic coordinates of the proteins whose structure has been obtained after their crystallization through x-ray and NMR analysis. In this context it is relevant to the macromolecular structure database (MSD), a European resource for data and software supporting structural studies (see the URL in Table 5.1).

5.10 GENOMIC DATABASES AND RESOURCES

The advent of the genomic era with complete genome mapping and sequencing of several organisms has required the design and implementation of genomic databases. Relevant examples of genomic databases from model organisms are FlyBase, the *Drosophila* Genome Database (FlyBase Consortium 1996), MGD, the Mouse Genome Database (Blake, Eppig et al. 2001), SGD, the *Saccharomyces* Genome Database (Cherry, Ball et al. 1997), and WormBase (Stein, Sternberg et al. 2001), the *C. elegans* database. All these databases integrate different types of data, ranging from maps of different types to EST data, genomic sequences, gene expression, gene structures, and protein functions (the URLs are in Table 5.1). A great effort has been made for design of the databases above. The curators of these databases have recently launched the Generic Model Organism Database (GMOD) project (see the URL in Table 5.1) to capitalize on the expertise of the groups above to support the developers of new genomic databases. Besides the genomic databases listed above, the great quantity of genomic data produced worldwide within coordinated genomic projects, particularly for the human genome project, is available at genomic sites developed at the EBI (see "EBI_genomes" in Table 5.1), and at the NCBI (see "NCBI_genomes" in Table 5.1), and through The Institute of Genomic Research (see "TIGR" in Table 5.1), the latter principally specialized in microbial genomes. The EBI genome site reports annotations about the complete genomes from eukaryotes, bacteria, archaea, viruses, viroids, phages, and organelles, and links to primary databases where the high-throughput data are stored and to relevant sites for genomic data such as Ensembl (see below) and Proteomes. Moreover, a monitoring page (GenomeMOT) about the status of a number of eukaryotic genome projects is available at the EBI. The NCBI genome site is not only a report and a "portal" of genomic data from the same class of organisms as those maintained at the genome EBI site, but also allows users to perform queries on these data through the Entrez program and to navigate in a much more complete way through the genomic data available worldwide. Relevant to both these sites is access to genome maps and hence the possibility to browse the data along the chromosomes. At the EBI, chromosome browsing is performed through Ensembl.

Ensembl (see the URL in Table 5.1) is a joint project between the EMBL-EBI and the Sanger Centre that provides a window into the draft human, mouse, zebrafish, and mosquito genomes, reporting a curated gene annotation as well as many other features, such as repetitive sequences, cytological bands, genetic markers, CpG islands, and UniGene clusters. At the Ensembl site the genomic data can be freely downloaded; moreover, it is possible to carry out similar searches, to browse chromosomes, to find genes, SNPs (single nucleotide polymorphisms), and matches to other genomes, to look for proteins and protein families, to find genomic sequences similar to a new protein sequence, to look up positional markers and analyze candidate disease genes in a region, to find the expression profile of a gene, and to use the Ensembl display to view personal genomic data. At present Ensembl provides identification of 90% of known human genes in the genome sequence and 10,000 additional genes, predicted *in silico* and supported by experimental evidence. A recent paper (Hogenesch, Ching et al. 2001) has reported results from a comparison of the Celera and Ensembl human genes predicted according to which it seems that little overlap exists in novel predicted genes. Probably this depends on the draft status of genome sequences and the prediction software used. An example of the use of Ensembl is reported in Fig. 5.4.

The University of California at Santa Cruz (UCSC) Human Genome Browser (see the URL in Table 5.1) also provides annotation of genome assemblies (known as a "golden path" assembly) displaying a variety of features, like those given by Ensembl, and additional ones such as the predictions obtained by several ab initio gene-finding programs (see Sec. 6.8.7). A similar tool is the NCBI Map Viewer (see the URL in Table 5.1), which using its own draft assembly (NCBI Human Contig Assembly) provides many connections to other NCBI data resources (e.g.,

LocusLink, RefSeq, etc.). Figure 5.5 shows the result of a query to the UCSC and NCBI genome browsers with the human p73 mRNA sequence (Y11416).

The TIGR site reports a collection of curate databases containing DNA and protein sequences, gene expression, cellular roles, protein families, and taxonomic data for microbes, plants, and humans. The comprehensive microbial resource available at TIGR (Peterson, Umayam et al. 2001) covers both bacterial genomes produced at TIGR and others produced at other centers. The microbial genome data annotation for the latter is derived from primary annotations thoroughly revised with the support of software available at TIGR. Finally, from the TIGR site it is possible to reach other genomic centers reporting data from specific organisms or from group of organisms.

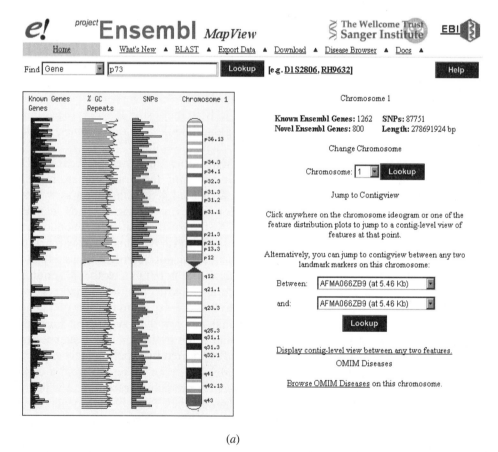

(*a*)

Figure 5.4. (*a*) Human chromosome 1 map view through Ensembl. A search function allows a user to browse the chromosome selected according to various criteria (marker, gene, SNP, family, disease, clone, etc). Shown here is an example of retrieval by gene name, the gene searched being p73. (*b*) Ensembl report for the p73 gene: gene properties, genome localization, prediction method, cross-referencing to EMBL and InterPro, link to SAGE expression profile, transcript sequence predicted, and graphical display of the gene structure of p73 and neighbor genes.

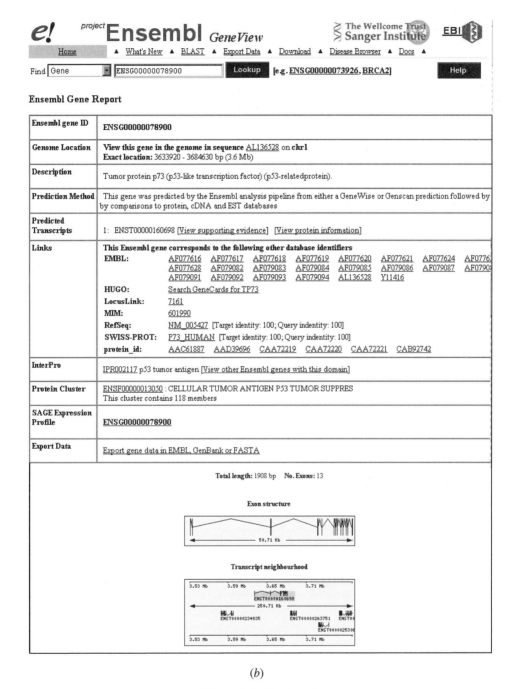

(b)

Figure 5.4. *Continued*

(a)

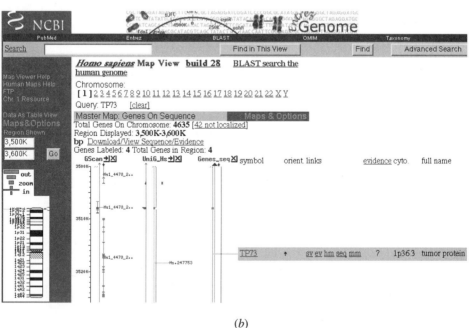

(b)

Figure 5.5. Result of a query to the UCSC (a) and NCBI (b) genome browsers with the p73 human mRNA sequence (Y11416).

Other genomic resources are:

- The *E. coli* genome resource GenProtEC (see the URL in Table 5.1), a database reporting information on *E. coli* genes and proteins resulting from both traditional experimental research and computational analysis. Interesting bioinformation available is the newly implemented classification system for cellular function and a report of paralogs for each gene and pattern and domain information for each protein.
- The Bioknowledge Library (see the URL in Table 5.1), a resource of databases for model organisms: *Schizosaccharomyces pombe* (PombePD), yeast (YPD), and *Caenorhabditis elegans* (WormPD) (Costanzo, Crawford et al. 2001). Here data can be retrieved using a quick or full search or a sequence search. Integration with human genome data and other mammalian genome data is available.
- RHdb (Rodriguez-Tome and Lijnzaad 2001; see the URL in Table 5.1) is a database of raw data used in constructing radiation hybrid maps. It includes STS data (see Sec. 6.7.4), scores, experimental conditions, and extensive cross-references.

5.11   GENE DATABASES AND RESOURCES

One of the unsolved problems related to the genome projects is gene finding, classification, and comparison. Much effort is devoted to this problem, and many resources have been developed to collect data and information on genes. Below is a description of gene databases and resources.

EuGENES (see the URL in Table 5.1) is a database collecting classified genes from eukaryotic organisms whose genomic projects have been completed or are in progress. For each organism the number of predicted and experimentally determined genes is reported; moreover, if available, the chromosome maps can be browsed; besides, genes can be searched for through different criteria such as gene name, gene function, phenotype and chromosomal location.

COG (Tatusov, Natale et al. 2001; see the URL in Table 5.1) is a database collecting clusters of orthologous groups (COGs) of proteins and represents an attempt at a phylogenetic classification of the proteins encoded in complete genomes. In October 2001, the database reported 3166 clusters derived from a comparison of 30 bacterial genomes and two eukaryotic genomes (*Drosophila melanogaster* and *Caenorhabditis elegans*). The clusters are produced through pairwise comparison of each protein in a genome against all the proteins in the other genomes and selecting the best hit as the corresponding orthologous gene. The procedure, described completely in Tatusov, Koonin et al. (1997), is not so trivial, due to the presence in the genomes of both orthologous and paralogous genes (see Sec. 8.3). The algorithm used for COG generation takes into account the problem related to the serious difficulty in distinguishing orthologous from paralogous genes, but to avoid missing orthologous genes in the cluster, it includes paralogous genes. The resulting clusters are classified in different functional categories and are analyzed with principal component analyses; the results are available at the COG site. The proteins in each cluster are multialigned and data can be downloaded easily.

New proteins can be compared against the COG database using the COGnitor software.

LocusLink (see the URL in Table 5.1) is a resource available at the NCBI, reporting accurate annotation for each genetic locus. The information stored in LocusLink for each gene is the official name, as stated by a committee, synonymous names, cross-referencing to primary databases (GenBank, EMBL, and DDBJ), phenotypes, EC numbers (if any), MIM numbers, cross-referencing to UniGene clusters, homologous data, and map locations. LocusLink can be queried through a simple Entrez-like search procedure. The resulting data are listed with a brief description, and clicking on a single locus ID or highlighting more locus IDs in the displayed list, a report for each locus ID is obtained. The report, in HTML format, allows navigation through several related resources and viewing of data graphically. LocusLink is strictly related to RefSeq (Pruitt, Katz et al. 2000; see the URL in Table 5.1) and UniGene (Schuler 1997) (see below). RefSeq contains references for genomic sequences, mRNAs, noncoding RNAs, and proteins to provide a stable basis for gene characterization, mutational analysis, expression analysis, and polymorphism discovery.

5.12   TRANSCRIPTOME DATABASES

A trascriptome can be defined as the entire set of transcripts produced by a given organism. Genome projects have opened doors to new technology and new perspectives, among them the study of the transcriptome through the production of ESTs, STSs (see Sec. 6.7.4), and the complete sequenced cDNA. dbEST (Boguski, Lowe et al. 1993; see URL in Table 5.1) collects expressed sequences derived through single-pass cDNA sequencing from a number of organisms. The information reported for each sequence includes tissue source, type (pathologic or normal), and map location. UniGene collects mRNA sequences into a nonredundant set of gene-oriented clusters. Each UniGene cluster that possibly represents a unique gene contains sequences obtained by either full-length cDNA sequencing or by EST projects. The current human UniGene release (build 151, May 2002) contains 101,698 clusters from 3,480,273 total sequences. The number of clusters is significantly higher than the total gene number predicted by both human genome draft sequences. This could be due to the fact that transcripts from alternatively spliced mRNAs have been cataloged in different clusters or to the artifacts occurring frequently in EST sequencing (Jongeneel 2000). UniGene is frequently used to select reagents for gene mapping projects and large-scale expression analysis.

More recently, full-length cDNA sequencing projects have been launched, such as the full-length mouse cDNA collection, which will provide a more comprehensive picture of the transcript inventory for a given species (Kawai, Shinagawa et al. 2001). Within the trascriptome context, the mRNA noncoding regions assume an important role at the level of functional studies. Of great support to these studies is the UTRdb database (Pesole, Liuni et al. 2000b; see the URL in Table 5.1), which collects a nonredundant set of 5′ and 3′ UTR (untranslated regions) eukaryotic mRNA sequences with careful annotations of functional elements localized in these regions. The database is updated starting from the EMBL database through an automatic procedure parsing EMBL Feature Tables. UTRdb may be very useful for annotation of the noncoding portion of mRNAs. Indeed, it is now well known that these regions contain sequence elements crucial for many aspects of gene regulation and expression. The database can be queried through SRS.

Transcript information is the necessary support to microarray technology, whose adoption in big genomic projects makes it possible to produce in a single experiment, microspot arrays for an entire transcriptome, which can be used to evaluate the expression level of each gene in connection with any perturbing effect. Hence, production of the microarray requires the use of bioinformatic tools for both their storage and their analysis. A list of microarray repositories is given in Sec. 4.2.6.

Indeed, being an emerging technology, the configuration of microarray databases is still a research topic, and hence it is more convenient to speak of repositories rather than of structured databases (but see also Sec. 4.2.6). However, some progress in this direction has been realized and here we can cite as a model the GXD mouse resource (Ringwald, Eppig et al. 2000, 2001; see the URL in Table 5.1). The Gene Expression Information Resource consists of three main components: Gene Expression Database (GXD), Anatomy Database, and 3D Atlas. The Gene Expression Database (GXD) contains mouse expression data and links to other relevant resources—first of all, the MGD database. The time and space for gene expression is described by a controlled Dictionary of Anatomical Terms that is part of the Anatomy Database, which provides the standard nomenclature for queries connecting gene expression to developmental anatomy. Finally, the 3D Atlas provides a high-resolution digital representation of mouse anatomy reconstructed from serial sections of single embryos at each representative developmental stage, enabling three-dimensional graphical display and analysis of in situ expression data.

5.13 METABOLISM DATABASES

The genome projects of several organisms are producing a great deal of information about genes and proteins. Although a lot of information related to them is stored in specialized databases available worldwide and in great part described here, it is absolutely necessary to collect all this information in an integrated manner in order to study and understand the molecular mechanisms with which these elements interact (i.e., to study metabolic and regulatory pathways and molecular assemblies). Resources in this direction are being designed and developed. Among them the EcoCyc ontology and KEGG resource are noteworthy.

The EcoCyc Ontology, an encyclopedia of *E. coli* genes and metabolism (Karp, Riley et al. 2000; see the URL in Table 5.1), is a database describing metabolic pathways of *E. coli* genes and some related microorganisms (*Bacillus subtilis, Haemophilus influenzae, Chlamydia trachomatis, Helicobacter pylori, Mycobacterium tuberculosis, Mycoplasma pneumoniae, Pseudomonas aeruginosa, Saccharomyces cerevisiae, and Treponema pallidum*). In particular, it is possible to select specific bacterial data sets and start from them to use the Pathway/Genome Navigator user interface to visualize the layout of genes within the bacterial chromosome, individual biochemical reactions, or complete biochemical pathways (with compound structures displayed) selected. EcoCyc is cross-referenced to a wide spectrum of biological databases.

KEGG (Kanehisa and Goto 2000; see the URL in Table 5.1), the *Kyoto Encyclopedia of Genes and Genomes*, is aimed to allowing the study of metabolic and regulatory pathways in all organisms whose genome sequencing has been completed or is in progress. Hence, KEGG collects all data related to the pathways besides a catalog of genes and chemical compounds involved in their pathways. Analysis tools allowing an estimation of new pathways are available at the same site. To summarize, KEGG consists of the following five types of data: pathway maps, ortholog group tables, molecular catalogs, genome maps, and gene catalogs. These classes of data are interlinked among them and with other relevant databases and can be queried through DBGET/LinkDB, an integrated database retrieval system. Moreover, a full text search can be performed using STAG software. Figure 5.6 shows a pathway obtained through the use of DBGET/LinkDB.

5.14 MUTATION DATABASES

The availability of the complete human genome sequence is only the starting point for a study of the molecular mechanisms related to variability of human phenotypes. Indeed, if the primary goal of the human genome project is to produce the complete map and sequence of one genome, for health studies it is crucial to map human genome variability. To this end, new technologies have been developed which are producing a great deal of data related to human variability: microsatellite maps, SNPs, maps, and phenotype-associated mutations. A resource collecting the majority of human mutation databases (MutRes; Lehvaslaiho 2000) is available at the EBI through the SRS Server. It includes locus and mutation databases. Locus databases collect phenotype-associated mutations occurring at a specific locus. At present, there are more than 40 locus databases implemented at the EBI, but the majority are not well structured. Among the locus databases worthy of mention is the p53 database (Beroud and Soussi 1998; see the URL in Table 5.1), collecting more than 15,000 p53 somatic mutations and about 200 germline mutations detected in human tumor cells and cell lines from a systematic search of reports published in the literature. As for SNP data, three databases have been developed: MITSNP, the dbSNP database at the NCBI, and the HGBASE database. The SNPs are single nucleotide polymorphisms that occur about once every 1000 bases in human genome.

MITSNP (see the URL in Table 5.1), the Whitehead Institute/MIT Center for Genome Research SNP database, reports (October 2001) 3241 candidate SNPs and a genetic map showing the location of 2227 of these SNPs in the human genome obtained by a combination of gel-based sequencing and high-density variation-detection DNA chips (Wang, Fan et al. 1998). The database can be queried by SNP name or by the corresponding STS name. The SNP map and chromosome distribution of human SNPs can be browsed.

dbSNP (Sherry, Ward et al. 2001; see the URL in Table 5.1), the NCBI SNP database, reports reference SNPs from human, rat, mouse, chimpanzee, gorilla, and *Plasmodium*. The database is linked to several Entrez resources and can be queried through diversified criteria. An SNP submission form is available at the same site. Data retrieved can be displayed in both text and graphical modality. Moreover, the entire database can be downloaded via FTP. This site is also linked to the DNA

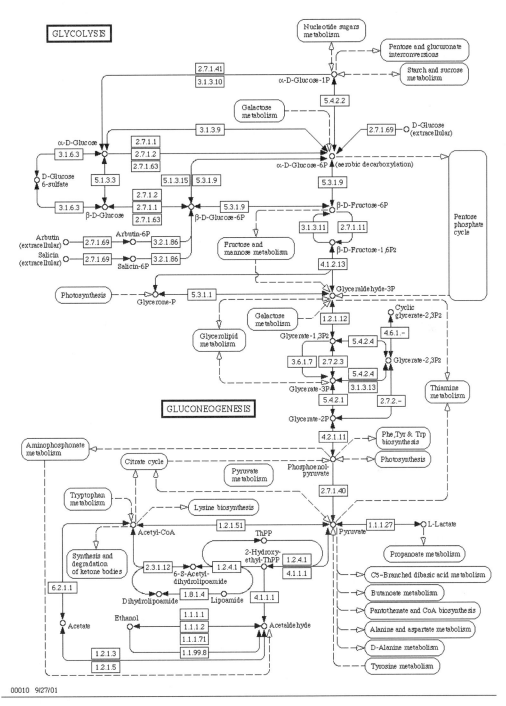

Figure 5.6. Glycolysis/gluconeogenesis reference pathway graphical display as it can be obtained through navigation among the KEGG data using the DBGEt/LinkDB retrieval system.

Polymorphism Discovery Resource at the National Human Genome Research Institute (see the URL in Table 5.1), where SNPs from 450 unrelated U.S. subjects have been determined and collected.

HGBASE (Brookes, Lehvaslaiho et al. 2000; see the URL in Table 5.1) aims to report all known sequence variations in the human genome. Sequence variations are presented with details of how they are physically and functionally related to the closest neighboring genes. Entries are classified in SNPs, indels, simple tandem repeats, and other variation types. These data are highly accurate and annotated with information about genome and chromosome location, related gene and protein information if known, and population and tissue source information. HGBASE is the product of a European consortium involving the Karolinska Institute (Sweden), the European Bioinformatics Institute (U.K.), and the European Molecular Biology Laboratory (Germany). Data retrieval is through SRS at the EBI. As in the majority of the databases described here, it is possible to perform, in addition to a text search, a sequence search applying the FASTA program. To avoid noisy results, the application of the RepeatMasker tool (see Sec. 6.8.3) to the database to be searched is highly recommended.

OMIMALLELE (see the URL in Table 5.1) contains alleles reported in OMIM, the *Online Mendelian Inheritance in Man*, a catalog of human genes and genetic disorders authored and edited by Victor A. McKusick and converted in a Web version by NCBI. OMIM contains textual information, pictures, reference information, and links to other resources. The *OMIM Morbid Map*, a catalog of genetic diseases and their cytogenetic map locations arranged alphabetically by disease, is now available.

5.15   MITOCHONDRIAL DATABASES AND RESOURCES

The great quantity of information related to the mitochondrion at the level of both mitochondrial DNA and nuclear proteins related to mitochondria for a large number of species is of great interest to the scientific community—hence the need to have these data available through specialized databases. The available mitochondrial data are distributed worldwide in primary and specialized databases. Following is a summary description of the mitochondrial specialized databases available worldwide.

GOBASE (Shimko, Liu et al. 2001; see the URL in Table 5.1), the organelle genome database, is a taxonomically broad organelle genome database that organizes and integrates diverse data related to organelles, mitochondria, and chloroplasts. Currently, the GOBASE database contains all mitochondrial nucleic acid sequences and protein data deduced, obtained from NCBI's Entrez database, taxonomic information extracted from NCBI's Taxonomy database, standardized gene names and product names assigned by the GOBASE biology experts, and various information about gene products. The sequences are extracted from GenBank through Entrez. Links to several resources and to the Organelle Genome Megasequencing Project (OGMP; see the URL in Table 5.1), the Fungal Mitochondrial Genome Project (FMGP; see the URL in Table 5.1), and the Protist Image Database (PID; see the URL in Table 5.1) are implemented. GOBASE allows sophisticated queries, including completely sequenced genomes and the use of search criteria concerning the general function of gene products. The majority of data contained in GOBASE have been verified and rectified by experts with respect to the consistency of gene and product nomenclature.

MITOMAP (Kogelnik, Lott et al. 1998; see the URL in Table 5.1), the Human Mitochondrial Genome Database, reports all variations detected in human mtDNA samples from subjects affected by mitochondrial pathologies and from subjects whose DNA has been sequenced for genetic population studies. The report is in HTML format and hence can be browsed using any Internet browser. Moreover, a query system has been implemented allowing one to search data in a very simplified mode; thus it does not allow one to navigate through the stored data. The database is updated to include the latest data available in the literature.

MITOP (Scharfe, Zaccaria et al. 2000; see the URL in Table 5.1), a database for mitochondria-related genes, proteins, and diseases, reports data on nuclear-encoded mitochondrial proteins for human, mouse, yeast, *Caenorhabditis elegans*, and *Neurospora crassa*. Each entry in MITOP is a protein. The following information is reported for each protein: function category, EC number, protein class, protein complex, PROSITE motifs, subcellular localization, molecular weight, isoelectric point, disease correlation, pathways, metabolism, and putative orthologs. The site where the database is available also contains the MITOPROT program, which identifies mitochondrial targeting sequences (see also Sec. 6.10.3). Direct links are provided to other databases, such as the Genome Data Base (GDB), PIR, Mouse Genome Database (MGD), Online Mendelian Inheritance in Man (OMIM), GenBank, and Medline.

MITODAT (see the URL in Table 5.1) is a database collecting nuclear genes specifying the enzymes, structural proteins, and other proteins, many still not yet identified, involved in mitochondrial biogenesis and function. MITODAT highlights predominantly human mitochondrial proteins, although proteins from other animals in addition to those currently known from yeast and other fungal mitochondria, as well as from plant mitochondria, are coded. The database consolidates information from various biological databases [e.g., GenBank, SWISS-PROT, Genome Data Base (GDB), Online Mendelian Inheritance in Man (OMIM)].

HvrBase (Burckhardt, von Haeseler et al. 1999; see the URL in Table 5.1) is a compilation of mtDNA control region sequences from primates. Sequences and related information on individuals, such as where the sequences were obtained from, is stored in three text files. Moreover, the collection is also available as a Mac/PC database application with a graphical user interface. It can be accessed on the Web.

MITOCHONDRIOME is a Web site of mitochondrial data and databases (see the URL in Table 5.1) that integrates mitochondrial databases developed by the Bari Bioinformatics Group, data derived from *in silico* analysis, and links to other mitochondrial sites and any relevant information. In particular, MITOCHONDRIOME gives access to the following databases: Human MitBASE, MITONUC, AMmtDB, and VARMmtDB. These databases can be queried through SRS.

Human MitBASE (Attimonelli, Cooper et al. 1999) contains data relevant to studies on DNA mutations that cause a modification of the genotype and/or phenotype and on their frequency. Data are derived from primary databases for genetic population studies and from the literature for pathological data. The overall aim of this database is to support phylogenetic and pathogenetic studies with updated information. Beside the variation data, information regarding geographical origin of the subjects whose DNA variation is reported, and clinical and biochemical data for pathological studies, are annotated.

MitoNuc (Pesole, Gissi et al. 2000a) is a database containing detailed information on sequenced nuclear genes coding for mitochondrial proteins in metazoans. Data derived from the EMBL and SWISS-PROT databases are thoroughly revised and annotated with value-adding information such as the metabolic pathways in which the protein product is involved; cellular and submitochondrial localization of the encoded proteins; the possible presence of tissue-specific isoforms, cross-references to the EMBL, SWISS-PROT/TREMBL, and UTRdb databases, comments about clinical data related to protein dysfunction, the location and description of relevant signatures in the protein; finally, homologous proteins in MitoNuc are flagged by cluster identifiers, which allows one to extract clusters of homologous proteins for further analyses.

AMmtDB (Lanave, Liuni et al. 2000) collects multialigned metazoan mitochondrial genes coding for proteins and tRNAs in class-specific clusters whose sequences are carefully multialigned. Both the nucleotide and amino acid multialignments can be downloaded. It also contains cross-referencing to intraspecies variants collected in the VARMmtDB database (unpublished).

PLMitRNA (Volpetti, Gallerani et al. 2000; see the URL in Table 5.1) is a database for tRNA molecules and genes identified in the mitochondria of all green plants (Viridiplantae). PLMItRNA reports information and multialignments of protein genes and tRNA molecules detected in higher plants and green algae. Retrieval of information or sequences can be accomplished according to several characteristics of the tRNA gene or molecule. A selection by single plant name or plant taxonomic group is also possible. For every sequence retrieved, a tRNA domain based on multialignment with homologous sequences in the database is reported.

CHAPTER 6

COMPUTATIONAL METHODS
FOR THE ANALYSIS OF
GENOME SEQUENCE DATA

6.1 INTRODUCTION

The development of rapid and economical sequencing techniques and the progress
of the huge number of large-scale projects for determination of the complete
genomic sequences of several species, from prokaryotes to humans, or the complete
repertorire of the transcripts expressed in various physiological or pathological conditions (Adams, Kelley et al. 1991), is producing an enormous flood of sequence
data, destined to increase ever faster in the coming years. Until the genetic message
encrypted in sequences is deciphered at least to some extent, any sequencing effort
will remain completely useless. To take advantage of the increasing amount of
biological information for the production of novel biological knowledge, adequate
tools are needed for their management and analysis, which are derived mainly from
information technology. This is the field of bioinformatics, an emerging discipline
involved in all aspects if database design, implementation, and maintenance, data
retrieval systems, and development of algorithms and software tools for the
functional and evolutionary analysis of newly determined sequences.

Sequence comparison is undoubtedly the first and probably the most informative step toward the functional characterization of sequence data under examination and has as its ultimate goal extensive definition of their structure–function
relationships. As the "Rosetta Stone," containing the same text written in different
languages, allowed hieroglyphic deciphering, so comparative analysis of genes and
genomes is the key to accessing the secrets of life as encoded in the informational
macromolecules. Comparative analysis can be carried out at several levels, including primary sequences, three-dimensional structures, and expression patterns up to
the level of whole metabolic pathways. The first analytical step in the characterization of a newly determined sequence is database searching to detect other sequences
or sequence regions showing a significant level of similarity with our query
sequence. If the function of any of the significantly matching sequences is known,
by analogy a similar function can be hypothesized for our sequence/gene. Indeed,

significant sequence similarity testifies to a common evolutionary descent and thus suggests a homology relationship. It is now worth clarifying that *homology* and *similarity* have different meanings (Reeck, de Haen et al. 1987). Indeed, two sequences are *homologous* if they share a common ancestry deriving from a speciation or gene duplication event, originating orthologous or paralogous sequences, respectively. Consequently, homology is a qualitative character in the sense that two sequences or genes may or may not be homologous, and it is incorrect to refer to percentages of homology. On the contrary, *similarity* is a quantitative character that measures the relatedness between sequences after their best alignment (see below).

Of course, two sequences sharing a significant degree of similarity are almost certainly homologous because it is extremely unlikely that two (or more) sequences acquire extensive similarity on their entire length just by chance. On the other hand, if two homologous sequences are highly divergent, due to a particularly fast evolutionary dynamics or to a very long time span since their common ancestor (or both), their homology relationship can no longer be recognized at the level of the primary sequence. In other words, two sequences may be homologous even if they do not show an appreciable degree of similarity (see also Sec. 8.3).

6.2 DOT-PLOT MATRIX

A very convenient way of identifying regions of similarity in large sequences, as well as features such as repeats and regions of sequence duplications, is provided by a dot-plot matrix. In a dot-plot matrix, two sequences are represented by the axes of a matrix. If we compare two sequences of length m and n, the matrix is made of m rows and n columns. At every cell i,j of such a matrix, the result of the comparison between the ith residue of the first sequence, a_i, and the jth residue of the second sequence, b_j, will be drawn in the form of a dot if $a_i = b_j$. A remarkable level of background noise is caused by matches that have occurred by chance. The level of noise will of course be much higher in the case of DNA sequence comparisons, which uses an alphabet of only four characters, than in the case of protein sequence comparisons, whose alphabet is made of 20 characters.

If we compare two unrelated sequences, the number of dots occurring just by chance will be given by

$$\sum_{i=1}^{D} A_i B_i$$

where A_i and B_i represent the frequencies of residues of type i present in sequences A and B, respectively, and D is the size of the alphabet: 4 for nucleotides and 20 for protein sequences. For example, if we compare two DNA or protein sequences, made of 100 residues, assuming equiprobability for each residue type, the total number of dots will be, on average, 2500 and 500 of the total 10,000 cells in the matrix for DNA and protein sequences, respectively.

It is clearly evident that the high background noise makes the dot-plot practically useless. For this reason, Maizel and Lenk (1981) suggested the use of filters to reduce noise. The idea is to compare "windows" of contiguous residues instead of single residues. In this way, a dot in the cell i,j is drawn only when between the two sequence strings $a_i a_{i+1} \ldots a_{i+w-1}$ and $b_j b_{j+1} \ldots b_{j+w-1}$ there are at least s matching residues out of the total w ($s \leq w$). The value of s defines the stringency parameter in the window of size w. In Fig. 6.1 the same dot-plot matrix is shown with and without filtering.

In the case of protein sequence comparisons, another way to filter matches is to give them a weight accounting for the chemical–physical similarities of the matching residues. A diagonal stretch of dots will indicate regions of sequence similarity. Vertical or horizontal breaks in the dot line will visualize insertion/deletion events, and parallel diagonal stretches will represent repeated subsequences. Indeed, other very useful applications of the dot-plot matrix allow the identification of repeated elements contained in the sequence under examination or of complementary oligonucleotides. In the case of repeated elements, the same sequence should be used on both axes. A perfect diagonal will be observed for the match of each base with itself, and repeated regions within the sequence will be represented as parallel lines. This type of dot-plot matrix will have identical halves above and below the main diagonal. Figure 6.2 shows a dot-plot matrix obtained by comparing a sequence with itself where a tandem repeated minisatellite can be identified. The size of the repeated unit is given by the distance between two parallel diagonal lines.

The identification of regions of self-complementarity within a sequence can also be obtained by a dot-plot matrix when the sequence on one axis is compared with its reverse complement sequence on the other axis. Diagonal lines in the upper left-to-lower right direction identify regions of self-complementarity which can contribute to the formation of specific secondary structures.

Dot-plot matrices can also be used to study gene conservation and order in different genomes and/or chromosomes (Tatusov, Mushegian et al. 1996; Wolfe and Shields 1997; Read, Brunham et al. 2000). In this case the horizontal and vertical axes report the ordered list of genes of two genome fragments being compared, and a dot is drawn if the similarity score calculated by comparing the corresponding protein sequences is above a predefined score. Observation of a main diagonal in the dot plot would indicate no major rearrangements in the genome tract under investigation and an overall degree of remarkable sequence conservation between homologous genes. Any diagonal stretch in the plot would indicate a specific synteny tract between the two genome regions, whereas inverted diagonals would indicate regions of large inversions. This approach has been used to compare gene order and content between two species of *Chlamydia* (Read, Brunham et al. 2000) or to visualize tracts of genes duplicated in yeast chromosomes (Wolfe and Shields 1997).

6.3 SEQUENCE PAIRWISE ALIGNMENT

To define the degree of similarity shared by two sequences, their *best alignment* must be determined. Computational tools are needed to align two sequences, since during evolution, homologous sequences may undergo several rounds of insertions and/or deletions (indels), which make them of different length. Indeed, the aim of best-alignment algorithms is to find the optimal arrangement of indels to maximize the overall similarity between sequences being compared. Alignment programs can find two types of alignment between sequences, global or local. *Local alignment algorithms* search for regions of local similarity between two sequences, and *global alignment algorithms* search for the best possible alignment between two sequences in

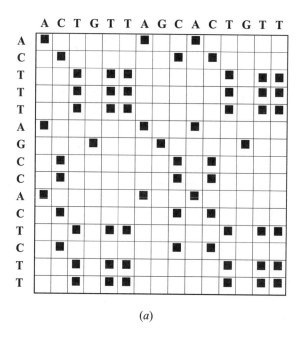

(a)

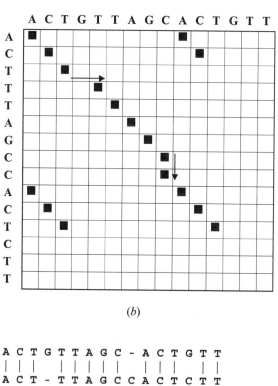

(b)

```
A C T G T T A G C - A C T G T T
| | |   | | | | |   | | |   | |
A C T - T T A G C C A C T C T T
```

(c)

Figure 6.1. Dot-plot matrix obtained by comparing two nucleotide sequences whose optimal alignment is shown in (c) and using window/stringency values of (a) $w = 1$ and $s = 1$ or (b) $w = 4$ and $s = 3$. The arrows represent diagonal displacements in the optimal alignment path corresponding to gaps.

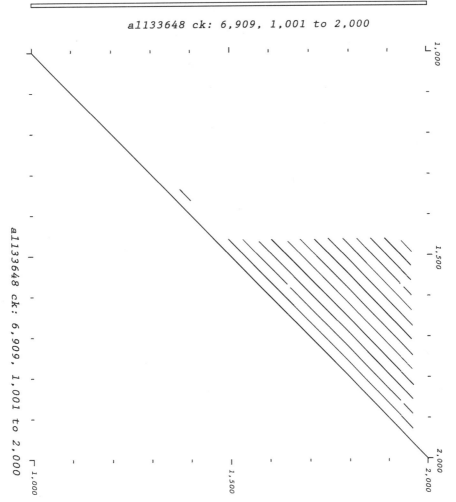

Figure 6.2. Dot-plot matrix obtained by comparing against itself a 1000-nt-long DNA sequence (region 1001–2000 in the EMBL entry AL133648) using a window of 21 nt and a stringency of 16 nt. The programs Compare and DotPlot of the GCG package were used to carry out the analysis.

their entirety by finding for each site of the first sequence a counterpart in the second, assuming that the two sequences are homologous.

6.3.1 Needleman–Wunsch Global Alignment Algorithm

The classical algorithm for the determination of global alignment between two nucleotide or protein sequences is that proposed by Needleman and Wunsch (1970). This method requires the construction of a two-dimensional array whose axes are the two sequences to be compared. To determine the best alignment between two

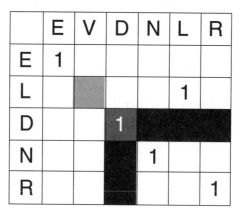

Figure 6.3. Untransformed 5×6 matrix showing the lower pathway for a cell (light gray) in dark gray and black. If a gap penalty is considered in the matrix transformation, it is applied only to cells not lying on the same diagonal (black cells).

sequences, $S_A = a_1 \ldots a_i \ldots a_n$ and $S_B = b_1 \ldots b_j \ldots b_m$, long n and m residues, respectively, a matrix **M** is constructed with m rows and n columns, respectively, whose elements identify all pairs of residues of the two sequences. The matrix is first initialized by assigning to each cell a predefined value as a function of the corresponding pairs of residues. In the simplest method, $M(i,j) = 1$ if $a_i = b_j$ and $M(i,j) = 0$ if $a_i \neq b_j$. This method for the first scoring of the matrix is adequate only when nucleotide sequences are compared; more appropriate values need to be used in the case of protein sequences based on the similarity of the chemicophysical properties of amino acid pairs (see below). The initialized matrix is then transformed starting from the bottom right-hand corner, which corresponds to the cell of coordinates m and n and proceeding toward the origin. The operating rule for the transformation consists of the addition to each cell value of the highest score in the cells lying in a lower pathway, [i.e., all the cells to its right in the row below or all the lower cells in the column to its right (see Fig. 6.3)]:

$$M(i,j) = M(i,j) + \max[M(i+1,j+1); M(i+1,j+2,\cdots,j_{max}) - \text{gap penalty};$$
$$M(i+2,\cdots,i_{max},j+1) - \text{gap penalty}]$$

Then starting from the cell at coordinated m, n the transformation procedure starts by adding to the value of cells in row $m - 1$ and/or column $n - 1$ the maximum value from the cells lying on a lower pathway. The same procedure is iterated for cells of indices $m - 2$ and $n - 2$ and continues until all cells in the matrix have been considered.

Every possible pathway through the array proceeding down from the upper left to the bottom right corner represents a possible global alignment (straight down, straight right and left paths are forbidden). The optimal global alignment is represented by the path totaling the highest score. If at a certain point this path moves off the diagonal, a gap is included in the best alignment to which a penalty factor is assigned, depending on size (see Fig. 6.3 and Sec. 6.3.2).

To clarify the method, let us make a simple example with the two sequences

$$S_A = \text{EVDNLRSDWDRMP}$$

$$S_B = \text{ELDLNRDYDRVM}$$

To initialize the matrix $\mathbf{M}_{AB}$ the simplest scoring system can be used that assigns 1 to identical and 0 to different residue pairs. The initialized matrix shown in Fig 6.4a is then transformed from the bottom rightmost cell as described above. From the transformed matrix, shown in Fig. 6.4b, the best alignment is given by the ridge line having the highest score, starting from the largest number in the first row and/or first column and ending at the last column and/or row of the matrix. The optimal global alignment between S_A and S_B, shown in Fig. 6.4c, is given by the path highlighted in Fig. 6.4b. The best matching path is the one producing the largest total score. If two or more paths reach the same largest score, one is chosen arbitrarily to represent the best global alignment. It is easy to see how the alternative paths shown by the black and gray arrows in Fig. 6.4b represent the equally optimal alignments shown in Fig. 6.4c.

It is interesting to note how the score in the cell at the top left-hand corner represents the total score of the pairwise alignment without the penalties for possible insertion/deletion events that result in lateral shifts of the optimal path from the current diagonal. In the case of gap penalty assignment, the total alignment score is decreased by a value proportional to the number and the length of gaps present in the alignment. For example, by assigning a penalty of one to each gap opening (see Sec. 6.3.2), the total score of both best alignments in Fig. 6.4c is 8 (match score) − 4 (gap penalties) = 4.

6.3.2 Smith–Waterman Algorithm for the Identification of Common Molecular Subsequences

In pairwise sequence comparisons, the identification of a subsequence of the first sequence having the highest degree of similarity with a subsequence of the second sequence is of the greatest importance. Smith and Waterman (1981) proposed an algorithm able to find a pair of segments, one from each of the two sequences compared, such that there is no other pair of segments with greater similarity. Given two sequences $S_A = a_1 a_2 \ldots a_n$ and $S_B = b_1 b_2 \ldots b_m$, a matrix $\mathbf{H} = \{H_{ij}\}$ is set up made of $n + 1$ rows and $m + 1$ columns, which is initialized conveniently by setting

$$H_{i,0} = H_{0,j} = 0 \qquad \text{for } 0 \le i \le n \quad \text{and} \quad 0 \le j \le m$$

To determine the value of each cell of $\mathbf{H}$ we need to assign a similarity score $\mathbf{s}$ to each pair of sequence elements a_i and b_j belonging to S_A and S_B, respectively. This can be done in a number of ways, also depending on whether nucleotide or amino acid sequences are being compared. It should be stressed that the alignment determined depends strictly on the scoring system used for sequence comparison. For the sake of simplicity, suppose that nucleotide sequences are compared and that the simplest possible scoring system is used:

$$\mathbf{s}(a_i, b_j) = 1 \text{ if } a_i = b_j \quad \text{and} \quad \mathbf{s}(a_i, b_j) = -\frac{1}{3} \text{ if } a_i \ne b_j$$

According to this scoring system, the average similarity score is zero between two long random sequences, with equiprobable occurrence of the four bases. We also

	E	V	D	N	L	R	S	D	W	D	R	M	P
E	1												
L					1								
D			1					1		1			
L					1								
N				1									
R						1					1		
D			1					1		1			
Y													
D			1					1		1			
R						1					1		
V		1											
M												1	

(*a*)

	E	V	D	N	L	R	S	D	W	D	R	M	P
E	8	7	6	6	5	4	4	3	3	2	1	0	0
L	7	7	6	6	6	4	4	3	3	2	1	0	0
D	6	6	7	6	5	4	4	4	3	3	1	0	0
L	6	6	6	5	6	4	4	3	3	2	1	0	0
N	5	5	5	6	5	4	4	3	3	2	1	0	0
R	4	4	4	4	4	5	4	3	3	2	2	0	0
D	3	3	4	3	3	3	3	4	3	3	1	0	0
Y	3	3	3	3	3	3	3	3	3	2	1	0	0
D	2	2	3	2	2	2	2	3	2	3	1	0	0
R	2	1	1	1	1	2	1	1	1	1	2	0	0
V	1	2	1	1	1	1	1	1	1	1	1	0	0
M	0	0	0	0	0	0	0	0	0	0	0	1	0

(*b*)

```
EVDNL-RSDWDR-MP
| | | | | || |
ELD-LNR-DYDRVM
```

```
EVD-NLRSDWDR-MP
| | | | | || |
ELDLN-R-DYDRVM
```
(*c*)

Figure 6.4. Needleman–Wunsch global alignment algorithm: (*a*) initialization of the M_{AB} matrix; (*b*) matrix transformation, no gap penalty is considered; (*c*) best global alignments. Black and gray arrows mark alternative paths leading to equally optimal alignments. [From Needleman and Wunsch (1970).]

need to assign a weight W to insertions/deletions depending on their length k as

$$W_k = \gamma + \delta(k-1)$$

W_k is a linear function assigning a constant penalty γ to the presence of insertion/deletion, also called *gap opening penalty*, plus a variable penalty proportional to the insertion/deletion length, also called *gap extension penalty*. The values we use here are $\gamma = 1$ and $\delta = \frac{1}{3}$.

The values of the matrix elements H_{ij} are determined according to the equation

$$H_{i,j} = \max\left\{ H_{i-1,j-1} + \mathbf{s}(a_i, b_j), \max_{k \geq 1}(H_{i,j,k} - W_k), \max_{l \geq 1}(H_{i,j-1} - W_l), 0 \right\}$$

where $1 \leq i \leq n$ and $1 \leq j \leq m$. The four terms of this formula are derived by considering the possible segment termination at any a_i and b_j:

1. a_i and b_j are associated.
2. a_i is at the end of a deletion of length k.
3. b_j is at the end of a deletion of length l.
4. Finally, a zero is inserted to prevent calculation of negative similarities.

The pair of segments with the highest degree of similarity is found by first locating the maximum value in **H**. A traceback procedure starting from the maximum H_{ij} element toward that element with the maximum value chosen between $H_{i-1,j-1}$, $H_{i-1,j}$ and $H_{i,j-1}$ determines the aligned segments as well as their alignment. In case the move is toward $H_{i-1,j}$ or $H_{i,j-1}$, a deletion is inserted in S_B or S_A, respectively. The traceback procedure ends when an element of **H** equal to zero is found backward in the main diagonal, in the cell of coordinates $i - 1, j - 1$. A simple example is given in Fig. 6.5 where the following sequences are considered:

$$S_A = \text{AAUGCCAUUGACGG}$$

$$S_B = \text{CAGCCUCGCUUAG}$$

The traceback path, shown shaded, starts from the maximal element 3.3 and determines the best local alignment, reported below, that contains both a mismatch substitution and a deletion.

```
G C C A U U G
| | |   | · |
G C C – U C G
```

The Smith–Waterman algorithm returns the single best local alignment.

However, two nucleotide sequences may share more than one common region. Indeed, the pairs of segments with the next-best local alignment can be determined by applying the same traceback procedure described above to the second-largest element of **H** not associated with the first traceback.

	0	C	A	G	C	C	U	C	G	C	U	U	A	G
0	0.0	0.0	0.0	0.0	0.0	0.0	0.0	0.0	0.0	0.0	0.0	0.0	0.0	0.0
A	0.0	0.0	1.0	0.0	0.0	0.0	0.0	0.0	0.0	0.0	0.0	0.0	1.0	0.0
A	0.0	0.0	1.0	0.7	0.0	0.0	0.0	0.0	0.0	0.0	0.0	0.0	1.0	0.7
U	0.0	0.0	0.0	0.7	0.3	0.0	1.0	0.0	0.0	0.0	1.0	1.0	0.0	0.7
G	0.0	0.0	0.0	1.0	0.3	0.0	0.0	0.7	1.0	0.0	0.0	0.7	0.7	1.0
C	0.0	1.0	0.0	0.0	2.0	1.3	0.3	1.0	0.3	2.0	0.7	0.3	0.3	0.3
C	0.0	1.0	0.7	0.0	1.0	3.0	1.7	1.3	1.0	1.3	1.7	0.3	0.0	0.0
A	0.0	0.0	2.0	0.7	0.3	1.7	2.7	1.3	1.0	0.7	1.0	1.3	1.3	0.0
U	0.0	0.0	0.7	1.7	0.3	1.3	2.7	2.3	1.0	0.7	1.7	2.0	1.0	1.0
U	0.0	0.0	0.3	0.3	1.3	1.0	2.3	2.3	2.0	0.7	1.7	2.7	1.7	1.0
G	0.0	0.0	0.0	1.3	0.0	1.0	1.0	2.0	3.3	2.0	1.7	1.3	2.3	2.7
A	0.0	0.0	1.0	0.0	1.0	0.3	0.7	0.7	2.0	3.0	1.7	1.3	2.3	2.0
C	0.0	1.0	0.0	0.7	1.0	2.0	0.7	1.7	1.7	3.0	2.7	1.3	1.0	2.0
G	0.0	0.0	0.7	1.0	0.3	0.7	1.7	0.3	2.7	1.7	2.7	2.3	1.0	2.0
G	0.0	0.0	0.0	1.7	0.7	0.3	0.3	1.3	1.3	2.3	1.3	2.3	2.0	2.0

Figure 6.5. Smith–Waterman local alignment algorithm between sequence A (vertical) and sequence B (horizontal). The traceback path (shaded) in the H_{ij} matrix constructed on the two sequences being compared determines their best local alignment. [From Smith and Waterman (1981).]

It is clear that the inferred local best alignment is dependent on the choice of the alignment parameters, including the score to be assigned to matches and mismatches and the gap penalty, the latter parameter being the most critical. To choose optimal values for gap penalty parameters, specific studies have been carried out to investigate frequency and length distribution of insertion/deletion events in large collections of homologous sequences (Vingron and Waterman 1994; Gu and Li 1995).

6.3.3   Alignment of cDNA and Genomic DNA Sequences

Most genes in common model organisms, such as human, mouse, or *Drosophila*, are known at the level of their mRNA (i.e., cDNA) sequence. Indeed, the numerous EST sequencing projects (see Sec. 4.2.2) have provided several million EST sequences from different species and various cellular lines corresponding to fragments of spliced mature transcripts. The rapid progress of genome sequencing projects, which provide the template from which transcripts are generated, makes it really useful to have tools available to align a spliced transcript efficiently and accurately with a genomic sequence containing that gene. Consider that a eukaryotic gene is generally made of several exons, some of them extremely small, which are joined to form the mature spliced transcript, and of the intervening intron sequences, some of them quite large. The alignment between the sequences of the gene and of the mature spliced transcript allows the detection of gene structure (i.e., the exact location of introns within the gene).

A computer program is then needed, capable of finding the best local alignment between a cDNA sequence and a genomic sequence where matching sequences are almost identical, assuming a relatively small number of sequencing errors, and long gaps are allowed in the cDNA sequence, corresponding to introns in the genomic sequence. The computer program Sim4 developed by Florea, Hartzell et al. (1998) addresses this specific problem and provides the coordinates of the matching

Genomic: gi|8217482|emb|AL136528.11|AL136528 Human DNA sequence from clone RP5-1092A11 on chromosome 1p36.2-36.33 Contains the gene for KIAA0495 protein, the TP73 (tumor protein p73) gene, a gene containing a WD repeat domain, ESTs, STSs, GSSs>

mRNA: gi|2370175|emb|Y11416.1|HSY11416 H.sapiens mRNA for P73

Alignment is on minus strand of genomic sequence and on plus strand of mRNA sequence
mRNA coverage: 100%
Overall percent identity: 100.0%

116590 ——— 36484

	Genomic coordinates	mRNA coordinates	length	identity	mismatches	gaps	Donor site	Acc.site
Exon 1	116590-116666	1-77	77	100.0%	0	0	d	
Exon 2	86801-86898	78-175	98	100.0%	0	0	d	a
Exon 3	86051-86171	176-296	121	100.0%	0	0	d	a
Exon 4	61440-61682	297-539	243	100.0%	0	0	d	a
Exon 5	47024-47210	540-726	187	99.5%	1	0	d	a
Exon 6	45762-45877	727-842	116	100.0%	0	0	d	a
Exon 7	42007-42116	843-952	110	100.0%	0	0	d	a
Exon 8	41461-41603	953-1095	143	100.0%	0	0	d	a
Exon 9	41014-41102	1096-1184	89	100.0%	0	0	d	a
Exon 10	39783-39904	1185-1306	122	100.0%	0	0	d	a
Exon 11	39083-39231	1307-1455	149	100.0%	0	0	d	a
Exon 12	38166-38304	1456-1594	139	100.0%	0	0	d	a
Exon 13	37675-37768	1595-1688	94	100.0%	0	0	d	a
Exon 14	35939-36484	1689-2234	546	100.0%	0	0		a

Figure 6.6. Output generated by the Spidey program by comparing the human p73 cDNA sequence (EMBL ID: Y11414) and a chromosome 1 genomic clone (EMBL ID: AL136528). The p73 gene structure identified consists of a total of 14 exons. Note the large size of the first, third, and fourth introns. The Spidey program is available on the Web at http://www.ncbi.nlm.nih.gov/IEB/Research/Ostell/Spidey/spideyweb.cgi. [From Wheelan, Church et al. (2001).]

segments and the corresponding identity percentages in the comparison between a cDNA sequence and a genomic sequence as long as 200 kbp. The program is designed so that the end of large gaps are trimmed in the attempt to fulfill the intron donor and acceptor consensuses matching either GT ... AG or CT ... AC. Another excellent tool for the mRNA to genomic alignment is Spidey (Wheelan, Church et al. 2001) that also allows interspecies alignment (i.e., mRNA and genomic sequences from two different species) and uses specific splice-site matrices for vertebrates, *Drosophila*, *C. elegans*, and plants. Figure 6.6 shows the output obtained by Spidey comparing the mRNA sequence coding for the p73 protein and the corresponding genomic clone.

6.3.4 Genome Alignment

The availability of complete genome sequences has generated the impelling need of tools specifically designed for the alignment of large genome regions or even complete genomes. It is particularly interesting to compare, at a genome level, different strains of the same bacterium showing pathogenic specificity. Existing algorithms are ineffective for such a task, due to the huge size of the sequences to be compared. This problem can be solved efficiently by using an efficient data structure known as a *suffix tree* (Gusfield 1997) when the genome sequences to be compared are very closely related or share very similar regions. The MUMer software (Delcher, Kasif et al. 1999) is able to detect maximal unique matches (MUMs) shared by two genomes in both strand orientations. After the detection of MUMs, which correspond to sequences occurring only once in the genome, alignment is completed by closing the gaps between the aligned MUMs, thanks to the local identification of large inserts, repeats, small mutated regions, tandem repeats, and SNPs using the Smith–Waterman best local alignment algorithm (see Sec. 6.3.2). Finally, an output is generated including all MUMs and the detailed alignment of the intervening regions which do not match exactly. Repeat sequences do not occur in the MUM alignment because, by definition, the latter includes only unique sequences in both genomes being compared.

For the identification of conserved sequences through the comparison of large homologous genome regions, the PipMaker software has been devised (Schwartz, Zhang et al. 2000; see the URL in the Appendix). PipMaker displays the percent identity plot (PIP) between the two sequences being compared and provides a graphical output where positions along the horizontal axis (corresponding to the first input sequence) can be labeled with features such as exons of genes and repetitive elements. PipMaker can be very useful in detecting similar noncoding regions in divergent species, probably endowed with significant functional activity. Alternatively, the VISTA software (see URL in Appendix) can be used that visualizes long sequence alignments of genomic DNA from two or more species with annotation information. Figure 6.7 shows the graphical and text output produced by MUMer in the alignment of *Mycoplasma pneumoniae* and *Mycoplasma genitalium* genomes and the output produced by VISTA in the comparison of two homologous genome regions from humans and mouse.

Figure 6.7. (*a*) Alignment of *Mycoplasma genitalium* (NC_000908) and *M. pneumoniae* (NC_000912) complete genomes by MUMer software. (i) Dot-matrix plot representation of the alignment, where the dots correspond to MUMs found provided by the TIGR facility (www.tigr.org). (ii) Sorted list of MUMs, where columns 1 and 2 are the position of MUMs in genomes 1 and 2, column 3 gives the MUM length, column 4 is the overlap with the preceding MUM (usually "none"), and columns 5 and 6 show the gap length between the end of the previous MUM in genomes 1 and 2, respectively. (iii) Smith–Waterman alignment of an intervening region between two consecutive MUMs. (*b*) Result of VISTA analysis in comparison of the human genome region containing the p73 gene and the 10 kbp upstream and downstream region against an homologous genome region in the mouse genome (ENSEMBL IDS: numer, ENSG00000078900; mouse, ENSMUS00000029026). Significant similarity is evident in the correspondence of exon, 3–14 and of some non coding regions. [(*a*) From Delcher, Kasif et al. (1999).]

(i)

```
> NC_000908.fna  Consistent matches
    3389    3412    26    none      -          -
    3698    3721    23    none     283        283         70
    4266    4289    24    none     545        545        120
    4812    4820    22    none     522        507        130
    5533    5541    20    none     699        699        162
    7039    7047    20    none    1486       1486        431
    7189    7197    20    none     130        130         33
   11525   11548    20    none    4316       4331       1463
```

(ii)

```
> NC_000908.fna  Consistent matches
    3389    3412    26    none      -          -
    3698    3721    23    none     283        283
    Errors = 70
T:  acaaacaaactgtaattgtaagcagactccagcaattagcttttttaaacaagggaataa
S:  acaaacaaacggtaattgctagtaggttgcaacaattagcctttcttaacaaagggatcc
              ^          ^^ ^ ^^ ^ ^         ^    ^ ^     ^    ^  ^^

T:  gaattgactttgttgataatcgtaaacaaaacccacagtctttttcttgaaaatatgatg
S:  aaattgactttgttgatgaacgccgtcaaaatccgcaaagcttttcttggaagtatgatg
    ^                 ^ ^  ^^^^  ^     ^  ^^^^      ^  ^  ^ ^

T:  ggggattggttgaatatatccaccacctaaacaacgaaaaagaaccacttttttaatgaag
S:  gtggcttagtccaatacatccaccacctcaacaacgaaaaggaacctttatttgaggaca
     ^ ^  ^  ^ ^^   ^           ^           ^      ^^ ^  ^ ^   ^^

T:  ttattgctgatgaaaaaactgaaactgtaaaagctgttaatcgtgatgaaaactacacag
S:  ttatctttggtgaaaaaaccgatactgttaaatcagttagccgtgatgagagctacacaa
        ^^^ ^^       ^ ^  ^    ^       ^

T:  taaaggttgaagttgcttttcaatataacaaaacatacaaccaatcaattttcagttttt
S:  ttaaggtggaagtggcctttcagtacaacaagacgtataaccaatcaatctttagttttt
     ^     ^     ^  ^   ^   ^ ^  ^ ^ ^  ^         ^   ^  ^

T:  gta
S:  gta
```

(iii)

(*a*)

p73 gene region

Alignment 1
Seqs: human/mouse
Criteria: 75%, 100 bp
Regions: 26

X-axis: human
Resolution: 39
Window size: 100 bp_

gene
exon
UTR
CNS

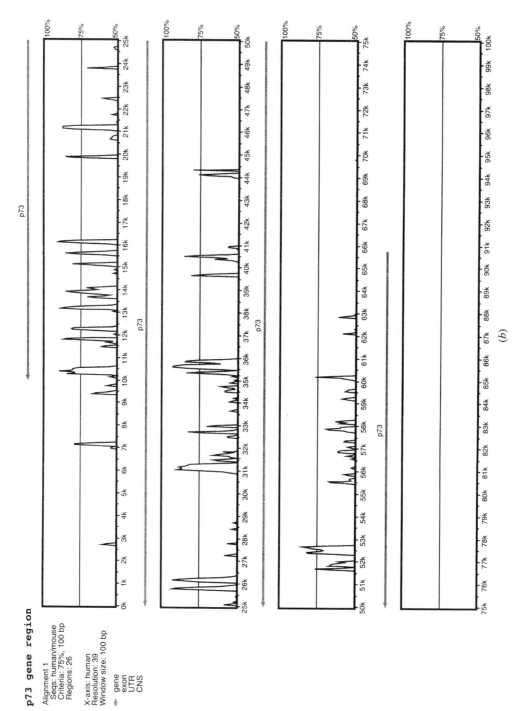

Figure 6.7. *Continued*

(b)

6.3.5 Cleanup of Sequence Databases from Redundancy

The issue of redundancy is a key concept in comparing sequence collections. Indeed, the production of sequence collections cleansed of redundancy is useful both in performing statistical analyses and in accelerating extensive database searches. Publicly available databases contain multiple entries of identical or almost identical sequences and, consequently, statistical analyses (e.g., searching for overrepresented oligonucleotides) on such biased data run the risk of misleading results. Since an unambiguous qualitative definition of redundancy is impracticable for biological sequence data, a quantitative definition can be used instead, based on a measure of sequence similarity. A sequence is considered redundant if it shows a degree of similarity and overlapping with a longer sequence in the database greater than a threshold fixed by the user.

To address this issue, the CleanUP algorithm (Grillo, Attimonelli et al. 1996) has been proposed, based on an approximate string matching procedure, which is able to determine the overall degree of similarity between each pair of sequences contained in a nucleotide or protein sequence database without performing the time-consuming task of pairwise global best alignment. The CleanUP application generates a "purified" dataset from the redundant sequence collection given as input, using cutoff parameters set by the user. Another useful application of CleanUP software, under the appropriate run option, is the search for highly similar shared segments between sequences contained in a data set.

A detailed description of the CleanUP algorithm and software is reported in Grillo, Attimonelli et al. (1996). Briefly, the basic idea of the CleanUP computational model is the generation of a set of highly descriptive indices for each w-gram in the sequences being compared containing information regarding both nucleotide composition and location. This data structure allows the user to detect very efficiently, both in terms of computational speed and memory allocation, any similarity match between sequence strings perturbed by up to $w - k$ differences (i.e., mutations, insertions, deletions). A suitable choice of k is fundamental to ensure a good compromise between sensitivity, selectivity, and efficiency. The quantity $w - k$ is defined as the *precision factor* ϕ, and $w = 8$ and $k = 7$ represent efficient default values for most applications. The sequences to be analyzed are sorted preliminarily according to their decreasing length so that the first sequence compared (i.e., the longest) is defined as the *primary sequence* and all others are *secondary sequences*.

The correct hooking of similar sequence shortmers in compared sequences is a crucial starting step for assessing the pattern of sequence similarity between compared sequences. If no suitable hooking point is found for a pair of sequences, the searching phase is skipped completely, which saves a considerable amount of computation time. Starting from the hooking pattern, the search proceeds until regional similarity drops below a cutoff score fixed by the user.

An example of CleanUP output is shown in Fig. 6.8. It has been applied to a data set of 362 sequences, accounting for all the human mitochondrial DNA sequences available in release 41 of the EMBL database and using a cutoff parameter of 95% for both overlapping and similarity. In the same format as the input sequences, CleanUP automatically generates a cleaned sequence collection by erasing all those sequences that are found to show a degree of similarity and overlapping with longer sequences above the user-fixed threshold.

```
                C L E A N U P  -  MAIN SEARCH FILE
========================================================================
Sequence 1  Sequence 2    L1    L2    P1    P2    OS1   OS2  %Ident.  %Overl.
------------------------------------------------------------------------
MIHSCG      HSCOXII     16569  708   7584    1    709   708   99.86   100.00
   .        HSMTALT1      .    360  16024    1    360   360  100.00   100.00
   .        HSMTNA02      .    360  16024    1    360   360   98.33   100.00
   .        HSMTNA03      .    360  16024    1    360   360   98.06   100.00
   .        HSMTNA04      .    360  16024    1    360   360   98.89   100.00
   .        HSMTNA05      .    360  16024    1    360   360   98.33   100.00
   .        HSMTNA06      .    360  16024    1    360   360   98.06   100.00
   .        HSMTNA07      .    360  16024    1    360   360   97.78   100.00
   .        HSMTNA08      .    360  16024    1    360   360   98.61   100.00
   .        HSMTNA09      .    360  16024    1    360   360   98.89   100.00
   .        HUMMTALT0     .    360  16024    1    360   360   99.72   100.00
   .        MIHS01        .    608    324    1    610   608   99.51   100.00  (c)
   .        MIHSAIA       .    360  16024    1    360   360   98.89   100.00
   .        MIHSAIAA      .    360  16024    1    360   360   99.44   100.00
   .        MIHSAIAB      .    360  16024    1    360   360   99.17   100.00
   .        MIHSAIB       .    360  16024    1    360   360   99.17   100.00
   .        MIHSAIC       .    360  16024    1    360   360   99.44   100.00
   .        MIHSAID       .    360  16024    1    360   360   99.17   100.00
   .        MIHSAIE       .    360  16024    1    360   360   98.06   100.00
   .        MIHSAIF       .    360  16024    1    360   360   98.61   100.00
   .        MIHSAIG       .    360  16024    1    360   360   98.06   100.00
   .        MIHSAIH       .    360  16024    1    360   360   98.33   100.00
   .        MIHSAII       .    360  16024    1    360   360   98.33   100.00
   .        MIHSAIJ       .    360  16024    1    360   360   98.33   100.00
   .        MIHSAIK       .    360  16024    1    360   360   98.61   100.00
   .        MIHSAIL       ,    360  16024    1    360   360   98.89   100.00
   .        MIHSAIM       .    360  16024    1    360   360   98.61   100.00
   .        MIHSAIN       .    360  16024    1    360   360   98.33   100.00
   .        MIHSAIO       .    360  16024    1    360   360   98.06   100.00
   .        MIHSAIP       .    360  16024    1    360   360   98.89   100.00
   .        MIHSAIQ       .    360  16024    1    360   360   98.89   100.00
   .        MIHSAIR       .    360  16024    1    360   360   98.89   100.00
   .        MIHSAIS       .    360  16024    1    360   360   98.61   100.00
   .        MIHSAIT       .    360  16024    1    360   360   98.61   100.00
   .        MIHSAIU       .    360  16024    1    360   360   99.17   100.00
   .        MIHSAIV       .    360  16024    1    360   360   99.17   100.00
   .        MIHSAIW       .    360  16024    1    360   360   98.89   100.00
   .        MIHSAIX       .    360  16024    1    360   360   98.89   100.00
   .        MIHSAIY       .    360  16024    1    318   318   98.43    88.33
   .        MIHSALT01     .    360  16024    1    360   360   98.61   100.00
   .        MIHSALT02     .    360  16024    1    360   360   98.89   100.00
   .        MIHSALT03     .    360  16024    1    360   360   99.17   100.00
   .        MIHSALT04     .    360  16024    1    360   360   98.61   100.00
   .        MIHSALT05     .    360  16024    1    360   360   98.61   100.00
   .        MIHSALT06     .    360  16024    1    360   360   99.17   100.00
   .        MIHSALT07     .    360  16024    1    360   360   98.89   100.00
   .        MIHSALT08     .    360  16024    1    262   262   98.47    72.78
   .          .           .       16287  264    97    97   98.97    26.94
   .        MIHSALT10     .    360  16024    1    360   360   98.61   100.00

                    ( ...... Continued .......)

========================================================================
```

Figure 6.8. Output produced by CleanUP application on a sample of 362 sequences extracted from release 41 of the EMBL database reporting pairwise % of identity and of overlapping. L1, L2, length of primary (L1) and secondary (L2) sequence; P1, P2, starting position of the similarity region in the primary (P1) and secondary (P2) sequence; OS1, OS2, length of the overlapping segment in the primary (OS1) and secondary (OS2) sequence; (c), similarity match found on the reverse/complementary strand of the secondary sequence.

6.3.6 Measure of the Similarity Degree between Homologous Sequences

Evolutionary relatedness between homologous sequences can be expressed in terms of their degree of similarity. This can be measured after determining the best alignment of homologous sequences so that each pair of aligned residues presumably has a common evolutionary descent. To this end, insertion/deletion events occurring after sequence divergence need to be accommodated accurately in the alignment. Since best alignment algorithms are designed to maximize the similarity between sequences being compared, the scoring of similarity is crucial and strongly influences the resulting alignment. The similarity degree S_{AB} calculated on the best alignment for a pair of sequences, A and B, of length L (aligned sequences have the same length) can be calculated as

$$S_{AB} = \sum_{i=1}^{L} s(a_i, b_i) - \sum_{j=1}^{NG} [\gamma + \delta(\text{len}(j) - 1)]$$

where to the first term of the expression $s(a_i, b_i)$, giving the similarity score for aligned residues a_i and b_i at the ith position of the alignment excluding gapped sites, is subtracted the *gap penalty*. The gap penalty is proportional to the number of gaps, NG, and to their length, $\text{len}(j)$, γ being a constant penalty for gap opening and $\delta(\text{len}(j) - 1)$ a variable penalty for gap extension proportional to its length $\text{len}(j)$ (see also Sec. 6.3.2). The similarity score can be obtained from similarity tables, where a specific score is assigned to each pair of residues.

The simplest possibility is the *unitary scoring matrix* (or *identity matrix*) which assigns the same score to all matches or mismatches [e.g., $s(a_i, b_i) = 1$ if $a_i = b_i$ and $s(a_i, b_i) = 0$ if $a_i \neq b_i$]. This scoring system is appropriate for DNA and RNA comparisons but not for protein sequences, where different matches or mismatches among the 210 possible amino acid pairs require different scores because of the specific structures and properties of the 20 amino acids. In the comparison of nucleotide sequences, the only possible variant of the unitary matrix is the use of a matrix that scores differently nucleotide substitutions in the class of transitions (purine $\leftrightarrow$ purine or pyrimidine $\leftrightarrow$ pyrimidine) and in the class of transversions (purine $\leftrightarrow$ pyrimidine and vice versa), where a higher similarity score is assigned to the former, due to their higher probability of occurrence during evolution (Topal and Fresco 1976). However, the broad observed variability of the transition/transversion ratio makes practically unfeasible the use of matrices that weight transitions and transversions differently.

In protein sequence comparison, the similarity score between any pair of residues is generally based on the expectation that the amino acid substitution observed is likely to preserve or disrupt the essential structural and functional features of the protein. Indeed, if interconverted amino acids possess similar physical and chemical properties, no drastic changes are expected in the structure and function of a given protein. Alternatively, the minimum mutation distance matrix has been proposed, which based on the genetic code, scores an amino acid substitution according to the number of nucleotide substitutions necessary to interconvert the codon coding for the first amino acid into the one coding for the second amino acid. According to this matrix, a higher similarity score is attributed to the change Phe → Leu, which involves a single nucleotide change in the third codon position (UUY

→ UUR), than to the change Asp → Arg, which always involves three nucleotide changes (GAY → AGR). Indeed, it is generally true, as in the example above, that interconversion between amino acids with similar features requires fewer nucleotide substitutions than between amino acids with drastically divergent features.

Feng, Johnson et al. (1984) proposed a matrix that takes into account both structural similarity and genetic interconvertibility of amino acids. A real improvement in the determination of similarity score matrices has been derived from empirical studies on the frequency of replacements of each amino acid into any other during evolution of homologous proteins. This approach was pioneered by M. Dayhoff in the 1970s, who incorporated generation and selection of mutations and has been successful in several sequence alignment applications.

PAM Matrices   The point accepted mutation (PAM) similarity matrix proposed by Dayhoff, Schwartz et al. (1978) is based on a compilation of amino acid substitutions observed on a collection of 71 families of homologous proteins. The amino acid substitutions observed are denoted as *point accepted mutations* because they refer to mutations that occurred in the gene template which have produced an amino acid replacement accepted by natural selection. This approach involves a careful alignment of proteins belonging to each family and construction of the relevant phylogenetic tree for each family. The principle of maximum parsimony (see Sec. 6.11.3) is used for construction of the phylogenetic tree, so that ancestral sequences are computed for each of the internal nodes in the tree, and amino acid replacements are counted along each branch of the tree.

To minimize the effect of multiple replacements in the same position, which by definition cannot be accounted for by maximum parsimony reconstruction for each protein family, only closely related sequences are to be considered (i.e., those showing a percent identity greater than 85%). A total of 1572 amino acid replacements were considered in the original formulation of the Dayhoff PAM matrix. Note that amino acid replacements calculated on the maximum parsimony tree are directional, from the ancestral to the descendant amino acid. From this 20×20 matrix of replacements observed, the symmetrical A_{ij} matrix is computed by averaging reciprocal entries. The A_{ij} matrix can be used to calculate the relative mutability of each amino acid m_j, given by

$$m_j = \sum_{\substack{i=1 \\ i \neq j}}^{20} \frac{A_{ij}}{f_j}$$

which corresponds to the ratio between the changes involving that amino acid and its average occurrence f_j. Given A_{ij} and m_j, a replacement probability matrix M_{ij} can be calculated that gives the probability that amino acid i is replaced by amino acid j during the evolution of two homologous sequences having length 100 amino acids and diverging for a given number of point accepted mutations:

$$M_{ij,i \neq j} = \frac{\lambda m_j A_{ij}}{\sum_i A_{ij}} \qquad M_{ii} = 1 - \lambda m_i$$

The normalization coefficient λ is chosen so that two homologous sequences 100 amino acids long diverge by only one PAM (PAM1 matrix). For the purpose of database searching or alignment of proteins containing divergent regions, it is better to use a matrix that reflects a greater number of mutations. To obtain such a matrix, the *n*th power of PAM1 matrix yields a matrix corresponding to amino acid replacement probabilities after *n* PAMs of evolutionary change. The PAM250 matrix is generally used, which corresponds to a pair of sequences that accepted 250 mutations per 100 residues, corresponding to an observed identity percentage of about 20%. In general, for a given PAM distance the percentage of amino acids that will be observed to change is given by

$$100\left(1 - \sum_i f_i M_{ii}\right)$$

The elements of the unsymmetrical M_{ij} matrix give the probability that amino acid *j* will change to *i* at any site of two related sequences whose divergence is given by the specific PAM distance. To assign a similarity score to each pair of residues, the symmetrical relatedness log-odds matrix is computed as

$$P_{ij}^{(d)} = 10\log_{10}\frac{M_{ij}^{(d)}}{f_i}$$

where *d* is the PAM distance. The division of each element M_{ij} by the frequency of the *i*th amino acid gives the probability of $i \rightarrow j$ replacement, which is normalized for the relative exposure to mutation of amino acid *i*. The log transformation of the odds matrix is taken to allow probabilities to be summed over all the alignment positions rather than requiring multiplication. The log-odds matrix PAM250 is shown in Fig. 6.9, and positive and negative scores denote amino acid replacements that occur, respectively, more or less often than in random sequences of the same composition. Positive scores denote conservative substitutions. Of course, the elements along the diagonal of the matrix will contain only positive scores, which, however, vary appreciably. This variation reflects both the average occurrence frequency of the relevant amino acid in proteins and the probability of observing its replacement by another amino acid. The frequent amino acids that can be conservatively replaced (e.g., Ala or Ser) have lower scores than those of rare amino acids (e.g., Trp), which are replaced infrequently.

BLOSUM Matrices The PAM family matrices are based on amino acid replacement probabilities derived from the alignment of proteins that are at least 85% identical. However, a higher efficiency of alignment algorithms is required to detect the best alignment between divergent sequences. For this reason, Henikoff and Henikoff (1992) derived substitution matrices denoted as BLOSUM (blocks substitution matrix). Their approach considered a large number (>2000) of ungapped local alignments of protein sequences (or blocks) as corresponding to the conserved regions of some 100 protein families. Several empirical tests have shown that BLOSUM matrices provide a more accurate estimate of the similarity degree between two homologous proteins and, consequently, perform better then the Dayhoff model in both best alignment and database-searching applications.

A	2																			
R	-2	6																		
N	0	0	2																	
D	0	-1	2	4																
C	-2	-4	-4	-5	12															
Q	0	1	1	2	-5	4														
E	0	-1	1	3	-5	2	4													
G	1	-3	0	1	-3	-1	0	5												
H	-1	2	2	1	-3	3	1	-2	6											
I	-1	-2	-2	-2	-2	-2	-2	-3	-2	5										
L	-2	-3	-3	-4	-6	-2	-3	-4	-2	2	6									
K	-1	3	1	0	-5	1	0	-2	0	-2	-3	5								
M	-1	0	-2	-3	-5	-1	-2	-3	-2	2	4	0	6							
F	-3	-4	-3	-6	-4	-5	-5	-5	-2	1	2	-5	0	9						
P	1	0	0	-1	-3	0	-1	0	0	-2	-3	-1	-2	-5	6					
S	1	0	1	0	0	-1	0	1	-1	-1	-3	0	-2	-3	1	2				
T	1	-1	0	0	-2	-1	0	0	-1	0	-2	0	-1	-3	0	1	3			
W	-6	2	-4	-7	-8	-5	-7	-7	-3	-5	-2	-3	-4	0	-6	-2	-5	17		
Y	-3	-4	-2	-4	0	-4	-4	-5	0	-1	-1	-4	-2	7	-5	-3	-3	0	10	
V	0	-2	-2	-2	-2	-2	-2	-1	-2	4	2	-2	2	-1	-1	-1	0	-6	-2	4
	A	**R**	**N**	**D**	**C**	**Q**	**E**	**G**	**H**	**I**	**L**	**K**	**M**	**F**	**P**	**S**	**T**	**W**	**Y**	**V**

Figure 6.9. PAM250 mutation score matrix. Positive values represent evolutionarily conservative replacements occurring more frequently than expected on the basis of the amino acid composition.

In a single protein block made of s sequences of w amino acids, the probability q_{ij} to observe the i,j amino acid pair is

$$q_{ij} = \frac{f_{ij}}{\sum_{i=1}^{20} \sum_{j=1}^{i} f_{ij}}$$

where f_{ij} is the frequency of the i,j amino acid pair observed for all the columns of all the blocks considered. The probability of occurrence of the ith amino acid in the pair i,j is

$$p_i = q_{ii} + \frac{\sum_{j \neq i} q_{ij}}{2}$$

Then the expected probability e_{ij} can be calculated for each pair of amino acids i,j equal to $p_i p_j$ for $i = j$ and $p_i p_j + p_j p_i = 2p_i p_j$ for $i \neq j$. The s_{ij} log odds ratio matrix is calculated in bit units as

	A	R	N	D	C	Q	E	G	H	I	L	K	M	F	P	S	T	W	Y	V
A	4																			
R	-1	5																		
N	-2	0	6																	
D	-2	-2	1	6																
C	0	-3	-3	-3	9															
Q	-1	1	0	0	-3	5														
E	-1	0	0	2	-4	2	5													
G	0	-2	0	-1	-3	-2	-2	6												
H	-2	0	1	-1	-3	0	0	-2	8											
I	-1	-3	-3	-3	-1	-3	-3	-4	-3	4										
L	-1	-2	-3	-4	-1	-2	-3	-4	-3	2	4									
K	-1	2	0	-1	-3	1	1	-2	-1	-3	-2	5								
M	-1	-1	-2	-3	-1	0	-2	-3	-2	1	2	-1	5							
F	-2	-3	-3	-3	-2	-3	-3	-3	-1	0	0	-3	0	6						
P	-1	-2	-2	-1	-3	-1	-1	-2	-2	-3	-3	-1	-2	-4	7					
S	1	-1	1	0	-1	0	0	0	-1	-2	-2	0	-1	-2	-1	4				
T	0	-1	0	-1	-1	-1	-1	-2	-2	-1	-1	-1	-1	-2	-1	1	5			
W	-3	-3	-4	-4	-2	-2	-3	-2	-2	-3	-2	-3	-1	1	-4	-3	-2	11		
Y	-2	-2	-2	-3	-2	-1	-2	-3	2	-1	-1	-2	-1	3	-3	-2	-2	2	7	
V	0	-3	-3	-3	-1	-2	-2	-3	-3	3	1	-2	1	-1	-2	-2	0	-3	-1	4
	A	**R**	**N**	**D**	**C**	**Q**	**E**	**G**	**H**	**I**	**L**	**K**	**M**	**F**	**P**	**S**	**T**	**W**	**Y**	**V**

Figure 6.10. BLOSUM62 mutation score matrix.

$$s_{ij} = \log_2 \frac{q_{ij}}{e_{ij}}$$

The BLOSUM matrix is derived by multiplying each element of the matrix by a scaling factor of 2 and rounding it to the nearest integer. As in PAM matrices, positive and negative values correspond to amino acid pairs more or less frequently than expected.

To reduce the contribution of amino acid pairs belonging to the most closely related members of a family, all sequences showing an identity percentage above a specified threshold value are clustered. When f_{ij} amino acid pairs are counted, each cluster is weighted as a single sequence. In this way, variation of the clustering threshold leads to a family of matrices. Figure 6.10 shows the BLOSUM62 matrix, the one most commonly used, which corresponds roughly to the PAM160 matrix.

There are several differences between BLOSUM and PAM matrices. The principal difference is that in PAM matrices mutation rates derive from an explicit evolutionary model where replacements are counted on the branches of a phylogenetic tree depicting relationships between closely related proteins and an extrapolation procedure is used to model distant relationships. BLOSUM matrices, instead, rely on amino acid replacement frequencies observed directly on the conserved blocks. Second, PAM matrices are based on mutations observed throughout global alignments, including both conserved and variable regions, whereas BLOSUM matrices are based only on highly conserved regions, where gaps are not allowed. This is probably an advantage, as database searches generally identify distantly related proteins from significant matches with the conserved regions of homologous proteins. Furthermore, BLOSUM matrices are based on a larger and more representative data set than that used to derive the Dayhoff matrices, although an updated substitution matrix has been derived by Jones, Taylor et al. (1992) by examining 2621 protein families.

Despite the differences noted above both PAM and BLOSUM matrices perform quite well in most cases. The ability of different scoring matrices to distinguish between chance and biologically meaningful alignments has been examined in detail (Altschul 1991, 1993; Henikoff and Henikoff 1993), and PAM120 or BLOSUM62 revealed the most effective choice in most cases.

More recently, other substitution matrices have been derived from protein pairs of sequence structures showing low or no sequence similarity, which proved to perform well for comparison of distantly related sequences (Prlic, Domingues et al. 2000) or were designed to model the substitution pattern of mitochondrial encoded proteins (mtREV matrix; Adachi and Hasegawa 1996). Being essentially membranous proteins, mitochondrial proteins present an amino acid composition quite different from that of nuclear-coded proteins and a specific substitution pattern. All matrices above can also be used to provide a state transition probability matrix to maximum likelihood reconstruction of phylogenetic trees (see also Sec. 6.11.3).

6.4   DATABASE SEARCHING

The most effective way to assign a putative function to a newly determined sequence is to determine similarity with other sequences of known function. This is generally done through a database searching approach, where the query sequence submitted is compared to all sequences in the database Indeed, if the unknown query sequence shares stretches of similar sequence (i.e., shows regions of local similarity) with one or more database entries whose function is already known, it may be inferred that the unknown sequence is likely to have a similar structure and probably the same or similar function or biochemical activity. This approach, allowing to establish the putative function of the query sequence by analogy, is by far the most widely used in bio-informatic analysis. In addition to the prediction of the biological function of an anonymous sequence, database searching can be very useful in detecting functional relationships between proteins of known function. For example, the recognition of sequence similarity through database searching has allowed recognition of novel protein families such as the tyrosin-kinase oncogenes, the growth hormone receptors, the G-protein coupled receptors, or the zinc-finger transcription factors.

The result of database searching is a ranked list reporting matches found between the query sequence and the sequences in the database searched. The matching hits are ranked according to the alignment score or to other statistics; generally, only hits having a score above a predefined threshold are listed. The comparison of a query sequence to all sequences in a specified database requires the implementation of specific algorithms responding essentially to four requisites: (1) speed, (2) coverage, (3) sensitivity, and (4) selectivity. *Speed* is fundamental to accomplish a large number of pairwise comparisons, depending on the size of the database, in a reasonably short computational time. *Coverage* can be defined as the ability of the searching algorithm to pick up all the appropriate database sequences actually homologous to the query sequence, irrespective of the alignment score measured. Full coverage is then reached if all homologous sequences are present in the top-ranking hit list. Coverage does not take into account whether the correct matches are statistically significant. If a specific threshold score is assumed, it is possible to rank all significant hits (i.e., above the threshold), denoted as predicted positives (PPs), as either true positives (TPs, homologous matching sequences) or false positives (unrelated matching sequences). If the actual number of homologous sequences contained in the searched database is known (i.e., APs, actual positives), the *sensitivity* and *selectivity* of the database searching analysis can be established as

$$\text{sensitivity} = \frac{\text{TP}}{\text{AP}}$$

$$\text{selectivity} = \frac{\text{TP}}{\text{PP}}$$

The hit threshold is usually based on the statistical significance of the sequence match (i.e., only those hits whose statistical significance is above the threshold will be listed in the database searching output). An ideal method should consider as statistically significant only sequence matches, with actual homologous sequences or domains thus maximizing coverage, sensitivity, and selectivity. However, it generally happens that an increase in sensitivity improves coverage but involves a decrease in selectivity. Thus an optimal trade-off between sensitivity (the ability to identify distantly related sequences) and selectivity (the avoidance of false positives, i.e., unrelated sequences with high statistically significant similarity) is the right choice.

Given that sequence similarity between the query sequence and the database entry may be limited to specific structural motifs or to the active site, a local alignment algorithm should be used to compare the user-submitted query sequence and all the subject entries of the database searched. To speed up sequence comparison, database searching methods apply the strategy of breaking sequences into short runs of nucleotides or amino acids called *words* and then comparing the vocabulary of words in the query sequence and in database entries under the basic assumption that two related sequences must share at least one word. The length w of the words to be used is depends strictly on the size of the alphabet D (four for DNA/RNA and 20 for protein sequences), where the expected frequency of word occurrence is given by $1/D^w$. This means that if we want less than 1 expected hit per 100,000 residues, $w = 4$ for proteins and $w = 9$ for nucleic acids should be used.

The most commonly used algorithms for database searching are FASTA (Pearson and Lipman 1988) and BLAST (Altschul, Gish et al. 1990), for which summary descriptions are given below.

6.4.1 FASTA

The FASTA algorithm (Pearson and Lipman 1988; Pearson 1990, 1996, 2000) has been the first widely used database searching tool. It detects sequence similarity between a query sequence and a group of sequences using a heuristic method across four consecutive phases. To speed up the search in the first phase, a lookup table is created with the position of all words in the query sequence and in all sequences in the database. The size of the word, here defined as the *ktup* parameter, is crucial to control the trade-off between speed and sensitivity.

For example, given the following query sequence **A** and database sequence **S**:

Sequence	1	2	3	4	5	6	7	8	9	10	11
A	Y	W	L	T	R	S	Y	W	A	L	
S	W	L	T	R	H	A	S	Y	W	G	A

the following lookup table for $w = 1$ is generated, where the offset reports the positional difference for each amino acid:

Amino Acid	Position in **A**	Position in **S**	Offset
Y	1, 7	8	−7, −1
W	2, 8	1, 9	1, −7, 7, −1
L	3, 10	2	1, 8
T	4	3	1
R	5	4	1
S	6	7	−1
H		5	
A	9	6, 11	3, −2

If the comparison is viewed as a dot-plot matrix with the query sequence as the vertical sequence and the database compared sequence as the horizontal sequence, the offset value defines the diagonal where a word is registered. If two sequences of length L1 and L2 are compared, a total of L1 + L2 − 1 diagonals are given, with the offset ranging between 1 − L2 and L1 − 1. In the first step, the diagonals (offset) having the largest number of word matches are recorded for each comparison (by default, the first 10). In the example given, the diagonals with offset +1 and −1 have the highest density of matches: four and three, respectively.

In the second step, these best similarity regions are re-scored using scoring matrices such as PAM (Dayhoff, Schwartz et al. 1978) and BLOSUM (Henikoff and Henikoff 1992) for protein comparison. The highest scoring value of these initial similarity regions, defined as the INIT1 score, is thus computed for each pairwise comparison between the query sequence and each of the database sequences. In the example above, the score for the two best initial similarity regions having offset +1 (WLTR) and −1 (SYW) using the PAM250 matrix is +32 and +29, respectively, thus making INIT1 = +32 for this comparison.

In the third step, the algorithm verifies if the initial similarity regions defined previously for each pairwise comparison can be joined to form a single alignment. The joining between two best initial similarity regions is made if the penalty score assigned to their intervening gap is lower than the score assigned to the joined segment. The following alignment can thus be derived for the sequence comparison in the example given:

```
A    W L T R - - S Y W
     | | | |     | | |
S    W L T R H A S Y W
```

The resulting score is called INITN and is used to rank all database matches in the FASTA output in decreasing order. The default gap penalties are −12 for gap creation and −4 for each further gap extension. The INITN score for the resulting alignment above is thus given by 45 = 32 (score for the best initial region) + 29 (score for the second-best initial region) − 16 (gap penalty).

Sequence matches with an INITN score above a fixed threshold (usually, 20) are optimized through a pairwise alignment procedure (Smith and Waterman 1981; Chao, Pearson et al. 1992) that finally provides the optimized OPT score. In our example, OPT = 46, given by the INITN score (45) plus the score (+1) of the Ala–Gly match included in the final optimized alignment:

```
A    W L T R - - S Y W A
     | | | |     | | | :
S    W L T R H A S Y W G
```

More recently, the FASTA program has been modified to provide the statistical significance of similarity hits found (Pearson 1998). A normalized z-score is calculated from the OPT score to remove its dependence on the natural log of the search set sequence length [see Fig. 1 in Pearson (1996)]. The distribution of z-scores should follow the extreme value distribution (Altschul, Boguski et al. 1994), so that an expectation value can be calculated for each alignment representing the expected number of random alignments with z-score greater than or equal to the observed value.

A typical FASTA output is shown in Fig. 6.11. The first part of the output file contains a histogram showing the number of matching regions between the query and the database sequence searched observed for a given z-score range. Each bin of the histogram reports the number of observed (=) and expected (*) alignments having a z-score ranging between the one reported in the leftmost column of the histogram and that reported above it. Below the histogram, FASTA displays a list of best-scoring alignments showing INITN, INIT1, and OPT scores as well as the relevant z-score and the expected (E) value that gives the number of alignments having the same z-score that one would expect to occur by chance in a database of the same size. Finally, the best scoring alignments are reported.

6.4.2 BLAST

The BLAST (Basic Local Alignment Search Tool) program employs a very fast algorithm to find sequences most similar to the user-submitted query sequences, if they are present in the database. The basic idea of this method is that the two homologous sequences, even if highly divergent, probably share high-scoring similarity segments that allow us to discriminate reliably between related and unrelated sequences. Maximal segment pair alignments have the very valuable property that the statistical distribution of chance high-scoring segments is well understood (Karlin and Altschul 1990, 1993), thus making possible computation of the statistical significance of maximal segment pair alignments. On the other hand, this implies

```
/free/Embnet/Programs/Fasta3-1/fasta3_t -n -w 80 -m 6 -q - %u

FASTA searches a protein or DNA sequence data bank
version 3.1t03 February, 1998
Please cite:
W.R. Pearson & D.J. Lipman PNAS (1988) 85:2444-2448

-: 50 nt
QUERY sequence
vs  UTRnr Untranslated Sequences library

          opt      E()
< 20 45695    0:===========================================================:
   22     0    0:                     one = represents 762 library sequences
   24     0    0:
   26     0    2:*
   28     0   19:*
   30     0  112:*
   32     0  435:*
   34     0 1179: *
   36     5 2421:=   *
   38    31 4001:=     *
   40  3311 5581:==== *
   42 15405 6822:=======*============
   44  8349 7525:=========*=
   46  4593 7665:======= *
   48  4869 7338:======= *
   50  3416 6696:=====  *
   52  3207 5887:===== *
   54  2483 5028:==== *
   56  1990 4200:=== *
   58  1586 3448:=== *
   60  1242 2793:== *
   62   990 2239:==*
   64   847 1781:==*
   66   661 1408:=*
   68   563 1107:=*
   70   475  868:=*
   72   363  678:*
   74   338  529:*
   76   268  411:*
   78   258  320:*
   80   185  248:*
   82   142  190:*
   84   134  150:*
   86   113  116:*
   88    91   90:*           inset = represents 2 library sequences
   90    87   70:*
   92    91   54:*      :==========================*=============
   94    69   42:*      :====================*===============
   96    55   32:*      :================*=============
   98    44   25:*      :============*=========
  100    35   19:*      :========*========
  102    27   15:*      :=======*======
  104    26   12:*      :=====*=======
  106    24    9:*      :====*=======
  108    20    7:*      :===*======
  110    19    5:*      :==*=======
  112    11    4:*      :=*====
  114    13    3:*      :=*=====
  116    10    2:*      :*====
  118     3    2:*      :*=
 >120    25    1:*      :*===========
37480376 residues in 102169 sequences
 statistics extrapolated from 50000 to 81597 sequences
 Expectation_n fit: rho(ln(x))= 1.2469+/-0.000532; mu= 30.3024+/- 0.032;
 mean_var=101.8200+/-16.522, Z-trim: 92 B-trim: 0 in 0/66
 Kolmogorov-Smirnov statistic: 0.4493 (N=27) at  86

FASTA (3.12 Oct., 1997) function (optimized, DNA matrix) ktup: 6
 join: 45, opt: 30, gap-pen: -16/ -4, width:  16 reg.-scaled
 Scan time: 13.283
```

Figure 6.11. FASTA output obtained in the comparison of a 50-bp-long query sequence against the UTRdb database. The distribution graph of similarity scores, where an asterisk signifies the expectation value, gives a qualitative view of how well the extreme value distribution (see Sec. 6.4.6) fits the similarity scores calculated by the program. Below the graph the parameters for the distribution function and for database searching (e.g., k-tup word size, gap penalty, etc.) are shown followed by the ranked list of optimal scores and the relevant alignments. In this example, only the first matching hit can be considered statistically significant based on a $E < 0.001$ threshold. [From Pesole, Liuni et al. (2000a).]

```
The best scores are:                                                initn init1 opt  z-sc  E(81597)
5HSA001299 Bb002455 5'UTR in Human nucleoporin 98 (NUP98) mRN ( 144)  250   250  250 261.6 1.2e-07
5RNO002179 Bb038197 5'UTR in Rattus norvegicus nucleoporin (N ( 155)   91    91  146 158.4 0.062
5HSA017572 Bb064274 5'UTR in Homo sapiens CREB-binding protei ( 818)   79    79  119 129.6 0.47
5MMU009814 Bb081387 5'UTR in Mus musculus zinc finger protein ( 521)   87    87  120 131.2 0.61
5MMU001977 Bb032625 5'UTR in Mus musculus neuronal dihydropyr (1166)   75    75  114 124.2 0.67
5HSA010729 Bb042776 5'UTR in Homo sapiens RalBP1-interacting  ( 588)   95    95  116 127.0 0.92
5GGA000863 Bb021070 5'UTR in Gallus gallus beta-1,4-galactosy ( 201)   88    88  123 135.3 0.93
5MMU000343 Bb030989 5'UTR in Mus musculus calcium-activated p ( 820)   65    65  112 122.7 1.2
5HSA002342 Bb003498 5'UTR in Human bcr mRNA (break point clus ( 596)   95    65  113 124.1 1.3
5HSA018763 Bb069819 5'UTR in Human DNA sequence from clone RP ( 309)   80    80  116 127.8 1.6
5MMU002136 Bb032784 5'UTR in Mouse fibrillin (Fbn-1) mRNA, co (2379)   60    60  102 111.4 1.7
5HSA011105 Bb043152 5'UTR in Homo sapiens CD24 gene promoter  (3397)   75    75   99 108.0 1.8
5CGR000148 Bb080648 5'UTR in Cricetulus griseus folate recept (  96)  122   122  122 135.2 2
5HSA008866 Bb010022 5'UTR in Human DP2 gene, partial cds. 12/ (1933)   66    66  102 111.7 2


>>>-, 50 nt vs %u library

>>5HSA001299 Bb002455 5'UTR in Human nucleoporin 98 (NUP98) mRNA, complete cd (144 nt)
 initn: 250 init1: 250 opt: 250 Z-score: 261.6 expect() 1.2e-07
 100.000% identity in 50 nt overlap

>5HSA00    1- 50:------------------------------------------------------------:

           10        20        30        40        50
QUERY  CGGTGGCAGGGGTGGTAGCGGCGGCGGCGACGGTTTCGTGGGGGCCGCGC
       ::::::::::::::::::::::::::::::::::::::::::::::::::::
5HSA00 CGGTGGCAGGGGTGGTAGCGGCGGCGGCGACGGTTTCGTGGGGGCCGCGCTGCTCTGTGAGCGGCGGGTGGCAGCAGG
           10        20        30        40        50        60        70        80

5HSA00 GGACTCCTGACACTTCCCCTTCCCCACCGAACCGCGCTTTCTGAAACAAAGACTCATTTTGAAG
              90       100       110       120       130       140


>>5RNO002179 Bb038197 5'UTR in Rattus norvegicus nucleoporin (Nup98) mRNA, co (155 nt)
 initn:  91 init1:  91 opt: 146 Z-score: 158.4 expect() 0.062
 85.000% identity in 40 nt overlap

>5RNO00   11- 50:           ------------------------------------------------------------:

                     10        20        30        40        50
QUERY           `CGGTGGCAGGGGTGGTAGCGGCGGCGGCGGCGACGGTTTCGTGGGGGCCGCGC
                 :::::  :::::::: :::::::::::: :::  :: :::::
5RNO00 GTCTACGCGCTGCCCCGAACGGCGGTCGGTGGCGGCGGCGGTGGCGACGGTTTCCTGGCGGTCGCGCGCTGCTCTGGTGA


/// deleted stuff here ///
```

Figure 6.11 *Continued*

that gaps are not allowed in the alignment, which is simply broken into two or more high-score segment pairs (HSPs).

In the initial scanning phase the BLAST program searches for short regions of a given length in the query sequence that align with sequences in the database totaling a similarity score above a certain threshold. The similarity score is determined by using the BLOSUM or PAM matrix, described previously. BLOSUM62 is the default matrix in the case of protein comparisons, whereas for DNA sequences a positive score (default is +1) is assigned to matches and a negative score (default is −3) is assigned to mismatches.

The length of the initial region (the *word*) is by default set at 3 for protein sequences and at 11 for DNA sequences. In the case of DNA sequences, a match is recorded if a 11-mer word is found in the database to match exactly a word in the query sequence. A word length of 11 should exclude almost all chance alignments. In the case of protein sequences a match is recorded if a similar word is found in

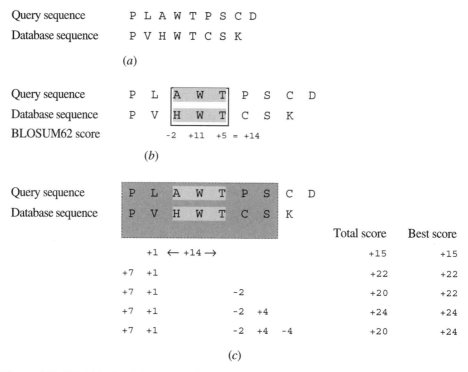

Figure 6.12. BLAST algorithm basics. (*a*) Query and database sequence; (*b*) search of similar "words" with similarity above threshold T ($T = 11$); (*c*) Extension of word in both directions to define the HSP (shaded region).

the database with a score equal to or greater than a threshold T (default is 11 using BLOSUM62).

One principal innovation introduced in the BLAST algorithm is that for each w-mer present in the query sequence, a list of neighborhood or synonymous words having score $\geq T$ is determined. For any word in a database sequence, the BLAST search finds its occurrence and position as well as that of any synonymous word in the query sequence. Figure 6.12 illustrates the basics of the BLAST algorithm. Given a query sequence and a database sequence nine and eight residues long, respectively (Fig. 6.12*a*), a search is done for synonymous words (i.e. having a similarity score above the threshold $T = 11$, Fig. 6.12*b*). Initial hits are extended in each direction until the score drops by a certain amount from its maximum (drop threshold is set by default at 22 for proteins and 20 for DNA sequences; Fig. 6.12*c*). At this point, the best-scoring extended region is designated HSP (high scoring pair) if the similarity score is greater than a cutoff value S (HSP is shaded in Fig. 6.12). A suitable S cutoff is chosen so that the expected times that a value greater than or equal to S can occur purely by chance is lower than E times for that particular database search (the default E is 10). In Fig. 6.12, one HSP of length 7 is found with a score of 24.

After the pairwise comparisons, the maximum-scoring HSPs, called *maximal scoring pairs* (MSPs), are listed and corresponding alignments are displayed in the output. The most critical parameter determining the speed and sensitivity of the BLAST search is T. If T is increased, the total number of word hits drops drastically and makes the search considerably faster but at the same time less sensitive, because more distant relationships are neglected. If two or more MSPs are found in the same pairwise comparison, all of them are reported and the higher significance of these multiple hits is modeled as a Poisson process. The Poisson P value shown in the BLAST output is the probability that at least as many HSPs can occur by chance, each with score at least as high as the lowest-scored member of the group.

The computation time requested by the BLAST program is proportional to the product of the length of the query sequence and of the database searched. Given the exponential growth of database sizes, the BLAST algorithm has recently been modified to enhance both speed and sensitivity. A two-hit method has now been introduced that requires the existence of two nonoverlapping word pairs on the same diagonal, whose distance is $\leq A$ (default $A = 40$ for comparisons at the protein level). At the same time, the threshold T can be decreased to increase the sensitivity.

A further improvement in the BLAST algorithm is the ability to provide a gapped alignment. In the heuristic approach used by the BLAST program, the gapped local alignment is built starting from a pair of aligned amino acids, called the *seed*, located in the central position of the 11 mer segment with the highest alignment score. Then the standard dynamic programming algorithm for pairwise sequence alignments proceeds in both directions until the alignment score drops to no more than Xg (default = 22) below the best score yet computed. The algorithm is thereby able to adjust to the data the region of the path graph explored, whose dimensions are typically the lengths of the two sequences being compared.

A typical BLAST output is shown in Fig. 6.13, where the highest MSPs are listed in decreasing order. It starts with some header information that lists the program used (e.g., BLASTP) and the features of the query sequence and of the database searched. Then a graphical representation of BLAST hits is reported with matching sequence colors, corresponding to a given alignment score range. Then an one-line description of database matches is shown with the database sequence identifier, the corresponding *definition line*, the score (in bits units), and the statistical significance (E value) (see Sec. 6.4.6 for their description). In the given example the first match is the actual query sequence. Afterward, the best alignments are shown with the query sequence at the top and the matching database sequence labeled "Subject." Identical residues are shown between the two sequences, whereas conservative changes are labeled "+." For each alignment the bits score and the raw score (in parentheses) are reported, as well as the percentage of identical residues ("Identities") and conservative changes ("Positives"). In the last section of the output the database search parameters are shown.

6.4.3 BLAST and FASTA Family of Programs

The versions of BLAST and FASTA available are listed in Table 6.1. The decision about which program to use depends basically on the type of data to be compared. Comparison at the protein level is generally more sensitive for detecting distant homologies. Thus, if a DNA sequence contains a region coding for proteins, the sequence comparison at the protein level is normally preferred for its higher sensitivity (i.e., two homologous genes differ more at the nucleotide level than at the protein level). However, there are a number of circumstances when protein comparison is not practicable: for example, when comparing non coding DNA sequences.

In case one is looking for homologous proteins in newly completed genomes, it is advisable to carry out the comparison not only against the annotated protein collection of that genome but also against all possible six-frame translations of the genome sequence (i.e., using TBLASTN or TFASTA). Indeed, it should be taken into account that some annotated genes may not be true functional genes but also that not all true functional genes might have been identified already.

BLAST is generally the first choice in database-searching experiments because it is certainly the fastest program, but FASTA is certainly more sensitive than BLAST in the comparison of DNA sequences. The gapped version of BLAST provides a single alignment block including gaps, but it could sometimes be more advantageous to use the ungapped version, as in the case of multiple matching blocks, when long gaps are requested in the alignment (e.g., the alignment of a intron-containing genomic sequence with the corresponding mRNA sequence). Furthermore, the gapped BLAST version can be inaccurate in the case of frameshift errors in the CDS portion of a DNA sequence when searching a protein database using BLASTX. The example in Fig. 6.14 shows clearly that the ungapped BLAST alignment performs better than gapped BLAST when detecting similarities between an EST sequence of *Arabidopsis thaliana* and its protein product.

Among other programs available to carry out similarity searches are SSearch (also included in the GCG package), MPSRCH, and Scanps, which perform a rigorous Smith–Waterman (Smith and Waterman 1981) comparison, thus increasing remarkably the sensitivity of the search but at the expense of increased computation time, and the MegaBLAST program, which implements a greedy algorithm for the DNA sequence gapped alignment search. Still implementing the basic BLAST algorithm, WU-BLAST offers added functionality and the setting of additional parameters. Very recently the BLAT (Kent 2002) and SSAHA (Ning, Cox et al. 2002) programs have been developed that perform searches on databases containing multiple gigabases of DNA much faster than BLAST or FASTA programs. These programs were specifically designed for rapid mRNA/DNA and cross-species alignments in vertebrate genomes. Table 6.2 lists main database searching programs that can be run over the Web.

6.4.4   Filtering Matches to Unwanted Sequences

Amino acid or nucleotide sequences may contain highly repetitive regions using a highly restricted variety of w-mer words. These regions, made up of homopolymers,

Figure 6.13. Typical BLAST output, showing a graphical representation of matching hits between the query and the database sequences using a color code for the similarity value, the list of matching hits reporting the similarity score (in bits) and the E value, and the relevant pairwise alignments. The search statistics and parameters are listed at the end of the output. (Output obtained by a database search carried out at the NCBI server *http://www.ncbi.nlm.nih.gov/BLAST.*)

BLASTP 2.1.2 [Nov-13-2000]

Reference:
Altschul, Stephen F., Thomas L. Madden, Alejandro A. Schäffer,
Jinghui Zhang, Zheng Zhang, Webb Miller, and David J. Lipman (1997),
"Gapped BLAST and PSI-BLAST: a new generation of protein database search
programs", Nucleic Acids Res. 25:3389-3402.

Query= gi|117086|sp|P13073|COX4_HUMAN CYTOCHROME C OXIDASE
POLYPEPTIDE IV PRECURSOR.
       (169 letters)
Database: nr
     597,014 sequences; 188,726,214 total letters

If you have any problems or questions with the results of this search
please refer to the **BLAST FAQs**

Taxonomy reports

Distribution of 35 Blast Hits on the Query Sequence

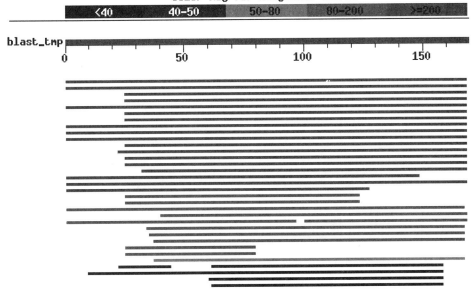

	Score	E
Sequences producing significant alignments:	(bits)	Value
ref|NP_001852.1| cytochrome c oxidase subunit IV [Homo sapi...	346	1e-94
sp|P00423|COX4_BOVIN CYTOCHROME C OXIDASE POLYPEPTIDE IV PR...	303	1e-81
sp|O46577|COX4_PANTR CYTOCHROME C OXIDASE POLYPEPTIDE IV >g...	299	1e-80
sp|O46578|COX4_GORGO CYTOCHROME C OXIDASE POLYPEPTIDE IV >g...	297	6e-80
sp|O46579|COX4_PONPY CYTOCHROME C OXIDASE POLYPEPTIDE IV >g...	289	1e-77

```
ref|NP_058898.1|  cytochrome c oxidase, subunit IV [Rattus n...     289   1e-7
sp|O46580|COX4_HYLAG  CYTOCHROME C OXIDASE POLYPEPTIDE IV >g...      287   4e-7
ref|NP_034071.1|  cytochrome c oxidase, subunit IV [Mus musc...     284   5e-7
gb|AAB02139.1|  (M37831) cytochrome c oxidase subunit IV [Mu...     282   2e-7
gb|AAF13254.1|AF198088_1  (AF198088) cytochrome c oxidase subu...   281   3e-7
sp|O46581|COX4_THEGE  CYTOCHROME C OXIDASE POLYPEPTIDE IV >g...      275   3e-7
pdb|2OCC|D  Chain D, Bovine Heart Cytochrome C Oxidase At Th...      264   4e-7
sp|O46584|COX4_AOTAZ  CYTOCHROME C OXIDASE POLYPEPTIDE IV >g...      261   4e-6
```

Alignments

```
ref|NP_001852.1| cytochrome c oxidase subunit IV [Homo sapiens]
Length = 169

 Score =  346 bits (887), Expect = 1e-94
 Identities = 169/169 (100%), Positives = 169/169 (100%)

Query: 1    MLATRVFSLVGKRAISTSVCVRAHESVVKSEDFSLPAYMDRRDHPLPEVAHVKHLSASQK 60
            MLATRVFSLVGKRAISTSVCVRAHESVVKSEDFSLPAYMDRRDHPLPEVAHVKHLSASQK
Sbjct: 1    MLATRVFSLVGKRAISTSVCVRAHESVVKSEDFSLPAYMDRRDHPLPEVAHVKHLSASQK 60

/////

emb|CAC18794.1|  (AL117381) dJ857M17.2 (novel protein similar to cytochrome c
            oxidase subunit IV (COX4)) [Homo sapiens]
        Length = 138

 Score =  166 bits (419), Expect = 2e-40
 Identities = 78/129 (60%), Positives = 98/129 (75%)

Query: 41   RRDHPLPEVAHVKHLSASQKALKEKEKASWSSLSMDEKVELYRIKFKESFAEMNRGSNEW 100
            +R +P+PE      L+A ++ALKEKEK SW+ L+  EKV LYR++F E+FAEMNR SNEW
Sbjct: 10   QRYYPMPEEPFCTELNAEEQALKEKEKGSWTQLTHAEKVALYRLQFNETFAEMNRRSNEW 69

/////

   Database: nr
     Posted date:  Dec 18, 2000 10:24 PM
   Number of letters in database: 188,726,214
   Number of sequences in database:  597,014

Lambda     K        H
  0.318    0.130    0.393

Gapped
Lambda     K        H
  0.267    0.0410    0.140


Matrix: BLOSUM62
Gap Penalties: Existence: 11, Extension: 1
Number of Hits to DB: 64618003
Number of Sequences: 597014
Number of extensions: 2418723
Number of successful extensions: 5422
Number of sequences better than 10.0: 35
Number of HSP's better than 10.0 without gapping: 32
Number of HSP's successfully gapped in prelim test: 3
Number of HSP's that attempted gapping in prelim test: 5388
Number of HSP's gapped (non-prelim): 35
length of query: 169
length of database: 188,726,214
effective HSP length: 53
effective length of query: 116
effective length of database: 157,084,472
effective search space: 18221798752
effective search space used: 18221798752
T: 11
A: 40
X1: 16 ( 7.3 bits)
X2: 38 (14.6 bits)
X3: 64 (24.7 bits)
S1: 41 (21.7 bits)
S2: 68 (30.8 bits)
```

Figure 6.13. *Continued*

TABLE 6.1. FASTA and BLAST Families of Programs

FASTA Program	BLAST Program	Comparison	Description
FASTA	BLASTN	DNA level	Compares a nucleotide query sequence against a nucleotide sequence database
FASTA	BLASTP	Protein level	Compares a protein query sequence against a protein sequence database
FASTX	BLASTX	Protein level	Compares the six-frame conceptual translation (both strands) of a nucleotide query sequence against a protein sequence database
TFASTA	TBLASTN	Protein level	Compares a protein query sequence against a nucleotide sequence database dynamically translated in the six reading frames
	TBLASTX	Protein level	Compares the six-frame conceptual translation (both strands) of a nucleotide query sequence against a nucleotide sequence database translated dynamically in the six reading frames
TFASTX		Protein level	Enhanced version of TFASTA, taking frameshift errors into account

TABLE 6.2. Major Database Searching Programs

Program	URL
FASTA	www.ebi.ac.uk/fasta33
BLAST	www.ncbi.nlm.nih.gov/BLAST
MPsrch	www.ebi.ac.uk/MPsrch
SSearch	www.ebi.ac.uk/bic_sw
WU-BLAST	www.ebi.ac.uk/blast2/blast.wustl.edu/
Scanps	www.compbio.dundee.ac.uk/Software/Scanps/scanps.html www.ebi.ac.uk/scanps
MegaBLAST	www.ncbi.nlm.nih.gov/BLAST
BLAT	genome.ucsc.edu/cgi-bin/hgBlat?command=start&db=hg11
SSAHA	www.sanger.ac.uk/Software/analysis/SSAHA

short-period repeats, or nonperiodic regions using only few types of characters, account for about 15% of total protein residues (Altschul, Boguski et al. 1994). Matches between such segments are more likely due to these local amino acid or nucleotide compositional bias than to actual evolutionary descent. Furthermore, regions defined as *low complexity* violate the assumptions of Karlin–Altschul

```
>sp|P25858|G3PC_ARATH GLYCERALDEHYDE 3-PHOSPHATE DEHYDROGENASE, CYTOSOLIC
 pir||JQ1287 glyceraldehyde-3-phosphate dehydrogenase (EC 1.2.1.12), cytosolic -
Arabidopsis thaliana


Length = 338

 Score =  104 bits (257), Expect = 3e-22
 Identities = 59/80 (73%), Positives = 65/80 (80%)
 Frame = +3


Query: 3    EIKKAIKEESEGKMKGILGYSEDDVVSTDFVGDNRSSIFDAKAGLHCIERQVCEVGVMVR 182
            EIKKAIKEESEGK+KGILGY+EDDVVSTDFVGDNRSSIFDAKAG+   ++ V  V
Sbjct: 261  EIKKAIKEESEGKLKGILGYTEDDVVSTDFVGDNRSSIFDAKAGIALSDKFVKLVS-WYD 319


Query: 183  QRMGLHSSRVVDLIVHMSKA 242
               G +SSRVVDLIVHMSKA
Sbjct: 320  NEWG-YSSRVVDLIVHMSKA 338
```

(a)

```
>sp|P25858|G3PC_ARATH GLYCERALDEHYDE 3-PHOSPHATE DEHYDROGENASE, CYTOSOLIC
 pir||JQ1287 glyceraldehyde-3-phosphate dehydrogenase (EC 1.2.1.12), cytosolic -
Arabidopsis thaliana


Length = 338

 Score =  101 bits (215), Expect(3) = 2e-36
 Identities = 42/45 (93%), Positives = 45/45 (99%)
 Frame = +3


Query: 3    EIKKAIKEESEGKMKGILGYSEDDVVSTDFVGDNRSSIFDAKAGL 137
            EIKKAIKEESEGK+KGILGY+EDDVVSTDFVGDNRSSIFDAKAG+
Sbjct: 261  EIKKAIKEESEGKLKGILGYTEDDVVSTDFVGDNRSSIFDAKAGI 305


 Score = 55.1 bits (114), Expect(3) = 2e-36
 Identities = 20/21 (95%), Positives = 21/21 (99%)
 Frame = +1

Query: 139  IALSDKFVKLVSWYDNEWGYT 201
            IALSDKFVKLVSWYDNEWGY+
Sbjct: 305  IALSDKFVKLVSWYDNEWGYS 325


 Score = 33.6 bits (67), Expect(3) = 2e-36
 Identities = 14/15 (93%), Positives = 15/15 (99%)
 Frame = +3

Query: 198  HSSRVVDLIVHMSKA 242
            +SSRVVDLIVHMSKA
Sbjct: 324  YSSRVVDLIVHMSKA 338
```

(b)

Figure 6.14. Gapped (*a*) and ungapped (*b*) BLASTX alignment between an *A. thaliana* EST sequence (Acc. No. Z17438) and its related protein (Acc. No. P2585). The ungapped alignments are able to detect the second similarity block missed by the gapped alignment due to a frameshift.

statistics (Karlin and Altschul 1990, 1993) that the probability of finding a residue at any particular position in a sequence is simply proportional to its composition. To avoid the problem that high similarity scores of matching low-complexity regions (LCRs) obscure other important similarities, it is worth filtering such regions in both query and database sequences. Such a practice, also known as *masking*, consists in converting residues belonging to such problematic regions to ambiguity characters ("X" for proteins and "N" for nucleotide sequences), thus preventing the contribution of such regions to the alignment score. Low-complexity regions can be identified by comparing the contrasting residue frequencies of specific clusters and those of the complete entry of a database. The program SEG (Wootton 1994) is most commonly used to detect low-complexity regions and is also employed by the BLAST program to mask low-complexity regions before executing a database search. The huntingtin protein (P42858) provides a good example of LCR-containing protein. The expansion of a glutamine repeat in its amino-terminal region is the cause of Huntington's disease. Following are the first 100 residues of the huntingtin protein before and after using the SEG program.

```
 1 MATLEKLMKA FESLKSFQQQ QQQQQQQQQQ QQQQQQQQQQ PPPPPPPPPP
   MATLEKLMKA FESLKSFXXX XXXXXXXXXX XXXXXXXXXX XXXXXXXXXX
51 PQLPQPPPQA QPLLPQPQPP PPPPPPPPGP AVAEEPLHRP KKELSATKKD
   XXXXXXXXXX XXXXXXXXXX XXXXXXXGP AVAEEPLHRP KKELSATKKD
```

SEG masks LCRs with **X**'s, thus preventing their contribution to sequence matches in database searching.

6.4.5   Filtering Matches to Repetitive Sequences

A large number of statistically significant but biologically uninteresting sequence matches can be obtained when the query sequence contains highly abundant repetitive sequences. To overcome this problem, prescreening the query sequence against a library of repetitive elements, such as RepBase (Jurka 2000), is usually carried out before executing the database search. BLAST has a built-in utility allowing masking of human repetitive sequences that is necessary in the case of database searching involving human genomic query sequences. For example, a database search between a genomic sequence and human ESTs, which may prove extremely useful in identifing actually transcribed exons, can be misleading when repetitive elements have not previously been masked. In such a case, most of the matches observed will correspond to repetitive elements (e.g., Alu's).

RepeatMasker is the most widely used software tool to accomplish this task. It compares the input sequence to a repository of repetitive elements by using a fast version of the Smith–Waterman algorithm (Smith and Waterman 1981). Repeat elements are masked when the Smith–Waterman similarity score is above a cutoff value, generally kept rather conservative to avoid false matches. The output reports a detailed annotation of the repeats that are present in the query sequence as well as a masked version of the query sequence, where all nucleotides belonging to mapped repeats are replaced by N's. This masked sequence can be used for further analyses. An example of the RepeatMasker output is shown in Fig. 6.15, where a human genome contig (AL136528) has been analyzed. For each matching element

```
Repeat sequence:
SW    perc perc perc  query       position in query    matching     repeat             position in repeat
score div. del. ins.  sequence  begin   end  (left)   repeat       class/family     begin   end  (left)

   21  0.0  0.0  0.0  AL136528    498   518 (139260) +  AT_rich     Low_complexity       1    21    (0)
  308 26.8 15.4  0.8  AL136528   2642  2764 (137014) C  MIR         SINE/MIR           (91)  171    31
   27  0.0  0.0  0.0  AL136528   3132  3158 (136620) +  GC_rich     Low_complexity       1    27    (0)
  273 37.5  5.0  0.0  AL136528   3678  3797 (135981) C  L2          LINE/L2           (563) 2750  2625
  217 19.1  0.0  4.4  AL136528   4137  4204 (135574) +  L2          LINE/L2           2925  2989  (283)
  253 22.3 18.4  1.0  AL136528   4286  4388 (135390) +  L2          LINE/L2           3138  3258   (14)
 2135 11.6  0.0  0.0  AL136528   5250  5541 (134237) +  AluJb       SINE/Alu             1   292   (20)
 1857 16.4  2.7  0.0  AL136528   7271  7568 (132210) +  AluJb       SINE/Alu             1   306    (6)
  361 32.6  7.4  3.9  AL136528  11765 11817 (127961) +  L1M4        LINE/L1           4618  4668 (1478)
  663 21.6  0.0  0.0  AL136528  11818 11942 (127836) +  FLAM_A      SINE/Alu             1   125    (8)
  361 32.6  7.4  3.9  AL136528  11943 12167 (127611) +  L1M4        LINE/L1           4668  4921 (1225)
 2291  8.0  0.7  0.0  AL136528  12168 12467 (127311) +  AluSp       SINE/Alu             1   302   (11)
  224 31.7  5.0  0.0  AL136528  15005 15165 (124613) +  (ACGTG)n    Simple_repeat        1   169    (0) *
  288 29.7  4.8  0.0  AL136528  15089 15253 (124525) +  (CGCGG)n    Simple_repeat        4   176    (0)
  919 23.2  4.2  1.9  AL136528  15839 16278 (123500) C  MER44C      DNA/MER2_type     (226)  502     4
   23  3.3  0.0  0.0  AL136528  19562 19591 (120187) +  GC_rich     Low_complexity       1    30    (0)
  575 17.6  6.1  1.4  AL136528  20001 20148 (119630) +  MER53       DNA                 20   174   (15)
  210 17.6  0.0  0.0  AL136528  20172 20222 (119556) +  (TATATG)n   Simple_repeat        4    54    (0)
  226 24.6  1.6  4.1  AL136528  22152 22273 (117505) +  (CCCG)n     Simple_repeat        4   122    (0)
 2044 12.5  0.3  1.3  AL136528  23600 23910 (115868) +  AluSx       SINE/Alu             1   308    (4)
  625 28.1  8.9  1.3  AL136528  24125 24359 (115419) +  MIR         SINE/MIR             4   256    (6)
 2093 12.7  0.0  0.0  AL136528  24816 25115 (114663) +  AluSx       SINE/Alu             1   300   (12)
 2446  4.7  0.7  0.0  AL136528  25607 25903 (113875) +  AluY        SINE/Alu             1   299   (12)
  657 30.2  1.9  0.0  AL136528  26146 26350 (113428) C  MIR         SINE/MIR            (2)  260    52
  991 27.4  5.4  0.3  AL136528  26493 26806 (112972) +  Charlie4a   DNA/MER1_type       30   359  (149)
 1163 24.0  2.2  4.2  AL136528  26821 27229 (112549) C  MLT1C       LTR/MaLR            (6)  460    60
```

(a)

```
====================================================
file name: al136528.tfa
sequences:           1
total length: 139778 bp
GC level:      57.33%
bases masked    33650 bp ( 24.07 %)
====================================================
             number of     length    percentage
             elements*    occupied   of sequence
----------------------------------------------------
SINEs:            65       14583 bp    10.43 %
       ALUs       43       11824 bp     8.46 %
       MIRs       22        2759 bp     1.97 %

LINEs:            19        4896 bp     3.50 %
       LINE1       6        1611 bp     1.15 %
       LINE2      11        1869 bp     1.34 %

LTR elements:     19        6432 bp     4.60 %
       MaLRs       5        1174 bp     0.84 %
       Retrov.     2        1918 bp     1.37 %
       MER4_group  2        2574 bp     1.84 %

DNA elements:      8        1887 bp     1.35 %
       MER1_type   4         859 bp     0.61 %
       MER2_type   2         801 bp     0.57 %
       Mariners    1          79 bp     0.06 %

Unclassified:      0           0 bp     0.00 %

Total interspersed repeats: 27798 bp    19.89 %


Small RNA:         0           0 bp     0.00 %

Satellites:        5        1231 bp     0.88 %
Simple repeats:   25        2743 bp     1.96 %
Low complexity:   14        1107 bp     0.79 %
====================================================
```

(b)

the percentage of divergence, insertions, and deletions is given with the RepBase reference element as well as the matching position, in the query sequence, either in the forward (+) or reverse (C) orientation, and in the reference element. Other output files provide a summary table of matching elements and the "masked" query sequence, with N corresponding to matching regions.

See Sec. 6.8.3 for a wider description of tools for repeat identification and masking in genomic sequences.

6.4.6 Statistical Significance of Alignment Scores

When we are faced with a sequence alignment showing a certain degree of similarity, the basic question is how great the likelihood is that the alignment score observed can occur just by chance. Any sequence alignment from database searching reports the similarity score and a Poisson probability (P) or expectation (E) value. The smaller the latter, the higher the statistical significance of the match and, consequently, the more reliable the assessment of a functional and/or evolutionary relationship between sequences being compared. A basic question then is: what is a reliable P or E threshold value that would make it possible to exclude a casual occurrence of the alignment score observed? A correct answer to this question is particularly relevant for a database search experiment carried out to try to establish the functional role of anonymous sequences, now produced in large amounts in megasequencing projects. The key issue in solving this problem is the availability of a suitable model for the distribution of unrelated sequence scores, an issue still unsolved. Two sequences can be defined as *unrelated* when they do not share any evolutionary ancestry. An empirical score distribution can be generated by comparing many unrelated sequences having the same length as the two sequences being compared. The distribution of unrelated sequence scores can also be obtained by comparing real sequences after a random shuffling of residues to preserve their overall compositional properties or just randomly generated sequences according to a specific DNA or protein model (e.g., Markov chain generators, see Sec. 6.8.1). From this distribution the statistical significance of the alignment score of interest can be measured by the number of standard deviations from the average score. A normal distribution of similarity scores should not be assumed; rather, such a distribution is generally unknown since real sequences show peculiar statistical properties, also depending on their functional role. This implies that an accurate estimate of the chance probability of a given score cannot easily be assumed (Smith, Waterman et al. 1985).

In the case of local alignment, a rather accurate description of the random scores can be obtained in the case of ungapped aligned segments (Karlin and Altschul 1990, 1993). Under reasonable assumptions, it has been demonstrated that the

◄────────────────────────────────

Figure 6.15. (*a*) RepeatMasker sample output listing low complexity, simple repeats, and interspersed repeats found in the query sequence. For each repeated element are reported the Smith–Waterman (SW) score, the percentage of divergence, deletions and insertions with the template repeat, the position in the query and in the repeat of the matching segment, the name of the repeat, and its class/family. (*b*) Summary table of repeats composition is also provided by the program.

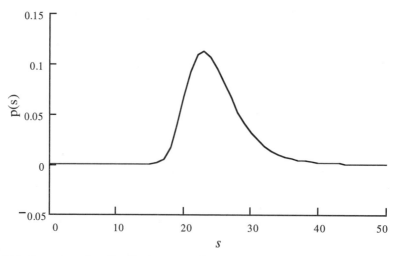

Figure 6.16. Extreme value distribution according to Altschul et al. describing the distribution $p(s)$ of similarity scores s of HSPs calculated by comparing protein sequence pairs. The distribution plotted here is obtained from the comparison of two 100-residue-long amino acid sequences with $K = 0.107$ and $\lambda = 0.305$ that are typically observed in real sequence comparisons. [From Altschul, Boguski et al. 1994).]

random score distribution for optimal ungapped alignments follows the extreme value distribution. The right-hand tail of this distribution decays exponentially in x rather than in x^2 as for the normal distribution (with x = score), so assuming normality tends to grossly overestimate alignment score significance. Although the same rigorous formal demonstration cannot be obtained for gapped alignments, it has been observed that their random scores fit the same type of distribution well (Altschul and Gish 1996).

According to the extreme value distribution, the expected number of HSPs having a score equal or greater than S is simply given by

$$E = Kmne^{-\lambda S}$$

where λ is the distribution decay constant, and K is another constant, dependent on the individual probabilities of the sequence residues and the scoring matrix used (Altschul and Gish 1996). The query and database sequence lengths, m and n, used in the formula correspond to the effective length obtained by subtracting from the actual length the length κ of a typical optimal subalignment (Altschul and Gish 1996).

Figure 6.16 shows the probability density function for the extreme value distribution in a comparison of two amino acid sequence having an effective length of 100 residues and assuming that $K = 0.107$ and $\lambda = 0.305$, typical values for real sequences. The raw score S used in the formula above depends on the scoring system used (e.g., a specific PAM or BLOSUM matrix). It can be normalized as S' by the following formula:

$$S'(\text{bits}) = \frac{\lambda S - \ln K}{\ln 2}$$

which provides the *bit score*. The expected (E) value corresponding to a given bit score is simply

$$E = mn \cdot 2^{-S'}$$

This makes E dependent only on the search space defined by mn. The λ, K, m, and n parameters are reported in the last section of the BLAST output (in the example in Fig. 6.13, they are given, respectively, by 0.267, 0.0410, 116, and 157,084,472).

As the number of HSPs with score $\geq S$ can be described by the Poisson distribution (Karlin and Altschul 1990), the probability of finding exactly q HSPs with score $\geq S$ is

$$e^{-E} \frac{E^q}{q!}$$

Then, provided that the probability of finding zero HSPs is e^{-E}, the probability of finding at least one such HSP is

$$P = 1 - e^{-E}$$

BLAST program reports E values rather than P values because they are more intuitively understandable. However, when $E < 0.01$, P and E values virtually coincide.

As a rule of thumb for alignment significance it has been shown that two protein sequences showing more than 25% identity over a length greater than 80 residues can be considered "bona fide" homologous. On the other hand, it has been demonstrated by computer simulation that at DNA level the percentage of identity is not a good parameter for determining whether or not a match is significant. Furthermore, influence from both the database searching algorithm and the query sequence length has been observed (Anderson and Brass 1998) to affect the alignment efficiency and the assessment of statistical significance. It has been shown that the best-performing algorithm is that of Smith and Waterman (1981) and matches are reliably significant for a z-score > 5 (see Sec. 6.4.1) or $E < 0.005$ (Anderson and Brass 1998) for a query sequence at least 200 bp long.

However, a lack of statistical significance of the similarity score calculated between two sequences does not necessarily imply that they are not correlated evolutionarily. Indeed, owing to very fast substitution rates, ancient common ancestry, and/or functional shifts in the biological activity, two homologous sequences can become so highly divergent that their common origin can no longer be inferred from similarity at the level of their primary structure. It is not unlikely that two homologous proteins, not showing significant similarity at the level of their primary structure, still maintain similar biological activity and a common protein folding. Indeed, the observation, when possible, of common folding is much more predictive of common function and of evolutionary ancestry than is sequence similarity. For example, even though the two ferredoxins from spinach (PDB entry: 1A70) and *Azotobacter vinelandii* (PDB entry: 7FD1) show a remarkably similar three-dimensional structure (Fig. 6.17*a*), the corresponding protein alignment does not show significant similarity (Fig. 6.17*b*).

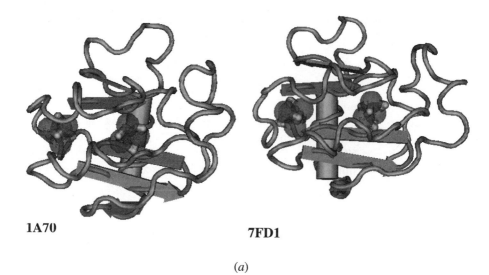

1A70   7FD1

(*a*)

```
  1 .AFVVTDNCIKCKYTDCVEV.CPVDCFYEGPNFLVIHPDECID...CALC 45
     |: ||      .. |: ||.|| |.:      :|   .:|.||   :. .
  1 XAYKVT....LVTPTGNVEFQCPDDVY.....ILDAAEEEGIDLPYSCRA 41
          .           .          .        .       .
 46 EPECPAQAIFSEDEVPEDMQEFIQLNAELAEVWPNITEKKDPLPDAEDWD 95
     :. :...: :..:.: :| |.|:: :.::.|.|. :|    |:.|.. .
 42 GSCSSCAGKLKTGSLNQDDQSFLD.DDQIDEGWV.LTCAAYPVSDVTIET 89
        .
 96 GVKGKLQHLER 106
     |:.|
 90 HKKEELTA... 97
```

(*b*)

Figure 6.17. (*a*) Three-dimensional folding of spinach and *Azotobacter vinelandii* ferredoxin; (*b*) optimal pairwise alignment of the corresponding primary sequences.

6.5 MULTIPLE ALIGNMENT

The simultaneous alignment of three or more nucleotide or amino acid sequences is one of the most commonly used techniques in sequence analysis. The profile of sequence conservation along the alignment allows us to define conserved patterns or domains, corresponding to functionally and structurally important regions. The basic information it provides may prove very useful in designing experiments to test and modify the function of specific proteins. Furthermore, it can also be used for predicting the secondary or tertiary structure of new protein sequences, to find diagnostic patterns for specific protein families, and to suggest the best candidates to be primers for PCR amplification. Finally, multiple alignments are the necessary input data for phylogenetic and evolutionary analyses. Multiple alignments are carried out on a set of homologous sequences generally across their entire length (global alignment). Deciding if a sequence is or is not related by homology relationship to a group of related sequences generally depends on the significance of its similarity to the other members of the group (see Sec. 8.3).

The Needleman and Wunsch (1970) algorithm for finding the best global alignment of two sequences can be in theory readily be extended to align multiple sequences, but this is computationally impracticle in most cases. Since a rigorous algorithm can be applied only to a small number of sequences, strategies have been developed to find optimal multiple global alignments in a reasonable amount of time. In particular, the *progressive method* of Feng and Doolittle (1987), now used in the great majority of multiple alignment algorithms, is based on the underlying idea that homologous sequences are phylogenetically related. According to this method, the Needleman and Wunsch (1970) method is first used to align all possible pairs of sequences [$n(n - 1)/2$ pairwise alignments for n sequences]. Then the pairwise similarity scores are used to construct a tree using the UPGMA or the neighbor-joining method (Swofford, Olsen et al. 1996) (see Sec. 6.11.3). Finally, this tree is used as a guide for progressive multiple alignment, which starts aligning the two most closely related sequences to make an alignment cluster where gaps are fixed for further alignments. Then the two sequences, the sequence and the cluster, or the two most closely related clusters are aligned and the procedure continues until all sequences are aligned in a single cluster. A possible limitation of the progressive alignment methodology is that the final alignment could be influenced by the alignment order. Indeed, any mistake that may be introduced during early stages of alignment cannot be corrected later as new sequence information is added.

The most commonly used method for progressive multiple alignment is implemented in the program ClustalW (Higgins, Thompson et al. 1996). The main feature of this program is that it allows different weights to sequences and to alignment parameters at different positions along the alignment. Sequence weights are determined so that they are higher for more divergent sequences and lower for more conserved sequences. In general, these weights are proportional to the length of the branch from the sequence under consideration down to the root divided by the number of sequences sharing that same branch (Higgins, Thompson et al. 1996). Figure 6.18a shows the multiple alignment determined for six mammalian p53 proteins by ClustalW using the guide tree shown in Fig. 6.18b. From branch lengths in Fig. 6.18b it is possible to assign sequence weights. For example, the weight assigned to the rat sequence, calculated as $0.048 + 0.050/2 + 0.040/3 = 0.086$, is lower than that assigned to the hamster sequence given by $0.102 + 0.040/3 = 0.115$.

In the ClustalW position, specific gap penalties are used to calculate the multiple alignment. Very simple heuristic rules are used that take into account the propensity of each amino acid residue to be adjacent to a gap, using lower gap penalties at positions where gaps already occur or in correspondence to stretches of hydrophilic residues most likely to correspond to loop regions. Users can choose default parameters selected empirically by the software authors by trial and error or define their own parameter set. A striking feature of ClustalW is that it allows one to add new sequences to an existing alignment or to align two existing alignments. An example is shown in Fig. 6.19, where the sheep p53 protein is added to the alignment shown in Fig. 6.18a. A suitable weighting scheme is also used in this case.

The alternatives to progressive multiple alignment are the use of hidden Markov models (HMMs) or the use of iterative methods that find optimal multialignments by maximizing their overall quality measured analytically using an objective function (OF). HMMs attempt simultaneously to find the best alignment and the most

```
COW__p53       MEESQAELNVEPPLSQETFSDLWNLLPENNLLSS---ELSAPVDDLLPY-TDVATWLDEC
CAT__p53       MQEPPLELTIEPPLSQETFSELWNLLPENNVLSS---ELSSAMNELPLS-EDVANWLDEA
HUMAN__p53     MEEPQSDPSVEPPLSQETFSDLWKLLPENNVLSP---LPSQAMDDLMLSPDDIEQWFTED
HAMSTER__p53   MEEPQSDLSIELPLSQETFSDLWKLLPPNNVLSTLP--SSDSIEELFLS-ENVAGWLEDP
MOUSE__p53     MEESQSDISLELPLSQETFSGLWKLLPPEDILP-----SPHCMDDLLLP-QDVEEFFEGP
RAT__p53       MEDSQSDMSIELPLSQETFSCLWKLLPPDDILPTTATGSPNSMEDLFLP-QDVAELLEGP
               *..   .  .* ******** ** ***   .*          . .*          .

///

COW__p53       KKGQSCPEPPPRSTKRALPTNTSSSPQPKKKPLDGEYFTLQIRGFKRYEMFRELNDALEL
CAT__p53       KKGEPCPEPPPGSTKRALPPSTSSTPPQKKKPLDGEYFTLQIRGRERFEMFRELNEALEL
HUMAN__p53     KKGEPHHELPPGSTKRALPNNTSSSPQPKKKPLDGEYFTLQIRGRERFEMFRELNEALEL
HAMSTER__p53   KKGEPCPELPPKSAKRALPTNTSSSPQPKRKTLDGEYFTLKIRGQERFKMFQELNEALEL
MOUSE__p53     KKEVLCPELPPGSAKRALPTCTSASPPQKKKPLDGEYFTLKIRGRKRFEMFRELNEALEL
RAT__p53       KKEEHCPELPPGSAKRALPTSTSSSPQQKKKPLDGEYFTLKIRGRERFEMFRELNEALEL
               **      *  ** *.*****  **..*  *.* ********.***  *.  **.***.****

COW__p53       KDALDGREPGESRAHSSHLKSKKRPSPSCHKKPMLKREGPDS
CAT__p53       KDAQSGKEPGGSRAHSSHLKAKKGQSTSRHKKPMLKREGLDS
HUMAN__p53     KDAQAGKEPGGSRAHSSHLKSKKGQSTSRHKKLMFKTEGPDS
HAMSTER__p53   KDAQALKASEDSGAHSSYLKSKKGQSASRLKKLMIKREGPDS
MOUSE__p53     KDAHATEESGDSRAHSSYLKTKKGQSTSRHKKTMVKKVGPDS
RAT__p53       KDARAAEESGDSRAHSSYPKTKKGQSTSRHKKPMIKKVGPDS
               ***          * ****  *.**   *  *   ** *  *   * **
```

(a)

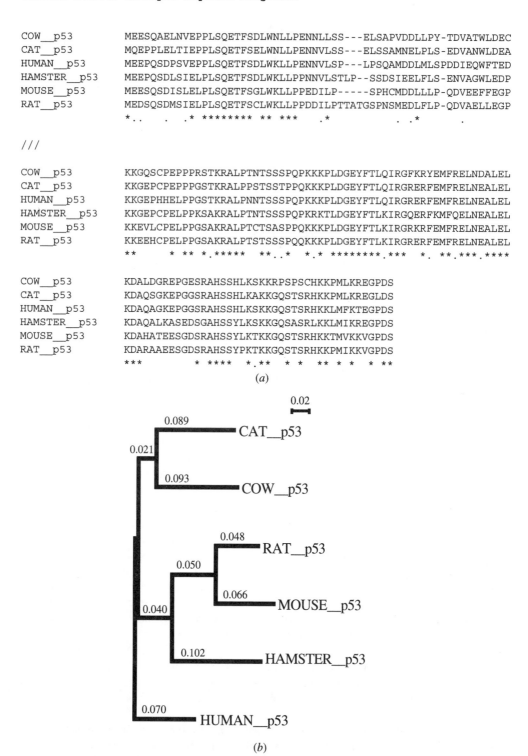

(b)

Figure 6.18. (a) Multiple alignment of six mammalian p53 proteins determined by ClustalW using the guide tree in (b), where the branch lengths inferred are also shown. An asterisk or dot below the alignment denotes identical or functionally similar amino acid residues, respectively.

```
**************************************************************
******** CLUSTAL W(1.60) Multiple Sequence Alignments *******
**************************************************************

     1. Sequence Input From Disc
     2. Multiple Alignments
     3. Profile / Structure Alignments
     4. Phylogenetic trees

     S. Execute a system command
     H. HELP
     X. EXIT (leave program)

Your choice: 3

****** PROFILE AND STRUCTURE ALIGNMENT MENU ******

     1.  Input 1st. profile
     2.  Input 2nd. profile/sequences

     3.  Align 2nd. profile to 1st. profile
     4.  Align sequences to 1st. profile (Slow/Accurate)

     5.  Toggle Slow/Fast pairwise alignments = SLOW

     6.  Pairwise alignment parameters
     7.  Multiple alignment parameters

     8.  Toggle screen display            = ON
     9.  Output format options
     0.  Secondary structure options

     S.  Execute a system command
     H.  HELP
     or press [RETURN] to go back to main menu

Your choice: 1
Enter the name of the sequence file: p53.aln

Sequence format is Clustal

No. of seqs=6

Sequences assumed to be PROTEIN

Sequence 1: COW__p53        402 aa
Sequence 2: CAT__p53        402 aa
Sequence 3: HUMAN__p53      402 aa
Sequence 4: HAMSTER__p53    402 aa
Sequence 5: MOUSE__p53      402 aa
Sequence 6: RAT__p53        402 aa


****** PROFILE AND STRUCTURE ALIGNMENT MENU ******

     1.  Input 1st. profile            (loaded)
     2.  Input 2nd. profile/sequences

     3.  Align 2nd. profile to 1st. profile
     4.  Align sequences to 1st. profile (Slow/Accurate)

     5.  Toggle Slow/Fast pairwise alignments = SLOW

     6.  Pairwise alignment parameters
     7.  Multiple alignment parameters

     8.  Toggle screen display            = ON
     9.  Output format options
     0.  Secondary structure options

     S.  Execute a system command
     H.  HELP
     or press [RETURN] to go back to main menu


Your choice: 2
Enter the name of the sequence file: sheep.p53

Sequence format is Pearson

No. of seqs in profile=1

Total no. of seqs      =7

Sequences assumed to be PROTEIN

Sequence 7: SHEEP__p53      382 aa

****** PROFILE AND STRUCTURE ALIGNMENT MENU ******
```

Figure 6.19. ClustalW run to add a new sequence (sheep p53) to an existing alignment (Fig. 6.15*a*). User-supplied answers are shown in boldface type.

consistent probabilistic model for substitutions, insertions, and deletions (see below). Practically, this approach is limited to cases with high number sequences (e.g., 100 or many). Iterative methods use deterministic or stochastic heuristics. In the MSA program (Lipman, Altschul et al. 1989) an attempt to define a relatively small solution space, where the best alignment is likely to be, is currently computationally unfeasible for more that seven or eight sequences. Among stochastic optimization methods are the *Gibbs sampler method* (Lawrence, Altschul et al. 1993), which allows us to find the best local alignment block with no gaps, and the *genetic algorithm* (GA) *strategy*, which involves the evolution of a population of alignments by crossover and mutation operators while the best-fitting alignment is determined as that maximizing the OF (i.e., the measure of alignment quality). The program SAGA (sequence alignment by genetic algorithm; Notredame and Higgins 1996) cannot be used for large data sets (>20 to 30 sequences not longer than 400 residues), but it has been shown to perform better than some of the widely used alternative alignment programs. The program RAGA (RNA alignment by genetic algorithm), mostly adapted from SAGA, performs pairwise alignments taking into account secondary structure information of one of the two RNAs (Notredame, O'Brien et al. 1997).

The OF used to measure the quality of the multiple alignment is generally obtained by summing up the costs calculated for all pairwise alignments weighted to correct for their unequal representation in the set of sequences to be aligned. The weighted sum-of-pair (WSP) score is given by the following equation:

$$\text{WSP score} = \sum_{i=1}^{N-1} \sum_{j=i}^{N} w_{ij} \text{COST}(A_{ij})$$

where w_{ij} is the weight given to the sequence pair i,j and $\text{COST}(A_{ij})$ is the alignment score between sequences A_i and A_j calculated using a substitution matrix (e.g., PAM or BLOSUM) and a scoring scheme for the gap penalty.

6.6 ALIGNMENT PROFILES TO RECOGNIZE DISTANTLY RELATED PROTEIN OR PROTEIN MODULES

The observation of a multiple alignment provides information on residues and/or domains more conserved than others and on the points where insertions and deletions occur with higher probability. Then the information contained into a multiple alignment can result in position-specific scoring matrices, called *profiles* (Gribskov, McLachlan et al. 1987; Gribskov and Veretnik 1996). Position-specific scoring information is derived by the frequency with which a given residue occurs in an alignment column. Indeed, the degree of conservation of residues along the alignment depends on their structural or functional relevance, with some sites strictly invariant and others highly variable. The occurrence of gaps is also not uniform, and thus a position-specific cost can be estimated for their insertion and extension.

The profile of a multiple alignment, which embodies the position-specific information for a given gene family alignment, allows a remarkable increase in the accuracy of the alignment between a sequence and a group of aligned sequences as well as an increase in both sensitivity and selectivity in database searching looking for distantly related homologous sequences or sequence blocks. A profile can be treated as a single sequence and aligned to any other sequence (or profile) using the profile substitution costs and penalties.

One of the main uses of alignment profiles is database searching of a new, distantly related member of a gene family while minimizing spurious matches with unrelated sequences. In this context, a key issue is the quality of the profile, which in turn depends on the sequences included in the alignment, on the overall accuracy of the multiple alignment, and on the method used for the transformation of multiple alignment into a profile, especially concerning the attribution of position-specific gap costs. The availability of sequences in the database is the criterion generally used for their inclusion in profile construction, although in some cases more appropriate criteria can be used (Neuwald, Liu et al. 1997). Figure 6.20 shows the profile calculated by ProfileMake (Genetics Computer Group 1998) on the alignment shown in Fig. 6.18. Each row represents a specific position of the multiple alignment, where substitution costs are indicated, based on the substitution matrix chosen, for any of the 20 amino acids. The first column reports the consensus residue (i.e., the one having the highest score).

An iterative profile database search has been included in the PSI-BLAST program (Altschul, Madden et al. 1997). When using the PSI-BLAST server, the query sequence is first searched against the database using the normal BLAST algorithm; then a profile of the matching alignments is generated and this profile is searched against the database to possibly gather more hits, which can be used in their turn for further profile refinement and new database searching. In contrast to profile construction based on multiple alignments, when each site is represented by the same number of sequences, PSI-BLAST profiles may have a variable number of sequences contributing to each site, depending on the matching aligned segments used for profile construction. Another serious pitfall of this kind of approach is that in contrast to multiple alignment profiles, where a curated set of related sequences is used, the PSI-BLAST profile may include some unrelated sequences, with borderline significant scores. In this case spurious matches will be recorded that may severely affect further PSI-BLAST profile search iterations.

A more efficient way to construct alignment profiles is based on the use of *hidden Markov models* (HMMs; Krogh, Brown et al. 1994), developed previously and extensively used for speech recognition. HMMs provide a statistical model representative of a given family of proteins or of a protein domain endowed with a specific functional or structural role. When applied to alignment profiles, the HMM is composed of a number of elements, corresponding to the columns of a multiple alignment that may exist in three different states: (1) match, (2) insert, and (3) delete. Transition probabilities that interconnect element states are generally derived from preexisting multiple alignments but can also be obtained from a set of unaligned sequences, producing a multiple alignment in the process.

Figure 6.21 shows a typical HMM state diagram corresponding to a simple multiple alignment of four positions. Each position of the multiple alignment can be modeled by the triplet of states above (match, insert, and delete), with three state transition probabilities (arrows) departing from each of the three states. Match and insert states have 20 symbol emission probabilities, one for each amino acid. The sequence of states is a first-order Markov chain (see Sec. 6.8.1) because the nature of the next state is dependent on the identity of the current state. The Markov model

```
(Peptide) PROFILEMAKE v4.40 of: p53_6.msf{*}     Length: 403
Sequences: 6  MaxScore: 416.92  December 21, 19100 12:41

                Gap: 1.00              Len: 1.00
                GapRatio: 0.33         LenRatio: 0.10

        p53_6.msf{cow}      From: 1   To: 403    Weight: 1.00
        p53_6.msf{cat}      From: 1   To: 403    Weight: 1.00
        p53_6.msf{human}    From: 1   To: 403    Weight: 1.00
        p53_6.msf{hamster}  From: 1   To: 403    Weight: 1.00
        p53_6.msf{mouse}    From: 1   To: 403    Weight: 1.00
        p53_6.msf{rat}      From: 1   To: 403    Weight: 1.00

Symbol comparison table: profilepep.cmp  FileCheck: 1254
  Relaxed treatment of non-observed characters
  Exponential weighting of characters
```

Cons	A	B	C	D	E	F	G	H	I	K	L	M	N	P	Q	R	S	T	V	W	Y	Z	Gap	Len
M	0	-30	-60	-40	-20	50	-30	-30	60	20	130	150	-30	-20	0	20	-30	0	60	-30	-10	-10	100	100
E	21	49	-43	70	102	-51	34	31	-15	23	-20	-13	35	9	57	3	12	12	-14	-75	-37	80	100	100
E	22	54	-43	76	105	-53	38	29	-14	22	-23	-16	38	7	51	0	15	15	-14	-80	-36	78	100	100
S	26	12	23	9	9	-29	26	0	-9	9	-20	-14	9	55	6	12	55	17	0	-14	-34	6	100	100
Q	17	33	-38	46	46	-57	15	47	-21	27	-9	-2	26	31	99	28	-3	-4	-12	-39	-45	72	100	100
S	29	11	11	7	9	-8	28	-11	2	6	-6	-3	12	19	-3	-1	65	16	4	11	-17	0	100	100
D	18	60	-32	83	70	-56	39	24	-12	18	-27	-21	39	6	43	0	12	12	-12	-67	-30	58	100	100
L	1	-17	-25	-18	-11	38	-17	-9	38	-8	57	52	-16	0	-3	-10	-11	1	37	2	2	-6	100	100
S	22	23	30	16	14	-19	32	-5	-5	14	-21	-15	27	20	-2	4	70	27	-5	6	-21	4	100	100
I	3	-13	-13	-13	-11	33	-9	-15	68	-11	49	38	-17	-6	-13	-17	-8	9	64	-24	3	-11	100	100
! 11																								
E	30	70	-60	100	150	-70	50	40	-20	30	-30	-20	50	10	70	0	20	20	-20	-110	-50	110	100	100
P	12	-11	-20	-11	-6	14	-6	0	17	-6	35	32	-11	35	6	-3	0	6	26	-9	-14	0	100	100
P	50	10	10	10	10	-70	30	20	-20	10	-30	-20	0	150	30	30	40	30	10	-80	-80	20	100	100
L	-10	-50	-80	-50	-30	120	-50	0	80	-30	150	130	-40	-30	-10	-40	-40	-10	80	50	30	-20	100	100
.	0	0	0	0	0	0	0	0	0	0	0	0	0	0	0	0	0	0	0	0	0	0	30	30
S	40	30	70	20	20	-30	60	-20	-10	20	-40	-30	30	40	-10	10	150	30	-10	30	-40	0	100	100
Q	20	50	-60	70	70	-80	20	70	-30	40	-10	0	40	30	150	40	-10	-10	-20	-50	-60	110	100	100
E	30	70	-60	100	150	-70	50	40	-20	30	-30	-20	50	10	70	0	20	20	-20	-110	-50	110	100	100
T	40	20	20	20	20	-30	40	-10	20	20	-10	0	20	30	-10	-10	30	150	20	-60	-30	10	100	100
F	-50	-70	-10	-100	-70	150	-60	-10	70	-70	120	50	-50	-70	-80	-50	-30	-30	20	130	140	-70	100	100

```
...
```

Figure 6.20. Alignment profile generated by ProfileMake from the multiple alignment in Fig. 6.15. It consists of a $L \times 22$ matrix (L = alignment length) where a position-specific substitution score is assigned to each of the 20 amino acid residues and to gap opening and extension. The first column of the matrix reports the consensus residue for each alignment position.

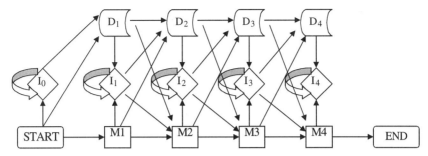

Figure 6.21. Hidden Markov model (HMM) state diagram for sequence analysis. Start and end nodes correspond to initial and final states. The nodes from M1 to M4 represent the states at four residue positions. The I_0–I_4 and D_1–D_4 nodes represent the states for insertion and deletion of residues, respectively. The arrows connecting nodes represent transition probabilities.

is defined as "hidden" because we do not observe the sequence state and the connecting probabilities, just the symbol sequence generated by the model.

Two major assumptions are made by HMMs. The first one is that possible long-range correlations between residues are ignored (i.e., an HMM is basically a primary structure model and consequently, cannot model conserved secondary structure sequence patterns). Second, HMMs do not take into account common evolutionary descent relating real biosequences that are considered to be generated independently using the model. Notwithstanding, HMMs provide a more consistent probabilistic theory for scoring insertions and deletions, which makes them perform much better than standard profile analyses.

Profile HMM software is well suited for modeling a particular sequence family of interest and finding additional remote homologs in a sequence database. In this case, the HMM profile built from a preexisting alignment can be searched against a protein database. This can be done, for example, by HMMER software (Eddy 1998), which provides a ranked list of hits between a query profile and a protein database, giving the corresponding E values. Another common use of HMM profiles is the search for one or more known protein domains in the query sequence of interest. This can be done by searching the query sequence against a library of HMM profiles constructed previously, starting from multiple alignments of common protein domains.

The two main collections of annotated HMM profiles currently available are the Pfam database (Bateman, Birney et al. 2000) and the ProSite profile database (Hofmann, Bucher et al. 1999). Pfam is a large collection of multiple sequence alignments and corresponding hidden Markov models that cover many common protein domains. The user can analyze a protein query sequence (or a DNA sequence translated in the six possible frames) to search Pfam domain matches. ProSite is a database of biologically significant sites and patterns described either as a regular expression syntax (see Sec. 5.8.4) or by profiles. Other protein signature databases include PRINTS (Attwood, Croning et al. 2000), ProDom (Corpet, Servant et al. 2000), BLOCKS (Henikoff, Greene et al. 2000), CDD (Marchler-Bauer, Panchenko et al. 2002), and EMOTIF (Huang and Brutlag 2001). PRINTS is a compendium of protein fingerprints, each represented by a group of nonoverlapping conserved

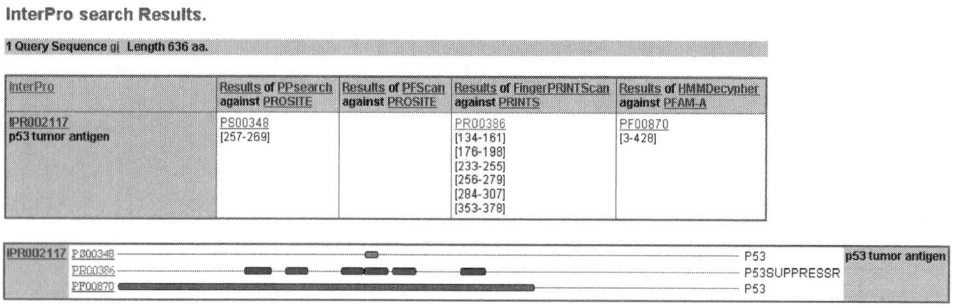

InterPro search Results.

1 Query Sequence gi Length 636 aa.

InterPro	Results of PPsearch against PROSITE	Results of PFScan against PROSITE	Results of FingerPRINTScan against PRINTS	Results of HMMDecypher against PFAM-A
IPR002117 p53 tumor antigen	PS00348 [257-269]		PR00386 [134-161] [176-198] [233-255] [256-279] [284-307] [353-378]	PF00870 [3-428]

IPR002117   PS00348 ———————————————————————————— P53   p53 tumor antigen
PR00386 ———————————————————————————— P53SUPPRESSR
PF00870 ———————————————————————————— P53

Figure 6.22. Output of an InterPro search carried out at the EBI Web server (*www.ebi.ac.uk/interpro/index.html*), showing domains detected in the p73 tumor suppressor protein (SwissProt ID: O15350).

motifs specific for a protein family. ProDom is a protein domain database containing an automatically generated compilation of homologous domains built by recursive PSI-BLAST searches. The BLOCKS database is derived basically from the information contained in ProSite and contains multiply aligned ungapped sequence segments corresponding to the most highly conserved regions of proteins. The Conserved Domain Database (CDD; Marchler-Bauer, Panchenko et al. 2002) is a compilation of multiple protein alignments representing protein domains derived mostly from SMART (see Sec. 5.9) and Pfam collections. The EMOTIF database is a collection of more than 170,000 highly specific and sensitive protein sequence motifs representing conserved biochemical properties and biological functions. Table 6.3 lists some of the available Internet resources for motif- or profile-based sequence searches.

The search of conserved profiles/patterns in uncharacterized proteins translated from genomic or cDNA sequences is a powerful tool to determine their possible function. To this end, the InterPro resource (Apweiler, Attwood et al. 2001) has been developed, which allows us to search unknown proteins against major protein signature databases. Figure 6.22 shows the result of the InterProScan software to the p73 gene product, a protein homologous to the p53 tumor suppressor protein.

6.7 METHODS FOR SEQUENCE ASSEMBLY

Sequence assembly is a process that involves comparison of sequences, finding overlapping fragment pairs, merging as many fragments as possible with contig construction, and creating a consensus sequence from the merged fragments. Such a process is essential in genome projects for reconstruction of the original DNA progenitor sequence and in EST projects for the production of a comprehensive gene catalog.

The assembly process consists of a complex procedure that can be split into three phases: cleaning, clustering, and construction of sequence alignment consensus. The last two steps can be performed separately with a single piece of software (i.e., the TIGR Assembler; see below) or with two or more distinct algorithms. The alignment stage adds two main benefits to the assembly process: generation of

TABLE 6.3. Internet Resources for Motif- or Profile-Based Sequence Searches

Software	URL	Description
BLOCKS	blocks.fhcrc.org	Collection of conserved protein family regions
EMOTIFS	motif.stanford.edu/emotif	Collection of protein sequence motifs representing conserved biochemical properties and biological functions
HMMER	hmmer.wustl.edu	Profile hidden Markov models for biological sequence analysis
HMMpro	www.netid.com	Commercial package implementing general-purpose HMM architectures for multiple alignment, database searches, and pattern discovery
InterPro	www.ebi.ac.uk/interpro/index.html	Integrated documentation resource for protein families, domains, and functional sites, which amalgamates the efforts of the ProSite, PRINTS, Pfam, and ProDom database projects
Meta-MEME	www.cse.ucsd.edu/users/bgrundy/metameme.1.0.html	Software toolkit for building motif-based hidden Markov models (HMMs)
Pfam	www.sanger.ac.uk/Software/Pfam	The Pfam protein family database and related software
PRINTS	http://www.bioinf.man.ac. uk/dbbrowser/PRINTS	Collection of protein family fingerprints
Probe	ncbi.nlm.nih.gov/neuwald/probe1.0	Implements models based on multiple ungapped HMM motifs, and includes an implementation of training models by Gibbs sampling
ProDom	www.toulouse.inra.fr/prodom.html	Automatically generated collection of protein domain families
Sequence Alignment and Modeling (SAM) system	www.cse.ucsc.edu/research/compbio/sam.html	Collection of flexible software tools for creating, refining, and using linear hidden Markov models for biological sequenceanalysis
SMART	smart.embl-heidelberg.de	Simple Molecular Architecture Search Tool

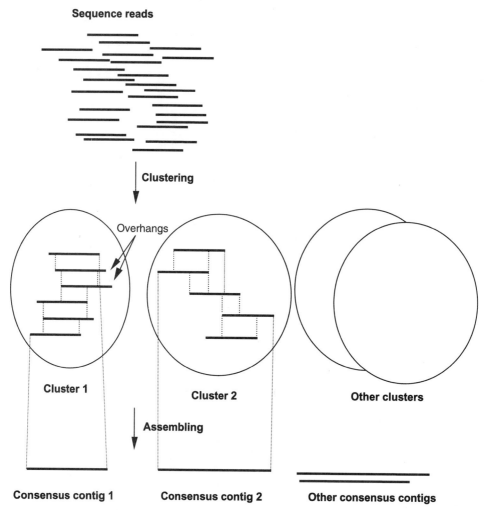

Figure 6.23. Major phases in sequence assembling and contig construction. Dashed lines delineate similar sequences.

consensus sequences with complete elimination of the sequence redundancy, and improvement in the length and quality of the sequence reconstructed. Figure 6.23 shows major phases in sequence assembling and contig construction. The protocols and software used in the assembly process have been improved and optimized following different criteria, depending on their use in genomic or EST projects.

6.7.1 Sequence Cleaning

Specific tools for identifying segments of vector, linker, adapter, or PCR primer origin are used prior to sequence analysis. For example, the VecScreen program (see the URL in the Appendix) is a system for quickly identifying vector sequences by launching a query against a specialized nonredundant vector database (UniVec).

The search uses BLAST with parameters preset for optimal detection of vector contamination.

Repeated regions also confound sequence assembly efforts; therefore, before running similarity seaarch methods, these regions are identified and masked. The process makes it possible to avoid false positive results and to accelerate downstream computational steps. An effective program to achieve such a goal is Repeat-Masker (see Sec. 6.4.5), which screens DNA sequences against a curated library of repetitive elements, as well as for low-complexity regions. Sequence comparisons are performed using the alignment program CrossMatch (see the URL in the Appendix), an efficient implementation of the Smith–Waterman–Gotoh algorithm. The program returns a masked query sequence in which all the annotated repeats have been replaced by X's and a table with a detailed annotation of the repeats is present in the query sequence.

6.7.2 Sequence Clustering

The clustering step is based on sequence similarity and overlapping criteria. It can be performed with several tools, using alignment- or word-based algorithms. The simplest alignment-based clustering method is to use scripting languages to run and parse sequence database searches by BLASTN or FASTA. Both algorithms are designed for local similarity searches, not for searching alignments consistent with overlap/contiguity; thus the script wrappers makes it possible to run the search, parse the results, and define rules for sequences to be considered as overlapping. Word-based algorithms try to estimate the significance of discovering DNA words shared by two sequences without the cost of a full alignment. Word-based clusterings are inherently fast because they omit all alignment steps but to the detriment of sensitivity and selectivity, unless specific refinements are made. Among word-based clusterings there is the D2_CLUSTER algorithm (Hide, Burke et al. 1994; Burke, Davison et al. 1999). It is a transitive closure method and uses an n-mer oligo frequency comparison between a query sequence and a compressed database. Sequences sharing a given region (window size) with at least a given percent of identity (stringency) or two sequences sharing similarity values with a third sequence are included in the same cluster.

The sequence similarity and overlapping criteria can also take into account sequence quality data and are commonly defined by the following parameters:

1. *Minimum overlap:* minimum length by which two DNA fragments must overlap to be included in an assembly.

2. *Minimum identity in overlap:* minimum percent identity that two fragments must achieve over their entire region of overlap to be included in an assembly. Sequence quality constraints can also be included. Indeed, TIGR assembler version 1.0 takes into account the positions of first and last nonvector good-quality base call, whereas TIGR assembler version 2.5 sets a different minimum percent identity in the various portions of overlap (beginning ending, and middle EST regions) because it takes into account that the ends of sequences are of lower quality, with frequent doubled base call errors.

3. *Maximum overhang:* maximum length at the end of a DNA fragment that does not match another overlapping DNA fragment (referred to as *overhang*) that

will allow such DNA fragments to become part of the same assembly (see Fig. 6.23).

4. *Maximum local errors:* maximum number of alignment errors (mismatches or gaps) allowed within a given window in the overlap region between two DNA fragments in the same assembly; meant to split apart splice variants in EST assembly.

Some of the more sophisticated programs make use of two classes of clustering criteria: (1) sequence similarity and overlapping criteria, and (2) ancillary information such as base-call probability estimates, clone annotation (forward/reverse directionality of two sequences derived from the same clone, clone insert-size range), DNA source (tissue, cell line, development stage, pathological or disease state of the DNA source), and sequence category (i.e., 3' EST derived from the 3' end of mRNA and 5' EST derived from the 5' end of mRNA).

6.7.3 Construction of Alignment Consensus

A number of programs have been developed for such a goal and the most widely used have been improved with the implementation of new program versions. An assembler is a sequence fragment assembly program that performs clustering and builds contigs and consensus sequences from small sequence reads. The TIGR assembler (Sutton, White et al. 1995; see the URL in the Appendix), originally designed for the *H. influenzae* genome assembly, has been used for many purposes, including the assembly of repeat-rich microbial genomes, of chromosomes from BAC sequences, and the construction of tentative consensus transcripts (TGI, TIGR Gene Indices) from EST data of various organisms (see the URL in the Appendix). The theoretical upper limit is beyond a 100-Mbp genome project. Different versions optimized for genome or EST assembly have been produced, and the last version also supports quality files. The TIGR assembler version 2.0 uses a greedy algorithm and heuristics to build contigs, find repeat regions, and target alignment regions. Sequence overlaps are detected and scored using a 32-mer hash. Sequence alignment and merging are done using a Smith–Waterman algorithm with gap penalties and score values, including quality values. In addition, the TIGR assembler (version 2.0) is the only program capable of assembling a mix of sequences with and without sequence quality values.

The following programs can only perform assembly of sequences already clustered.

1. The PHRED/PHRAP/CONSED software package (see the URL in the Appendix) has been widely used in shotgun sequencing projects. PHRED (Phragment Reading Program) reads DNA sequence trace data, calls bases, and assigns quality values to the bases as log-transformed error probability, writing the sequence and relative quality values to output files (Ewing and Green 1998; Ewing, Hillier et al. 1998). PHRAP (Phragment Assembly Program) assembles shotgun DNA sequence data allowing the use of base quality values. The program clips 5' and 3' low-quality regions of reads and uses base quality values in the evaluation of overlaps and generation of contig sequences. The consensus sequence is extracted by

using a mosaic approach, where the best read in any multiple alignment is copied directly to the result. CONSED (Gordon, Abajian et al. 1998) is a graphical tool for viewing and editing PHRAP assemblies. The editing can be guided by base error probability, and an autofinish option helps to choose experimental solutions to complete the sequencing project (i.e., to close gaps and to improve the error rate).

2. The CAP program (Contig Assembly Program) has been improved through three consecutive versions (Huang 1992, 1996; Huang and Madan 1999). The last CAP version, CAP3, is able to clip 5' and 3' low-quality regions of reads and uses base quality values produced by PHRED in the computation of overlaps between reads, sequence alignment, and consensus generation. Moreover, forward–reverse constraints (which specify strand and range of distance between two sequence reads obtained from the same clone) are used to correct assembly error and link contigs.

3. GAP4 (Genome Assembly Program; see the URL in the Appendix; Bonfield, Smith et al. 1995) is suitable for large and small projects and a variety of sequencing strategies. One of the most powerful features of GAP4 is its highly interactive graphical user interface, which enables data to be viewed and manipulated at several levels of resolution, making it easier for the user to interpret and check difficult assemblies. The program also includes several functions for finding problems and suggesting possible experimental solutions. Preassembly steps plus quality clipping, sequencing vector, and cosmid vector removal are controlled by the PREGAP script. Several modes of assembly, including CAP2, CAP3, PHRAP, and others, can be used. The original assembly algorithm compares new reads against contig consensus sequences looking for hash hits that extend the contigs in a greedy fashion. The GAP4 program is included in the Staden (1996) package.

4. Arachne is a new public-domain software (see the URL in the Appendix) for assembling genome sequences from whole genome shotgun reads obtained by sequencing clone ends. It requires as input the base calls and associated quality scores of each read, as well as its relevant ancillary information. The outputs consists of a list of supercontigs ("scaffolds"), each of which has an ordered list of forward-oriented contigs and includes the estimates for the gaps between them in the supercontig.

Performance analyses of CAP3 and PHRAP on BAC data using quality values have shown that PHRAP produces longer contigs than CAP3 but that CAP3 often produces fewer errors in consensus sequences than PHRAP (Huang and Madan 1999). The relative performance of CAP3, PHRAP, and old and new TIGR assemblers versions 1.0 and 2.0 in reconstructing transcripts from ESTs and gene fragments has been tested using simulated sequences (Liang, Holt et al. 2000b). The best assembler for EST analysis must be fairly tolerant to sequence discrepancies; that is, a balance between discrepancies tolerance and resolution is required. Indeed, such sequence discrepancies can be due to sequencing errors, polymorphisms, or transcripts belonging to the same gene family. The assembler has to discard sequencing errors, still discriminating between closely related but distinct transcripts. CAP3 produces the fewest high-quality assemblies from single genes and is tolerant of random errors, while maintaining the ability to discriminate between related genes; thus it has been chosen to construct the most recent release of the Human Gene Index (Liang, Holt et al. 2000). PHRAP produces accurate consensus sequences using sequence quality values. When quality values are not available and sequence quality is low, as for ESTs, the assembling procedure for the construction of a consensus is problematic. In this case the consensus contains a large number of insertions and other errors, up to a 5% error rate. Moreover, the total error rate, particularly insertions and substitutions, increase as the depth of coverage increases, because the program has a tendency to retain the cluster sequence variability, introducing them as insertions or substitutions in the consensus (Liang, Holt et al. 2000a). Both TIGR assemblers are less tolerant of sequence discrepancies. Such programs tend to split cluster sequences into singletons or separate contigs as the error rate increases and fail to produce a consensus if the error rate is high. The discrimination between gene family members is very low for TIGR assembler version 1.0, whereas TIGR assembler version 2.0 provides the greatest discrimination among the four programs tested (Liang, Holt et al. 2000b).

6.7.4 Sequence Mapping by Electronic PCR

The clone-by-clone sequencing strategy (see Sec. 4.1.1) requires the availability of a high-resolution physical map of the genome of interest to get the ordered list of clones. Landmarks used to construct genomic physical maps are the sequence-tagged sites (STSs) that correspond to unique, or presumably unique, genome subsequences. Operationally, an STS is defined by a pair of oligonucleotide primers that through a PCR assay detect its presence or absence in the genome clone under investigation. STSs for different genomes have been determined within several mapping projects. STSs are collected in the UniSTS database (see the URL in Table 5.1) that integrates marker and mapping data from several public resources, such as RHdb, GDB, Genethon genetic map, and so on. At present (March 2002) the species with the highest number of available STSs are human (133,202), rat (29,808), mouse (26,618), and zebrafish (22,391).

STSs are defined by the sequence; thus their identification is possible by searching for subsequences that match the corresponding couple of primers whose order, orientation, and spacing is consistent with the STS-specific PCR product. To accomplish this task an electronic PCR procedure (e-PCR) has been developed (Schuler 1997b, 1998) that given an input query sequence, provides the list and the position of all STSs detected. Figure 6.24 shows the output obtained by e-PCR software on a human genomic clone from chromosome 1.

6.7.5 Sequence Assembly in Genome and EST Projects

Large-scale eukaryotic genome sequencing uses bacterial artificial chromosomes (BAC) as the main substrate. Mapped BACs are subcloned, sequenced, and assembled into a complete and finished sequence, essentially one BAC at a time. Microbial genomes have been sequenced to completion successfully employing a total-shotgun strategy, aided by end-sequencing of phage clones. Combinatorial PCR reactions to align sequence contigs and to aid in the reduction of gaps has often been used to finish the work. Physical maps constructed from restriction digests or hybridization can play an important role in the final assembly and verifi-

Electronic PCR

Query sequence: gi|8217482|emb|AL136528.11|AL136528, 139823 bases

Human DNA sequence from clone RP5-1092A11 on chromosome 1p36.2-36.33 Contains the gene for KIAA0495 protein, the TP73 (tumor protein p73) gene, a gene containing a WD repeat domain, ESTs, STSs, GSSs and CpG Islands, complete sequence [Homo sapiens]

Site (bases)	Marker	Chr.	Organism
9563..9812	D1S1403		Homo sapiens
12074..12212	stSG62963	1	Homo sapiens
31734..31873	stSG31866	1	Homo sapiens
33028..33130	A009E17	1	Homo sapiens
33028..33130	A009E17		Homo sapiens
33183..33303	N27761	1	Homo sapiens
34573..34764	D1S1310		Homo sapiens
98986..99310	SHGC-147180	1	Homo sapiens
98986..99297	SHGC-147179	1	Homo sapiens
100656..100838	D1S468	1	Homo sapiens
100658..100790	SHGC-1560	1	Homo sapiens
122018..122158	SHGC-32565	1	Homo sapiens
127568..127701	SHGC-80737	1	Homo sapiens
138197..138364	stSG49557	1	Homo sapiens

Figure 6.24. e-PCR output obtained from the NCBI Web server (*www.ncbi.nlm.nih.gov/genome/sts/epcr.cgi*) on a genome contig of human chromosome 1 (EMBL ID: AL136528).

cation of the finished sequence data. In all cases the genome has been sequenced redundantly to ensure maximal coverage and sequence accuracy.

In higher organisms only a minor part of the genome is represented by protein-coding sequences (about 1.5% in human) and the mRNA/cDNA molecules represent a direct source of spliced and coding sequences of genes. Thus the genome sequencing projects have been complemented by the cDNA sequencing project, which analyses the transcribed portion of the genome, also referred to as *transcriptome*, to enhance the interpretation and annotation of genomic sequences.

The EST (expressed sequence tag) sequences are short single-pass sequence reads of cDNA clones randomly selected from cDNA libraries. They represent partial cDNA sequences and generally correspond to 3′ untranslated regions (UTRs) of mRNAs, but they can also represent 5′ UTRs or internal protein-coding regions, depending on cDNA synthesis, library construction, and sequencing strategies. ESTs are low quality, poorly annotated, and highly redundant sequences, due to the method used to produce them. Moreover, contaminants (pre-mRNA, intron, vector, and bacterial sequences) and chimeras (artifactual fusion of cDNA derived from unlinked genes) are also quite common.

Both genomic and cDNA sequencing projects include as a major step the assembly of short reads into long sequences for the reconstruction of complete genomes and full-length transcripts, respectively. The clustering and assembly step present both common and distinct computational problems for the genome and EST projects, and in general assembling gene transcript is more complex than assembling shotgun clones of genomic DNA.

In genomic projects, the sequence trace data are known and base quality values can be assigned. Single reads are mostly of good quality (0.1% errors), with low insertion and deletion rates. A global sequencing coverage higher than twofold ensures the good quality of the final sequences. Information on cloning and sequencing strategies are stored together with the sequences and can help in the reconstruction of original DNA scaffolds. Moreover, the sequenced DNA arises from a single clone as a DNA source; thus, sequences sharing less than 98% identity can be assumed to come from different copies of a repeated element or paralogous genes.

In contrast, the single-pass sequencing of ESTs implies relatively inaccurate sequences (about 2% errors) with a high frequency of insertions and deletions. Because ESTs are a snapshot of genes expressed in a given tissue/cell line/developmental stage, there is a high redundancy of sequences derived from highly expressed genes and a low representation of low-expressed genes. Thus, compared with the genome projects, the sequence fold coverage and resulting sequence accuracy are low and highly variable between genes. Chimeric clones due to artifactual fusion of cDNA derived from unlinked genes and reversal clones due to failure in the directional cloning procedure can also occur, producing chimeric sequences and erroneous clone annotations. The EST data derive from a wide variety of sources because the cDNA molecules are extracted from different tissues, cell lines, and development stages of an organism, thus they represent the spectrum of polymorphisms in the original samples. Splice variants, often tissue-specific, and transcripts belonging to the same or related gene families further confuses the pattern of overlapped sequences. Therefore, the degree of identity for sequence overlapping has to be lower in EST assembly than in genome assembly to take into account high sequence errors, polymorphisms, and alternative splicing events. Finally, the EST base-call quality values are available through Washington University for only a low fraction of EST data; thus they lack this important information used in most assembly programs. Moreover, the errors generated in automated DNA sequencing are concentrated at the start and end of sequence reads (Adams, Kerlavage et al. 1995). In genomic sequence assembly, this is mitigated by the random distribution of sequence start points, whereas in ESTs, errors are positionally biased, most of them being clustered at 3′ UTRs of mRNAs because of the used cDNA synthesis, cloning, or sequencing strategy.

In conclusion, programs used in EST assembly should be relatively tolerant of random errors; they should also be able to separate EST from distinct but closely related transcripts (i.e., derived from paralogous genes, alternative splicing, or alternative polyadenilation sites).

6.7.6   Sequence Assembly for Gene Index Construction

The gene catalogs now available have been constructed using only ESTs or combining EST and mRNA sequence data derived from GenBank and EMBL. Among the efforts to form such gene catalogs can be mentioned UniGene (Boguski and Schuler 1995; Schuler, Boguski et al. 1996; see the URL in Table 5.1), TIGR Gene Indices (see the URL in the Appendix), STACK (Sequence Tag Alignment and Consensus Knowledge Base; Christoffels, van Gelder et al. 2001; see the URL in the Appendix), GeneExpress (Houlgatte, Mariage-Samson et al. 1995; see the URL in the Appendix), and Merck Gene Index (Aaronson, Eckman et al. 1996). The input data and building procedure depends on the principal use of the database and on the features of clustered-reconstructed sequences that the authors wish to stress.

We discuss the TIGR Gene Index, UniGene, and STACK databases briefly as a case study for the EST clustering approach. The sequence cleaning strategy is common to all three projects, whereas the clustering–assembling protocol and strategy differ in stringency. The first versions of the TIGR Gene Index used the TIGR Assembler as the clustering–assembling tool. By contrast, the current TIGR Gene Indices, UniGene, and STACK databases have been constructed using distinct programs for clustering and assembly.

The goal of the TIGR Gene Index is the EST classification in distinct transcripts rather than distinct genes. The stringent grouping strategy allows the identification and partition of alternative transcript isoforms and of distinct members of related gene families. Cluster variants are defined as splice variants only if they match fully sequenced genes with known isoforms in the related EGAD database (Expressed Gene Anatomy Database) of well-characterized genes (White and Kerlavage 1996). In this way, chimerism is prevented; however, this approach results in a more fragmented representation and in the retention of most sequence redundancy. In addition to clustering, the construction of tentative consensus (TC), representing virtual transcripts, improves the length and quality of sequence reconstructions beyond those available from any one EST and simplifies its functional annotation. The TIGR Gene Index treats ESTs as elements of a transcriptome shotgun sequencing project (Adams, Kerlavage et al. 1995). Only stringent sequence similarity and overlapping criteria are used to construct clusters and to produce a set of unique, high-fidelity, full-length or partial virtual transcripts.

The current build procedure (Quackenbush, Liang et al. 2000) is as follows:

1. Vector, linker, bacterial, ribosomal, mitochondrial, polyA/polyT, and low-quality sequences are removed. Only ESTs longer than 100 nt and with no more than 3% undefined nucleotides are retained.

2. A pairwise comparison is made of ESTs with the EGAD database (White and Kerlavage 1996) and of ESTs with ESTs using the WU-BLAST program (see Sec. 6.4.3). Sequences sharing a minimum of 95% identity over regions at least 40 nt in length with unmatched overhangs of less than 20 nt long are included in the same cluster.

3. Each cluster is assembled separately using the CAP3 program (Huang and Madan 1999). One or more consensus sequences can be produced for each cluster. Chimeric and low-quality sequences are discarded during assembly.

4. A second round of clustering and assembly is carried out using only the newly constructed TC sequences, to further eliminate redundancy.

UniGene is an experimental system for automatic partitioning of ESTs and full-length mRNAs from characterized genes into a nonredundant set of gene-oriented clusters. Each cluster is very likely to represent a distinct gene and is 3′ anchored; that is, it includes a 3′ mRNA terminus. A single cluster can contain several splice forms of the same transcript, but chimeras and other artifacts may be included incorrectly. Different from the TIGR Gene Index, UniGene does not attempt to reconstruct contigs; the longest and highest-quality sequence within each cluster, instead, is reported as representative of the entire cluster. Indeed, multiple assemblies can be obtained per cluster without knowing whether these assemblies represent splice variants, partially divergent members of closely related families, or just plain artifacts, such as chimeras resulting from "dirty" EST data.

The looser clustering strategy allows distinct transcript isoforms to be included in the same cluster and the elimination of (probably) all sequence redundancy. Algorithms and procedures for clustering are experimental and still under development; anyway, the clustering procedure makes use of sequence similarity criteria and clone information. Indeed, 3′ and 5′ EST reads from the same cDNA clone are included in the same cluster, to avoid mRNA regions not sampled by EST sequencing separate clusters belonging to the same gene. The UniGene build procedure (see the URL in the Appendix) proceeds in several steps:

1. Contaminants (vector, ribosomal, and mitochondrial sequences) repeats, and low-complexity regions are removed. Only sequences with more than 100 informative sites are retained.
2. Comparisons are made of ESTs with characterized genes and of ESTs with ESTs using the WHALE (previous UniGene versions) or MEGABLAST program (Zhang, Schwartz et al. 2000) (see Sec. 6.4.3). The similarity threshold is set to retain sequence pairs with a minimum of 95% identity over 70% coverage of alignable regions.
3. Only clusters containing a sequence with a polyadenylation signal or two labeled 3′ ESTs are retained (3′ anchored clusters), and clone-based edges are added. Due to frequent imperfect clone labeling, two clones with overlapping at both the 3′ EST and 5′ EST ends are needed to merge previous clusters.
4. ESTs that do not belong to an anchored cluster are rechecked at a lower level of stringency and can be added to a cluster as a guest member.
5. Clusters with only one EST (singletons) are rechecked at a lower stringency against all the clusters and eventually merged.

The goal of the STACK database is the detection and visualization of putative human transcripts in the context of an expression state. Transcript variability is then analyzed in the context of developmental and pathological states (Christoffels, van Gelder et al. 2001). The STACK database uses as input only EST sequences and is organized via tissue subdivision as a whole body index. Thus, sequence redundancy is retained in the tissue subdivision, but is mostly eliminated in the whole body index. The building procedure is a loose clustering approach followed by a postassembly step to highlight the intracluster sequence diversity. Allowing simultaneous viewing of inconsistencies within a cluster, the CRAW program (Burke, Wang et al. 1998) facilitates discrimination of sequence variants that can arise from alternative splicing or polymorphism of the same gene, aberrant transcripts associated with diseases, genes belonging to the same families, or sharing a common motif or even chimeric clones and other artifacts. Sequence similarity and overlapping criteria are used for clustering, whereas additional information such as a clone source, or normal and disease-related tissue source, are used in postclustering steps. The STACK database is generated using the following procedure (Miller, Christoffels et al. 1999; Christoffels, van Gelder et al. 2001):

1. Arbitrary partitioning of ESTs into tissue/state subdivisions.
2. Masking and/or removal of repeated elements, low-complexity sequences, and vector, mitochondrial, and ribosomal sequences (RepeatMasker and CrossMatch programs). Sequences shorter than 50 nt are eliminated.
3. EST clustering using D2_CLUSTER (Burcke, Davison et al., 1999), a word-based clustering program. Sequences sharing a 150-nt window with at least 96% identity fall into the same cluster.
4. Assembly of each cluster using the PHRAP program and no sequence quality values (trace information is not available for all human ESTs).
5. Multiple alignments verification using CRAW (Burke, Wang et al. 1998; Chou and Burke 1999) and CONTIGPROC (Miller, Christoffels et al. 1999) programs. CRAW provides a simple means to view clusters, partition them in subclusters and maximize the consensus length. A subcluster is generated if at least 50% of a 100-nt window differs from the remaining sequences of the cluster, excluding the initial 100 nt of any read. CONTIGPROC partitions and ranks the multiple consensuses that can be generated from a single cluster. In addition, the clusters are 3' or 5' oriented.
6. Joining of clusters containing ESTs originated from the same cDNA clone and extension of the linked consensus entries.

6.8 LINGUISTIC ANALYSIS OF BIOSEQUENCES

Biosequences (i.e., DNA, RNA, and proteins), can be considered and treated as a written text that contains the information needed for their biological function. In particular, a genome can be described as a very long string made up of four letters, the nucleotides. Indeed, linguistic metaphors, such as transcription, translation, editing, and so on, have always been used extensively to describe the features of genetic material. As with any language, the letters of the genetic language are not distributed at random but have to follow defined rules to make sense (i.e., they should be organized in words). In the context of a genome, we can define as a word any sequence endowed with a specific biological function.

Despite the several features resembling natural languages, the genetic language has some peculiar features, such as the lack of fixed ends, and punctuation and spaces that make the definition of words much more difficult, although we can easily observe that only a small number of the various possible combinations of the four

TABLE 6.4. Single-Letter Code Recommendations for DNA Sequences

Symbol	Meaning	Origin of Designation
G	G	Guanine
A	A	Adenine
T	T	Thymine
C	C	Cytosine
R	G or A	puRine
Y	T or C	pYrimidine
M	A or C	aMino
K	G or T	Keto
S	G or C	Strong interaction (three H bonds)
W	A or T	Weak interaction (two H bonds)
H	A or C or T	not-G, H follows G in the alphabet
B	G or T or C	not-A, B follows A
V	G or C or A	not-T (not-U), V follows U
D	G or A or T	not-C, D follows C
N	G or A or T or C	aNy

nucleotides in a sequence can actually be found in nature. However, this number is large enough to allow the same or similar encoded information to be expressed in multiple ways. This feature reflects one of the fundamental properties of genetic information, and consequently of the resulting living beings, that is, the capability to evolve.

The fact that only a small fraction of all possible sequences can actually be found in nature depends on constraints, most of them still unknown, operating on the information content of the genetic material (i.e., a sort of "grammar" whose rules describe the possible relationships between "words" to assure meaningfulness of the encoded information). Many linguistic approaches have been proposed for the analysis of sequence information that revealed powerful tools for a number of purposes, including the determination of gene structure, the identification of sequence motifs having a functional role, and the establishment of functional correlations between sequences (i.e., evolutionary analyses; see Chapters 7 and 8). As in natural languages, an alphabet may be defined for the genetic language. In the case of nucleic acids (DNA/RNA), the alphabet is made up of four elements (the four nucleotides A, C, G, T), whereas 20 elements are used for proteins (the amino acids). Given the possible ambiguites in nucleotide sequences, the IUPAC–IUB nomenclature commission defined the 15-item alphabet in Table 6.4. Similarly, the protein alphabet is shown in Table 6.5.

An alphabet is generally used to compose words. The term *word* is commonly used interchangeably with other terms, such as *string*, *motif*, or *pattern*. For the sake of clarity, we propose a more accurate definition of these terms, keeping in mind that in the literature they can be used as synonyms. We define as a *string* any subsequence of length equal to w. Then a sequence of length L will contain $L - w + 1$ strings of length w. The number of strings w nucleotides long obtainable with an alphabet of size D is thus equal to D^w. Hence, the occurrence probability of a string w nucleotides long in a random text where the different symbols are used with the same frequency should be equal to $1/D^w$. Indeed, the basic feature of the DNA lan-

TABLE 6.5. Alphabet of Amino Acids Encoded by mRNAs

Trivial Name	Three-Letter Symbol	One-Letter Symbol
Alanine	Ala	A
Arginine	Arg	R
Asparagine	Asn[a]	N[a]
Aspartic acid	Asp[a]	D[a]
Cysteine	Cys	C
Glutamine	Gln[b]	Q[b]
Glutamic acid	Glu[b]	E[b]
Glycine	Gly	G
Histidine	His	H
Isoleucine	Ile	I
Leucine	Leu	L
Lysine	Lys	K
Methionine	Met	M
Phenylalanine	Phe	F
Proline	Pro	P
Serine	Ser	S
Threonine	Thr	T
Tryptophan	Trp	W
Tyrosine	Tyr	Y
Valine	Val	V
Unspecified amino acid	Xaa	X

[a] Asx denotes Asp or Asn; B denotes N or D.
[b] Glx and Z represent glutamic acid or glutamine.

guage is its base composition (i.e., the frequency of the four nucleotides). Grantham, Gautier et al. (1980) were the first to discover that each genome has its own peculiar distribution of bases, usually expressed in terms of G + C content (%G + C) in the case of a homogeneous distribution of the complementary nucleotides between the two strands. This property was called the *genome hypothesis*. It is equivalent to saying that any genome has a personal language (see also Sec. 7.3).

The linguistic approach to the analysis of sequence information could be a powerful tool in interpreting and deciphering the message contained in biological texts (Trifonov 1989, 1999; Pietrokovski, Hirshon et al. 1990; Popov, Segal et al. 1996) and to assess evolutionary relationships. Several methods and approaches have been proposed for pattern recognition and, consequently, for the construction and comparison of genetic vocabularies (Brendel, Beckmann et al. 1986; Mengeritsky and Smith 1987; Hertz, Hartzell et al. 1990; Pevzner 1992). This genetic text can be more or less complex, and its complexity can be determined by information theory parameters as described by Shannon (Ebeling and Jimenez-Montano 1980) or as a function of the richness of its vocabulary. Vocabulary-based complexity depends on the number of different strings adopted in the text with respect to the maximum number of strings that can theoretically be present in a sequence of L length with an alphabet of D number of characters ($D = 4$ for DNA and $D = 20$ for proteins). A more detailed description of various complexity measures for genetic texts is given in Sec. 6.8.2.

Average genome complexity is generally not very informative. The variation of complexity along the genome could be a useful measure to detect, for instance, lateral transfer events or tandem repeat-containing regions (see below). Another approach to linguistic analysis consists in the discovery of novel regulatory oligonucleotide patterns based on the observation that they are overrepresented in a group of functionally related sequences (e.g., promoter sequences or a subset of promoter sequences for a specific collection of functionally equivalent genes). Such patterns, which can be considered as reliable candidates of some regulatory activity, can be used later, after experimental validation of their activity, for functional annotation of novel sequences. The main problem for pattern discovery algorithms is that generally, biological functional patterns often tolerate few mutations, insertions, and/or deletions without losing their biological activity.

6.8.1 Biosequences as Markov Chains

The linguistic methods developed for the analysis of sequence information are mostly based on the assumption that nucleotide sequences can be represented as a Markov chain (Almagor 1983). A Markov chain sequence can be regarded as a sequence of symbols generated according to a principle whereby the chain has a finite memory h. This means that the probability of having the nucleotide k at position i depends on the h immediately preceding nucleotides. This concept can be formalized by the following conditional probability equation:

$$p(a_i|a_{i-1} \quad a_{i-2} \quad \cdots \quad a_1) = p(a_i|a_{i-1} \quad a_{i-2} \quad \cdots \quad a_{i-h})$$

This rule allows us to calculate the occurrence probability for a given nucleotide string according to different statistics, depending on the value of h, which determines the "order" of the Markov chain. Assuming that $h = 0$, the probability of having one of the four nucleotides at position i does not depend on the preceding ones, and thus the occurrence probability of a string is given simply by the product of the occurrence probability of the component nucleotides. For example,

$$p(\text{ACGT}) = p(\text{A})p(\text{C})p(\text{G})p(\text{T})$$

Assuming a Markov chain h greater than zero, the expected occurrence E of the string $S = a_1a_2 \ldots a_n$ is given by

$$E(S = a_1a_2 \cdots a_n) = \frac{f(a_1 \cdots a_{h+1})f(a_2 \cdots a_{h+2}) \cdots f(a_{n-h} \cdots a_n)}{f(a_2 \cdots a_{h+1})f(a_3 \cdots a_{h+2}) \cdots f(a_{n-h} \cdots a_{n-1})}$$

The observation of a significant deviation between the expected (E) and the observed (O) value of a given string that can be expressed by an O/E ratio significantly divergent from unity or by a high standard deviation value [SD(S), otherwise defined as the *contrast value* (Brendel, Beckmann et al. 1986)]:

$$\text{SD}(S) = \frac{f(S) - E(S)}{\max[\sqrt{E(S)}, 1]}$$

As stated previously, statistically significant oligomers are defined here as *motifs* or *patterns*.

As a simple example, let us calculate the expected occurrence of the tetramer ACGT in the human mitochondrial genome (AC V00662, 16,569 nt). Assume the following occurrences observed for the relevant mono-, di-, and trinucleotides:

A: 5123	C: 5178	G: 2176	T: 4094
AC: 1493	CG: 436	GT: 418	
ACG: 119	CGT: 78		

If we use a zero-order Markov chain ($h = 0$), the ACGT expected occurrence is

$$E(\text{ACGT}) = p_A p_C p_G p_T = 0.31 \cdot 0.31 \cdot 0.13 \cdot 0.25 \cdot 16,568 = 51.9$$

for $h = 1$,

$$E(\text{ACGT}) = \frac{f(\text{AC})f(\text{CG})f(\text{GT})}{f(\text{C})f(\text{G})} = 24.1$$

and for $h = 2$,

$$E(\text{ACGT}) = \frac{f(\text{ACG})f(\text{CGT})}{f(\text{CG})} = 21.2$$

The discrepancy between the expected values obtained with $h = 0$ and those with $h = 1$ or $h = 2$ is clearly evident. This is probably due to the fact that a zero-order Markov chain does not take into account the remarkable frequency biases generally observed at the dinucleotide level. In this particular case, all three dinucleotides embedded in the tetramer under analysis are avoided [i.e., $O/E < 1$; $E(\text{AC}) = 1600$; $E(\text{CG}) = 680$; $E(\text{GT}) = 537$]. Given that $O(\text{ACGT}) = 21$ in the human mitochondrial genome, we would have considered the ACGT tetranucleotide significantly underrepresented by assuming a zero-order Markov chain ($O/E = 0.4$ or $SD = -4.3$), whereas it is not for $h = 1$ or $h = 2$.

An uneven distribution of dinucleotides along the genetic text was noted in the early 1980s (Nussinov 1980, 1981, 1987) and received further support from the analyses of Kozhukhin and Pevzner (1991), who showed that the inhomogeneity was determined primarily by WW and SS dinucleotides (see Table 6.4). They observed that WW and SS dinucleotides are generally strongly underrepresented compared to their rate of expected occurrence.

6.8.2 Linguistic Complexity of Biosequences

In the context of the linguistic analysis of biosequences, the information content of a text can be evaluated in terms of *complexity* or *entropy* of the language used. These parameters may provide an appropriate measurement of the quality of information stored in a biosequence and of the grammar rules this message obeys. Ebeling and Jimenez-Montano (1980) have shown that polynucleotide sequences have a lower degree of complexity than protein sequences. This can be explained as the result of a higher number of constraints (i.e., grammar rules) acting on nucleotide sequences than on proteins. The lower complexity of DNA sequences, or conversely, their higher redundancy, may also reflect the necessity of DNA to accommodate multiple messages at the same location. Indeed, in DNA sequences there are several examples of code overlapping (i.e., the simultaneous involvement of the same nucleotide in several messages; Trifonov 1989). Multiplicity and overlapping of the genetic code poses serious problems to biomathematicians, who attempt to determine biologically significant oligomers (DNA motifs) using statistical methods.

The complexity K_n of a sequence S can be measured using several methods; the most relevant is illustrated in what follows (a different n subscript for K will be associated with each method). The simplest way to measure linguistic complexity of nucleotide sequences is derived from Shannon's theory of information systems. Such a complexity measure, also implemented in the SEG program developed by Wootton and Federhen (1996), has been used to detect low-complexity regions in protein or nucleotide sequences (see Sec. 6.4.4 for its use in database searching).

Assuming a zero-order Markov chain, sequence entropy, S, can be defined as

$$K_1(S) = -\sum_{i=1}^{N} p_i \log_D p_i$$

When a first-order Markov chain is assumed, entropy is

$$K_2(S) = -\sum_i \sum_j p_i p_{ij} \log_D p_{ij}$$

D being the logarithm base equal to the size of the alphabet and p_i and p_{ij} being the probability observed for the ith nucleotide and the (i,j)th dinucleotide, respectively (Gatlin 1966, 1968). According to these equations, the sequence complexity value ranges between zero (lowest complexity, e.g., the sequence AAAA) and 1 (highest complexity, e.g., the sequence ACGT).

An alternative way to measure sequence complexity (or entropy) is based on the combinatorial enumeration of possible states according to Boltzmann theory whereby the complexity status of a sequence or subsequence is represented by a list of numbers defined as a *complexity state vector* of length equal to the size of the alphabet (D) that summarizes its composition. For example, the complexity state vector of the pentanucleotide ACGAT is given by

$$\{2, 1, 1, 1\}$$

as A is present twice and all other nucleotides once. The number of sequences that can be generated having exactly the same composition is given by

$$\Omega = \frac{L!}{\prod_{i=1}^{N} n_i!}$$

where L is the sequence length and n_i are the numbers in the complexity state vector. In our particular example ($S = \text{ACGAT}$), we have $\Omega = 5!/2! = 60$. The sequence complexity can be thus defined as

$$K_3(S) = \frac{1}{L} \log_N \Omega$$

$K_3(S)$ converges to $K_1(S)$ or $K_2(S)$ at large sequence length L.

Sequence complexity measured previously is based on residue composition, regardless of the possible periodicity patterns. In this way the four-gram periodic sequence

$$\text{ACGT–ACGT–ACGT–ACGT}$$

would have the same linguistic complexity as the random sequence:

$$\text{ACAGCTATGTCATGCG}$$

The previous measure is evidently inadequate, as it can reveal low-complexity regions owing to compositional bias but not to some periodic structure. To overcome this problem, alternative methods have been proposed that measure vocabulary richness used in the sequence as compared to the maximal possible vocabulary size (e.g., the number of different four-gram strings used in the actual sequence or subsequence as compared to a random sequence of the same length and composition). The elements of the vocabulary are here defined as contiguous oligonucleotide k-grams of various sizes. The greater the number of different strings used in the sequence as compared to the D^w total possible strings of a given length w ($D = 4$ for DNA/RNA and 20 for proteins), the more complex the genetic text.

The linguistic complexity of a nucleotide sequence S of length L can thus be described according to the simple empirical formula (Trifonov, unpublished)

$$K_4(S) = \prod_{k=1}^{L-1} U_k$$

where

$$U_k = \sum_{j=1}^{4^k} \frac{x_{kj}}{L-k+1} \max\left(1, \frac{L-k+1}{4^k}\right)$$

where $x_{kj} = 1$ if the jth string of length k is present at least once in the sequence under examination, $x_{kj} = 0$ if it is missing. For example, if we consider the two tetramers GGGG and ACGT, their linguistic complexity is

$$K_4(\text{GGGG}) = U_1 U_2 U_3 = 1/4 \cdot 1/3 \cdot 1/2 = 0.04$$
$$K_4(\text{ACGT}) = U_1 U_2 U_3 = 4/4 \cdot 3/3 \cdot 2/2 = 1.00$$

Indeed, as in a sequence data set, the sequence length can vary considerably. It is recommended that complexity be calculated on a moving window of size w and

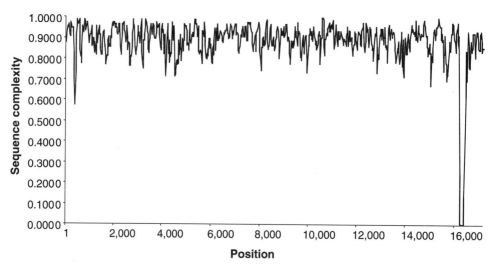

Figure 6.25. Plot of sequence complexity calculated for a window of 100 bp moving along the rabbit complete mitochondrial genome (EMBL ID: AJ001588) generated by the program Complex. Word sizes used ranged between 5 and 7 bp and the moving step was of 20 bp. (From Pesole et al., unpublished; EMBOSS package.)

using k-grams within a given size range. Choosing the window size w so that $w \ll 4^k$ (for all k values used in the complexity measure) to avoid vocabulary richness becomes trivial due to too large a window (e.g., all 16 different dinucleotides are expected to be found in a window of size much larger than 4^2). For instance, using $w = 100$, a suitable complexity calculation can be made varying k from 5 to 7 nt.

The formula above can be used to calculate the complexity profile of a sequence plotting the window complexity as a function of its position in the sequence and of the global sequence complexity, which is the complexity value averaged over all windows scanned. The Complex software (available in the EMBOSS package; Pesole et al., unpublished) is able to calculate sequence complexity according to the formula above but also provides a measure based on the ratio between the richness of the k-gram vocabularies of the actual sequence against that of a random sequence of the same length generated by a Markov chain of order h. Figure 6.25 shows the plot obtained by Complex software applied to the complete mitochondrial genome of rabbit. A low-complexity peak corresponding to the short tandem repeat region present in the D-loop region is evident (nt 16,224 to 16,423, a 20-nt-long repeat unit), but no complexity decrease is observed for a long tandem repeat region (nt 16,610 to 17,221, a 153-nt-long repeat unit). This is due to the specific k-gram size range, from 5 to 7 nt, chosen in this complexity analyis.

A completely different way of measuring sequence complexity is the *algorithmic entropy* proposed independently by Kolmogorov (1965) and Chaitin (1969). According to this definition, if a sequence S of length $L(S)$ is random, there does not exist any shorter sequence Q, μ characters long with $\mu < L$, which permits construction of sequence S. On the other hand, if there is any regularity, it is possible a shorter representation of the sequence [i.e., $L(Q) < L(S)$]. The entropy of the sequence is defined as

$$K_5(S) = \sum_\sigma K(\sigma \rightarrow Q) + I_F$$

where $K(\sigma \rightarrow Q)$ represents the character length of the terms of the production rule and I_F is the length of the final sequence. A measure of sequence complexity is given by $K_5(S)/K_{max}$, where K_{max} is given by $L(S)$. To clarify, let us give a simple example. Given the sequence

$$S = \text{ACACACACACACA} \qquad L(S) = K_{max} = 13$$

If we pose the production rule $\sigma_1 = \text{ACAC}$, we obtain a shorter sequence Q_1:

$$Q_1 = \sigma_1 \sigma_1 \sigma_1 A$$

Given the character length of the production rule $K(\sigma_1 \rightarrow Q) = 4$, we have

$$K_5(S) = K(\sigma_1 \rightarrow Q) + L(Q_1) = 8$$

The complexity can be measured as $K_5(S)/K_{max} = 8/13 = 0.61$.

This approach is generally used in many compression algorithms, and the higher the compression ratio (i.e., the ratio between the size of the uncompressed sequences and that of the compressed one), the lower the sequence complexity. Sequence complexity can also be used to reduce disk space occupied by biosequence databases on the computer. Gusev, Nemytikova et al. (1999) have proposed a novel complexity measure that takes into account typical regularities observed in DNA such as direct, symmetric, and complemented repeats. This method also takes into account inexact repeats that can be regarded as exact matching segments separated by mismatching symbols. Quite interestingly, by analyzing a large number of single base-pair substitutions in regulatory regions of 65 different human genes involved in inherited diseases, Krawczak, Chuzhanova et al. (2000) observed that changes in DNA sequence complexity induced by mutation were correlated to phenotypic consequences of mutations. In particular, mutations leading to an increase in complexity are more likely to have pathological relevance than mutations leading to decrease or no change in complexity.

6.8.3 Identification of Repeats in Genomic Sequences

Repetitive DNA that represents a large fraction of the eukaryotic genome can also be found in lower amounts in prokaryotic genomes. It is generally located in the noncoding portion of the genome and may consist of tandem repeated sequences (micro- and minisatellites) or of interspersed repeats (e.g., SINEs and LINEs). For known repeat families, the repeat annotation can be carried out by comparing the sequences under investigation against a database of repetitive elements, as done by RepeatMasker (see Sec. 6.4.5), which is able to identify simple sequences (i.e., low-complexity sequences such as a homopolymeric DNA sequences), microsatellites (short tandem repeats with a monomeric unit up to five nucleotides long), and interspersed repeats stored in the RepBase reference repeat database (Jurka 2000). Other tools need to be used to identify tandem repeats with longer repeat units (e.g., minisatellites) or unknown interspersed repeats. The dot-plot matrix (see Sec. 6.2) can be used efficiently to detect tandem or interspersed repeats, but it becomes impracticable if large genome sequences have to be analyzed.

The program TRF (Tandem Repeat Finder; Benson 1999) is able to identify tandem repeat sequences without the need to specify neither the pattern nor the pattern size. Figure 6.26 shows the result of TRF application on a transcript encoded by a human gene located on chromosome 1 and flanking the p73 tumor suppressor gene. The REPUTER program, implementing a suffix tree representation of the sequence searched (Kurtz and Schleiermacher 1999), allows us to determine exact and slightly degenerate repeated substrings contained in DNA sequences as large as complete microbial genomes. The search is allowed for direct, reversed, complementary, or palindromic repeats of maximal length above the threshold chosen, also providing an estimate of statistical significance. Figure 6.27 shows the list of repeated sequences, longer than 50 nucleotides, found by the REPUTER software tool in the entire *E. coli* genome.

6.8.4 Pattern Searching in Biosequences

Biological processes are the result of a specific pattern of gene expression in both temporal and spatial frameworks. Transcriptional or posttranscriptional control of gene expression involves short DNA or RNA tracts (cis-acting elements), respectively, interacting with specific binding proteins (trans-acting factors). Interaction of RNA-binding proteins may also require the corresponding cis-acting elements to be folded into conserved secondary structures to be specifically recognized.

When analyzing unknown nucleotide sequences, we can follow two different approaches. The first one is to apply methods able to detect sequence patterns already known to be endowed of some biological function. In this case the structure of the functional pattern needs to have been defined previously by comparative analysis of a set of functional patterns whose activity was verified experimentally and whose structure has been investigated carefully by site-specific mutagenesis or chemical probing. The second approach consists in the discovery of novel regulatory patterns based on the observation that they are common to a group of functionally related sequences.

Cis-acting regulatory elements generally consist of subtle sequence patterns where each particular position has a different functional relevance and thus may present a variable degree of conservation. This is the reason why they are commonly represented by the consensus sequence, where ambiguous characters can be used

---

Figure 6.26. Output of the TRF program used on a transcript sequence encoded by an unknown human gene located in chromosome 1 (EMBL ID: AB007964). The parameters used in the run, reported in the output, are match weight, 2; mismatch penalty, 7; indel penalty, 7; match probability, 80%; indel probability, 10%; minimum alignment size to report, 50 nt; minimum period size to report, 500 nt; A notable minisatellite with repeat size 36 nt and 27.2 copy number has been identified by TRF in this transcript.

Tandem Repeats Finder Program written by:

            Gary Benson
    Department of Biomathematical Sciences
        Mount Sinai School of Medicine

Version 2.02

Sequence: AB007964 Ab007964 Homo sapiens mRNA, chromosome 1 specific transcript
KIAA0495. 8/1998

Parameters: 2 7 7 80 10 50 500

Pmatch=0.80,Pindel=0.10
tuple sizes 0,4,5,7
tuple distances 0, 29, 159, MAXDISTANCE

Length: 6357
ACGTcount: A:0.20, C:0.25, G:0.29, T:0.25

Indices	Period Size	Copy Number	Consensus Size	Percent Matches	Percent Indels	Score	A	C	G	T	Entropy (0-2)
2029--2101	36	2.0	36	97	0	137	28	35	27	8	1.86
2154--3129	36	27.2	36	90	1	1386	20	11	37	30	1.89

    Indices: 2029--2101 Score: 137
    Period size: 36 Copynumber: 2.0 Consensus size: 36

      2019 TGGTTAGCGT

      2029 GGACATCAGCCAGCACTGCACAGACAGCTGCGACCA
         1 GGACATCAGCCAGCACTGCACAGACAGCTGCGACCA

             *
      2065 GGACGTCAGCCAGCACTGCACAGACAGCTGCGACCA
         1 GGACATCAGCCAGCACTGCACAGACAGCTGCGACCA

      2101 G
         1 G

      2102 AGACTGGAGG

Statistics
Matches: 36, Mismatches: 1, Indels: 0
      0.97   0.03   0.00

Matches are distributed among these distances:
  36   36   1.00

ACGTcount: A:0.29, C:0.36, G:0.27, T:0.08

Consensus pattern (36 bp):
GGACATCAGCCAGCACTGCACAGACAGCTGCGACCA

Indices: 2154--3129 Score: 1386

    Period size: 36 Copynumber: 27.2 Consensus size: 36

      2144 CACTGTGAAA

```
# repfind -f -l 100 -h 5 U00096.fna
# 4639221 -5 100 U00096.fna
3107 3616822 F 3107 3759809 -5 0.00e+00
3104 3616840 F 3104 3759827 -5 0.00e+00
3095 3616900 F 3095 3759887 -5 0.00e+00
2985 4166150 F 2985 4207552 -5 0.00e+00
2958 4166062 F 2958 4207464 -5 0.00e+00
2952 4166197 F 2952 4207599 -5 0.00e+00
1952 4032758 F 1952 4163876 -5 0.00e+00
1842 2725475 F 1842 3422691 -5 0.00e+00
1757 4164087 F 1757 4205574 -5 0.00e+00
1738 4032972 F 1738 4205577 -5 0.00e+00
1660  226300 F 1660 3941868 -5 0.00e+00
1621 4164227 F 1621 4205714 -5 0.00e+00
1603 4033109 F 1603 4205714 -5 0.00e+00
1458  226022 F 1458 4166483 -5 0.00e+00
1458  226022 F 1458 4207885 -5 0.00e+00
1456  226166 F 1456 3941734 -5 0.00e+00
1394  227488 F 1394 4209351 -5 0.00e+00
1367   15374 F 1367 2512280 -5 0.00e+00
1364  226604 F 1364 3942172 -5 0.00e+00
1359   15375 F 1359  607218 -5 0.00e+00
1358  607219 F 1358 2512282 -5 0.00e+00
1343  380477 F 1343 3184105 -5 0.00e+00
1340 1465927 F 1340 2994376 -5 0.00e+00
1338  380476 F 1338 4495741 -5 0.00e+00
1338 3184104 F 1338 4495741 -5 0.00e+00
1337 1465932 F 1337 2066961 -5 0.00e+00
1336 2066957 F 1336 2994377 -5 0.00e+00
1324 3941731 F 1324 4166624 -5 0.00e+00
1324 3941731 F 1324 4208026 -5 0.00e+00
1289 3618715 F 1289 3761702 -5 0.00e+00
1282  225686 F 1282 4207549 -5 0.00e+00
1279  225689 F 1279 4166150 -5 0.00e+00
1269  390922 F 1269 1093457 -5 0.00e+00
1268  565995 F 1268 2168188 -5 0.00e+00
1264  314452 F 1264 2168195 -5 0.00e+00
1262  314450 F 1262  566000 -5 0.00e+00
1258 4036971 F 1258 4209493 -5 0.00e+00
1255  227628 F 1255 4036969 -5 0.00e+00
1250  225736 F 1250 4166197 -5 0.00e+00
1250  225736 F 1250 4207599 -5 0.00e+00
1223  269762 F 1223 1467317 -5 0.00e+00
1217  687069 F 1217 3649661 -5 0.00e+00
1211  573802 F 1211 3363179 -5 0.00e+00
1211  573808 F 1211 3649660 -5 0.00e+00
1210  573802 F 1210  687062 -5 0.00e+00
1210  687062 F 1210 3363179 -5 0.00e+00
1210 2099762 F 1210 3363182 -5 0.00e+00
1209  273169 F 1209  687064 -5 0.00e+00
1209 3363180 F 1209 3649655 -5 0.00e+00
1208  573805 F 1208 2099762 -5 0.00e+00
```

to represent variable positions following the rules in Table 6.4. Typically, a consensus sequence is derived from the alignment of regulatory patterns known to be functional and choosing each position to be represented by the most used residue. Indeed, the consensus representation would be adequate only if it was able to provide an exhaustive representation of all functional sequences. This is the case of recognition sites for restriction enzymes, where, for example, the sequence GTYRAC fully describes the HindII site. In contrast, cis-acting elements recognized by regulatory proteins may have a variable level of degeneracy and may allow some mismatches at specific positions. A possible representation of these elements is the regular expression pattern which provides a deterministic definition of the pattern, which may include ambiguous characters, wildcard characters, and specific pattern constraints. Such regular expression syntax is used for the definition of patterns to be searched by the FindPatterns program (Genetics Computen Group 1998) and is also the notation used in most ProSite patterns (Hofmann, Bucher et al. 1999). As an example, the zinc-finger domain signature of the C_2H_2 class (ProSite entry PS00028), where two pairs of cysteine and histidines coordinate a zinc ion, can be defined using the following regular expression:

$$C-X-(2,4)-C-X(3)-[LIVMFYWC]-X(8)-H-X(3,5)-H$$

where the single-letter amino acid alphabet is used (Table 6.5), the numbers in parentheses define the minimum and maximum number of allowed repetitions of the preceding residues (including the wildcard residue, X) and brackets contain residues allowed at a given position.

A more complex pattern syntax can be used by the PatSearch program (Pesole, Liuni et al. 2000a), where a sequence pattern is made of a number of pattern units that may include paired complementary regions (this applies to RNA sequences) and permits up to a fixed amount of mismatches, insertions, and deletions. As an example, the consensus structure and the corresponding pattern syntax defining the *iron responsive element* (IRE) is shown in Fig. 6.28. IRE is a specific hairpin structure located in the 5′ or the 3′ UTRs of various mRNAs coding for proteins involved in cellular iron metabolism. The PatSearch software also allows postprocessing of first-pass matching sequences to allow further constraints to be applied at the pattern searched. As an example, we report below the PatSearch syntax for the p53 recognition element that is made of four tandem repeated decamers that can be separated by up to 13 nucleotides:

```
p1=rrrcwwgyyy[2,0,0]  p1:(p2=nnncnngnnn)  0...13
p3=rrrcwwgyyy[2,0,0]  p3:(p4=nnncnngnnn)  0...13
p5=rrrcwwgyyy[2,0,0]  p5:(p6=nnncnngnnn)  0...13
   p7=rrrcwwgyyy[2,0,0]  p7:(p8=nnncnngnnn)
```

◀──

Figure 6.27. Output of the RepFind software of the Reputer package, showing direct repeats at least 100 nucleotides long, with up to five mismatches allowed, identified in the *E. coli* complete genome sequence (EMBL ID: U00096). The parameters used in the search are: –f, direct repeats; –l 100, minimum repeat length of 100 nt; –h 5, maximum of 5 allowed mismatches. The output consists of lines, each listing, respectively, the first repeat length, its position in the sequence, the orientation, the second repeat length, its position in the sequence, the distance between repeats (negative for mismatch repeats), and expected value of the repeat.

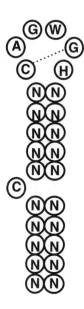

PatSearch pattern:

r1={au,ua,gc,cg,gu,ug}
(p1=2...8 c p2=5...5 CAGWGH r1~p2 r1~p1 |
p3=2...8 nnc p4=5...5 CAGWGH r1~p4 n r1~p3)

Figure 6.28. Hairpin structure of the iron responsive element with the regular expression PatSearch pattern. The r1 pattern unit defines the base pairings allowed, including non-canonical GU–UG pairings. The p1 pattern unit represents any sequence of length varying between two and eight nucleotides, whereas p2 is a pattern unit of exactly five nucleotides, r1~p1 and r2~p2 represent oligonucleotides complementary to p1 and p2, respectively, following the pairing rules defined in the r1 pattern unit. The "|" symbol is the "either/or" pattern unit that allows matching between alternative consensus structures, which in this case differentiate in the bulge loop, which may be a single C or the trinucleotide NNC. [From Pesole, Liuni et al. (2000).]

The [2,0,0] qualifier of the p1, p3, p5, and p7 pattern units means that up to two mismatches, but no insertions or deletions, are permitted in each decamer, while the p1: (p2 = nnncnngnnn) postprocessing command (as well as p3:p4, p5:p6, and p7:p8) imposes that C4 and G7 are strictly conserved in each decamer. The definition of a regular expression pattern allows this pattern to be searched over an entire genome or in a large set of sequences. The problem arises of assessing the statistical significance of the matched sequences. The simplest way to accomplish this task is to carry out the same search on a set of random sequences generated by a Markov chain simulator (Almagor 1983) (see Sec. 6.8.1). In this way the number of hits expected by chance gives a measure of the statistical significance of hits found in real sequences.

In consideration of the high degeneracy of most regulatory protein-binding oligonucleotides, an alternative is to use a statistical description of such regulatory elements: namely, the *base frequency matrix*. In contrast to the deterministic description of a pattern, such as that given by the general regular expression pattern that gives a simple "yes" or "no" answer on the occurrence of a pattern in a given position along the sequence, base frequency matrices allow a quantitative score to be calculated for each match in a sequence to which a level of statistical significance can usually be associated.

A base frequency matrix can be constructed when a sufficiently large collection of experimentally tested regulatory sites is known. If all DNA/RNA regulatory elements endowed of a specific function have the same length L a base frequency matrix will be made of four rows and L columns. The (i,j)th element of the matrix simply contains the number of times the ith nucleotide is found at position j in the set of aligned functional patterns. If functional regulatory elements are not of the same length, they need to be aligned previously, and five rows are used to construct the matrix to include the gap character. The base frequency matrices can conveniently be transformed in weight matrices, also called *position-specific weight matrices* (PWMs) by using a log-odd (lod) transformation where the lod-score w_{ij} assigned to nucleotide i at position j is

$$w_{ij} = \log\left(\frac{q_{ij}}{f_i} + s\right)$$

where q_{ij} is the observed frequency of nucleotide i at position j with f_i its background frequency in the sequence under investigation, and s is a smoothing percentage to prevent the argument of the logarithm to be zero. PWMs assign a weight to each possible nucleotide at each position within the matrix that reflects the frequency with which the given nucleotide occurs at that specific position. This is consistent with the observation that changes at different positions of the regulatory element described by the matrix may have drastically different effects on their activity, critical sites generally being the most conserved. The local goodness of fit between a target sequence starting at position k and the signal recorded in the matrix can be measured as

$$S_k = \sum_{k}^{k+w-1} w_{ij}$$

with $S_k = 0$ if a perfect match with the consensus sequence is found and $S_k < 0$ in the case of one or more changes with respect to the consensus sequence. A match between the matrix and the scanned sequence is considered significant if the lod score is above a threshold that can be defined to optimize the trade-off between sensitivity and selectivity (Bucher 1990). Alternatively, the quality of a match between the matrix and the sequence scanned can be measured by the similarity score, ranging between zero and 1, given by the ratio between the matrix score computed and its maximum value. MatInspector program (Quandt, Frech et al. 1995) detects consensus matches in nucleotide sequences and may use a different threshold similarity for the core region of the matrix, including most constrained sites, and the entire matrix. Table 6.6 shows the base frequency matrix and the PWM for the TATA-box promoter element as reported in the TRANSFAC database (entry V$TATA_01). An optimal cutoff value of −8.16 has been suggested for this specific matrix (Bucher 1990).

TABLE 6.6. *(a)* **TATA-Box Consensus Matrix;** *(b)* **Lod-Score Matrix**[a]

(a)

Base	Position														
	1	2	3	4	5	6	7	8	9	10	11	12	13	14	15
A	61	16	352	3	354	268	360	222	155	56	83	82	82	68	77
C	145	46	0	10	0	0	3	2	44	135	147	127	118	107	101
G	152	18	2	2	5	0	20	44	157	150	128	128	128	139	140
T	31	309	35	374	30	121	6	121	33	48	31	52	61	75	71
Consensus	S	T	A	T	A	A	A	W	R	N	N	N	N	N	N

(b)

Base	Position														
	1	2	3	4	5	6	7	8	9	10	11	12	13	14	15
A	-0.65	-2.52	1.86	-4.61	1.87	1.47	1.89	1.20	0.68	-0.77	-0.21	-0.23	-0.23	-0.50	-0.32
C	0.59	-1.05	-6.64	-3.16	-6.64	-6.64	-4.61	-5.03	-1.11	0.48	0.61	0.40	0.29	0.15	0.07
G	0.65	-2.36	-5.03	-5.03	-4.03	-6.64	-2.21	-1.11	0.70	0.63	0.41	0.41	0.41	0.53	0.54
T	-1.60	1.67	-1.43	1.95	-1.65	0.33	-3.80	0.33	-1.52	-0.99	-1.60	-0.88	-0.65	-0.36	-0.43

[a] *(a)* TRANSFAC entry V$TATA_01; calculated on 389 TATA-box elements (Bucher 1990). *(b)* Base 2 logarithms; calculated assuming a homogeneous background frequency (0.25) and $s = 0.01$.

TABLE 6.7. Specialized Databases Collecting DNA-Binding Sites[a]

Database	URL	Description
COMPEL	compel.bionet.nsc.ru	Composite regulatory elements
DBTSB	elmo.ims.u-tokyo.ac.jp/dbtbs	*Bacillus subtilis*–binding factors and promoters
EPD	www.epd.isb-sib.ch	Eukaryotic pol II promoters with experimentally determined transcription start sites
RegulonDB	www.cifn.unam.mx/ Computational_Biology/regulondb	*E. coli* transcriptional regulation and operon organization
TRANSFAC	transfac.gbf.de/TRANSFAC	Transcription factors and binding sites
OoTFD	www.ifti.org	Transcription factors and gene expression
SCPD	cgsigma.cshl.org/jian	*Saccharomyces cerevisiae* promoter database

[a] Recognized by prokaryotic and eukaryotic transcription.

6.8.5 Identification of Promoter Regions in Chromosomal Sequences

Transcription initiation is generally thought to be the most important point in gene expression regulation, although many other possible levels of regulation have been described, mostly in eukaryotes, including chromatin packaging (Kingston, Bunker et al. 1996), polyadenylation status (Wahle and Keller 1996), splicing (McKeown 1992), mRNA stability (Wickens, Anderson et al. 1997), and translation initiation (Kozak 1999). Correct recognition of the promoter region may greatly improve gene identification and provide a better description of the gene expression context. Transcription initiation involves the cooperative binding of a number of proteins to DNA; therefore, the computational approach to promoter recognition first tries to identify individual binding sites and then to find some general structural rules on the relative arrangement and spacing of the various individual sites, which is crucial for specific promoter activity.

Binding sites can be described by consensus sequences, reporting the most preferred bases at each position within a site or by the most informative PWM, assigning a weight to each possible nucleotide at each position of a putative binding site and giving a site score equal to the sum of weights (see Sec. 6.8.4) that may approximate the energy of the protein binding (Berg and von Hippel 1988).

A number of specialized databases have been constructed that collect DNA-binding sites recognized by transcription factors in both prokaryotes and eukaryotes, which are listed in Table 6.7. In addition, several promoter prediction tools have been devised that use data stored in the databases noted above [see Fickett and Hatzigeorgiou (1997) for a review]. Promoter prediction tools can be divided roughly into three main groups: (1) heuristic approaches; (2) approaches that attempt to recognize core promoter elements such as TATA boxes, CAAT boxes, and transcription initiation sites (INRs); (3) approaches that attempt to use the entire ensemble of elements (TF-binding sites, oligonucleotide frequencies in true promoters, etc.). However, the selectivity of such methods is generally low; they can predict one promoter per kilobase pair against the estimated one per 30 to 40 kbp

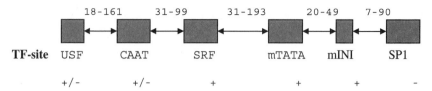

Figure 6.29. Basic muscle-specific promoter model of mammalian actin genes consisting of six TF-binding sites. The + and − signs indicate strand orientation. [From Werner (2000).]

(Fickett and Hatzigeorgiou 1997). It should be also assumed that core promoter elements such as TATA boxes are neither necessary nor sufficient for promoter function. This makes prediction methods essentially useless for automatic annotation of promoters in large sequence contigs.

To overcome this problem, a possibility is to restrain promoter analysis to specific regions that are more likely to correspond to promoter regions for their peculiar features (Scherf, Klingenhoff et al. 2000) and then carry out more accurate analyses on these restricted regions. Alternatively, the search for combination of individual regulatory elements can be carried out, usually called *promoter modules*. A promoter module is made of two or more TF-binding sites whose functional activity is dependent on their sequential order, strand orientation, and distance range. Some of these modules have already been described in the literature; an example is the actin promoter shown in Fig. 6.29, which is highly specific and can allow selective database scanning.

Specific computational approaches can be used to predict the matrix attachment regions (MARs), also called scaffold attachment regions (SARs), which are sequence regions responsible for the attachment of genomic DNA to the nuclear matrix or scaffold. Indeed, transcription absolutely requires anchorage of genomic DNA to the nuclear matrix (Singh, Kramer et al. 1997). Some Web-based tools, such as Mar-Finder (see the URL in the Appendix) and Mar-wiz (see the URL in the Appendix), have been designed specifically to predict these regions based on the co-occurrence of several patterns known to occur in the neighborhood of MARs. The density of such patterns is used to associate a matrix attachment potential with a given region of the DNA sequence. Figure 6.30 shows the Mar-wiz prediction on a genomic contig (AL136528) where a potential MAR region is found close to the promoter region of the p73 gene. Regulatory MARs facilitate the access of transcription activators to the gene promoter via the formation of a ternary complex constituted of a MAR, a matrix, and activators (Bode, Benham et al. 2000).

6.8.6 Pattern Discovery for the Identification of Gene Regulatory Elements and of Protein Motif Models

Understanding the control mechanisms ruling coordinated expression of genes in time and space as a result of the physiological requirements of the cell or in response to environmental stimuli is one of the major challenges of modern biology. Genetic information for the regulation of gene expression is stored mainly in noncoding regions of the genome and controls conversion of genetic information from gene to

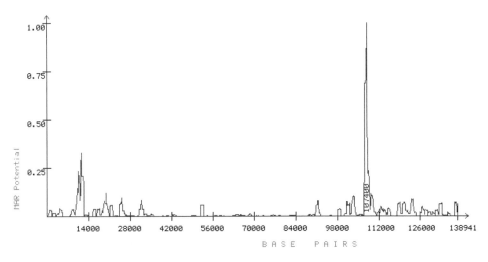

MAR-Analysis Summary Report

Sequence Description: p73 contig
Sequence Length: 138941
Maximum and Minimum Potential = [9.9913e-07 .. 115.646]

High Scoring Regions with threshold = 0.6
Region   Average Strength   Integrated Strength
107000 ... 107800 0.750232   600.936

Frequency of Rule Matches
Number of Rules Selected = 6
Rule Name   Forward Strand Match Count   Reverse Strand Match Count
ORI Pattern   370   340   View Detail Locations
TG-Richness Pattern   13   25   View Detail Locations
Curved DNA Pattern   68   68   View Detail Locations
Kinked DNA Pattern   114   114   View Detail Locations
Topo II Pattern   223   197   View Detail Locations
AT-Richness Pattern   484   468   View Detail Locations

Questions or Comments...Please send E-mail to
Dr. Gautam B. Singh, *gbs@futuresoft.org*

Figure 6.30. Prediction of a potential MAR region in a genomic contig of chromosome 1 (EMBL ID: AL136528) carried out with Mar-wiz software (*www.futuresoft.org*). The MAR region predicted is located in the promoter region of the p73 oncosuppressor gene.

protein at the transcriptional level via promoters, enhancers, MARs/SARs (see Sec. 6.8.5), locus control regions (LCRs), and at the posttranscriptional level via regulatory elements mostly embedded in the untranslated regions of mRNA (Pesole, Grillo et al. 2000).

Microarray gene expression profiles usually define groups of genes with a similar expression profile in specific cellular conditions. These data need to be interpreted by recognition of the specific transcription factor-binding sites responsible for gene coregulation. The fact that regulatory elements are relatively short sequence motifs (5 to 25 nt in length), usually highly degenerate, with few conserved sites, makes local alignment algorithms generally inadequate for their detection. A further complication in searching for common oligonucleotide motifs in the regulatory regions (e.g., promoters) of a set of coregulated genes derives from the fact that functional motifs are of unknown size and that expression data clustering algorithms very often mistakenly group unrelated genes within the same cluster.

The problem of finding common elements in a set of functionally related sequences can be tackled using two different approaches: alignment methods and enumerative or exhaustive methods. *Alignment methods* try to identify significantly shared signals by performing local multiple alignments of all sequences, whereas *enumerative* of *exhaustive methods* examine all oligomers of fixed length and report those statistically more represented than expected, with expectations calculated from the background sequence composition of the analyzed sequences. Alignment methods are particularly suitable for the analysis of protein sequences for the availability of quite accurate models of amino acid interconversion probabilities as provided by PAM or BLOSUM family matrices (see Sec. 6.3.6). By contrast, enumeration methods are clearly better suited to nucleotide sequences that use a much smaller alphabet than that used by protein sequences.

From a practical point of view, the most remarkable difference between alignment and enumerative methods is the presentation of results. The alignment approach represents conserved patterns as a weighted matrix built from the optimal local alignment. Enumerative methods provide a lot of over represented oligomers possibly ranked based on their statistical significance.

Local multiple alignments are computationally very expensive; therefore, various other strategies are used. For example, the *consensus algorithm* (Hertz and Stormo 1999) aligns sequences one by one, optimizing the weight matrix construction from the alignment. Other alignment algorithms use a statistical approach that considers the start positions of the motifs in the sequences to be unknown and adopts a local optimization strategy to detect which positions correspond to the most conserved motifs.

The *Gibbs sampler* (Lawrence, Altschul et al. 1993; Neuwald, Liu et al. 1995) and the expectation maximization strategy implemented in the MEME system (Bailey and Elkan 1994) are the most commonly used methods of this type. Two different implementations of the Gibbs sampler may be used: the *site sampler*, which addresses the problem of finding motifs when the number of occurrences of each motif in each sequence is assumed (Lawrence, Altschul et al. 1993), and the *motif sampler*, which addresses the problem of detecting motifs when little prior information about the number of occurrences of each motif is known (Neuwald, Liu et al. 1995). Both the Gibbs sampler and the MEME system provide a measure of statistical significance of the motifs detected.

The alignment methods described above can be used valuably to detect locally conserved motifs common to a set of functionally related (not necessarily homologous) protein sequences. Gibbs sampling has been used to detect subtle conserved patterns in helix-turn-helix proteins, lipocalins, phenyltransferases (Lawrence, Altschul et al. 1993), and in outer membrane bacterial proteins (Neuwald, Liu et al. 1995).

Enumeration methods may search for exact or slightly degenerate oligo-nucleotide patterns. The WordUP method (Pesole, Prunella et al. 1992) is able to find oligonucleotide patterns shared significantly by the collection of sequences under examination, where the statistical significance of each pattern is estimated by comparing with a simple chi-square test the observed and expected number of sequences containing that pattern. WordUP calculates expected values by assuming that pattern occurrences are Poisson distributed with pattern probabilities derived from a first-order Markov chain (see Sec. 6.8.1). More recent enumerative methods are able to extract gapped substring patterns or flexible patterns of unrestricted length with predefined ambiguous character positions and wildcards of fixed length (Brazma, Jonassen et al. 1998; van Helden, Andre et al. 1998; Pevzner and Sze 2000; van Helden, Rios et al. 2000).

Figure 6.31 illustrates the output obtained by the consensus algorithm (Hertz and Stormo 1999) and the Gibbs motif sampler (Neuwald, Liu et al. 1995) by analyzing a data set of 10 random sequences each containing a common 15-mer with up to two mismatches allowed. A very conservative way to represent and visualize biological information contained in a weighted matrix is by use of sequence logos (Schneider and Stephens 1990), which provide a graphical representation of the frequencies of bases at each position and of the degree of sequence conservation measured in bits of information (Fig. 6.31*c*).

6.8.7 Gene Prediction

The initiation of large-scale genome sequencing projects has provided a strong impetus for the development of computational methodologies for gene discovery. These methodologies are particularly useful when no identifiable homolog matches are found in the primary databases. Computational methods for gene finding use two essentially different approaches, based on either general sequence statistical features or on the presence of specific local signals. One approach tries to identify gene coding regions based on their specific content features by simply searching for open-reading frames (ORFs) longer than expected, where nucleotide frequencies show dependency from the triplet codon structure of the coding regions. However, it is not unlikely to find quite long unfunctional ORFs, particularly in the antisense strand of the expressed ORFs (Silke 1997).

The TestCode program developed by Fickett (1982) helps identify potential protein-coding regions in nucleic acid sequences using a statistics based on the observation that protein-coding sequences show a period 3 bias absent in non-coding sequences. The period 3 asymmetry for each base is calculated as

$$\frac{\max(N_1, N_2, N_3)}{\min(N_1, N_2, N_3) + 1}$$

where $N_1, N_2,$ and N_3 are the frequencies of each nucleotide ($N = A, C, G, T$) at positions $(1, 4, 7, \ldots), (2, 5, 8, \ldots),$ and $(3, 6, 9, \ldots)$, respectively, within the settled analysis window, which should be at least 100 nt long. The weighted combination of TestCode model parameters, with optimal weights derived by empirical statistics

```
MATRIX 1
number of sequences = 10
width = 16
crude information = 12.194
unadjusted information = 17.157
sample size adjusted information = 14.421
ln(p-value) = -125.39   p-value = 3.49834E-55
ln(expected frequency) = -56.3275   expected frequency = 3.44587E-25

    1|5    :    1/14     CGAGGCCTTTCCTCCC
    2|7    :    2/157    CGAGGCCTTCCCCCCC
    3|3    :    3/419    CGAAGCCTCTCCTCCC
    4|2    :    4/684    CGAGGCCTTCCCTCCC
    5|6    :    5/95     GGAAGCCTCTCCTCCC
    6|10   :    6/284    AGAGGCCCTACCTCCC
    7|4    :    7/997    CGAAGCCTCTCCTCCC
    8|9    :    8/184    GGAGGCCTTTCTTCCC
    9|1    :    9/340    CGAAGCCATTCCTCCC
   10|8    :   10/821    AGAGGCCTGTCGTCCC
```

(a)

```
1-1     15   tcttgatcac GAGGCCTTTCCTCCC aggggtcgt    29  (1.00)
2-1    158   ataggacatc GAGGCCTTCCCCCCC ggaaccagga   172  (0.94)
3-1    420   gaaaggcaac GAAGCCTCTCCTCCC caggatattg   434  (1.00)
4-1    685   cggtatccgc GAGGCCTTCCCTCCC acgcattaga   699  (1.00)
5-1     96   cattagactg GAAGCCTCTCCTCCC gtcttacaag   110  (1.00)
6-1    285   cacttaagga GAGGCCCTACCTCCC agacctgggg   299  (1.00)
7-1    998   gagcttaacc GAAGCCTCTCCTCCC ttt         1012  (1.00)
8-1    185   ctgtacgtag GAGGCCTTTCTTCCC ccctattaac   199  (0.95)
9-1    341   gtaactacgc GAAGCCATTCCTCCC gcagtgatgc   355  (0.99)
10-1   822   cgccgtaaaa GAGGCCTGTCGTCCC agagtagcga   836  (0.96)
sites:                 ******  ******
                        5    10    15
          (10 sites in 10 sequences)

model map = 107.057; betaprior map = -71.1967
MAP = 35.859964
```

(b)

(c)

carried out on coding and noncoding sequences, gives a measure of the coding potential for each of the sliding windows moving along the sequence analyzed. The statistics is independent of the reading frame and has been shown to perform well for both prokaryotic genomes and eukaryotic cDNA sequences. Following a similar approach, the similarity between ORF triplet frequency and the organismal codon usage strategy is computed that provides a reliable gene predictor mostly when a marked preference in the use of synonymous codons is observed.

In the case of microbial genomes where genes generally saturate over 90% of the genome sequences and have no introns, the identification of significantly long ORFs gives quite a reliable proof of a gene coding region. To this aim the ORF Finder tool (see the URL in the Appendix) at NCBI can find all open-reading frames of a selectable minimum size in a user's sequence or in a sequence already in the database. However, the tight packaging of prokaryotic genes, frequently overlapping with each other, makes exact predictions of prokaryotic genes (i.e., detection of translation initiation and termination sites) quite difficult. To improve prediction accuracy in microbial genomes, several algorithms have been developed, such as Glimmer (Delcher, Harmon et al. 1999), and GeneMark (Shmatkov, Melikyan et al. 1999) that use a combination of Markov models in their implementation and reach over 97% of prediction accuracy.

The gene predictors described above, also called *content sensors*, are clearly inadequate for eukaryotic gene prediction because of their discontinuous exon/intron structure. In this case, in addition to content sensors, signal sensors have to be used, which predict start codon, donor, and acceptor splice sites as well as polyadenylation sites. Although content sensors may provide a rough indication of exon locations in a genomic sequence, they cannot set boundaries that are better defined by signal-specific motifs such as the start codon, the donor and acceptor splice contexts, and the polyadenylation site. Table 6.8 shows base frequency matrices for start codon context, donor, and acceptor splice sites calculated on a large collection of genes, from mammals to plants (Senapathy, Shapiro et al. 1990). These motifs can be searched by PWMs (see Sec. 6.8.4) or neural networks (Brunak, Engelbrecht et al. 1991) but may produce too many false positives to be considered effective predictors.

Gene-finding tools integrating both content and signal sensors performed better, particularly when adopting hidden Markov models (HMMs), which apply probabilistic models to sequence content (e.g., modal length, nucleotide and oligonucleotide composition for 5′ UTRs, initial exon, intron, internal exon, final exon, and 3′ UTR) and boundary signals (e.g., transcription start site, donor and acceptor splicing sites, start codon context, and polyadenylation signal).

Figure 6.31. Output produced by (*a*) the Consensus program executed at the Pasteur Web site (*bioweb.pasteur.fr/seqanal/interfaces/consensus-simple.html*) and (*b*) the Gibbs motif sampler on a data set of 10 random sequences 1000 nt long where the 15-mer oligonucleotide GAGGCCTTTCCTCCC containing up to two mismatches has been inserted artificially. (*c*) Sequence logo of the motif detected by the Gibbs motif sampler. [(*a*) from Hertz and Stormo (1999); (*b*) from Lawrence, Altschul et al. (1993) and Neuwald, Liu et al. (1995); (*c*) from Schneider and Stephens (1990).]

TABLE 6.8. Nucleotide Percentages at Splice Sites[a]

| Location | Frequency of Occurrence (%) at Sequence Position: | | | | Consensus |
	A	C	G	T	
	5′ Splice Sites (3724 Total Sequences)				
−3	32	37	18	13	
−2	60	13	12	15	A
−1	9	5	79	7	G
+1	0	0	100	0	G
+2	0	0	0	100	T
+3	59	3	35	3	AG
+4	71	9	11	9	A
+5	7	6	82	6	G
+6	16	16	18	50	T
	3′ Splice Sites (3683 Total Sequences)				
−14	11	29	14	46	T
−13	11	33	12	44	TC
−12	10	30	10	50	TC
−11	8	30	10	57	TC
−10	11	32	9	48	TC
−9	10	34	11	45	TC
−8	11	37	10	42	TC
−7	11	38	9	43	TC
−6	7	39	7	47	TC
−5	8	36	6	51	TC
−4	25	26	26	23	N
−3	3	75	1	21	C
−2	100	0	0	0	A
−1	0	0	100	0	G
+1	27	14	49	10	G
	Branch Point				
−3	1	76	2	21	C
−2	0	8	0	91	T
−1	39	15	42	4	AG
0	99	1	0	0	A
1	11	45	6	38	CT

Source: Data from Senapathy, Shapiro et al. (1990).

[a] Calculated on a collection of genes from several taxa, from mammals to plants.

An excellent review of available gene-finding programs can be found in Stormo (2000) and an up-to-date compilation of the relevant bibliography is maintained by Wentian Li (see the URL in the Appendix). More recent gene-finding methods, designed specifically for eukaryotic genes, which use generalized hidden Markov models (HMMs; see Sec. 6.6) include Genie (Reese, Kulp et al. 2000), used for gene prediction of the *Drosophila melanogaster* genome, HMMgene (Krogh 1997), and the Genscan program (Burge and Karlin 1997). The latter two revealed the best predictive accuracy among available gene-finding methods, above 91% at nucleotide level, at least for mammalian genes (Rogic, Mackworth et al. 2001). The GenomeScan program (Yeh, Lim et al. 2001) combines the exon–intron and splice signal models used in Genscan predictions with similarity to known protein sequences in an integrated model. Figure 6.32 shows the output generated by Genscan program on a human genomic contig, where the genes predicted actually correspond to transcribed regions coding for protein products showing similarity with homologous proteins of other organisms. This allows a more accurate gene structure prediction and lower rate of false positives.

To increase gene prediction accuracy further, some auxiliary information can be used, such as the presence of CpG islands (see Sec. 6.8.8) in the upstream gene region, expressed sequence tag (EST) matches to exons predicted, and significant matches between either the protein predicted or the six-frame translation of the genomic region under investigation and protein database entries (Krogh 2000; Yeh, Lim et al. 2001). It should be noted that all computationally based gene-finding tools are unable to predict noncoding exons, thus implying that even if gene recognition methods are rather sensitive, the accuracy of gene boundaries is still rather inaccurate. A list of major gene-finding programs is shown in Table 6.9.

6.8.8 Identification of CpG Islands in Genomic Sequences

The dinucleotide CpG is greatly underrepresented in human and other vertebrate genomes. It accounts for about one-fifth of the roughly 4% expected, calculated based on the 42% G + C content of human genome. This pattern is determined by the high susceptibility of the cytosine base to be methylated with methyl-C residues deaminating spontaneously to give rise to thymine residues. The CpG dinucelotides are thus depleted by mutation to TpG. However, there are some genome regions where CpG dinucleotides are poorly methylated, and thus their observed frequency is close to the expected one based simply on the local C + G content. These regions, denoted as CpG islands, are particularly interesting because they are probably associated with the 5′ end of genes. The identification of CpG islands can easily be performed by computer programs that scan the genomic sequence with a moving window of fixed size (usually, >200 nt) measuring for each window the %GC and the CpG *O/E* ratio.

A CpG island can be defined operationally as a genomic sequence region, at least 200 nt long having GC content greater than 50% and CpG *O/E* ratio greater than 0.6 (Gardiner-Garden and Frommer 1987). The latter parameter is calculated as

$$\text{CpG observed/expected} = \frac{f(\text{CG})}{f(\text{C})f(\text{G})} L$$

with L the window size and $f(\text{CG})$, $f(\text{C})$, and $f(\text{G})$ the observed frequencies of CG, C, and G, respectively. The program CPGPLOT of the EMBOSS package (Rice, Longden et al. 2000) can be used to detect CpG islands in genomic sequences.

```
GENSCAN 1.0      Date run: 5-Mar-101     Time: 13:17:04

Sequence gi : 31874 bp : 44.58% C+G : Isochore 2 (43 - 51 C+G%)
Parameter matrix: HumanIso.smat
Predicted genes/exons:

Gn.Ex Type S .Begin ...End .Len Fr Ph I/Ac Do/T CodRg P.... Tscr..
----- ---- - ------ ------ ---- -- -- ---- ---- ----- ----- ------

 1.04 Intr -   3644   3554   91  0  1   97   94    26 0.075  3.15
 1.03 Intr -   9283   9136  148  1  1   55   85    43 0.368  0.51
 1.02 Intr -   9526   9402  125  2  2   88   73   124 0.632 11.20
 1.01 Init -  18880  18562  319  1  1   63   91   176 0.864 12.90
 1.00 Prom -  19283  19244   40                              -4.66

 2.00 Prom +  19738  19777   40                              -4.46
 2.01 Init +  26255  26490  236  1  2   50   72   130 0.496  3.21
 2.02 Intr +  28041  28137   97  0  1   84   70    64 0.719  4.21
 2.03 Intr +  28774  28863   90  0  0   59   61   125 0.992  7.09
 2.04 Term +  30023  30178  156  1  0   79   38   182 0.948 10.23
 2.05 PlyA +  31229  31234    6                               1.05
```

Predicted peptide sequence(s):

```
>gi|GENSCAN_predicted_peptide_1|228_aa
MSRESDVEAQQSHGSSACSQPHGSVTQSQGSSSQSQGISSSSTSTMPNSSQSSHSSSGTLSSLETVSTQELYSIPEDQEPEDQEPEEPT
PAPWARLWALQDGFANLECVNDNYWFGRDKSCEYCFDEPLLKRTDKYRTYSKKHFRIFREVGPKNSYIAYIEDHSGNGTFVNTELVGKK
RRPLNNNSEIALSLSRNKVFVFFDLTVDDQSVYPKALRDEYIMSKTLGX

>gi|GENSCAN_predicted_peptide_2|192_aa
MWRGRAGALLRVWGFWPTGVPRRRPLSCDAASQAGSNYPRCWNCGGPWGPGREDRFFCPQCRALQAPDPTRDYFSLMDCNRSFRVDTAK
LQHRYQQLQRLVHPDFFSQRSQTEKDFSEKHSTLVNDAYKTLLAPLSRGLYLLKLHGIEIPERTDYEMDRQFLIEIMEINEKLAEAESA
AMKEIESIVKGER
```

Explanation

```
Gn.Ex : gene number, exon number (for reference)
Type  : Init = Initial exon (ATG to 5' splice site)
        Intr = Internal exon (3' splice site to 5' splice site)
        Term = Terminal exon (3' splice site to stop codon)
        Sngl = Single-exon gene (ATG to stop)
        Prom = Promoter (TATA box / initation site)
        PlyA = poly-A signal (consensus: AATAAA)
S     : DNA strand (+ = input strand; - = opposite strand)
Begin : beginning of exon or signal (numbered on input strand)
End   : end point of exon or signal (numbered on input strand)
Len   : length of exon or signal (bp)
Fr    : reading frame (a forward strand codon ending at x has frame x mod 3)
Ph    : net phase of exon (exon length modulo 3)
I/Ac  : initiation signal or 3' splice site score (tenth bit units)
Do/T  : 5' splice site or termination signal score (tenth bit units)
CodRg : coding region score (tenth bit units)
P     : probability of exon (sum over all parses containing exon)
Tscr  : exon score (depends on length, I/Ac, Do/T and CodRg scores)
```

Comments
The SCORE of a predicted feature (e.g., exon or splice site) is a log-odds measure of the
quality of the feature based on local sequence properties. For example, a predicted 5'
splice site with score > 100 is strong; 50-100 is moderate; 0-50 is weak; and
below 0 is poor (more than likely not a real donor site).

The PROBABILITY of a predicted exon is the estimated probability under
GENSCAN's model of genomic sequence structure that the exon is correct.
This probability depends in general on global as well as local sequenc
properties, e.g., it depends on how well the exon fits with neighboring
exons. It has been shown that predicted exons with higher probabilities
are more likely to be correct than those with lower probabilities.

TABLE 6.9. Major Gene-Finding Programs

Program	URL
GENEID	www1.imim.es/geneid.html
SELFID	igs-server.cnrs-mrs.fr/~audic/selfid.html
MZEF	sciclio.cshl.org/genefinder
GeneParser	beagle.colorado.edu/~eesnyder/GeneParser.html
WebGene	www.itba.mi.cnr.it/webgene
GeneMark	opal.biology.gatech.edu/GeneMark
Genemark.hmm	opal.biology.gatech.edu/GeneMark/hmmchoice.html
Framed [(G + C)-rich procaryotic sequences]	www.toulouse.inra.fr/FrameD/cgi-bin/FD
EUGENE (*A. thaliana* DNA sequences)	www-bia.inra.fr/T/EuGene
Glimmer (microbial DNA)	www.tigr.org/~salzberg/glimmer.html
VEIL	www.tigr.org/~salzberg/veil.html
Genie	www.fruitfly.org/seq_tools/genie.html
Genscan	genes.mit.edu/GENSCAN.html
GenomeScan	genes.mit.edu/genomescan
GRAIL 1.3	compbio.ornl.gov/Grail-1.3
GENLANG	www.cbil.upenn.edu/~sdong/genlang_home.html
HMMgene	www.cbs.dtu.dk/services/HMMgene

6.8.9 Analysis of Codon Use Strategy

The triplet organization of the genetic code has been the first linguistic code to be deciphered in DNA sequences. The degeneracy of the genetic code implies that most amino acids are encoded by several *synonymous codons* that define 21 disjoint codon families for the 20 amino acids and a termination signal. As soon as the first nucleotide sequences were published, it clearly emerged that use of synonymous codons was not random. Indeed, the vast majority of prokaryotic and eukaryotic organisms show codon use bias that may be related to the specific organism (Grantham, Gautier et al. 1980), to the level of expression (Bennetzen and Hall 1982; Gouy and Gautier 1982), or to the genomic compartment where the gene is located (Bernardi 2000a,b).

Species specificity in codon use strategy is particularly evident in bacterial genomes with extreme base composition, such as *Streptomyces* sp., which because they have a genomic GC content of about 70%, use G- and C-ending codons almost exclusively (Wright and Bibb 1992). When a systematic similarity in codon use between genes of the same organism is observed, the detection of atypical codon use can help to recognize genes probably acquired by horizontal gene transfer (Moszer, Rocha et al. 1999).

◀───────────────────────────────────────

Figure 6.32. Genscan output obtained by its application on a human genome contig of chromosome 22 (EMBL ID: AL117330). The structure of two predicted genes having opposite orientation, made each of four exons, is provided with the corresponding peptides predicted. [From Burge and (Karlin 1997).]

Preferential use of specific codons in unicellular prokaryotic and eukaryotic organisms is determined primarily by natural selection toward the use of a restricted set of codons, providing optimal translation efficiency. This selection is most pronounced in highly expressed genes whose codon choice may be correlated with tRNA abundance or tRNA-binding affinity. In this case, codon use analysis can be used profitably to identify highly expressed genes (Sharp, Tuohy et al. 1986).

A codon use bias is also observed in mammalian genes but seems not to be related to their function or level of expression; instead, it is connected to the compositional features of the genomic context. In mammalian genomes, regional differences in base composition have been recognized with GC-rich and GC-poor segments, also called *isochores*, where each segment is hundreds of kilobase pairs long and shows a rather homogeneous GC content (Bernardi, Olofsson et al. 1985; Bernardi 2000b) (see Sec. 7.3). The higher the average GC content of the isochore where a specific gene is located, the higher the observed preference toward GC-ending codons. Indeed, it is not clear whether the codon use bias depends on natural selection or on directional mutational bias. A compilation of codon use tables for protein coding sequences from the taxonomic divisions of the GenBank/ EMBL/DDBJ databases has been collected in the CUTG database (Nakamura, Gojobori et al. 2000).

The preference of a specific codon within its family can be expressed by the *relative synonymous codon use* (RSCU), given by

$$\mathrm{RSCU}_{ij} = \frac{n_{ij}}{(1/n_i)\sum_{j=1}^{n_i} n_{ij}}$$

where n_{ij} is the number of occurrences of the jth codon for the ith amino acid and n_i is the number of synonymous codons belonging to the ith family. The simplest measure of codon use bias, which quantifies how far the codon use of a gene departs from equal use of synonymous codons, is given by the effective number of codons used in a gene, N_c (Wright 1990). This index can easily be calculated from codon use data alone and is independent of gene length and amino acid composition. N_c values range from 20 in the case of extreme bias, where one codon is used exclusively for each amino acid, to 61 where the use of alternative synonymous codons is equally possible.

When preliminary information on codon use strategy in the species being analyzed is known, other indices can be calculated that provide a measure of the frequency of optimal codons (F_{op}) or of the Codon Adaptation Index (CAI). The index F_{op} requires the identification of the optimal codons for the species, and the CAI requires that codon use of highly expressed genes be identified. The CAI (Sharp and Li 1987) ranges between 0 and 1 and provides the degree of fitness between codon use of the gene under investigation and the average codon use of highly expressed genes. The analysis of codon use can be also as a tool for gene classification. This type of analysis, which requires multivariate statistic techniques such as correspondence analysis or hierarchical clustering, is able to identify discrete groups of genes each having homogeneous codon use. The program CodonW (see the URL in the Appendix) is able to perform correspondence analysis (COA) for codon use, RSCU, and amino acid use. COA creates a series of orthogonal axes to identify trends that explain data variation, with each subsequent axis explaining a decreasing amount of variation. CodonW also implements computation of the major codon use indexes, such as F_{op}, CAI, and N_c. A hierarchical cluster analysis is implemented in the CodonTree program (Pesole, Attimonelli et al. 1996) that is able to determine gene relationships on the basis of codon use similarity. The result of clustering is represented graphically by a binary tree so that one or more groups of genes using a similar codon use strategy may be distinguished. Codon use analysis can also be used fruitfully to construct reliable codon use tables to be incorporated in backtranslation techniques aimed at designing suitable oligonucleotide probes from protein sequences.

6.9 PREDICTION OF RNA SECONDARY STRUCTURES

Beyond their major role in the transmission of genetic information encoded in their primary structure, RNA molecules, carry out a broad spectrum of biological functions that require specific pairing interactions involving complementary bases that produce a precise two-dimensional folding, also termed *secondary structure*. A large number of RNA structural motifs have been recognized to be capable of catalyzing numerous biochemical transformations. In particular, specific RNA folding motifs are involved in peptidyltransferase activity of ribosomes (Noller, Green et al. 1995), RNA splicing (Scott 1998), or translocation of proteins across the plasma membrane (Stroud and Walter 1999). In mRNA transcripts, specific structural motifs, generally located in the 5′ and 3′ untranslated regions, regulate posttranscriptionally gene expression in a variety of ways through the control of alternative mRNA splicing, translation efficiency, and stability (Mignone, Gissi et al. 2002).

RNA secondary structures are further organized in the space to assume a tridimensional folding that in some cases has been resolved by x-ray crystallographic studies. Intramolecular interactions responsible for RNA secondary structure are generally much more stable than those involved in the formation of tertiary folding, thus implying that RNA secondary structures form rapidly and precede the three-dimentional packaging of RNA in its tertiary structure. Positively charged ions such as Mg^{2+} have a larger general effect on RNA folding, but mostly at the level of the tertiary structure (Tinoco and Bustamante 1999). The RNA secondary structure can be defined as a combination of only a few elementary structural motifs, such as helices, loops, bulges, and junctions, each contributing independently to the global folding free energy. More formally, the secondary structure of an RNA is a collection of the base pairs, S_{ij}, between the ith and jth nucleotides, with $i < j$ by convention. Figure 6.33 shows the fundamental elements of RNA secondary structure in their typical two-dimensional appearance.

RNA stems generally form A-form duplexes, where base pairs are approximately coplanar and generally stacked onto other base pairs. In addition to canonical Watson–Crick (WC) base pairs (A–U and G–C), noncanonical base pairs also occur, particularly the G–U pair, which is almost as stable as the canonical WC pairings. Indeed, it is striking to note that most structures solved to date contain one or more non-WC paired bases (Wahl and Sundaralingam 1995); some functionally active stem–loop structures located in mRNA UTRs such as the selenocysteine insertion sequence (SECIS; Fagegaltier, Lescure et al. 2000) contains four consecutive non-WC

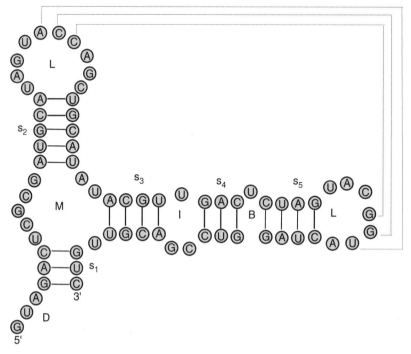

Figure 6.33. Typical elements of RNA secondary structure: s1–s6, helices or stems; L, loops; M, multibranched loops; I, internal loops; B, bulge loops; D, dangling ends. A pseudoknot pairing is also indicated between the two loops.

paired bases. The loops are single-stranded subsequences bounded by base pairs and can be classified as *hairpin loops*, *internal loops*, or *multibranched loops*, depending on one, two, or more stems radiating from them, respectively. Single-stranded bases occurring on only one side within a stem are called *bulges* or *bulge loops*. Unstructured single-stranded regions are called *junctions*. Base pairs that occur between different single-stranded regions, called *pseudoknots*, are often excluded by the definition of secondary structures essentially because common predictive methods cannot deal with them.

To predict the secondary structure of RNAs, an algorithm is needed that is able to assign a score to all possible structures and possibly assign the highest score to the correct structure. Indeed, the number of possible structures increases exponentially with sequence length. A possible approach is to find the structure with the highest number of base pairs (Comay, Nussinov et al. 1984). A simple dot-plot matrix can be used to visualize all possible pairs by comparing the sequence on one axis with its complementary sequence on the other axis. In this matrix, stem regions appear as dot segments, perpendicular to the diagonal from the bottom left to the top right of the plot (see Fig. 6.35).

RNA folding is dictated primarily by thermodynamics rather than by counting and maximizing base pairs. Zuker (2000) and colleagues thus introduced an RNA secondary structure prediction algorithm assuming that the structure most likely to be correct is the one with the lowest free energy (ΔG). The main assumption of this

TABLE 6.10. Base-Pairing Energies (Kcal/Mol) of Structural Elements of RNA Secondary Structures

(a) Stacking energies (UG = GU)

	5'-				
-3'	GU	AU	UA	CG	GC
GU	−0.5	−0.5	−0.7	−1.5	−1.3
AU	−0.5	−0.9	−1.1	−1.8	−2.3
UA	−0.7	−0.9	−0.9	−1.7	−2.1
CG	−1.9	−2.1	−2.3	−2.9	−3.4
GC	−1.5	−1.7	−1.8	−2	−2.9

(b) Bulge loop destabilizing energies by size of loop

L																	
1	2	3	4	5	6	7	8	9	10	12	14	16	18	20	25	30
3.2	5.2	6	6.7	7.4	8.2	9.1	10	10.5	11	11.8	12.5	13	13.6	14	15	15.8

(c) Hairpin loop destabilizing energies by size of loop

	L																	
Closing	0.1	0.2	0.3	0.4	0.5	0.6	0.7	0.8	0.9	1	1.2	1.4	1.6	1.8	2	2.5	3
CG	99.9	99.9	7.4	5.9	4.4	4.3	4.1	4.1	4.2	4.3	4.9	5.6	6.1	6.7	7.1	8.1	8.9
AU	99.9	99.9	7.4	5.9	4.4	4.3	4.1	4.1	4.2	4.3	4.9	5.6	6.1	6.7	7.1	8.1	8.9

(d) Interior loop destabilizing energies by size of loop

	L																	
Closing	0.1	0.2	0.3	0.4	0.5	0.6	0.7	0.8	0.9	1	1.2	1.4	1.6	1.8	2	2.5	3
CG-CG	99.9	0.8	1.3	1.7	2.1	2.5	2.6	2.8	3.1	3.6	4.4	5.1	5.6	6.2	6.6	7.6	8.4
CG-AU	99.9	0.8	1.3	1.7	2.1	2.5	2.6	2.8	3.1	3.6	4.4	5.1	5.6	6.2	6.6	7.6	8.4
AU-AU	99.9	0.8	1.3	1.7	2.1	2.5	2.6	2.8	3.1	3.6	4.4	5.1	5.6	6.2	6.6	7.6	8.4

Source: Data from Freier, Kierzek et al. (1986).

algorithm is that the contribution of the various secondary structure elements (i.e., stem, loops, etc.) is additive. Indeed, good agreement is usually found between predicted structure calculated by energy minimization and experimentally tested structures (Mathews, Sabina et al. 1999). In contrast to the simple Nussinov algorithm (Comay, Nussinov et al. 1984) process, the stem energy is calculated by adding stacking energies between neighboring base pairs assumed to be independent:

$$E\begin{pmatrix} 5' & ACTAG & 3' \\ 3' & TGATC & 5' \end{pmatrix} = E\begin{pmatrix} 5' & AC & 3' \\ 3' & TG & 5' \end{pmatrix} + E\begin{pmatrix} 5' & CT & 3' \\ 3' & GA & 5' \end{pmatrix} + E\begin{pmatrix} 5' & TA & 3' \\ 3' & AT & 5' \end{pmatrix} + E\begin{pmatrix} 5' & AG & 3' \\ 3' & TC & 5' \end{pmatrix}$$

In this way, the energy of a stem of N bases is given by the sum of $N - 1$ base stacking terms. Base-pair stacks take into account both hydrogen bonds and stacking effects. Table 6.10 shows ΔG parameters for RNA structure prediction ob-

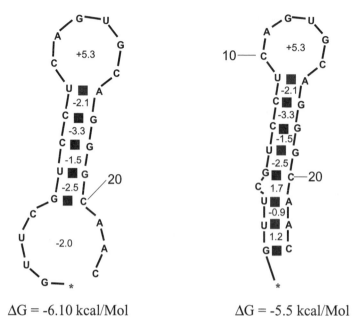

$$\Delta G = -6.10 \text{ kcal/Mol} \qquad \Delta G = -5.5 \text{ kcal/Mol}$$

Figure 6.34. Alternative foldings of the iron responsive element located in human mRNA coding for the 5-aminolevulinate synthase (EMBL ID: X60364).

tained by Freier, Kierzele et al. (1986) from experimental thermodynamic studies carried out on small RNA molecules at a fixed temperature (usually, 37°C) and ionic strength (usually, 1M NaCl). These include parameters for stacking energy, hairpin loops, bulge loops, and internal loops. It can be noted that free energies are not symmetric; for example, GA–CU and AG–UC provide different stability contributions.

The original Zucker algorithm (Zuker and Stiegler 1981; Zuker 1989a), as implemented in the FOLDRNA program of the GCG package, predicts a single optimal secondary structure for an RNA molecule. However, it may happen that the biological functional structure is not coincident with the optimal structure but, rather, has a structure whose free energy is slightly different, above a few percent, from the minimum energy. The program MFOLD (Zuker 1989b, 2000) is able to predict suboptimal structures within a few percent of the calculated energy minimum using an updated set of energy parameters obtained by additional experiments as described by Mathews, Sabina et al. (1999).

Figure 6.34 shows the two foldings predicted for the iron responsive elements (IREs) located in human mRNA coding for 5-aminolevulinate synthase (EMBL ID: X60364). It is striking to note that in this case, the correct folding, with the C-bulge, does not correspond to the structure with minimum energy. The program MFOLD, also available through a Web server (see the URL in the Appendix), makes it possible to set specific constraints forcing or preventing predetermined bases from pairing or remaining single-stranded.

In addition to the planar graph representation in Figs. 6.33 and 6.34, RNA secondary structures can be drawn in a number of ways (e.g., circular or linear graphs). Particularly useful are the Energy DotPlot and the P-Num Plot representations. The Energy DotPlot is a dot-plot matrix where the two axes represent the same RNA molecule in the direct and complementary orientations, respectively, with dots indicating base pairs of all the bases involved in optimal and suboptimal secondary structures within the energy range specified. The P-Num Plot may reveal how well determined the various motifs of the secondary structure are. For each position of the sequence reported on the horizontal axis, the number of times that a particular base is in a paired state for all P-optimal foldings is plotted on the vertical axis, where P represents the percentage increment of the minimum energy, including suboptimal foldings. Figure 6.35 shows the Energy DotPlot and the P-Num Plot calculated for the human 5S rRNA (EMBL ID: V00589) by the MFOLD program.

It is relatively common for homologous RNAs to have a common secondary structure without sharing significant sequence similarity. In this case, base substitutions at one side of the stem are compensated by a base substitution on the other side that maintains base-pairing complementarity. For this reason, in a structurally correct multiple alignment of RNAs, compensatory mutations are usually found in pairs of aligned positions corresponding to paired bases. Then, comparative sequence analysis may recognize probable base pairs, thus validating structures predicted based on the observation of a substitution covariation pattern [see Zuker (2000) for a review]. Such comparative analyses can be carried out with programs calculating the mutual information content among different positions along a multiple alignment (e.g., the MatrixPlot software; see the URL in the Appendix).

An important issue in RNA secondary structure prediction concerns the assignment of a confidence level to minimum energy folding predictions. The simplest way to accomplish this task is to fold a number, perhaps 10, of random sequences having the same compositional features (e.g., GC content) as the sequence under analysis and then comparing the resulting folding energies with that of our particular sequence. Specific software has been developed to fold RNAs in moving windows to detect regions having a significantly high or low folding energy, also called *unusual folding regions* (UFRs; Le and Maizel 1997). Another indication of significant folding derives from the observation of well-definedness of a particular folding prediction as inferred by observation of the Energy DotPlot (Zuker and Jacobson 1998) or by computation of base-pair probabilities of base pairings (Fekete, Hofacker et al. 2000).

Specific prediction tools have been developed for the prediction of tRNAs in eukaryotic and prokaryotic genomes. TRNAs are short RNA molecules (usually 75 to 95 nt long) with relatively little shared primary sequence identity but a highly conserved "cloverleaf" secondary structure. This makes the introduction of covariance models particularly effective for improving tRNA prediction efficiency. The tRNAscan-SE program (Lowe and Eddy 1997), implementing covariance models derived from the analysis of a large number of aligned tRNAs, turned out to be the best performing, with a negligible proportion of false positives (less than 1 per 15 billion random nucleotides). It also implements the search for eukaryotic tRNA pol III promoters (Pavesi, Conterio et al. 1994). Figure 6.36 shows the tRNAscan

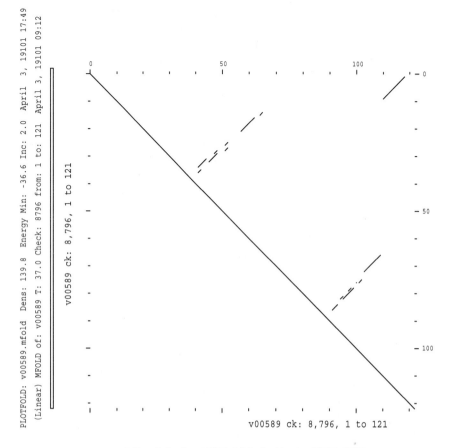

v00589 ck: 8,796, 1 to 121

P-Numplot of: v00589mfold April 3, 19101 17:35
Min Energy: -36.6 Energy Inc: 2.0 Density: 105.22
(Linear) MFOLD of: v00589 T: 37.0 Check: 8796 from 1 to: 121 April 3, 19101 09:12

(a)

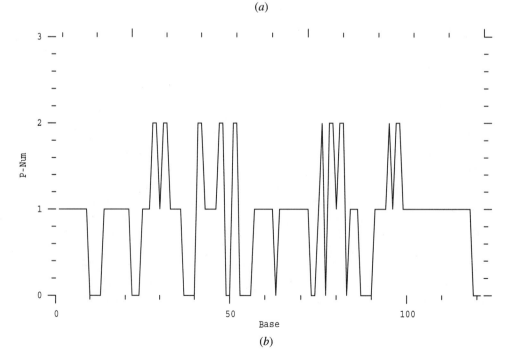

(b)

tRNAscan-SE Secondary Structure Prediction

Sequence: gi|4104639|gb|AF037471.1|AF037471
tRNA #: 1
Isotype: Asp
Anticodon: GUC
Cove Score: 60.58 bits

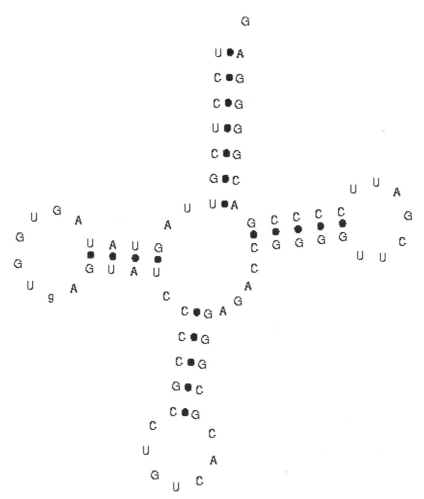

Figure 6.36. tRNAscan prediction carried out on a DNA sequence (EMBL ID: AF037471) containing a tRNA[Asp] sequence. The tRNAscan URL is *http://www.genetics.wustl.edu/eddy/tRNAscan-SE.* [From Lowe and Eddy (1997).]

Figure 6.35. Energy DotPlot (*a*) and P-Num Plot (*b*) calculated for the human 5S rRNA (EMBL ID: V00589) by the MFOLD program. [From *http://bioinfo.math.rpi.edu/ ~mfold/rna/form1.cgi.*)]

prediction output obtained on a DNA sequence (EMBL ID: AF037471) containing a tRNA^{Asp} sequence.

6.10 PROTEIN SEQUENCE ANALYSIS

As a result of large-scale sequencing projects, marked by the successful completion of numerous genome projects ranging from prokaryotes to human, the sequence of the complete protein set for several organisms (proteome) has become available for analysis. In this respect, a cautionary note to be recalled is that most protein sequences derive from the conceptual translation of coding sequences annotated in the nucleotide sequence entries. Then it cannot be excluded that some of them could be defined incorrectly (e.g., by the use of a different AUG start codon) or be completely nonexistent, especially when no clear homologs are detected. On the other hand, it is also possible that even if the complete genome sequence is available, not all protein coding genes have been identified.

Most of the protein sequences predicted, even if conserved evolutionarily across species, have no documented functional role. The use of suitable bioinformatic methods, including statistical and comparative analyses, is thus of utmost importance to predict structural or other properties which may reveal protein function. These methods can also contribute to the much more ambitious task of elucidating the cellular context in which proteins operate, the pathways in which they are involved, as well as their interactions with other cellular components.

The determination of protein three-dimensional structures is certainly crucial for defining accurately its functional activity. However, theoretical tools that analyze protein primary structure may contribute greatly to uncovering functional and structural features of proteins whose three-dimensional structure remains to be resolved experimentally. Although experimental determination of protein three-dimensional structure based on conventional crystallographic or NMR spectroscopic techniques has become much more efficient (Jones 2000), the gap between the number of known protein sequences and the number of known structures is steadily increasing. The pioneer work of Anfisen and co-workers has revealed that most of the relevant information for protein function resides is its primary structure. Indeed, they demonstrated, through denaturation/renaturation experiments on bovine ribonuclease (Anfinsen 1973), that the monodimensional information encoded in the primary sequence fully specifies protein conformation that is at a free-energy minimum. This is generally true with some exceptions when accessory proteins, known as *chaperones*, are needed in the folding process or for correcting misfolds (Grantcharova, Alm et al. 2001). In any case, the problem of three-dimensional structure prediction starting simply from the known amino acid sequence has represented, and still represents, a major challenge in modern biology.

The first step for the functional annotation of a novel protein sequence is usually a database search using BLAST or similar programs (see Secs. 6.4.2 and 6.4.3). If a clear similarity is observed with a well-characterized protein spanning the complete length of the query sequence, the sequence function can be inferred by analogy. The most frequent result of database searching is a list of significant matches between one or more segments of the query sequence and a variety of proteins, most of them still not characterized. This is due to the modular structure of proteins, made of distinct structural domains each folding independently in the relevant three-dimensional structure and harboring its specific independent function.

The worst but improbable scenario is that database searching shows no significant hit. In this case the search can be carried out for a specific domain or for family protein signatures through the application of pattern recognition methods (see also Secs. 5.9 and 6.8.4). Pattern databases of protein functional signatures have become a vital tool for the identification of specific functional activities in novel sequences. Currently, the most commonly used protein pattern databases include ProSite (Hofmann, Bucher et al. 1999), PRINTS (Attwood, Croning et al. 2000), Pfam (Bateman, Birney et al. 2000), and ProDom (Corpet, Servant et al. 2000) (see Secs. 5.9 and 6.6). To create a single coherent resource for automatic annotation of protein sequence data and for proteome analysis, the InterPro resource has been developed (Apweiler, Attwood et al. 2001), which includes databases noted above. SMART (Simple Modular Architecture Research Tool) is a similar tool, which allows identification and annotation of genetically mobile domains and the analysis of domain architecture (Schultz, Copley et al. 2000). All these methods allow assignment to a known family of the protein submitted, or identification of a specific biological activity, if available.

Figure 6.37 shows the InterPro output obtained by analyzing the human p73 tumor protein (entry O15350) and the InterPro entry for the sterile alpha motif (SAM). If no significant match is found against the InterPro or SMART database, other prediction methods can be used that, by exploiting the fairly well known physical and chemical properties of amino acids, can provide useful information about protein structure, function, and classification (see below). Furthermore, other powerful prediction methods have recently been devised, not based on sequence or structure similarity, that analyze patterns such as domain fusion, conserved gene position, and gene co-inheritance and co-expression. These nonhomology methods, based primarily on comparative evolutionary genomic analyses, may allow the functional classification of an unknown protein by the discovery of some kind of relationship with other known proteins (see below) (Marcotte 2000; Pellegrini 2001).

6.10.1 Analysis of Protein Primary Sequences

Analysis of the amino acid composition of an unknown protein may provide several physical properties, such as the isoelectric point (pI), molecular weight (MW), and occurrence of charged amino acid runs and/or of repetitive structures. The PEPSTATS program of the EMBOSS package or the SAPS (statistical analysis of protein sequences; Brendel, Bucher et al. 1992) Web resource (see Table 6.9) output a large amount of chemical and physical information based solely on what can be inferred from the protein primary sequence. Figure 6.38 shows samples of PEPSTATS and SAPS outputs obtained by analyzing human cytochrome c (Acc. No. P00001).

The ProtScale Web resource (see Table 6.11) allows us to compute and represent the profile produced by a number of amino acid scales on a selected protein. The scales used most frequently are the hydrophobicity or hydrophilicity scales, the secondary structure parameter scales, and many other scales based on different chemical and physical properties of amino acids. Plotting the average of these scale values computed on amino acid windows moving over the entire length of a protein may

(*a*)

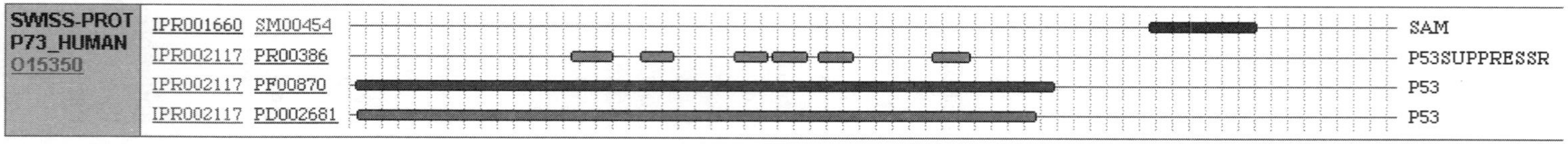

(*b*)

SAM domain (Sterile alpha motif)

Database	InterPro
Accession	IPR001660; SAM (matches 290 proteins)
Name	SAM domain (Sterile alpha motif)
Type	Domain
Dates	08-OCT-1999 (created) 29-NOV-2000 (last modified)
Signatures	PS50105; SAM_DOMAIN (232 proteins) PF00536; SAM (192 proteins) SM00454; SAM (284 proteins)
Secondary no.	IPR000964
Abstract	The sterile alpha motif (SAM) domain is a putative protein interaction module present in a wide variety of proteins [1] involved in many biological processes. The SAM domain that spreads over around 70 residues is found in diverse eucaryotic organisms [2]. SAM domains have been shown to homo- and hetero-oligomerize [3], nevertheless with a low affinity constant [4], and to mediate specific protein-protein interactions. Structural analyses show that the SAM domain is arranged in a small five-helix bundle with two large interfaces [3]. In the case of the SAM domain of EphB2, each of these interfaces is able to form dimers. The presence of these two distinct intermonomers binding surface suggest that SAM could form extended polymeric structures [4].
Examples	• Q64028 EDR1_MOUSE: Drosophila polyhomeotic (ph) and Sex comb on midleg (Scm) and mouse Rae28, member of the Polycomb group of genes, implied in the transcriptional repression of the homeotic genes • P39769 PHP_DROME • O75334 O75334: Human liprin alpha, interact with LAR transmembrane protein tyrosine phosphatase • Q24191 Q24191 • P23561 ST11_YEAST: Yeast Ste 11, serine/threonine protein kinase which participate in mating pheromone response pathways • O35867 NEB1_RAT: Mammalian Neurabin actin filament binding protein, involved in neurite formation • P21709 EPA1_HUMAN: Vertebrate Eph receptor protein tyrosine kinase. Involved in developmental regulation • P09759 EPB1_RAT View examples
References	1. Shultz J., Ponting C.P., Hofmann K., Bork P. *SAM as a protein interaction domain involved in developmental regulation.* Protein Sci. 6: 249-253(1997). [MEDLINE:97160498] [PUB00005047] 2. Stapleton D., Balan I., Pawson T., Sicheri F. *The crystal structure of an Eph receptor SAM domain reveals a mechanism for modular dimerization.* Nat. Struct. Biol. 6: 44-49(1999). [MEDLINE:99101382] [PUB00003966] 3. Peterson A.J., Kyba M., Bornemann D., Morgan K., Brock H.W., Simon J. *A domain shared by the Polycomb group proteins Scm and ph mediates heterotypic and homotypic interactions.* Mol. Cell. Biol. 17(11): 6683-6692(1997). [MEDLINE:98001597] [PUB00006191] 4. Thanos C.D., Goodwill K.E., Bowie J.U. *Oligomeric structure of the human EphB2 receptor SAM domain.* Science 283(5403): 833-836(1999). [MEDLINE:99132419] [PUB00006192]
Database links	PROSITE predoc; PDOC50105
Matches	Table all Graphical all

Figure 6.37. (*a*) Output generated by the InterPro Web server after submission of the human p73 tumor protein (O15350); (*b*) example of the fully annotated InterPro entry IPR001660.

```
PEPSTATS of P00001 from 1 to 104

Molecular weight = 11617.50          Residues = 104
Average Residue Weight  = 111.707    Charge   = 10.5
Isoelectric Point = 10.1635
```
(a)

```
SWISS-PROT ANNOTATION:
ID   sp|P00001|CYC_HUMAN
DE   sp|P00001|CYC_HUMAN (CYC)CYTOCHROME C.[Homo sapiens], 104 bases, C485894F checksum

number of residues: 104;   molecular weight:  11.6 kdal

        1   GDVEKGKKIF IMKCSQCHTV EKGGKHKTGP NLHGLFGRKT GQAPGYSYTA ANKNKGIIWG
       61   EDTLMEYLEN PKKYIPGTKM IFVGIKKKEE RADLIAYLKK ATNE


------------------------------------------------------------------------------
COMPOSITIONAL ANALYSIS (extremes relative to: swp23s.q)

A  :  6( 5.8%); C  :  2( 1.9%); D  :  3( 2.9%); E  :  8( 7.7%); F  :  3( 2.9%)
G  : 13(12.5%); H  :  3( 2.9%); I  :  8( 7.7%); K  : 18(17.3%); L  :  6( 5.8%)
M  :  3( 2.9%); N  :  5( 4.8%); P  :  4( 3.8%); Q  :  2( 1.9%); R  :  2( 1.9%)
S  :  2( 1.9%); T  :  7( 6.7%); V  :  3( 2.9%); W  :  1( 1.0%); Y  :  5( 4.8%)


KR      :   20 ( 19.2%);     ED     :   11 ( 10.6%);     AGP     :   23 ( 22.1%);
KRED    :   31 ( 29.8%);     KR-ED  :    9 (  8.7%);     FIKMNY  :   42 ( 40.4%);
LVIFM   :   23 ( 22.1%);     ST     :    9 (  8.7%).


------------------------------------------------------------------------------


CHARGE DISTRIBUTIONAL ANALYSIS
 1  0-0-+0++00 00+0000000 -+00+0+000 0000000++0 0000000000 00+0+00000
61  --000-00-0 0++00000+0 00000+++-- +0-00000++ 000-

A. CHARGE CLUSTERS (None)
B. HIGH SCORING (UN)CHARGED SEGMENTS (None)
C. CHARGE RUNS AND PATTERNS (None)

DISTRIBUTION OF OTHER AMINO ACID TYPES
1. HIGH SCORING SEGMENTS.
High scoring hydrophobic segments: None
High scoring transmembrane segments: None
2. SPACINGS OF C: H2N-13-CSQC-87-COOH

REPETITIVE STRUCTURES
A. SEPARATED, TANDEM, AND PERIODIC REPEATS: amino acid alphabet.
Repeat core block length:   4
Aligned matching blocks:
[   3-   6]   VEKG
[  20-  23]   VEKG
B. SEPARATED AND TANDEM REPEATS: 11-letter reduced alphabet (i= LVIF; += KR; -= ED; s=
AG; o= ST; n= NQ; a= YW; p= P; h= H; m= M; c= C)

MULTIPLETS (None)

PERIODICITY ANALYSIS (nothing found)
```
(b)

Figure 6.38. Sample output generated by PepStats (*a*) and SAPS (*b*) programs obtained by analyzing human cytochrome *c* (SwissProt entry P00001).

TABLE 6.11. Web Resources for Protein Primary Structure Analysis

Program	URL
PepStats	www.uk.embnet.org/Software/EMBOSS
SAPS	www.isrec.isb-sib.ch/software/SAPS_form.html
ProtScale	www.expasy.ch
Antigenic	www.uk.embnet.org/Software/EMBOSS
NetOglyc	www.cbs.dtu.dk/services/NetOGlyc
NetPhos	www.cbs.dtu.dk/services/NetPhos

TABLE 6.12. Amino Acid Scales for Hydrophobicity

	KD[a]	NCHS[b]	Polarity[c]	Percent Buried[d]
Ala	1.8	0.6	8.1	11.2
Arg	−4.5	−2.5	10.5	0.5
Asn	−3.5	−0.8	11.6	2.9
Asp	−3.5	−0.9	13.0	2.9
Cys	2.5	0.3	5.5	4.1
Gln	−3.5	−0.9	10.5	1.6
Glu	−3.5	−0.7	12.3	1.8
Gly	−0.4	0.5	9.0	11.8
His	−3.2	−0.4	10.4	2
Ile	4.5	1.4	5.2	8.6
Leu	3.8	1.1	4.9	11.7
Lys	−3.9	−1.5	11.3	0.5
Met	1.9	0.6	5.7	1.9
Phe	2.8	1.2	5.2	5.1
Pro	−1.6	0.1	8.0	2.7
Ser	−0.8	−0.2	9.2	8
Thr	−0.7	−0.1	8.6	4.9
Trp	−0.9	0.8	5.4	2.2
Tyr	−1.3	0.3	6.2	2.6
Val	4.2	1.1	5.9	12.9

[a] Proposed by Kyte and Doolittle (1982).
[b] Normalized consensus hydrophobicity scale, proposed by Eisenberg, Schwarz et al. (1984).
[c] Proposed by Grantham (1974).
[d] Amino acid propensity to be buried in protein hydrophobic core, calculated by Janin (1979) as percentage occurrence in proteins with known structure.

provide important information about its structure. Predictions of the relative propensity of some amino acid to bury in the protein hydrophobic core or to form a transmembrane segment can be derived. Table 6.12 lists the most widely used scales for amino acid hydrophobicity, polarity, or propensity to be buried in a hydrophobic core. These amino acid properties can also be used to predict antigenic regions (Hopp 1986; Jameson and Wolf 1988; Kolaskar and Tongaonkar 1990) whose sequence should correspond to the most likely synthetic peptides for making antibodies (see, e.g., the programs Antigenic in the EMBOSS package or PeptideStructure in the GCG package). More recently, the prediction of antigenic peptides for humans has been improved greatly, taking into account the requirements for MHC–peptide interactions (Hammer 1995), and several computational tools have been developed [see Sturniolo, Bono et al. (1999) and the references therein].

Other programs, based only on the analysis of protein primary structure, may predict posttranslational modification. NetOglyc (see the URL in Table 6.11; Hansen, Lund et al. 1995) predicts *O*-glycosilation at serine and threonine residues by using a neural network trained on eukaryotic *O*-glycoproteins. The NetPhos server (see the URL in Table 6.11; Blom, Gammeltoft et al. 1999) produces neural network predictions for serine, threonine, and tyrosine phosphorylation sites in eukaryotic proteins. Figure 6.39 shows NetPhos prediction carried out on human cyclin-dependent kinase 2 (CDK2, P24941) which correctly predicts phosphorilation at sites Thr14, Thr160, and Tyr15.

When performing protein sequence analysis, the prediction of false positives and false negatives should also be taken into consideration. Such figures are discussed routinely when methods are carefully validated when implemented and described in the literature, and may help in the evaluation of prediction when a new target sequence is submitted to the server electronically for predicting the feature at hand.

6.10.2 Prediction of Transmembrane Protein Helices

Membrane proteins constitute a relevant and heterogeneous class and are responsible for several important functions, such as signal trasduction, transport phenomena, and energy conversion. One interesting problem is the prediction of the location (topography) of transmembrane helices (TMHs) and the topology (the location of the N and C termini with respect to the membrane plane) of the transmembrane protein based on its primary sequence.

Transmembrane segments of crystallized membrane proteins are typically apolar helices made up of 17 to 25 residues perpendicular or partially tilted with respect to the membrane bilayer. These hydrophobic helices are separated by polar connection loops with a preponderance of positively charged residues on the cytoplasmic side of the transmembrane segment (interior). The topology pattern, determined statistically for both procaryots and eukaryots, is known as the *positive inside rule* (Sipos and von Heijne 1993) and is used to predict the location of inner and outer loops. A number of algorithms designed to identify putative transmembrane segments have been developed. They are commonly based on the analysis of the primary amino acid sequence or on amino acid multiple alignments. In the latter case, a higher level of prediction accuracy was obtained (Rost, Casadio et al. 1995).

The availability of prediction techniques to recognize the topology of transmembrane regions is particularly important because structure determination of membrane proteins at high resolution is still a difficult task. Table 6.13 lists some of the algorithms available for detection of transmembrane protein domains. Among those most frequently quoted in the literature, MemSat (Jones, Taylor et al. 1994) is a statistical method based on the propensity for a given residue to be located in a specific position along the transmembrane helix, together with dynamic programming to maximize the propensity during the predictive phase and to assign the topology. PHDhtm (Rost, Casadio et al. 1995; Rost, Fariselli et al. 1996) is a neural network predictor trained on some 150 membrane proteins whose alpha-helix

```
Phosphorylation sites predicted:    Thr: 5  Tyr: 2
```

```
                    Threonine predictions

   Name           Pos    Context   Score  Pred
                            v
   gi_116051_s      14    IGEGTYGVV  0.768  *T*
   gi_116051_s      26    RNKLTGEVV  0.201   .
   gi_116051_s      39    IRLDTETEG  0.930  *T*
   gi_116051_s      41    LDTETEGVP  0.113   .
   gi_116051_s      47    GVPSTAIRE  0.068   .
   gi_116051_s      72    DVIHTENKL  0.088   .
   gi_116051_s      97    ASALTGIPL  0.098   .
   gi_116051_s     137    LLINTEGAI  0.020   .
   gi_116051_s     158    VPVRTYTHE  0.127   .
   gi_116051_s     160    VRTYTHEVV  0.801  *T*
   gi_116051_s     165    HEVVTLWYR  0.036   .
   gi_116051_s     182    KYYSTAVDI  0.103   .
   gi_116051_s     198    AEMVTRRAL  0.369   .
   gi_116051_s     218    RIFRTLGTP  0.036   .
   gi_116051_s     221    RTLGTPDEV  0.987  *T*
   gi_116051_s     231    WPGVTSMPD  0.682  *T*
   gi_116051_s     290    FQDVTKPVP  0.036   .
                            ^
```

```
                    Tyrosine predictions

   Name           Pos    Context   Score  Pred
                            v
   gi_116051_s      15    GEGTYGVVY  0.856  *Y*
   gi_116051_s      19    YGVVYKARN  0.102   .
   gi_116051_s      77    ENKLYLVFE  0.330   .
   gi_116051_s     107    LIKSYLFQL  0.006   .
   gi_116051_s     159    PVRTYTHEV  0.013   .
   gi_116051_s     168    VTLWYRAPE  0.011   .
   gi_116051_s     179    LGCKYYSTA  0.129   .
   gi_116051_s     180    GCKYYSTAV  0.039   .
   gi_116051_s     236    SMPDYKPSF  0.936  *Y*
   gi_116051_s     269    QMLHYDPNK  0.419   .
                            ^
```

Figure 6.39. Output generated by NethPhos prediction server for serine, threonine, and tyrosine phosphorylation sites on human cyclin-dependent kinase 2 (cdk2, P24941). Thr14, Tyr15, and Thr160 sites, which are actually phosphorylated, are shown in boldface type.

transmembrane domains had been detected experimentally. Evolutionary information in the form of a sequence profile computed by a multiple alignment generated by the program automatically is used as input to the networks. Originally, the method yielded an overall accuracy of about 95% when tested with a cross-validation procedure to predict membrane protein topography (Rost, Casadio et al. 1995). The topology is predicted essentially by implementing a variant of the positive inside rule, and the accuracy was 86% (Rost, Fariselli et al. 1996). PHDhtm, cor-

TABLE 6.13. Software for the Prediction of Location and Topology of Transmembrane Helices in Membrane Proteins

Program	Description	References
DAS	Prediction of transmembrane regions in prokaryotes using the dense alignment surface method	Cserzo, Wallin et al. (1997)
HMMTOP	Prediction of transmembrane helices and topology of proteins	Tusnady and Simon (1998)
PredictProtein (PHDhtm)	Prediction of transmembrane helix location and topology	Rost, Casadio et al. (1995); Rost, Fariselli et al. (1996)
MemSat	Prediction of the secondary structure and topology of integral membrane proteins based on the recognition of topological models	Jones, Taylor et al. (1994)
IRENE	Neural network predictor	www.biocomp.unibo.it
SOSUI	Prediction of transmembrane regions	Hirokawa, Boon-Chieng et al. (1998)
TMAP	Transmembrane detection based on multiple sequence alignment	Persson and Argos (1994)
TMHMM	Prediction of transmembrane helices in proteins	Krogh, Larsson et al. (2001)
TopPred 2	Topology prediction of membrane proteins	Claros and von Heijne (1994)

rectly, does not predict transmembrane domains of porins, as they are made up of beta-strands forming a barrel, comprising an even number of strands. To this end, selective tools, also based on neural networks, have been developed, starting from a database of nonredundant beta barrel proteins solved at atomic resolution and present in the outer membrane of gram-negative bacteria (Rost, Fariselli et al. 1996).

The DAS (dense alignment surface) method (Cserzo, Wallin et al. 1997), based on low-stringency dot plots of the query sequence against a collection of non-homologous membrane proteins, scores as high as PHDhtm when topography is predicted. However, the method relies on the selection of threshold values whose selection may affect the number of transmembrane helices of unknown proteins. A similar method (TopPred), also implementing the positive inside rule, makes possible prediction of topological models.

The recently developed TMHMM (Krogh, Larsson et al. 2001) is a membrane topology prediction method based on hidden Markov models that correctly predicts 97 to 98% of transmembrane helices. HMMs based on the probability that a given model is suited to describe the sequence at hand can discriminate between soluble and membrane proteins with both sensitivity and specificity higher than 99% (excluding signal peptides containing proteins). This high degree of accuracy made possible reliable identification of integral membrane proteins in organism proteomes inferred from the genomes sequenced. Based on these predictions, Krogh, Larsson et al. (2001) estimated that 20 to 30% of total genes in most genomes encode membrane proteins and that their preferred topology is N(interior)–C(interior). More recently, another predictor based on neural networks became available (IRENE; see Table 6.13). This predictor is focused only on membrane proteins whose atomic structure is presently available in the database.

Figure 6.40 shows the output generated using TMHMM, PHDhtm, DAS, and IRENE on the human cytochrome *b* mitochondrially encoded subunit (P00156) of the cytochrome bc1 complex. Prediction accuracy can be evaluated through comparison with the resolved structure of the entire complex that identifies eight trans-

```
TMHMM result

# gi_117863_sp_P00156_CYB_HUMAN Length: 380
# gi_117863_sp_P00156_CYB_HUMAN Number of predicted TMHs 9
# gi_117863_sp_P00156_CYB_HUMAN Exp number of AAs in TMHs: 196.7936
# gi_117863_sp_P00156_CYB_HUMAN Exp number, first 60 AAs:  22.34318
# gi_117863_sp_P00156_CYB_HUMAN Total prob of N-in:        0.09002
# gi_117863_sp_P00156_CYB_HUMAN POSSIBLE N-term signal sequence
gi_117863_sp_P00156_CYB_HUMAN    TMHMM2.0       outside      1     32
gi_117863_sp_P00156_CYB_HUMAN    TMHMM2.0       TMhelix     33     55
gi_117863_sp_P00156_CYB_HUMAN    TMHMM2.0       inside      56     75
gi_117863_sp_P00156_CYB_HUMAN    TMHMM2.0       TMhelix     76     98
gi_117863_sp_P00156_CYB_HUMAN    TMHMM2.0       outside     99    112
gi_117863_sp_P00156_CYB_HUMAN    TMHMM2.0       TMhelix    113    135
gi_117863_sp_P00156_CYB_HUMAN    TMHMM2.0       inside     136    139
gi_117863_sp_P00156_CYB_HUMAN    TMHMM2.0       TMhelix    140    158
gi_117863_sp_P00156_CYB_HUMAN    TMHMM2.0       outside    159    177
gi_117863_sp_P00156_CYB_HUMAN    TMHMM2.0       TMhelix    178    200
gi_117863_sp_P00156_CYB_HUMAN    TMHMM2.0       inside     201    228
gi_117863_sp_P00156_CYB_HUMAN    TMHMM2.0       TMhelix    229    251
gi_117863_sp_P00156_CYB_HUMAN    TMHMM2.0       outside    252    287
gi_117863_sp_P00156_CYB_HUMAN    TMHMM2.0       TMhelix    288    310
gi_117863_sp_P00156_CYB_HUMAN    TMHMM2.0       inside     311    322
gi_117863_sp_P00156_CYB_HUMAN    TMHMM2.0       TMhelix    323    340
gi_117863_sp_P00156_CYB_HUMAN    TMHMM2.0       outside    341    349
gi_117863_sp_P00156_CYB_HUMAN    TMHMM2.0       TMhelix    350    372
gi_117863_sp_P00156_CYB_HUMAN    TMHMM2.0       inside     373    380
```

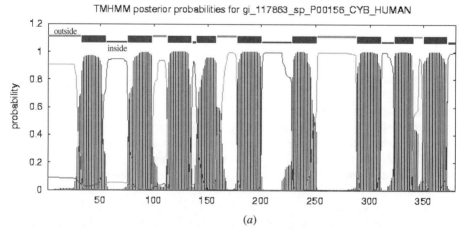

(a)

Figure 6.40. Helical transmembrane domains predicted by TMHMM (*a*), PHDhtm (*b*), DAS (*c*), and IRENE (*d*) programs on human cytochrome *b* (SwissProt entry P00156). TMHMM, PHDhtm, and IRENE also provide topology prediction.

```
PHDhtm Helical transmembrane prediction
        note: PHDacc and PHDsec are reliable for water-
              soluble globular proteins, only.  Thus,
              please take the  predictions above with
              particular caution wherever transmembrane
              helices are predicted by PHDhtm!


  PHDhtm
  ---
  --- PhdTopology REFINEMENT AND TOPOLOGY PREDICTION: SYMBOLS
  --- AA           : amino acid in one-letter code
  --- PHD htm      : HTM's predicted by the PHD neural network
  ---                system (T=HTM, ' '=not HTM)
  --- Rel htm      : Reliability index of prediction (0-9, 0 is low)
  --- detail       : Neural network output in detail
  --- prH htm      : 'Probability' for assigning a helical trans-
  ---                membrane region (HTM)
  --- prL htm      : 'Probability' for assigning a non-HTM region
  ---        note: 'Probabilites' are scaled to the interval
  ---              0-9, e.g., prH=5 means, that the first
  ---              output node is 0.5-0.6
  --- subset       : Subset of more reliable predictions
  --- SUB htm      : All residues for which the expected average
  ---                accuracy is > 82% (tables in header).
  ---        note: for this subset the following symbols are used:
  ---              L: is loop (for which above ' ' is used)
  ---              '.': means that no prediction is made for this,
  ---                residue as the reliability is:  Rel < 5
  --- other        : predictions derived based on PHDhtm
  --- PHDFhtm      : filtered prediction, i.e., too long HTM's are
  ---                split, too short ones are deleted
  --- PHDRhtm      : refinement of neural network output
  --- PHDThtm      : topology prediction based on refined model
  ---                symbols used:
  ---              i: intra-cytoplasmic
  ---              T: transmembrane region
  ---              o: extra-cytoplasmic
  ---
  --- PhdTopology REFINEMENT AND TOPOLOGY PREDICTION
                  ...,....1....,....2....,....3....,....4....,....5....,....6
          AA      |MTPMRKINPLMKLINHSFIDLPTPSNISAWWNFGSLLGACLILQITTGLFLAMHYSPDAS|
          PHD htm |                         TTTTTTTTTTTTTTTTTTTTTTTTTTTTTTTTTTT|
          Rel htm |99999999999999999999999988750234567777777777777777766543322|
  detail:
          prH htm |000000000000000000000000001246677888888888888888888887776666|
          prL htm |99999999999999999999999998753322111111111111111111223333|
          PHDRhtm |                         TTTTTTTTTTTTTTTTTTTTT              |
          PHDThtm |iiiiiiiiiiiiiiiiiiiiiiiiiiiiTTTTTTTTTTTTTTTTTTTTTooooooooo|
  subset: SUB htm |LLLLLLLLLLLLLLLLLLLLLLLLLLL...HHHHHHHHHHHHHHHHHHHHHHHHHH....|
                  ...,....7....,....8....,....9....,...10...,....11...,....12
          AA      |TAFSSIAHITRDVNYGWIIRYLHANGASMFFICLFLHIGRGLYYGSFLYSETWNIGIILL|
          PHD htm |T                    TTTTTTTTTTTTTTTTTTTTTT        TTTTTTTTT|
          Rel htm |10244565334466542113024667777877777777777754201344310346778887|
  detail:
          prH htm |543222123322112234434678888889888888888877644322345678889998|
          prL htm |456777876677887765565321111110111111111122355677654321110001|
          PHDRhtm |                     TTTTTTTTTTTTTTTTTTTT          TTTTTTT|
          PHDThtm |ooooooooooooooooooooooTTTTTTTTTTTTTTTTTTTTTTTiiiiiiiiiiiiiiiTTTTTTT|
  subset: SUB htm |...LLLLL..LLLLLL......HHHHHHHHHHHHHHHHHHHHHH....LL....HHHHHHHH|
                  ...,....13...,....14...,....15...,....16...,....17...,....18
          AA      |LATMATAFMGYVLPWGQMSFWGATVITNLLSAIPYIGTDLVQWIWGGYSVDSPTLTRFFT|
          PHD htm |TTTTTTTTTTTTTTTTTTTTTTTTTTTTTTTTTTTTTTTTTT        TTTTTTTTT|
          Rel htm |77777777777666666655556667777777776666666554210011111343467 7|
  detail:
          prH htm |888888888888888887777888888888888888888888877765544444567 67888|
          prL htm |111111111111111122221111111111111111111122234455555543232111|
          PHDRhtm |TTTTTTTTTTTT        TTTTTTTTTTTTTTTTTT                   TT|
          PHDThtm |TTTTTTTTTTTTTooooooooooTTTTTTTTTTTTTTTTTTTTTTiiiiiiiiiiiiiiiiiiiiiiiTT|
  subset: SUB htm |HHHHHHHHHHHHHHHHHHHHHHHHHHHHHHHHHHHHHHHHHHH...........H.HHHH|
                  ...,....19...,....20...,....21...,....22...,....23...,....24
          AA      |FHFILPFIIAALATLHLLFLHETGSNNPLGITSHSDKITFHPYYTIKDALGLLLFLLSLM|
          PHD htm |TTTTTTTTTTTTTTTTTTTTT                   TTTTTTTTTTTTTTTTTTTTT|
          Rel htm |777777777778887776403578899999999999999886303457778777777888|
```

Figure 6.40. *Continued*

membrane helices in the protein from bovine heart mithocondria (Iwata, Lee et al. 1998). Interestingly, both IRENE and DAS (used with the strongest threshold) assign the same number of helices (eight), in approximately the same position, contrary to the other two predictors.

6.10.3 Identification of Protein Signal Peptides and Prediction of Their Subcellular Location

A large number of eukaryotic proteins synthesized in the citosol need to be further sorted to several subcellular compartments, such as the mitochondrion or the chloroplast, the vescicles of the secretory system, the endoplasmic reticulum, the Golgi apparatus, and so on. Protein sorting usually relies on the presence of N-terminal targeting sequences recognized by translocation machinery (Schatz and Dobberstein 1996). The targeting sequence is generally removed proteolitically during or after compartment sorting. Only weak consensus sequences have been found for these *presequences*, although some specific features can be identified.

Proteins destined for the endoplasmic reticulium, Golgi compartment, lysosome, plasma membrane, and exterior of the cell contain an N-terminal signal peptide (SP) of about 15 to 25 residues. SPs show a tripartite structure with a positively charged

```
detail:
        prH htm |8888888888899988887532100000000000000000001356778889888888999|
        prL htm |11111111111000111124678999999999999999999864322110111111000|
        PHDRhtm |TTTTTTTTTTTTTTT              TTTTTTTTTTTTTTT|
        PHDThtm |TTTTTTTTTTTTTTTTToooooooooooooooooooooooooooooooTTTTTTTTTTTTTTTT|
subset: SUB htm |HHHHHHHHHHHHHHHHHHHH..LLLLLLLLLLLLLLLLLLLLL...HHHHHHHHHHHHHHHHH|
                ....,....25...,....26...,....27...,....28...,....29...,....30
        AA      |TLTLFSPDLLGDPDNYTLANPLNTPPHIKPEWYFLFAYTILRSVPNKLGGVLALLLSILI|
        PHD htm |TTTTTT                    TTTTTTTTTTTTTTTTTTTTTTTTTTTTTTTTTTT|
        Rel htm |877652035788999999999999998753113567777777766666566777777777777|
detail:
        prH htm |988876432100000000000000001234567888888888888788888888888888|
        prL htm |0111235678999999999999999987654321111111111111121111111111111|
        PHDRhtm |TTTTT              TTTTTTTTTTTTTTTT    TTTTTTTTTTTTTT|
        PHDThtm |TTTTTiiiiiiiiiiiiiiiiiiiiTTTTTTTTTTTTTTTTTTTTTooooTTTTTTTTTTTT|
subset: SUB htm |HHHHH...LLLLLLLLLLLLLLLLLLL....HHHHHHHHHHHHHHHHHHHHHHHHHHHH|
                ....,....31...,....32...,....33...,....34...,....35...,....36
        AA      |LAMIPILHMSKQQSMMFRPLSQSLYWLLAADLLILTWIGGQPVSYPFTIIGQVASVLYFT|
        PHD htm |TTTTTTTTT              TTTTTTTTTTTTTTTTTTTT     TTTTTTTTTTTTTT|
        Rel htm |888776531146777764113577888788877764203344203577888888888|
detail:
        prH htm |9998887654211111112456788999889998888765332235678899999999999|
        prL htm |0001112345788888887543210001000111234667764321000000000000|
        PHDRhtm |TTTTT              TTTTTTTTTTTTTTTTTTT    TTTTTTTTTTTTTT|
        PHDThtm |TTTTTTiiiiiiiiiiiiiiiiiiTTTTTTTTTTTTTTTTTToooooooooTTTTTTTTTTTTT|
subset: SUB htm |HHHHHHH...LLLLLLLLL...HHHHHHHHHHHHHHHHH....LL...HHHHHHHHHHHHH|
                ....,....37...,....38...,....39...,....40...,....41...,....42
        AA      |TILILMPTISLIENKMLKWA|
        PHD htm |TTTTTTTTTTTTT|
        Rel htm |88888888775202456777|
detail:
        prH htm |99999999887643221111|
        prL htm |00000000112356778888|
        PHDRhtm |TTTTTTTTTT          |
        PHDThtm |TTTTTTTTTTTiiiiiiiiii|
subset: SUB htm |HHHHHHHHHHH...LLLLLL|
---
--- PhdTopology REFINEMENT AND TOPOLOGY PREDICTION END
---
                                (b)
```

Figure 6.40 *Continued*

Potential transmembrane segments

Start	Stop	Length	~	Cutoff
35	51	17	~	1.7
37	48	12	~	2.2
88	99	12	~	1.7
89	98	10	~	2.2
115	129	15	~	1.7
116	126	11	~	2.2
145	153	9	~	1.7
179	199	21	~	1.7
181	197	17	~	2.2
229	245	17	~	1.7
230	244	15	~	2.2
275	280	6	~	1.7
290	307	18	~	1.7
291	306	16	~	2.2
326	337	12	~	1.7
327	336	10	~	2.2
350	370	21	~	1.7
355	368	14	~	2.2

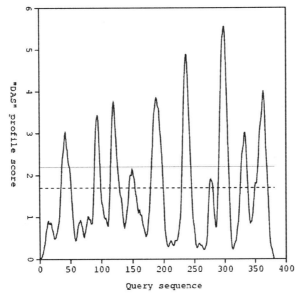

"DAS" TM-segment prediction

loose cutoff ----
strict cutoff _____

(c)

Figure 6.40 *Continued*

Transmembrane Helix Predictions of Proteins

DB /DB/SW/sprot
The Prediction of your sequence

```
MTPMRKINPLMKLINHSFIDLPTPSNISAWWNFGSLLGACLILQITTGLFLAMHYSPDAS
TAFSSIAHITRDVNYGWIIRYLHANGASMFFICLFLHIGRGLYYGSFLYSETWNIGIILL
LATMATAFMGYVLPWGQMSFWGATVITNLLSAIPYIGTDLVQWIWGGYSVDSPTLTRFFT
FHFILPFIIAALATLHLLFLHETGSNNPLGITSHSDKITFHPYYTIKDALGLLLFLLSLM
TLTLFSPDLLGDPDNYTLANPLNTPPHIKPEWYFLFAYTILRSVPNKLGGVLALLLSILI
LAMIPILHMSKQQSMMFRPLSQSLYWLLAADLLILTWIGGQPVSYPFTIIGQVASVLYFT
TILILMPTISLIENKMLKWA
```

TOPOLOGY (N-Terminus) = IN

Predicted Segments

	Start	-	End
TM (1)	29	-	57
TM (2)	79	-	103
TM (3)	111	-	131
TM (4)	174	-	201
TM (5)	224	-	246
TM (6)	288	-	308
TM (7)	318	-	338
TM (8)	349	-	370

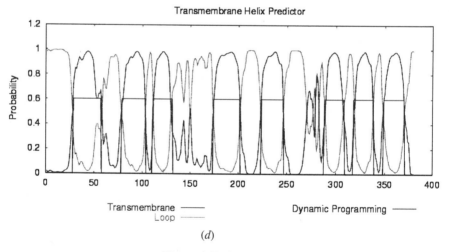

Figure 6.40 *Continued*

N-terminal region (n-region), a 7- to 15-residue-long hydrophobic core (h-region), and a polar 3- to 7-residue C-terminal region (c-region) leading to the signal peptidase cleavage site. Mitochondrial targeting peptides (mTPs) consist of an N-terminal presequence 20 to 80 residues long enriched with the amino acids Arg, Ala, and Ser, depleted of acidic residues (Asp and Glu), and believed to form amphiphilic alpha-helices. Amphiphilic helices, which separate hydrophobic and hydrophilic regions in a protein or mediate protein–membrane interactions, show distinct features on the two opposite sides of the helix, the one more hydrophobic and the other more hydrophilic. mTPs can be predicted by calculating the hydrophobic moment (Eisenberg, Schwarz et al. 1984) or plotting helical wheel graphs along a peptide window scanning along the sequence. The Moment and HelicalWheel programs of the GCG package can be used for such prediction.

Chloroplast targeting peptides (cTPs), with a length ranging from 20 to more that 100 residues, are mainly enriched in serine and other hydroxylated amino acids. cTPs seem to be organized in three domains, with an N-terminal part uncharged and poor in Gly and Pro, a central part rich in Ser, and a C-terminal part enriched in Arg and forming an amphiphilic beta-strand. Several methods have been developed for the prediction of targeting peptides based on statistical methods or neural network approaches [see Claros, Brunak et al. (1997) for a review].

SignalP (Nielsen, Engelbrecht et al. 1997) is a neural network–based program suited to predict signal peptides in both prokaryotes and eukaryotes. The PSORT resource (Nakai and Horton 1999) has been developed for predicting the localization compartment of eukaryotic proteins. Among 14 subcellular locations, the most likely localization, with about 65% accuracy, is predicted by discriminant functions trained on a set of proteins with known subcellular localization. A higher level of accuracy is reached using methods devised to recognize a specific subset of targeting signals.

The MitoProt program (Claros and Vincens 1996) applies discriminant analysis, using a large number of parameters, to identify mitochondrial proteins, reaching about 80% accuracy. TargetP (Emanuelsson, Nielsen et al. 2000) is a neural network–based tool for the determination of proteins destined to the mitochondrion, the chloroplast, or the secretory pathway, with a success rate of about 90%. It also predicts the cleavage site, but at a lower level of accuracy. Figure 6.41 shows the MitoProt and TargetP predictions carried out on human subunit 4 of cytochrome *c* oxidase. It shows clearly that both methods accurately predict the protein as mitochondrial, as well as the site of the cleavage removing the 22 amino acid mTPs.

6.10.4 Prediction of Protein Secondary Structure

The secondary structure of a protein describes the regular close arrangement of amino acids in the protein, involving hydrogen bonds between the backbone carbonyl of one amino acid and the backbone NH of another amino acid. Major elements of secondary structure are the alpha-helix and the beta-sheet. The *alpha-helix* is a tightly coiled rodlike structure with an average of 3.6 amino acids per turn, stabilized by hydrogen bonds between residues i and i + 4 along the chain with R side groups sticking out from the structure in a spiral arrangement. The *beta-sheet* is a more extended structure composed of two or more straight beta-strand chains with side-by-side hydrogen bonding. The sheets can be parallel or antiparallel, depending on whether beta-strands run in the same or in opposite directions as defined by the amino and carboxy termini of component beta-strands. Sheets may be formed from a single chain if they contain a beta-turn, which forms a sharp hairpin reversion of the chain, producing an antiparallel orientation of beta-strands. Protein regions not belonging to any secondary structure element are generally classified as *random coils*.

The prediction of secondary structures may provide important structural infor-

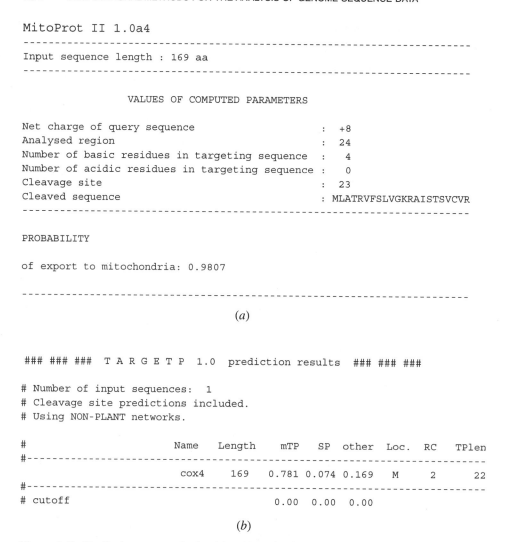

```
MitoProt II 1.0a4
--------------------------------------------------------------------
Input sequence length : 169 aa
--------------------------------------------------------------------

                    VALUES OF COMPUTED PARAMETERS

Net charge of query sequence                        :  +8
Analysed region                                     :  24
Number of basic residues in targeting sequence   :   4
Number of acidic residues in targeting sequence :    0
Cleavage site                                       :  23
Cleaved sequence                                    : MLATRVFSLVGKRAISTSVCVR
--------------------------------------------------------------------


PROBABILITY

of export to mitochondria: 0.9807


--------------------------------------------------------------------
```

(*a*)

```
### ### ###   T A R G E T P  1.0  prediction results  ### ### ###

# Number of input sequences:  1
# Cleavage site predictions included.
# Using NON-PLANT networks.

#                         Name   Length   mTP   SP  other  Loc.  RC  TPlen
#------------------------------------------------------------------------
                          cox4    169    0.781 0.074 0.169   M    2    22
#------------------------------------------------------------------------
# cutoff                                 0.00  0.00  0.00
```

(*b*)

Figure 6.41. Prediction output obtained by MitoProtII (*a*) and TargetP (*b*) on human subunit 4 of cytochrome *c* oxidase (P13073). MitoProtII calculates the overall probability that the protein submitted is exported to mitochondria. TargetP gives the probability that the protein submitted contains a target peptide for the mitochondrion (mTP) or for the secretory pathway (SP), assessing the predicted location (M, mitochondrion), and providing a reliability class (RC) determined on the difference between the highest and second-highest score. The length predicted for the presequence (Tplen) is also indicated.

mation on a protein in case its three-dimensional structure has not been resolved experimentally (e.g., using NMR or x-ray crystallography). If the secondary structure of a protein is determined by prediction methods, it is possible to derive a relatively small number of possible tertiary structures using knowledge about the ways secondary structural elements can pack. Additionally, considering that protein structure is conserved more extensively than primary structure (see also Sec. 6.4.6 and Fig. 6.17), more reliable alignments can be obtained under the guidance of shared secondary structure regions. The accuracy of the multiple alignment between the target protein and the template homologous proteins whose three-dimensional structure has been determined experimentally greatly affects the performance of homology modeling methods (see Sec. 6.10.6). Then, alignment refinement based on the conservation of secondary structure elements may considerably extend the applicability range of homology modeling to cases of distantly related templates.

The local accuracy of secondary structure prediction can be obtained by measuring the number of residues predicted correctly in each of the three states—alpha-helix, beta-strand, and coil—for proteins whose three-dimensional structures have been determined experimentally. The programs DSSP (Kabsch and Sander 1983) and STRIDE (Frishman and Argos 1995) are currently used to assign a secondary structure state to each residue from atomic coordinates of known protein structures.

The parameter most used to estimate the global accuracy of the prediction is Q3, defined by the equation

$$Q3 = \frac{P_{\text{alpha}} + P_{\text{beta}} + P_{\text{coil}}}{T}$$

where P_{state} is the number of residues predicted in a given state that actually are in that state (T = total number of residues). As the accuracy measure of single-residue predictions does not provide a reliable assessment of prediction quality, other measures have been proposed, based on the number of segments in a protein and on their length. Indeed, a prediction where all secondary structure elements are identified, although not precisely in their start and end positions, is certainly more accurate than a prediction characterized by a similar Q3 value but where some elements are missed or wrongly assigned.

Early methods for secondary structure prediction were designed for single protein sequences and were based on simple residue statistics. These methods measured the relative propensity for each amino acid to stay in a specific secondary structure element based on information derived from known structures. For example, it is well known that alanine, glutamate, and methionine are commonly found in alpha-helices, whereas glycine and proline are not: instead, frequently being found in beta-turns. Table 6.14 shows the amino acid scales proposed by Deleage and Roux (1987), Chou and Fasman (1978b), and Levitt (1978), reporting the relative propensity of each amino acid to stay in an alpha-helix, a beta-sheet, or a beta-turn.

Among these early methods, those used most extensively were those of Chou and Fasman (1978a), Garnier, Osguthorpe et al. (1978), and Lim (1974), with prediction accuracies of 50 to 60%. These early methods suffered from a lack of data, as there were very few known three-dimensional structures from which to derive parameters, even if their major limitation is the fact that they are based essentially on single residue statistics (i.e., do not take into consideration the local environment of a given residue). The second generation of prediction tools improved accuracy by a combination of a larger database of protein structures and the use of statistics based on local and nonlocal interactions among residues (e.g., SOPM and GORII programs; see Table 6.15). All these methods were limited by prediction accuracy levels slightly higher than 60%, and although performing better than the first-

TABLE 6.14. Amino Acid Relative Propensity Scales for Secondary Structure Elements[a]

AA	Alpha-Helix			Beta-Sheet			Beta-Turn		
	DR	Levitt	CF	DR	Levitt	CF	DR	Levitt	CF
Ala	1.489	1.290	1.420	0.709	0.900	0.830	0.788	0.770	0.660
Arg	1.224	0.960	0.980	0.920	0.990	0.930	0.912	0.880	0.950
Asn	0.772	0.900	0.670	0.604	0.760	0.890	1.572	1.280	1.560
Asp	0.924	1.040	1.010	0.541	0.720	0.540	1.197	1.410	1.460
Cys	0.966	1.110	0.700	1.191	0.740	1.190	0.965	0.810	1.190
Gln	1.164	1.270	1.110	0.840	0.800	1.100	0.997	0.980	0.980
Glu	1.504	1.440	1.510	0.567	0.750	0.370	1.149	0.990	0.740
Gly	0.510	0.560	0.570	0.657	0.920	0.750	1.860	1.640	1.560
His	1.003	1.220	1.000	0.863	1.080	0.870	0.970	0.680	0.950
Ile	1.003	0.970	1.080	1.799	1.450	1.600	0.240	0.510	0.470
Leu	1.236	1.300	1.210	1.261	1.020	1.300	0.670	0.580	0.590
Lys	1.172	1.230	1.160	0.721	0.770	0.740	1.302	0.960	1.010
Met	1.363	1.470	1.450	1.210	0.970	1.050	0.436	0.410	0.600
Phe	1.195	1.070	1.130	1.393	1.320	1.380	0.624	0.590	0.600
Pro	0.492	0.520	0.570	0.354	0.640	0.550	1.415	1.910	1.520
Ser	0.739	0.820	0.770	0.928	0.950	0.750	1.316	1.320	1.430
Thr	0.785	0.820	0.830	1.221	1.210	1.190	0.739	1.040	0.960
Trp	1.090	0.990	1.080	1.306	1.140	1.370	0.546	0.760	0.960
Tyr	0.787	0.720	0.690	1.266	1.250	1.470	0.795	1.050	1.140
Val	0.990	0.910	1.060	1.965	1.490	1.700	0.387	0.470	0.500

[a] Proposed by Deleage and Roux (DR, 1987), Levitt (1978b), and Chou and Fasman (CF, 1978).

generation methods, suffered from the same limitations: beta-strands were predicted poorly, the lengths predicted for helices and strands were too short.

A significant breakthrough in prediction accuracy was obtained with the third-generation prediction tools, using evolutionary information embedded in the multiple alignment of all proteins belonging to the same family of the target protein. Indeed, amino acid changes occurring during evolution are probably selected to preserve protein function. This evolutionary pressure also leads to the conservation of protein structure, which then becomes much more highly conserved than the sequence. When accurate multiple alignments of the members of these families can be determined, the evolutionary information present in the alignment can be used to determine critical residues for the folding and function of the protein. Furthermore, protein multiple alignment data implicitly also contain information about long-range residue interactions. The combination of additional information provided by protein family alignments with sophisticated computing techniques such as neural networks has lead to an accuracy level presently well above 75%, as shown at the recent CASP 3 and 4 meetings (Moult, Fidelis et al. 2001; Venclovas, Zemla et al. 2001). These predictors correctly determine the protein core structural features.

Table 6.15 lists some of the methods currently in use for secondary structure prediction. Among secondary structure predictors with more than 70% accuracy are Predator (Frishman and Argos 1997), PSIpred (Jones 1999) and PHD (Rost, Sander et al. 1994). Predator (Q3 = 75%) is based on the calculated propensities of all pos-

TABLE 6.15. Software Used in Protein Secondary Structure Prediction

Program	URL	References
AGADIR	www.embl-heidelberg.de/Services/serrano/serrano/agadir/agadir-start.html	Munoz and Serrano (1997)
CHOU-FASMAN	fasta.bioch.virginia.edu/fasta/chofas.htm	Chou and Fasman (1978a)
DSC	npsa-pbil.ibcp.fr/cgi-bin/npsa_automat.pl?page=/NPSA/ npsa_dsc.html	King and Sternberg (1996)
GORI	npsa-pbil.ibcp.fr/cgi-bin/npsa_automat.pl?page=npsa_gor.html	Garnier, Osguthorpe et al. (1978)
GORII	npsa-pbil.ibcp.fr/cgi-bin/npsa_automat.pl?page=npsa_gib.html	Gibrat, Garnier et al. (1987)
GORIV	npsa-pbil.ibcp.fr/cgi-bin/npsa_automat.pl?page=npsa_gor4.html	Garnier, Gibrat et al. (1996)
HNN	npsa-pbil.ibcp.fr/cgi-bin/npsa_automat.pl?page=npsa_nn.html	Guermeur, Geourjon et al. (1999)
Jpred	jura.ebi.ac.uk:8888	Cuff, Clamp et al. (1998); Cuff and Barton (2000)
nnPredict	www.cmpharm.ucsf.edu/~nomi/nnpredict.html	Kneller, Cohen et al. (1990)
PHD	cubic.bioc.columbia.edu/predictprotein	Rost, Sander et al. (1994)
Predator	www.embl-heidelberg.de/cgi/predator_serv.pl	Frishman and Argos (1997)
Prof	www.aber.ac.uk/~phiwww/prof	Ouali and King (2000)
PSA	bmerc-www.bu.edu/psa/request.htm	White, Stultz et al. (1994)
PSIpred	insulin.brunel.ac.uk/psipred	Jones (1999)
SOPM	npsa-pbil.ibcp.fr/cgi-bin/npsa_automat.pl?page=npsa_sopm.html	Geourjon and Deleage (1994)
SOPMA	npsa-pbil.ibcp.fr/cgi-bin/npsa_automat.pl?page=npsa_sopma.html	Geourjon and Deleage (1995)

sible 400 amino acid pairs to interact inside an alpha-helix or on one of three types of beta-bridges. It also uses propensities for the alpha-helix, beta-strand, and coil structure states derived from a nearest-neighbor approach (Salamov and Solovyev 1995) and incorporates nonlocal interaction statistics. PSIpred (Q3 = 76.1 – 78.3%) uses a two-stage neural network based on the position-specific scoring matches generated by PSI-BLAST (Altschul, Madden et al. 1997). After automatically generating its own multiple alignment with the sequence submitted, PHD (Q3 = 72.2) uses profile-based neural networks trained previously on proteins of known structure. Its output assigns to each residue the relative probability that it belongs to any of the three secondary structure states (alpha-helix, beta-strand, and coil), thus allowing the user to identify more reliably predicted structural elements.

Combining predictions from several different methods may provide a further increase in accuracy. The Jpred server (Cuff, Clamp et al. 1998; Cuff and Barton 2000) returns a consensus prediction obtained from six secondary structure prediction algorithms, exploiting evolutionary information from multiple sequences. Figure 6.42 shows the secondary structure prediction obtained by Jpred and PSIpred servers on spinach ferredoxin (PDB entry 1A70), both complying closely to the known structure derived from atomic coordinates. Several secondary structure predictors, such as PHD and Jpred, may additionally predict solvent accessibility for each residue. Solvent accessibility can be described in several ways (Rost and Sander 1994). The simplest is a two-state description distinguishing between residues that are buried (relative accessibility <16%) and exposed (>16%). Jpred predicts solvent accessibility at three thresholds, 25%, 5%, and 0%, with the latter value given to those residues that are likely to be deeply buried (see Fig. 6.42). Methods developed to predict solvent accessibility are based on the observation that this property is conserved evolutionarily at each position within sequence families (Rost and Sander 1994). This property has been used to develop methods for predicting accessibility using multiple alignment information (Rost and Sander 1994; Wako and Blundell 1994) that provide more than 75% prediction accuracy.

6.10.5   Prediction of Coiled-Coil and Helix-Turn-Helix Structures

A specific type of secondary structure element is represented by *coiled coils*, a ubiquitous protein motif that is often used to control oligomerization. It is found in many types of proteins, including transcription factors, where it generally assumes a leucine-zipper structure (Leu-X6-Leu-X6-Leu-X6-Leu-X6-Leu, where X is any amino acid residue), certain tRNA synthetases, G protein beta subunits, and so on. Coiled coils involve two to five alpha-helices, from the same or different proteins, wound around each other in a highly organized manner. Most coiled-coil sequences contain heptad repeats—seven residues, denoted, abcdefg—in which the core residues a and d are generally hydrophobic. The coiled coil is formed by component helices that twist to coil around each other to bury their hydrophobic seams. Helices generally run in the same direction, although antiparallel orientation may also occur.

Figure 6.42. Protein secondary structure prediction generated by Jpred on the spinach ferredoxin (PDB entry 1A70). Jpred shows also on the bottom the residue solvent accessibility at three different thresholds (0, 5 and 25%).

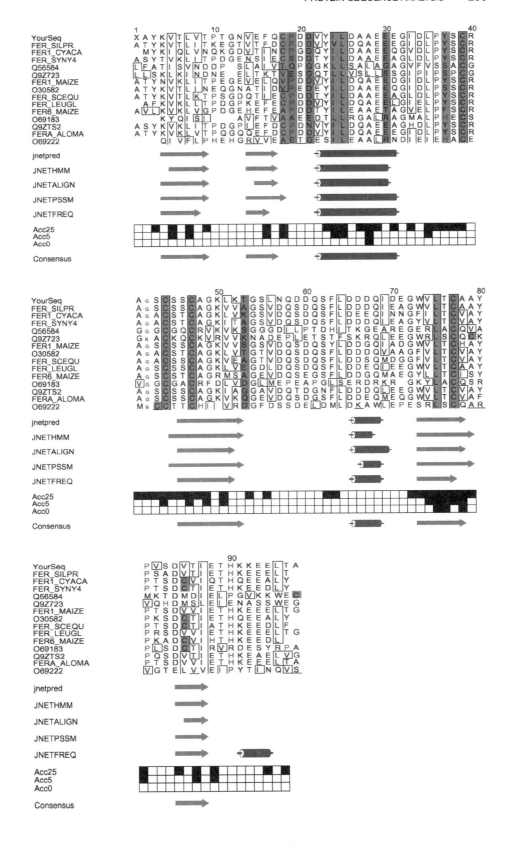

A number of programs are available that predict coiled coils in protein sequences, such as Coils (Lupas 1997), ParCoil (Berger, Wilson et al. 1995), and Multicoil (Wolf, Kim et al. 1997) (see the URLs in the Appendix). To identify coiled-coil motifs in protein structures in PDB (Berman, Westbrook et al. 2000), the program Socket (Walshaw and Woolfson 2001) has been developed, whose results may be useful for training coiled-coil predictors and to gain insight into new coiled-coil design rules.

The *helix-turn-helix* (HTH) *motif*, originally identified as the DNA-binding domain of phage repressors, is indicative of DNA-binding domains widely distributed in nature and often associated with gene regulation. The HTH minimal motif, made of about 20 amino acids, consists of two short helices connected by a tight turn and forming an angle of 90 to 120°. Among programs able to predict these motifs are HTHscan (GCG package), which uses a log-odds position weight matrix derived from three different families of HTH proteins (lysR, araC, and homeobox) and HelixTurnHelix (EMBOSS package; see the URL in the Appendix) that uses the method of Dodd and Egan (1990), also based on a scoring matrix derived from a HTH reference set.

6.10.6 Prediction of Protein Tertiary Structure

Determination of the tertiary structure of a protein is a crucial step in determining its function and mechanism of action. Despite the general improvement in x-ray crystallography the expression, purification, and crystallization of the target protein is still a severe rate-limiting step in experimental structure determination. Furthermore, NMR techniques, which avoid some of the limitations noted above, are still limited by protein size. For this reason, the gap between known sequences (>600,000 in the SWISSALL database on May 2001) and the number of known structures (about 15,000 in the PDB database on May 2001) is increasing rapidly, and its increase will speed up further with large-scale sequencing projects that are disclosing a huge number of genes and hence of new proteins.

The most successful tool for protein three-dimensional structure prediction is *homology modeling*, which allows us to build an approximate three-dimensional model for a sequence of unknown structure (target) if it has significant similarity to a protein of known structure (template). The basic assumption of homology modeling is that target and template proteins have identical backbones, whereas side-chain conformation could be predicted by energy-minimization molecular dynamics techniques. The accuracy of homology modeling is proportional to the similarity between the target and the template. When the sequence identity is above 90%, homology modeling is as accurate as experimental determination of protein structure. For lower levels of pairwise identity, homology modeling is able, at most, to produce an overall picture of the structure, which can be viewed as a ribbon plot (see the structure in Fig. 6.17). However, as structure is more conserved than sequence, even if a lower similarity between target and template reduces the overall accuracy of the prediction, the assignment of the target protein fold to a known class is extremely informative for predicting its function and activity. Again, the evolutionary information derived from the multialignment of homologous proteins can greatly improve three-dimensional prediction because the pattern observed for amino acid substitutions that occurred during evolution has probably preserved protein structure and function.

Successful homology modeling for proteins very distantly related to structure templates (<25% pairwise sequence identity) is hampered by three main obstacles: (1) detection of the remote homolog, (2) correct alignment between target and template, and (3) probable divergence between the three-dimensional structures of very distantly related proteins. In this case the possibility exists of using *threading methods*, whose basic idea is to "thread" the sequence of the target protein into the backbone of the template, to evaluate the fitness of the sequence for the structure (Sippl and Flockner 1996). However, the overall accuracy of such predictions is rather limited (Lemer, Rooman et al. 1995) since a correct or partially correct prediction is expected for less than 10% of the proteins analyzed.

Several programs are available that perform homology modeling predictions, such as Modeller (Sali and Blundell 1993), What-If (Vriend 1990), Insight II and SwissModel (Guex and Peitsch 1997) (see the URLs in the Appendix). Figure 6.43 shows a homology modeling prediction carried out on guinea pig p53 (SwissProt entry Q9WUR6), where two different template sets are used: 1TSR (chain A) for region 87–292 and 1A1U (chain A) for region 312–363 of the target protein, corresponding to the tetramerization domain and DNA-binding domain of p53 protein, respectively. Indeed, it is quite common that proteins contain multiple domains associated with different specific functions. The result of SwissModel analysis, which includes the atomic model of the submitted target protein, is sent by e-mail and can be visualized by any of several programs, such as SwissPDBviewer (Guex and Peitsch 1997), Rasmol (Sayle and Milner-White 1995; Bernstein 2000), and Molscript (Esnouf 1997), which also allow us to manipulate, analyze, and compare protein structures.

In the absence of known structures with significant identity to the protein of interest, one should apply procedures based on fold recognition (threading). The performances of these methods is evaluated periodically during Critical Assessment of Structure Prediction (CASP) meetings, where automatic prediction methods attempt to build models for newly solved structures not yet published. Although important progress has been made during the past few years, these methods are still not sufficiently accurate to be used reliably on a routine basis: for example, in the context of large-scale genome annotation. According to CASP competitions, the best-performing fold recognition methods are GeneThreader and 3D-PSSM (see the URLs in the Appendix). These servers return putative protein folds for a target sequence based on the notion that a sequence can be scored against different folds and that the score can be optimized.

An alternative to this procedure is *ab initio prediction*. Several methods have been proposed so far that are based on statistics- or physics-based potentials (Osguthorpe 2000). At CASP4, the Rosetta method was outperforming other methods, being consistent and sufficiently accurate in determining the backbone conformation of at least 25 targets. The method is based on alignment of the target sequence with all possible three- and nine-residue segments of a fragment library from the protein structure database (see the URL in the Appendix). A Monte Carlo procedure is then used to search the conformational space. The method correctly folds different structure types of chains 150 to 250 residues long with a root-mean-square deviation ranging from 0.3 to 1 nm. Although it is still far from being fully satisfactory, it is presently the best-performing method in its category. In conclusion, after four decades of research, it is still not possible to predict protein structure from sequence, but the increased data deposited in databanks and the

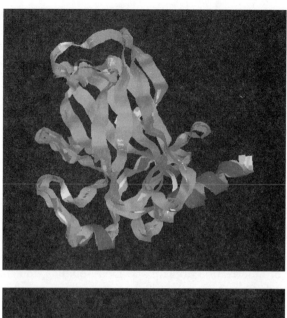

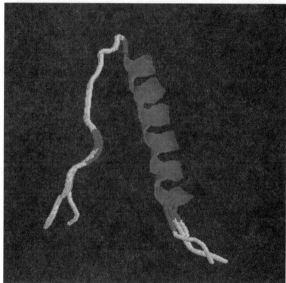

Figure 6.43. Homology modeling prediction carried out by SwissModel on the guinea pig p53 protein (SwissProt entry Q9WUR6), where two structures have been generated corresponding to two different PDB templates: 1TSR for region 87–292 and 1A1U for region 312–363. [From Guex and Peitsch (1997).]

refinement of prediction tools make it possible today to derive a lot of useful information for experiment design.

6.10.7   Protein Fold Recognition and Classification

Several research groups have tried to classify protein sequences into families and superfamilies. In particular, classification based on the similarity of specific protein modules at the level of the three-dimensional structure may make possible recognition of a common three-dimensional fold in two or more proteins without apparent sequence similarity, thus suggesting a shared functional activity. The SCOP (Structural Classification of Protein) database (Lo Conte, Ailey et al. 2000; see the URL in the Appendix) aims to providing a detailed and comprehensive description of the structural and evolutionary relationships between proteins whose structures have been determined experimentally.

The CATH database (Pearl, Todd et al. 2000; see the URL in the Appendix) provides a new hierarchical classification of protein domain structures using four major levels: Class (C), which is derived from secondary structure content; Architecture (A), which describes the gross orientation of secondary structures independent of connections; Topology (T), which clusters structures according to their topological connections; and Homologous superfamily (H), which groups proteins with highly similar structure and function. For example, the CATH classification for pig myoglobin (PDB ID: 1MWD) is given by four digits, corresponding to the four CATH levels: 1.10.490.10 (i.e., Class = 1, mainly alpha protein; Architecture = 10, nonbundle protein; Topology = 490, globinlike protein; Homologous superfamily = 10).

The Impala software (Schaffer, Wolf et al. 1999) can be used to compare a single query sequence with a database of PSI-BLAST (see Sec. 6.6)-generated position-specific score matrices (PSSMs) for the structural domains classified in the CATH database. In this way, probable structural and functional assignments may be inferred for regions of the query sequence that significantly match any of the PSSMs stored.

A comparison of three-dimensional structures may reveal biologically interesting similarities which are not detectable through sequence comparison. To identify the structural neighbors of a known protein structure, the VAST (Vector Alignment Search Tool) algorithm (Hogue, Ohkawa et al. 1996) or the DALI server (Holm and Sander 1998) can be used. Precomputed structure neighbors for entries already present in PDB have been collected in the FSSP database (Holm and Sander 1998) of structurally aligned protein folds and families.

6.10.8 Comparative Evolutionary Genomic Tools for Predicting Protein Function

The function of a protein is basically inferred through biochemical or molecular biology studies providing information on its structure and biological activity. During recent years several computational methods have been developed for predicting the functional role of novel proteins, based essentially on observation of global or local similarities with known proteins and/or modules or on the statistical analysis of their sequences. The rapid accumulation of new genes and proteins from genomic projects has spurred the development of novel computational methods relying on properties shared by functionally related proteins other than sequence or structural similarity.

The observation that some pairs of proteins, which are not homologous to one another, have homologs in other organisms that are fused into a single protein chain suggests their functional relationship. The fusion protein is named the *Rosetta Stone protein* for its ability to reveal the functional relationship between the component peptides (Fig. 6.44*a*) (Marcotte 2000). Using this method, Marcotte (2000) and colleagues found a total of 6809 putative protein interactions in *E. coli* and 45,502 in yeast. Many of the predicted interactions were confirmed functionally, although several false predictions also occurred due to *promiscuous domains* (e.g., SH2 and SH3 domains) present in a large number of otherwise different proteins. To reduce errors in predicting protein–protein interactions, the promiscuous domains can be identified and filtered out for domain fusion analysis.

The analysis of orthologous genes across genomes allows the construction of *phylogenetic profiles* (Fig. 6.44*b*) (Pellegrini, Marcotte et al. 1999). The basic assumption of this method is that proteins that operate in the same biochemical pathway in a cell are generally inherited in a correlated way. In the costruction of a phylogenetic profile for each protein, a binary vector is constructed that records its presence (1) or absence (0) across several genomes. Protein groups with similar inheritance patterns (profiles) are predicted to be linked functionally and may take part in the same structural complex or biochemical pathway. Analogously, two proteins can be inferred to be related functionally if their genes are found repeatedly as neighbors on genomes of different organisms. The conservation of relative gene position (Fig. 6.44*c*) probably derives from the early evolution of operon organization of genes encoding for proteins involved in closely related functions—still observed in microbial genomes. The application of this method to eukaryotic genomes, which usually do not have operons, is questionable and should then be restricted to the analysis of bacterial genomes.

All the methods described above require the accurate assessment of evolutionary orthology between genes analyzed from different organisms. Unfortunately, the definition of sets of orthologous genes has a remarkable degree of uncertainty, due to gene loss, gene duplication, or lateral gene transfer events (see Sec. 8.3), which may be sources of error and in any case, introduce a remarkable level of background noise. As a result of the exceptional development of experimental techniques to measure the concentration of cellular mRNA, due primarily to the explosion of DNA chip technology, it is now possible, simultaneously, to determine in a single experiment the expression level of thousands of genes. The result of such experiments is that to each gene an expression vector can be associated that reports the relative mRNA concentration of that gene in different physiological conditions and at different times. The application of clustering techniques (see Sec. 4.2.6) may allow the recognition of groups of genes having a similar expression profile and thus possibly being related functionally. Use of this method permitted the identification of genes co-expressed in the various phases of the yeast cellular cycle (Spellman, Sherlock et al. 1998) and of those up- or down-regulated from a specific stimulus (e.g., drug administration) or by knocking out a gene (Wilson, DeRisi et al. 1999; Hughes, Marton et al. 2000).

Figure 6.44. (*a*) Rosetta Stone method. When two separate proteins in one genome exist as single fused proteins in another genome, the two component proteins are functionally linked. (*b*) Phylogenetic profile. The five genomes (G1 to G5) code for proteins p1 to p8. The phylogenetic profile records the presence (1) or absence (0) of the proteins across several genomes. Identical profiles are clustered and establish a functional link between corresponding proteins. (*c*) Gene synteny. Chromosomal proximity of genes coding for proteins p1 and p2 conserved across several genomes (G1–G5) suggests their functional relatedness.

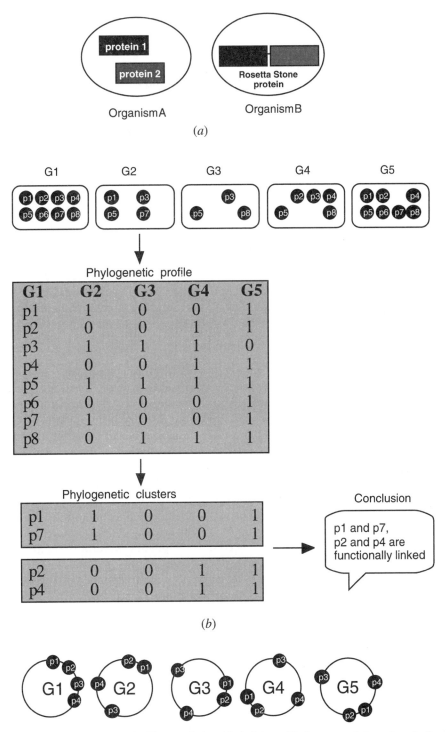

(a)

Phylogenetic profile

Phylogenetic clusters

Conclusion

p1 and p7,
p2 and p4 are
functionally linked

(b)

Conclusion: A statistically significant relationship is inferred between proteins p1 and p2

(c)

6.11   EVOLUTIONARY AND PHYLOGENETIC ANALYSIS

Molecular evolution encompasses two main areas of study: (1) the study of evolutionary dynamics of genetic material and its products, aimed at shedding light on rates and patterns of changes occurring in DNA and proteins; and (2) reconstruction of the evolutionary history of genes and organisms (molecular phylogeny). In the last few years, remarkable progress has been made in these two areas, which are strongly interconnected, due to great improvement in mathematical models used to study molecular evolution and phylogeny, but principally because of the exceptional growth of genomic sequence data.

The major steps in evolutionary analysis of homologous biosequences (DNA and proteins) are their multiple alignment, determination of the substitution pattern that occurred since their divergence, and finally, the building of a phylogenetic tree depicting sequence relationships. The multiple alignment of homologous sequences is the necessary input data for any kind of evolutionary analysis, and thus its accuracy is a prerequisite for any phylogenetic inference. The methodological approach to be used for data analysis should fit as closely as possible the actual evolutionary process the sequences being analyzed have gone through, without imposing unverified a priori assumptions that could lead systematically to erroneous results.

6.11.1   Estimating Genetic Distances between Homologous Sequences

Accurate determination of the number of substitutions occurring between two homologous sequences after their divergence is necessary to estimate the substitution rate as well as to reconstruct reliable phylogenetic trees. The number of differences observed is the simplest measure of the distance between two sequences. However, this measure underestimates the actual number of substitutions, and the extent of this underestimate is correlated with the amount of divergence. Multiple substitutions or backmutations may occur on the same site, thus implying the necessity of using stochastic models to provide accurate estimates of the number of substitutions between two homologous sequences. In this context, accurate modeling of the process of evolutionary change is practically feasible only in the case of DNA sequences, which are made up of only four characters, whereas for protein sequences, which use a much more complex alphabet, several assumptions are needed (see below).

A number of stochastic models to estimate genetic distances between DNA sequences have been proposed, some of which are illustrated in Table 6.16. The major drawback of most of these methods is that they are based on a number of a priori assumptions which could not be fulfilled by real data. Some of the most common a priori assumptions are (1) that all nucleotide sites change independently, (2) that all sites vary equiprobably, (3) that the substitution rate is constant over time in different evolutionary lineages, (4) that the base composition is at equilibrium ($f_i = \frac{1}{4}$, $i = A, C, G, T$), and (5) that all sites are equally variable. In general, the simpler a model, the greater the number of assumptions. In the simplest model, that of Jukes and Cantor (1969), all five assumptions above are accounted for, and only one parameter, the homologous substitution rate α, is inferred from the data.

To make models more advanced, some more parameters were introduced, including (1) base composition parameters, (2) base exchangeability parameters, and

(3) rate heterogeneity parameters. The model proposed by Kimura (1980) allows for different rates between transitions (purine–purine or pyrimidine–pyrimidine changes) and transversions (purine–pyrimidine changes). The method of Tamura (1992), in addition to the transition/transversion bias, also takes into account the G + C content (parameter θ) of compared sequences. The Felsenstein (1981) model extends the Jukes–Cantor model by taking into account the equilibrium frequencies of nucleotides derived simply from the base composition of the DNA sequences under study. The HKY model of Hasegawa, Kishino et al. (1985) incorporates the transition/transversion parameter and the base composition of sequences being compared.

The most general model, first proposed by Lanave, Preparata et al. (1984) and called the *general time reversible* (GTR or REV) model by Yang (1994), permits a different rate for each of the six types of base change: two transitions, A $\leftrightarrow$ G and C $\leftrightarrow$ T, and four transversions, A $\leftrightarrow$ C, A $\leftrightarrow$ T, C $\leftrightarrow$ G, and G $\leftrightarrow$ T. In all models listed in Table 6.16, the reversibility of the rate matrix is assumed (e.g., A $\leftrightarrow$ G is equivalent to G $\leftrightarrow$ A). A very basic prerequisite for the applicability of all the models above is that sequences under analysis should have, within statistical fluctuations, the same base composition at sites being compared. This condition, known as *stationarity* (Saccone, Lanave et al. 1990), should be checked carefully in advance because the inclusion of compositionally divergent sequences in the data set to be analyzed could lead to erroneous results. Stationarity condition can be verified by a standard chi-square test checking compositional homogeneity among all aligned sequences at sites analyzed. This test is implemented in Markov (Saccone, Lanave et al. 1990) and Tree-Puzzle (Strimmer and Von Haeseler 1996) programs (see the URLs in the Appendix).

An additional parameter to be considered in modeling the sequence evolutionary process is rate heterogeneity among sites. In the case of DNA sequences coding for proteins, the evolutionary analysis could be carried out separately on the three codon positions, where different constraints operate at the third codon positions, whose changes are mostly synonymous (i.e., do not change the amino acid coded), and which evolve much faster than the first and second codon positions, whose changes are almost always nonsynonymous and thus can also be analyzed together. In general, the most widespread approach for modeling rate heterogeneity among sites is the use of a gamma distribution, in combination with the parameters described above.

To preserve functional activity, some sites cannot change, thus the evolutionary model should incorporate an additional parameter that establishes the proportion of invariable sites. It has been shown that the estimates of genetic distances inferred are strongly dependent on the a priori assumptions superimposed in the model. In general, the larger the number of a priori assumptions not fulfilled by actual data, the more inaccurate the results. Another important aspect to be taken into account is the estimate of the extent of statistical fluctuations. These have to be considered in order to assess the statistical significance of inferred estimates. However, as the extent of fluctuations is usually correlated with the number of parameters, the use of simpler models may misleadingly produce more precise results than more general models. Indeed, statistical tests (see below) almost invariably show that more general models provide a significantly better description of the evolutionary process.

TABLE 6.16. Transition Probability Matrices Assumed in Different Models of DNA Substitutions[a]

Model	Transition Probability Matrix	Stationary Nucleotide Frequencies $f_i^\infty, i = A, C, G, T$	Number of Parameters
Jukes and Cantor (1969)	$\begin{matrix} p_{11} & \alpha & \alpha & \alpha \\ \alpha & p_{22} & \alpha & \alpha \\ \alpha & \alpha & p_{33} & \alpha \\ \alpha & \alpha & \alpha & p_{44} \end{matrix}$	$\left[\dfrac{1}{4}, \dfrac{1}{4}, \dfrac{1}{4}, \dfrac{1}{4}\right]$	1
Kimura (1980)	$\begin{matrix} p_{11} & \beta & \alpha & \beta \\ \beta & p_{22} & \beta & \alpha \\ \alpha & \beta & p_{33} & \beta \\ \beta & \alpha & \beta & p_{44} \end{matrix}$	$\left[\dfrac{1}{4}, \dfrac{1}{4}, \dfrac{1}{4}, \dfrac{1}{4}\right]$	2
Tamura (1992)	$\begin{matrix} p_{11} & \theta\beta & \theta\alpha & (1-\theta)\beta \\ (1-\theta)\beta & p_{22} & \theta\beta & (1-\theta)\alpha \\ (1-\theta)\alpha & \theta\beta & p_{33} & (1-\theta)\beta \\ (1-\theta)\beta & \theta\alpha & \theta\beta & p_{44} \end{matrix}$	$\left[\dfrac{1-\theta}{2}, \dfrac{\theta}{2}, \dfrac{\theta}{2}, \dfrac{1-\theta}{2}\right]$	3
Felsenstein (1981)	$\begin{matrix} p_{11} & \pi_C\alpha & \pi_G\alpha & \pi_T\alpha \\ \pi_A\alpha & p_{22} & \pi_G\alpha & \pi_T\alpha \\ \pi_A\alpha & \pi_C\alpha & p_{33} & \pi_T\alpha \\ \pi_A\alpha & \pi_C\alpha & \pi_G\alpha & p_{44} \end{matrix}$	$[\pi_A, \pi_C, \pi_G, \pi_T]$	4
Hasegawa, Kishino et al. (1985)	$\begin{matrix} p_{11} & \pi_C\beta & \pi_G\alpha & \pi_T\beta \\ \pi_A\beta & p_{22} & \pi_G\beta & \pi_T\alpha \\ \pi_A\alpha & \pi_C\beta & p_{33} & \pi_T\beta \\ \pi_A\beta & \pi_C\alpha & \pi_G\beta & p_{44} \end{matrix}$	$[\pi_A, \pi_C, \pi_G, \pi_T]$	5
Yang (1994); Saccone, Lanave al. (1990); Lanave, Preparata et al. (1984)	$\begin{matrix} p_{11} & \pi_C\beta_1 & \pi_G\alpha_1 & \pi_T\beta_2 \\ \pi_A\beta_1 & p_{22} & \pi_G\beta_3 & \pi_T\alpha_2 \\ \pi_A\alpha_1 & \pi_C\beta_3 & p_{33} & \pi_T\beta_4 \\ \pi_A\beta_2 & \pi_C\alpha_2 & \pi_G\beta_4 & p_{44} \end{matrix}$	$[\pi_A, \pi_C, \pi_G, \pi_T]$	9

[a] Labels for the four rows and the four columns are the nucleotides A, C, G, and T, respectively. For each matrix the nucleotide composition assumed at equilibrium and the number of parameters are also shown.

Several software packages to carry out evolutionary analyses have been developed that implement the methods described above to estimate genetic distances and to obtain phylogenetic reconstructions (see below). Among these, PHYLIP and PAUP* (see the URLs in the Appendix) certainly represent the most widely used multipurpose packages. A very comprehensive list of phylogeny programs can be found at the PHYLIP Web site, where for each program a short description is provided together with information on how to obtain or run the various software.

Figure 6.45 shows the pairwise genetic distances calculated between six primate ψη-globin sequences using PAUP* and Tree-Puzzle programs assuming, respectively, GTR and HKY models for DNA substitution. In the case of protein comparison, in contrast to DNA substitution models, the amino acid replacement transition probabilities are computed empirically simply by counting amino acid replacements observed in large sequence data sets. In this way, Dayhoff, Schwartz et al. (1978) have derived a relationship between observed (p) and actual pairwise distance (d) between homologous aligned protein sequences. Kimura (1983) produced an approximate formula to estimate d:

$$d_K = -\ln(1 - p - 0.2p^2)$$

that gives a good approximation for $p \leq 0.7$.

Updated matrices for amino acid replacement probabilities have been devised using larger protein data sets by Jones, Taylor et al. (1992; JTT matrix) or on specific collections of mitochondrial (Adachi and Hasegawa 1996) or chloroplast (Adachi, Waddell et al. 2000) encoded proteins. Figure 6.46 shows pairwise distances observed and calculated on a set of five homologous vertebrate beta-globins.

6.11.2 Molecular Phylogeny

A study of molecular phylogeny consists of reconstructing the evolutionary history of genes and organisms. The most commonly used visual representation of phylogenetic relationships are *phylogenetic trees* (Fig. 6.47), graphs composed of nodes and branches in which only one branch connects two adjacent nodes. The nodes represent the taxonomic units, and the branches define the relationships among the units in terms of descent and ancestry. The terminal nodes of the tree, corresponding to the actual compared homologous sequences, are called *operational taxonomic units* (OTUs). Ancestral taxonomic units are represented by internal nodes. The branching pattern of a tree is called the *topology* and branch length usually represents the number of changes (i.e., the genetic distance) occurring along that branch. When a tree is constructed from a set of orthologous genes from different species (gene tree), the resulting phylogeny is expected to reproduce the phylogenetic relationships among species (species tree). However, the gene tree does not necessarily coincide with the species tree, owing to several circumstances, such as erroneous assessment of orthology relationships or undetected lateral gene transfer events, or others.

In a phylogenetic tree, a *monophyletic group* or *clade* has a unique ancestral taxon and includes all its descendants; a *paraphyletic group* has a unique ancestor but does not include all descendant taxa; and a *polyphyletic group* has multiple evolutionary origins and does not originate from a common ancestor. Phylogenetic trees can be either rooted or unrooted. In a *rooted tree* there exists a particular node called the *root* which represents the common ancestor of all other nodes in the tree, from which a unique path leads to any other node. An *unrooted tree* specifies the relationships among the OTUs but does not define the evolutionary path from ascendant to descendant. An unrooted tree is also called an *additive tree*, in the sense that all pairwise distances should be equal to the sum of branches connecting the corresponding OTUs. A further constraint of an additive tree is *ultrametricity*. An *ultrametric tree* must satisfy the criterion that the distances between any two taxa

Uncorrected ("p") distance matrix

```
                    1         2         3         4         5         6
1 CHIMP            -
2 HUMAN           0.01422       -
3 GORILLA         0.01961   0.01471       -
4 ORANG           0.03627   0.03137   0.03725       -
5 MACAQUE         0.07696   0.07157   0.07549   0.07549       -
6 OWL MONKEY      0.10833   0.10490   0.10784   0.10784   0.12157       -
```

General time-reversible distance matrix

```
                    1         2         3         4         5         6
1 CHIMP            -
2 HUMAN           0.01440       -
3 GORILLA         0.01995   0.01490       -
4 ORANG           0.03744   0.03223   0.03846       -
5 MACAQUE         0.08204   0.07587   0.08028   0.08031       -
6 OWL MONKEY      0.11890   0.11474   0.11829   0.11817   0.13512       -
```

(*a*)

SEQUENCE ALIGNMENT
Input data: 6 sequences with 2040 nucleotide sites
Number of constant sites: 1678 (= 82.3% of all sites)
Number of site patterns: 88
Number of constant site patterns: 4 (= 4.5% of all site patterns)

SUBSTITUTION PROCESS
Model of substitution: HKY (Hasegawa et al. 1985)
Transition/transversion parameter (estimated from data set): 2.54 (S.E. 0.29)
Nucleotide frequencies (estimated from data set):

pi(A) = 29.5%
pi(C) = 19.1%
pi(G) = 23.7%
pi(T) = 27.7%

Expected transition/transversion ratio: 2.50
Expected pyrimidine transition/purine transition ratio: 0.76

SEQUENCE COMPOSITION (SEQUENCES IN INPUT ORDER)

```
               5% chi-square test   p-value
CHIMP              passed            99.19%
HUMAN              passed            99.92%
GORILLA            passed            95.69%
ORANG              passed            95.63%
MACAQUE            passed            97.86%
OWL_MONKEY         passed            96.98%
```

The chi-square tests compares the nucleotide composition of each sequence
to the frequency distribution assumed in the maximum likelihood model.

IDENTICAL SEQUENCES

The sequences in each of the following groups are all identical. To speed
up computation please remove all but one of each group from the data set.

 All sequences are unique.

MAXIMUM LIKELIHOOD DISTANCES

Maximum likelihood distances are computed using the selected model of
substitution and rate heterogeneity.

```
  6
CHIMP        0.00000   0.01439   0.01993   0.03748   0.08297   0.12035
HUMAN        0.01439   0.00000   0.01489   0.03228   0.07680   0.11614
GORILLA      0.01993   0.01489   0.00000   0.03851   0.08129   0.11974
ORANG        0.03748   0.03228   0.03851   0.00000   0.08142   0.12007
MACAQUE      0.08297   0.07680   0.08129   0.08142   0.00000   0.13727
OWL_MONKEY   0.12035   0.11614   0.11974   0.12007   0.13727   0.00000
```

Average distance (over all possible pairs of sequences): 0.07290

(*b*)

and their ancestor are the same. This implies a linear relationship between time and genetic distance (i.e., the molecular clock) and tree rooting so that all OTUs are equidistant from the root. In an additive tree for n taxa there are $2n - 3$ independent branches, reducing to $n - 1$ in an ultrametric tree.

For computer programs to read trees, these can be represented by a textual format with a series of nested parentheses, enclosing taxon names, separated by commas. This type of representation, called the *Newick format*, was adopted by the Society for the Study of Evolution in 1986. Branch lengths may also be included immediately after the descending group and separated by a colon (see the example in Fig. 6.49). The number of possible trees grows exponentially with the number of taxa. Given n OTUs, the number of different rooted (N_R) or unrooted (N_U) trees is given by

$$N_R = \frac{(2n-3)!}{2^{n-2}(n-2)!} \qquad N_U = \frac{(2n-5)!}{2^{n-3}(n-3)!}$$

For example, for six OTUs, as in the tree in Fig. 6.47, there are 954 rooted and 105 unrooted trees.

Three types of methods are currently used for phylogenetic reconstruction: maximum parsimony, distance matrix, and maximum likelihood. In *maximum parsimony* the character states (i.e., the nucleotides or the amino acids at a specific site of the multiple alignment) are considered and the tree(s) requiring the minimum number of changes is (are) considered as optimal. Maximum parsimony is a fully deterministic method using only the fraction of sites in the multiple alignment which can be considered as *informative*. A site is defined as informative if at least two different character states are represented, each one occurring at least twice (Fig. 6.48). Determination of the most parsimonious tree, which is, by definition, an unrooted tree, through the exhaustive evaluation of all possible distinct trees, is quite a simple task when a small number of taxa are considered but becomes practically unfeasible for large-sequence data sets.

A suitable approximation that in any case guarantees determination of the maximum parsimony tree is the *branch-and-bound algorithm* (Hendy and Hendy 1982), which ignores entire classes of trees because they are certainly suboptimal compared to an already determined parsimonious tree. For very large data sets, even the branch and bound algorithm is computationally impracticable; approximation techniques have to be used, which, however, do not guarantee the optimal solution. In this case, a heuristic search is carried out whereby a quick initial approximation of the optimal tree is obtained using several methods, such as neighbor-joining (see below) or stepwise addition; branch swapping, a refinement procedure, follows. Stepwise addition, which consists of the addition of taxa, one by one, to a growing tree, is generally more accurate than neighbor joining for very large data sets.

Figure 6.45. Genetic distances calculated on six ψη-globin sequences (human, X02133; chimp, X02135; gorilla, X02134, orang, M18038; macaque, AF012469; owl monkey, X02142): (*a*) GTR distances (subs/site) observed and calculated by PAUP* software; (*b*) HKY distances calculated by Tree-Puzzle software. The chi-square test for compositional stationarity is also reported.

Distance Matrix

Uncorrected for Multiple Substitutions
Gap weighting is 0.000000

1	2	3	4	5			
0.00	20.41	31.97	34.69	55.78		mouse	1
	0.00	27.89	32.65	53.74		rabbit	2
		0.00	30.61	51.02		opossum	3
			0.00	49.66		chicken	4
				0.00		frog	5

(a)

Distance Matrix

Using the Kimura correction method
Gap weighting is 0.000000

1	2	3	4	5			
0.00	23.88	41.58	46.36	96.66		mouse	1
	0.00	34.88	42.75	90.33		rabbit	2
		0.00	39.28	82.52		opossum	3
			0.00	78.86		chicken	4
				0.00		frog	5

(b)

```
5
mouse      0.00000   0.23181   0.41478   0.45224   0.97955
rabbit     0.23181   0.00000   0.33645   0.39879   0.88573
opossum    0.41478   0.33645   0.00000   0.39417   0.88531
chicken    0.45224   0.39879   0.39417   0.00000   0.82014
frog       0.97955   0.88573   0.88531   0.82014   0.00000
```

(c)

Figure 6.46. Protein distance matrix (subs/site·100) calculated between pairs of beta-globins from five vertebrates (mouse, rabbit, opossum, chicken, and frog): (*a*) uncorrected distances; (*b*) distances corrected using the Kimura method (Kimura 1980) calculated with the DistMat program of the EMBOSS package (see the URL in the Appendix); (*c*) distances corrected using Dayhoff method (Dayhoff, Schwartz et al. 1978) calculated with ProtDist program of the PHYLIP package.

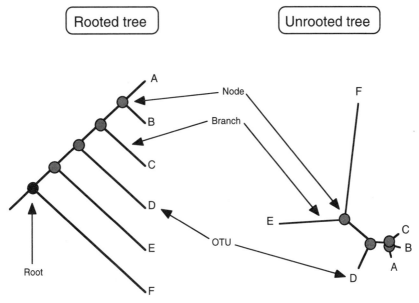

Figure 6.47. Graphical representation of rooted and unrooted phylogenetic trees, indicating nodes, branches, and operational taxonomic units (OTUs) corresponding to the extant analyzed sequences. The root is a special node representing the common ancestor for all the sequences considered.

Branch swapping involves the rearrangement of the branches on the initial tree to generate topologically related trees that are as good as or more parsimonious solutions.

The major drawback of maximum parsimony is that it lacks an explicit model of evolution and fails to detect parallel, convergent, reversed, or multiple changes, which become more and more important as sequence divergence increases. The impact of this limitation is much higher for DNA sequences, where the limited size of the alphabet makes homoplasic changes much more likely than in protein sequences. Furthermore, when in a tree there are both very short and very long branches, the *long-branch attraction phenomenon* (Felsenstein 1978) may occur, leading to an (erroneous) closer relationship between long branches. Another disadvantage of maximum parsimony is that multiple parsimonious solutions are not unique and multiple trees may result that are equally parsimonious, without any useful criterion to discriminate among them. The several pitfalls, illustrated above, make parsimony phylogenetic inferences misleading for rather divergent sequences, particularly in the case of rate heterogeneity among different lineages.

Among *distance-matrix methods*, the most commonly used are the unweighted pair group method with arithmetic mean (UPGMA), the neighbor-joining (NJ), and the minimum evolution (ME) methods. The *UPGMA method* (Sokal and Michener 1958) uses a sequential clustering algorithm where the two taxa, say i and j, having the smallest distance are clustered together first to form a new composite OTU that replaces the ith and jth OTUs. A new distance matrix is then constructed where the

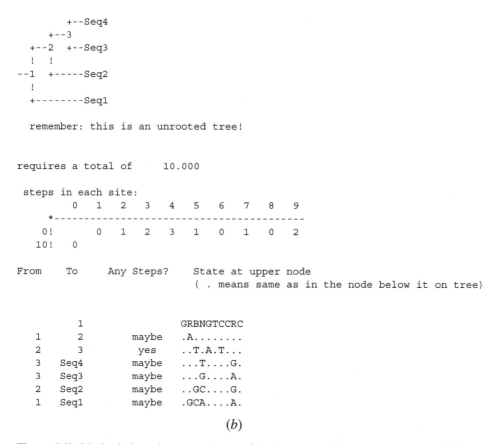

Figure 6.48. Method of maximum parsimony: (*a*) alignment of four sequences, each 10 characters long, where informative sites are highlighted; (*b*) maximum parsimony tree of the aligned sequences in (*a*) determined with the DNAPARS program of the PHYLIP package. The minimum number of changes required at each site and the probable ancestral sequence at each node is also shown in the output.

distance of the composite OTU from other OTUs is calculated as the average of all possible pairwise distances of each of the composite OTU elements. The same procedure is repeated until all OTUs have been clustered together. UPGMA assumes that all taxa are extant and that the rates of change are equal in all lineages. The latter assumption, better known as the *molecular clock hypothesis*, was first put forward by Zuckerkandl and Pauling (1965a). The molecular clock hypothesis is a general evolutionary rule which if actually fulfilled, enables us to calculate times of divergence between species and/or genes provided that a calibration time for a pair of these is known from other sources (e.g., paleontology). Given that under the assumption of the molecular clock the genetic distances are directly proportional to the evolutionary time, the UPGMA method determines an ultrametric rooted tree where all the tips are equidistant from the root of the tree.

The *neighbor-joining* (NJ) *method* (Saitou and Nei 1987) builds an unrooted tree starting from a fully unresolved "star" tree and then inserting branches representing the distance between pairs of neighbor taxa and all remaining taxa until all internal branches have been determined. Branch lengths are calculated to adjust for differences in the evolutionary rates of each taxon.

Unlike UPGMA and NJ, which build phylogenetic trees using a clustering algorithm, the minimum evolution (ME) *method* uses an optimality criterion whereby the overall length of the tree is minimized. The length L of a tree of n sequences having $2n - 3$ branches each of length b_i is given by

$$L_{\text{tree}} = \sum_{i-1}^{2n-3} b_i$$

The optimal minimum evolution solution must satisfy the two constraints that all branch lengths are nonnegative and that the distances computed on the tree are always equal to or greater than the distances observed.

Figure 6.49 shows the phylogenetic tree calculated on six primate $\psi\eta$-globin nucleotide sequences using UPGMA and ME algorithms. It is striking to note that in the UPGMA tree, it may happen that distance between two OTUs is smaller that the one observed, and this is quite unrealistic because it would imply that less evolutionary change took place than we actually observed. For example, as compared to the observed distance of 0.1961 between gorilla and chimp (see Figure 6.45), the corresponding UPGMA tree distance is only 0.01740.

The considerations above would suggest caution in the use of UPGMA for tree reconstruction. This method is better suited to cases where the hypothesis of the molecular clock can be proved (or cannot be rejected) by a suitable statistical test (see below). The most statistically robust approach to achieving phylogenetic reconstructions is the method of *maximum likelihood* (ML), which tries to infer the phylogenetic tree that is most consistent with the sequence data observed. To calculate the likelihood of an ML tree, an explicit stochastic model of sequence evolution is assumed such as those described in Sec. 6.11.1. The ML approach permits us not only to infer phylogenetic trees but also to derive evolutionary model parameters and to perform statistical tests to evaluate competing phylogenetic hypotheses. However, ML needs to examine each possible tree topology and to estimate the highest likelihood parameters (e.g., the nucleotide substitution rate

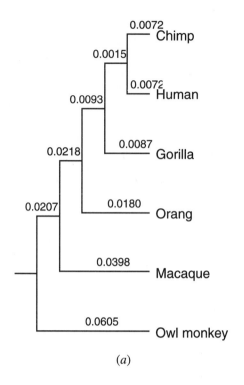

(a)

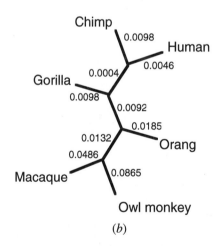

(b)

```
((((CHIMP:0.009774,HUMAN:0.004625):0.000443,GORILLA:0.009785):
0.009222,ORANG:0.018465):0.013215,MACAQUE:0.048612,
OWL_MONKEY:0.086509);
```

(c)

Figure 6.49. Phylogenetic tree of six primate ψη-globin sequences (see the legend of Fig. 6.45) calculated by the UPGMA (a) and ME (b) algorithms on distance matrices determined assuming the GTR model for DNA substitutions. (c) Newick format for the UPGMA/ME trees.

matrix, the branch lengths, or the proportion of invariable sites), which makes this method computationally very intensive and impracticable for a large number of sequences.

The quartet puzzling (QP) technique (Strimmer and Von Haeseler 1996) approximates the determination of ML trees on large-sequence data sets. According to this method, first all possible four-taxa ML trees are calculated from the sequences analyzed and then a combinatorial algorithm is applied to reflect inferred quartet phylogenies. Since the QP method is very sensitive to long branch attraction, it is usually less efficient that the traditional phylogenetic reconstruction approach based on general ML or pairwise evolutionary distances (Ranwez and Gascuel 2001).

More recently, a novel approach based on Bayesian inference has been proposed which is able to find the tree with the maximum posterior probability (Huelsenbeck, Ronquist et al. 2001). This method, implemented in MrBayes software (see the URL in the Appendix), proved to be highly accurate and allowed analysis of rather large data sets not tractable with the usual ML techniques.

Rooting of Unrooted Trees   Most methods for phylogeny reconstruction yield unrooted trees, and just looking at the tree topology, it is not possible to recognize the branch on which to place the root. Indeed, tree rooting is essential to establish which of the OTUs branched off before all the others. The simplest way to place a root in an unrooted tree is to add an *outgroup* to the data set. An outgroup is an OTU for which external information (e.g., paleontology) is available that indicates that the outgroup branched off before all other taxa represented in the data set. For example, the rooting of eutherian mammals (placental) can be provided by methaterian mammals (marsupials) or prototherian mammals (monotremes). A suitable outgroup should not be very distantly related to the taxa under analysis. This means that it is much better to root mammals with marsupials than with fish or amphibians. Indeed, in the latter case, serious topological errors may occur due to the degradation of the phylogenetic signal when comparing very distant species as a consequence of the saturation of nucleotide substitutions. Furthermore, the use of more than one outgroup generally improves the accuracy of the inferred tree topology.

An alternative way to introduce an outgroup is to use two sets of duplicated paralogous genes both existing in all OTUs under analysis. In this way the genes of the first orthologous gene group may outgroup the genes of the second. This rooting system has been used to identify the first branching event in the "tree of life" where the root can be placed on each of the three branches leading to bacteria, archaebacteria, and eukaryotes, respectively. For example, Iwabe, Kuma et al. (1989), using genes coding for elongation factors, EF-Tu and EF-G, present in all prokaryotes and eukaryotes, placed the root of the tree of life in the branch leading to Bacteria, with Eukarya and Archea appearing as sister groups. When using a single outgroup, the root can be placed somewhere in the branch leading to the outgroup. Otherwise, if multiple outgroups are used, the root has to be placed on the branch joining the outgroup with ingroup species (Fig. 6.50).

In the absence of a good outgroup, the root can be placed approximately at the midpoint of the longest branch between two OTUs. This method of rooting, called *midpoint rooting*, is fairly accurate if the rates over all the branches of the tree are approximately the same and the branch length between the real outgroup and the other taxa is not too small.

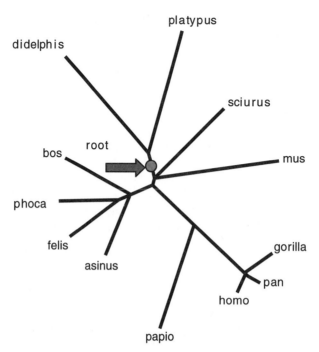

Figure 6.50. Outgroup rooting of a phylogenetic tree. The tree shown depicts phylogenetic relationships between mammals as inferred by analysis of the complete mitochondrial genome [see the sequence data in Reyes, Gissi et al. (2000)]. The root can be placed in the branch joining the two outgroups (marsupials didelphis and platypus) with the 10 ingroup placental mammalian species.

Statistical Test to Evaluate the Accuracy of Phylogenetic Reconstructions

There are two major prerequisites for the accuracy of any phylogenetic inference: the suitability of molecular data for the purpose of the analysis, and the fitting between the model used and the molecular data considered. Suitability of input sequence data not only depends on the type of data considered (i.e., truly orthologous genes) and on the quality of the multiple alignment, but also on taxon sampling and sequence length. For example, the reconstruction of phylogenetic relationships among mammalian orders requires each order to be represented by an adequate number of taxa (at least two), and the sequences to be analyzed are long enough to provide an acceptable degree of resolution in depicting phylogenetic relationships. Indeed, phylogenetic inferences provided by stochastic models have an inherent degree of uncertainty (i.e., the extent of statistical fluctuations, proportionally correlated to the number of parameters adopted by the model and to the length of analyzed sequences). In other words, if the sequences are too short, it is very likely that many phylogenetic relationships remain unresolved, thus producing trees with *polytomies*, nodes from which more than two branches depart. For this reason, a commonly used approach, first proposed by Saccone, Lanave et al. (1990), is to represent a single OTU by multiple genes concatenated to form a single sequence.

TABLE 6.17. Statistical Test of Competing Model of Nucleotide Substitution for a Set of Six Primate ψη-Globin[a]

Model	Number of Parameters	−ln(likelihood)	2Δ[b]	Prob(χ²)[c]
JK	1	4909.0	297.2	$<10^{-10}$
K2P	2	4797.7	74.6	$<10^{-10}$
F81	4	4879.0	237.2	$<10^{-10}$
HKY	5	4765.7	10.6	0.10
GTR	9	4760.7	0.6	0.74
GTR + I[d]	10	4760.4	0.0	1.00
GTR + I + Γ[e]	11	4760.4	best	—

[a] See the legend of Fig. 6.45. JK, Jukes and Cantor (1969); K2P, Kimura (1980); F81, Felsenstein (1981); HKY, Hasegawa, Kishino et al. (1985); GTR, Lanave, Preparata et al. (1984) and Saccone, Lanave et al. (1990); I, a proportion of sites is assumed to be invariant; Γ, rate heterogeneity is assumed following a discrete four-category gamma distribution.
[b] The twofold difference between the ln(likelihood) of the given model and that of the optimal one.
[c] Assuming a number of degrees of freedom equal to the difference between the number of parameters of models being compared.
[d] The proportion of variable sites estimated is given by 12.4%.
[e] The α parameter estimated for the gamma distribution is infinity.

To test the suitability of an evolutionary model, a likelihood framework can be used to compare the fitness of two or more competing models with the sequence data to be analyzed. In this context the best-fitting model is the one that obtains the highest likelihood value. Table 6.17 shows the maximum likelihood value, determined by PAUP* software (Swofford 2000), on a set of six primate ψη-globin sequences using different nucleotide substitution models. A *likelihood ratio test* (LRT) can be performed to compare two or more competing models by calculating the value 2Δ, given by

$$2\Delta = 2 \frac{\max[\text{likelihood}(H_0|\text{data})]}{\max[\text{likelihood}(H_1|\text{data})]}$$

where H_0 and H_1 represent the competing models. The quantity 2Δ approximately follows a χ^2 distribution with degrees of freedom equal to the difference in the number of free parameters between the alternative models.

When the models to be compared are not nested (i.e., the simplest model is a special case of the more complex one and is derived by fixing certain free parameters), a more reliable distribution for 2Δ can be obtained using a procedure known as Monte Carlo simulation or parametric bootstrapping (Huelsenbeck and Rannala 1997). *Parametric bootstraping* differs from nonparametric bootstraping (see below) because the former uses simulated replicates of the data rather than pseudoreplicates. In the parametric bootstraping simulation, replicated data are generated according to a specific model of nucleotide substitution and to a given tree topology estimated from the original data. To generate such simulated sequence data, SEQGEN software can be used (Rambaut and Grassly 1997) to define a DNA substitution model, including the gamma correction for the among-site rate heterogeneity, and to define a specific tree topology. After the simulation, the likelihood ratio is calculated for each simulated data set and the proportion of replicates in which the likelihood ratio calculated on the original data is exceeded for the simulated data represents the significance of the test [see Fig. 6.17 in Page and Holmes (1998)].

In the specific example shown in Table 6.17, the HKY model represents an adequate description of the evolutionary process, as its outcome cannot be rejected by a χ^2 test. Indeed, as expected in the case of a pseudogene evolution, no rate heterogeneity is observed resulting from the estimated value of the α parameter of the gamma distribution at infinity.

Statistical Test of Tree Topologies   The assessment of a confidence measure of a phylogenetic hypothesis is a crucial issue in molecular phylogenetics. The most popular way to assess the robustness of a tree is the method of *nonparametric bootstrapping* (Felsenstein 1985). According to this method a series of N multiple alignments are simulated (usually, 500 to 1000) by resampling with replacement the L sites of the original multiple alignment. Consequently, each simulated multiple alignment of length L may contain sites that are represented more than once and sites that are not represented at all. Each simulated or bootstrapped multialignment is used to construct a tree with the same method used for the inferred tree. In this way, a score for each node of the tree can be defined, denoted as the *bootstrap value*, given by the percentage of bootstrap replicates supporting a monophyly of descending OTUs, and a consensus tree can be drawn where only branching patterns supported by the majority of bootstrap replicates are shown collapsing the others into polytomies. The consensus tree, so determined, is called a *majority-rule consensus tree*. Figure 6.51 shows the majority rule consensus tree generated by PAUP* (Swofford 2000) on a set of six primate $\psi\eta$-globin sequences after 1000 bootstrap replicates. The chimp–human clade is supported by 60.5% of bootstrap replicates, whereas chimp–gorilla and human–gorilla are supported by 26.2% and 13.3% of bootstrap replicates, respectively.

A resampling technique alternative to the bootstrap is the *jackknife*. In this case, instead of generating a number of simulated alignments having the same size as the original alignment, with site replacement and duplication, smaller data sets are generated, randomly dropping a certain percentage of sites for each replicate without duplication. This technique, implemented in PAUP* and PHYLIP packages, is expected to give statistical support values very similar to those calculated by bootstrap, although the latter is the most widespread method.

Although bootstrap analysis is commonly used to represent the statistical confidence of group monophyly, it is not adequate to assess the overall confidence of a tree. Indeed, if there are three different clades, each having for example 70% bootstrap value, assuming their independence, the chance that all are correct should be as low as $(0.7)^3 \cdot 100 = 34\%$. As more clades are considered simultaneously, these values decrease further and the overall probability quickly becomes meaningless. A comparison of competing phylogenies can be obtained with the LRT method described previously for comparing competing models of DNA substitution. These tests began to be used when Kishino and Hasegawa (1989) devised a nonparametric test (KH test) to compare the merit of different phylogenetic hypotheses. A problem with this approach is that the topologies to be tested have to be selected a priori, thus leading to overconfident use of the wrong trees. The SH test

```
Bootstrap method with neighbor-joining search:
  Number of bootstrap replicates = 1000
  Starting seed = 1960379486
  Ties (if encountered) will be broken randomly
  Distance measure = general time-reversible
  (Tree is unrooted)

  Time used for neighbor-joining bootstrap = 1.78 sec

Bootstrap 50% majority-rule consensus tree

/---------------------------------------------------------------------- OWL MONKEY(6)
|
|                                               /------------------ CHIMP(1)
|                              /--------60--------+
|                              |                  \------------------ HUMAN(2)
|              /-------100-------+
|              |                 \---------------------------------- GORILLA(3)
+-------100--------+
|              \---------------------------------------------------- ORANG(4)
|
\---------------------------------------------------------------------- MACAQUE(5)

Bipartitions found in one or more trees and frequency of occurrence (bootstrap support
values):

123456      Freq       %
-----------------------
....**      1000 100.0%
...***      1000 100.0%
..****       605  60.5%
.*.***       262  26.2%
.**...       133  13.3%
```

Figure 6.51. Majority-rule consensus tree generated by PAUP* on a set of six primate ψη-globin sequences (see the legend of Fig. 6.45) after 1000 bootstrap replicates using the neighbor-joining method on distance matrices generated by the GTR substitution model.

(Shimodaira and Hasegawa 1999), derived from the KH test, addresses this problem, providing a more accurate evaluation of the relative significance of different phylogenetic hypotheses, much as the SOWH test (Goldman, Anderson et al. 2000), which uses a parametric bootstrap approach instead. Figure 6.52 shows the results obtained by the SH test carried out on all 105 possible unrooted trees for the six-taxa ψη-globin data set. Although the most likely topology corresponds to that now generally accepted (i.e., human closer to chimp than to gorilla), using the sequence data considered, it is not possible to rule out the other five competing topologies, which cannot be rejected at the 5% level.

Several other biological questions involving phylogeny can be addressed by LRTs, such as the test of constancy of substitution rates among lineages (i.e., the test of the existence of a molecular clock), the congruency of phylogenies estimated from different data (Huelsenbeck and Bull 1996), and the consistency between host and parasite phylogenies (Huelsenbeck, Rannala et al. 1997). Figure 6.53 shows the result of an LRT analysis to verify the existence of the molecular clock in the evolution of the six primate ψη-globin genes considered earlier. In this case, the LRT 2Δ value, calculated from the difference between the log(likelihhood) estimated

Shimodaira-Hasegawa test:
SH test using RELL bootstrap (one-tailed test)
Number of bootstrap replicates = 1000

Tree	-ln L	Diff -ln L	P
1	4760.79253	(best)	
2	4760.79253	0.00000	0.950
3	4760.79253	0.00000	0.920
4	4794.85435	34.06182	0.054
5	4794.85435	34.06182	0.054
6	4794.85435	34.06182	0.054

(*a*)

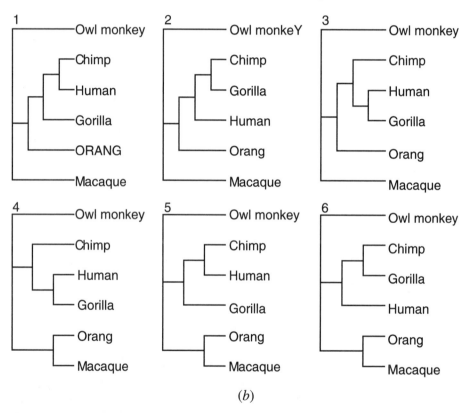

(*b*)

Figure 6.52. Result of the SH test comparing the likelihood of all 105 possible tree topologies for the six-taxa ψη-globin sequence data set carried out using PAUP* with GTR as the DNA substitution model. The result of the test (*a*) and the corresponding trees (*b*) are shown for topologies that cannot be rejected at the 5% level.

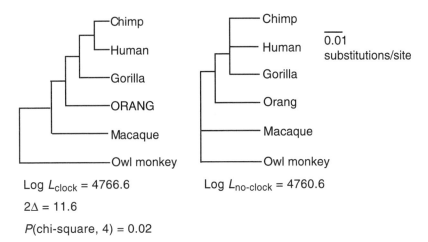

Figure 6.53. Maximum likelihood trees calculated assuming the GTR model of DNA substitution for the six-primate ψη-globin sequences with and without the molecular clock assumption. Using LRT, the molecular hypothesis cannot be rejected at the 1% confidence level.

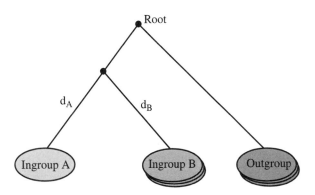

Figure 6.54. Relative rate test. The relative rate ratio (d_A/d_B) between the ingroup OTUs (A and B), given the outgroup OTUs (O), is given by $(d_{AB} + d_{AO} - d_{BO})/(d_{AB} - d_{AO} + d_{BO})$.

with and without assuming the clock, is assumed to be χ^2 distributed with $n - 2$ degrees of freedom, where n is the number of OTUs. In this case, the degrees of freedom correspond to the difference in the number of independent branches between an additive and an ultrametric tree, given by $(2n - 3) - (n - 1) = n - 2$. A necessary prerequisite for using such a test is the fact that the rooted and unrooted trees have the same tree topology. If the molecular clock hypothesis is rejected, use of the relative rate test (RRT; see Fig. 6.54) can provide information on the relative rate of several OTUs, provided that an outgroup reference OTU is available (Muse and Weir 1992; Muse and Gaut 1994).

PART III

COMPARATIVE GENOMICS

CHAPTER 7

MOLECULAR EVOLUTION

7.1 INTRODUCTION

In Parts I and II our goal has been to provide the reader with an overview of the general features of genomes based on sequencing data of complete genomes (Part I) and the methodologies and tools most used in genomic research (Part II). In this part, we wish to emphasize the comparative analysis that allows us to infer a number of general principles and mechanisms of evolution at a genome level, providing a degree of knowledge from which this new discipline, *evolutionary genomics*, may be built. Indeed, the aspect of gene origin timing and evolution is crucial to understanding the properties of the living matter and the relationships between genotype and phenotype. The major challenge in this field is to understand the mechanisms at the base of evolution of the genome as a whole. We are aware that there is still a long way to go and time is needed, but we are ready to start. We would probably be able to trace only the path along which this discipline, as predicted, may proceed and shortly explode.

Comparative genomics can reveal the extraordinary unity and the striking diversity of living matter, as genomes are shown to have great plasticity, variability in shape, size, and gene organization. Hence the first issue addressed is how objects that are so different can be compared. There are cases in which, owing to their close relationships, genomes have a high degree of similarity and may be aligned, as described in the methodological section. However, more frequently, genome structures and organizations are very divergent, thus requiring specific strategies.

Genomes may be compared using many different criteria; results, although all equally valid, may lead to different evolutionary implications and interpretations. For example, coding and noncoding portions have different constraints and follow different rules in evolution. Evolutionary changes can be qualitative and also quantitative; How can they be evaluated in terms of genetic distance? In addition, genomic evolution may often be species-specific.

This clearly demonstrates that data need to be interpreted and integrated in a global vision—only such an approach can provide reliable answers to our questions. Indeed, it is a fact that genomics is changing biology (Brent 2000): not only our understanding and its application to important fields, such as agricolture and medicine, but also our way of doing research. An important aspect of genomics is the production of a huge amount of varied information. More than ever before, this requires efficient management and research organization and involves multidisciplinary competence.

Our aim in this chapter is to raise and discuss points related to the evolutionary processes acting at the molecular level on the genome as a unit. Obviously, we should limit our attention to a few issues. Some of these issues, which can be considered as general, include fundamental questions still under discussion, or background concepts; in particular, we discuss extensively only a few issues, because our aim is to introduce the reader to this new discipline, and because this is a field in development, as evidenced by the newly produced genomic data. More specifically, we deal with the evolution of the three types of genomes whose general features were described earlier: prokaryotic, eukaryotic, and organelle genomes.

7.2 EVOLUTION OF GENOME SIZE

The genome sizes of some prokaryotic and eukaryotic genome are listed in Tables 1.1 and 2.3. How genome size has evolved is an old problem, yet never as new and dramatic as presently, because of genome sequencing. Genome complexity may be defined as the amount of information encoded by the DNA molecule and thus the number of genes together with the accessory regulatory elements necessary for their replication and expression. We already know that the degree of genome complexity, from unicellular to multicellular eukaryotes with specialized functions, increases in a way certainly not correlated directly with genome size and, more particularly, to the coding portion of genomes, whose extension cannot explain the numerous new functions arising in a given organism (phylum, lineage, tissues, etc.).

From the draft sequence of the human genome as contrasted to those of invertebrates (e.g., fruit fly and worm), we know that humans probably have only about twice the number of genes identified in the fly or worm. However, genes in humans are spread out in a much larger region of DNA, and in addition, they are used to create, by alternative splicing, a lot of new transcripts. In proteins, functions are determined by domains whose number may not be so different even in distantly related organisms (e.g., invertebrates and vertebrates). However, new proteins (and their relevant functions) may emerge from new *architectures*, linear arrangements of domains in a polypeptide. New architectures may be created by deletions, additions, or rearrangements of domains. Because of the molecular setup of living organisms, we are faced with the following fundamental questions:

1. How is complexity generated starting from a number of building blocks (genes), that are not very different among organisms?
2. On which molecular similarities is life based?
3. What makes a species or an individual unique?

TABLE 7.1. Examples of Paralogous Gene Displacements Observed Comparing
H. influenzae **and** *M. genitalium* **Gene Sets**

Function	Missing Orthologs
Translation	Aminoacyl-tRNA synthetases (Pro, Gly, Gln)
Replication	RNAse H
Nucleotide biosynthesis	Nucleotide diphosphate kinase
Glycolysis	Phosphoglyceromutase
Coenzyme biosynthesis	Lipoate-protein ligase

Source: Data from Mushegian (1999).

To understand the degree of complexity of a given genome, we should be able to estimate the minimal set of genetic elements sufficient to build a cellular organism. In other words, a minimal set of proteins should be established as necessary for the survival of the organism and thus the corresponding essential set of genes (minimal genome concept) necessary to code them. Attempts to reach this goal have already been made, but with questionable success. A minimal genome set of about 300 genes has been suggested by comparing the complete protein lists of two bacterial genomes, *H. influenzae* and *M. genitalium*, trying to identify all orthologous proteins (Mushegian 1999). A previous comparative analysis of these two genomes (Mushegian and Koonin 1996) detected a set of 244 proteins; but the first very remarkable result was the discovery that the same biochemical function might be performed in two genomes by proteins which, according to the current definition, cannot be considered as orthologous proteins. This happens because a process called *gene displacement*, in which the function of one gene is taken by another, nonorthologous gene, is occurring very frequently in genome evolution. This process takes place at a very consistent rate, around 5%, and demonstrates clearly that the functional equivalence of proteins requires neither primary nor tertiary structural similarity.

In Table 7.1 some examples of gene displacement are shown. Taking into account nonorthologous gene displacements and the very important aspect of gene duplication and gene loss, which implies a continuous gain and loss of genetic material in the genome, it has been possible to postulate a minimal genome concept. For this, instead of single orthologs, clusters of orthologous groups (COGs), including various paralogs, have been considered. In comparing various genomes, 327 COGs were found to include representatives of all three superkingdoms (i.e., bacteria, archea, eukaryotes), thus defining a minimal universal set of genes.

Another important aspect regarding genome size is the degree of redundancy of genomes, particularly of the largest ones, which is equivalent to the *level of paralogy*, defined as the fraction of genes having at least one paralog in the genome. This gene fraction has been found to change in phylogenetically closed organisms approximately by the square root of the gene number ratio. Then the relative level of paralogy (RLP) between two genomes containing N_1 and N_2 genes, respectively, is given by

$$\text{RLP}_{12} = \sqrt{\frac{N_1}{N_2}}$$

For example, the paralogy level in *E. coli* and *H. influenzae*, calculated to be 50% and 33%, respectively (RLP = 1.52), corresponds to a gene number ratio of 2.5.

There is increasing evidence (Nowak, Boerlijst et al. 1997) that inactivation of some very important genes has no apparent effect on the phenotype or survival of an organism. This is equivalent to admitting that the function of a gene, although very central, can be taken over by another gene. In other words, the genome as a whole may react even to the most drastic situations with a reassessment of its functional capacity to have devices for survival in an emergence situation. This observation is very important in some practical applications of genomics: for example, in the use of animal model systems, where the knocking out of genes can produce different effects, depending on their genetic background (Pearson 2002). The presence of such devices, which may be specific for a given genome, should be taken into consideration very seriously before extrapolating the results when the same conditions are applied to another genome. This process may be considered equivalent in functional terms to gene displacement, which is the result of natural selection.

Other attempts to define a minimal genome are always based on prediction of genes and their functions as well as on the ability to reconstruct the pathway followed by evolution. The task is particularly difficult because we do not know all the gene functions in a genome. As discussed in Chapter 2, this is one of the major challenges in the field of genomics. If genomic size and complexity do not require a great increase in the number of genes—indeed very simple organisms such as amoeba possess more DNA than humans—but are based on noncoding and repetitive ("extra") DNA, it should be understood which evolutionary forces are acting at the genome level to produce vast amounts of noncoding DNA. This is particularly relevant in the case of eukaryotes, which, as already discussed, posses various types of repetitive elements. In other words, we should be able to establish the function(s) of noncoding DNA and the reasons why it is perpetuated during evolution. In this context, there are two theories. The first, called the *adaptive hypothesis*, postulates an adaptive function of "extra" DNA on the phenotype. In this context, genome size would be essential for efficient natural selection in different organisms. The second hypothesis considers the vast abundance of extra DNA as "junk DNA" or "selfish DNA"—in other words, as useless.

As for the evolutionary implications, recent studies have focused attention on how important genetic drift is in genome-size evolution (Petrov 2001). It is fundamental to ascertain whether mutational rates of DNA addition and loss vary among organisms independent of natural selection for small or large genome size, and furthermore, if there is a correlation among *mutational rates* and genome size. For example, the Japanese pufferfish (*Fugu rubripes*) has the smallest vertebrate genome known to date (400 Mbp; see Table 2.3), but it maintains a gene set similar to that of other vertebrates. This is because *F. rubripes* probably has a mutational bias toward DNA elimination, and its size is likely to be close to the "minimal" vertebrate genome (Venkatesh, Gilligan et al. 2000). Figure 7.1 depicts the forces acting on the evolution of genome size, some having additive effects (Petrov 2001).

At the present stage of knowledge it is difficult to evaluate the effects of any single component, and even more so, their combinatorial effect. Data are limited or absent and systematic studies on the correlation of different components at comparative levels are still lacking. We should also once again emphasize that for better,

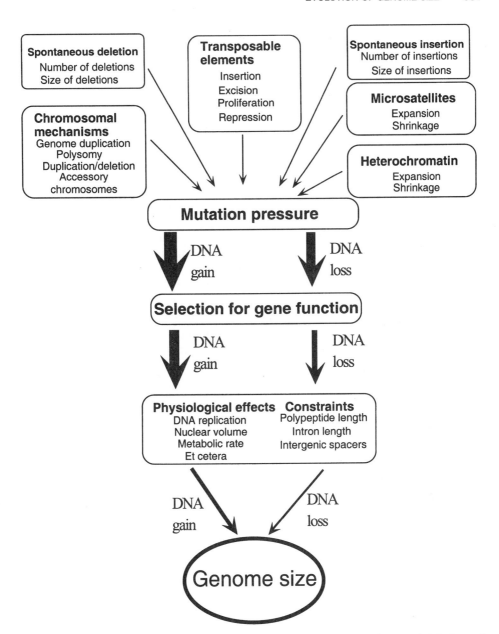

Figure 7.1. A large variety of mutational mechanisms generate insertions and deletions of different sizes. Some of these mutations are bound to have strong deleterious effects on gene function and will quickly be eliminated from the population. The remaining length mutations will be further affected by weaker natural selection acting either at the level of constraints on gene, intron, and intergenic DNA length or at the level of physiological effects modulated by overall genome size. The combined force of all selective effects change the number and the size of length mutations that will fix in the genome. The combined effect of mutational and selective biases produces a force toward either growth, shrinkage, or maintenance of genome size. (Courtesy of Dmitri Petrov.)

more complete analysis, we need not only molecular data but also approaches from other disciplines, such as cytogenetics, developmental biology, population genetics, and others.

7.3 ROLE OF BASE COMPOSITION IN EVOLUTION

Base composition is an extremely variable parameter in all types of genomes. Indeed, genome plasticity is expressed first through variations in base composition. These, observed already in prokaryotes, are amplified dramatically in higher eukaryotes, both animals and plants. Before the sequencing of the genomes, simply on the basis of indirect and crude measurements, much attention has been given to the evolutionary forces acting on base composition. Now we have complete primary structures of the genomes, base composition has obviously become one of the parameters most studied. The pioneer work of Grantham beginning in 1980 (Grantham, Gautier et al. 1980) has demonstrated that each genome has its own base composition feature, putting forth the *genome hypothesis* theory and various methodologies have been proposed to study the base composition of genomes.

As mentioned previously, in all genomes, but particularly in smaller ones such as those of prokaryotes or organelles, we can use a linguistic approach to analyze base composition. This methodological approach is described in Sec. 6.8. Variations in language analysis is very useful in detecting genomic regions of external sources, as in the case of lateral gene transfer. In the context of language analysis it is possible to discover some signatures. For example, the term *genome signature* has been defined by the Karlin group (Campbell, Mrazek et al. 1999) as the ratio between the dinucleotide frequencies observed and the frequencies expected on the basis of a random distribution. The authors claim that each genome has its own signature, which is rather constant along the genome length. It is remarkable that plasmids and their hosts have similar signatures, whereas in eukaryotes, there are large differences between mitochondrial and nuclear signatures. Another interesting observation is that archaea appear as a noncoherent clade with respect to this property. Comparison of signatures between genomes may be used in molecular taxonomy to estimate the phylogenetic relationship between organisms. This methodology has the advantage of being independent of the ability to align homologous sequences to measure their degree of similarity.

Not only is the average value of base composition different in the genomes of the various lineages, but there is variability in base composition along the genome length (see Sec. 1.5). There are different views and controversial data about the extent, meaning, and origin of such regional variability. One type of variability regards the distribution of complementary bases (A–T or C–G) between the two strands. Asymmetrical strand distribution, called *skew,* was defined previously (see Sec. 1.5).

It is interesting to note that in whole genomes, at least in those completely sequenced, there is no total base asymmetry between strands, in the sense that the parity rule A = T and G = C in each strand seems to be respected. Figure 7.2 shows the GC skew calculated on complete genomes from representative species of prokaryotes, eukaryotes, and viruses. There are, however, only some localized skewed regions, which in bacteria are in general found at the level of the replica-

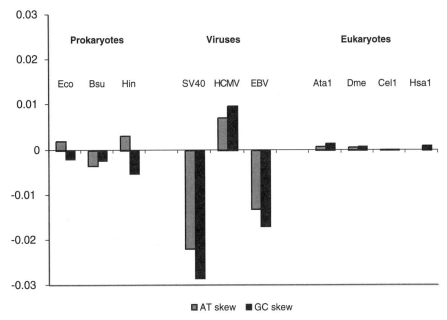

Figure 7.2. GC skew calculated on complete genomes from representative species of prokaryotes (Eco, *E. coli*; Bsu, *B. subtilis*; Hin, *H. influenzae*), viruses (SV40, simian virus 40; HCMV, human cytomegalovirus; EBV, Epstein–Barr virus), and eukaryotes (Ata1, *A. thaliana* chromosome 1; Dme, *D. melanogaster*; Cel1, *C. elegans* chromosome 1; Hsa1, *H. sapiens* chromosome 1).

tion origins, being due to the asymmetric nature of this process. In contrast, in this respect mitochondrial DNA of vertebrates represents an important exception, because the skew in this case involves the entire genome. As shown in Chapter 3, the skew is linked to the replication mechanism, whose asymmetry involves not a short genome tract but two-thirds of the entire genome.

A regional variation in the relative nucleotide abundance along genomes has been found in most organisms, both prokaryotic and eukaryotic. The extent of the DNA tracts that have such a peculiarity, and their boundaries, often depend on the method used for their detection, in particular on the window width used for the measurement. Their meaning has been interpreted in various ways. However, the existence of long tracts having homogeneous compositions, thus creating real genome compartments, seems a property of only some high eukaryotes such as warm-blooded vertebrates. Such compartments were demonstrated experimentally in 1985 by Bernardi, who proposed that the nuclear genome of warm-blooded vertebrates can be described as a mosaic of *isochores* (i.e., equal regions), very long DNA segments (>300 kbp) in which the base composition can be regarded as homogeneous (Bernardi, Olofsson et al. 1985). The isochore structure is correlated with various genomic features.

1. *Cytogenetics.* A striking correlation between isochores and chromosomal bands has been demonstrated by in situ hybridization of DNA from different iso-chore families in metaphase chromosomes. H2 and H3 isochores hybridize on a

small set of R(everse) bands, whereas GC-poor isochores are mainly concentrated in G(iemsa) bands (Saccone, De Sario et al. 1993; Bernardi 2001). The GC-rich isochores have higher transcription activity and open chromatin structure than do the GC-poor isochores, which show a more compact chromatin structure.

2. *Repetitive elements.* Alu and LINEs, which are widespread classes of repeats, are very unevenly distributed in the human genome, with the majority of GC-poor LINEs residing in the GC-poor isochores and the GC-rich Alu in the GC-rich (Pavlicek, Jabbari et al. 2001).

3. *Methylation.* GC-rich isochores display a lower relative level of methylation (methyl group/acceptor sites) than GC-poor isochores, as shown by the more pronounced shortage of the CpG dinucleotide in GC-poor isochores and the higher abundance of CpG islands in GC-rich isochores (Bernardi 2000b).

4. *Replication.* GC-rich isochores replicate earlier than GC-poor isochores and early/late transition regions occur at positions identical to or nearby GC% transitions (Tenzen, Yamagata et al. 1997).

5. *Recombination.* GC-rich isochores show a recombination rate much higher than that of GC-poor isochores (Bernardi 2000b).

6. *Gene density.* Gene concentration is much higher in GC-rich than in GC-poor isochores. The *genome core* formed by isochore families H2 and H3 in the human genome (see Sec. 2.4), which makes up 12% of the genome contains about 54% of the genes, the remaining 46% of the genes being located in the *empty space* (isochore families L and H1), accounting for 88% of the genome.

GC-rich isochores are found in mammals and birds but not in Anura (frogs and toads), fishes, and invertebrates. In reptiles, ultracentrifugation analyses provide evidence that compositional heterogeneity does exist in crocodile and turtle but not in snake genomes.

There are two alternative hypotheses to explain the presence of compositional compartments, one *selectionist* and another *neutralistic.* According to the selectionist hypothesis, followed by Bernardi, isochores are formed to increase the fitness of the organism to environmental factors: for example, to body temperature increase in homeothermic organisms. In this vision the presence of isochores in reptiles is a particularly important issue for supporting such a hypothesis. According to Hughes, Zelus et al. (1999), the presence of GC-rich isochores in cold-blooded vertebrates seems to rule out the homeothermy hypothesis, even if other relationships between isochores and organism fitness can be found.

The alternative hypothesis is that isochores result from processes that are of neutral nature. The possible causes may be mutational bias, and according to Eyre-Walker and Hurst (2001), biased gene conversion (BGC). In the context of mutational bias a model has been proposed relating the variation of mutational bias to the difference in replication time of the various isochore classes and/or to different nucleotide pools (Wolfe, Sharp et al. 1989). In particular, the pattern of misincorporation during DNA replication should be affected by the concentrations of free nucleotides, which vary during the cell cycle. Since some regions of the genome are replicated early and others are replicated late, regions of the genome that replicate at different times should have different mutation patterns and thus different base compositions. Indeed, even if some correlation has been observed between replication time and GC content, with early replicating regions more GC- and gene-rich than late-replicating regions, this mechanism has not been demonstrated to occur in germ lines. Also the view that the efficiency of DNA repair is always biased along the genome, as suggested by Filipski (1987), falls in the category of neutral evolution. The same holds for the cytosine deamination mechanism put forth by Frixell and Zuckerkandl (2000), according to which deamination of cytosine and methyl cytosine is expected to take place more in A–T-rich regions which are more unstable than in G–C-rich regions. If, for example, a DNA tract becomes more G–C-rich because it codes for a protein, a reduction in cytosine deamination should occur in that area and would be the cause of isochore formation.

In the domain of neutral evolutionary processes, there is the biased gene conversion. It is due to homologous recombination, which produces hetheroduplex DNA and, in turn, base mismatches. These mismatches are repaired by the DNA repair machinery, which tends to be biased. Such hypothesis is based on the observed correlation between the frequency of recombination and the GC content and the finding that almost all Y-linked genes, not subject to recombination, have lower GC content than X-linked paralog. However, this hypothesis has not been proved adequately to definitely establish causation of isochore formation. In conclusion: So far, the neutralistic hypothesis does nor rely on strong experimental evidence. The data produced are controversial and the predictions often have been proven incorrect. Thus this issue remains open.

Having the complete human genome sequence available, however, it has been natural to search directly for the presence of isochores. In the draft map of the International Human Genome Sequencing Consortium (Lander, Linton et al. 2001), the strict notion of isochores as compositionally homogeneous compartments has been rejected and it has been affirmed that isochores do not appear to merit the prefix "iso". This statement is based on analyses that deserve serious comment. The definition of *isochores* sensu stricto (i.e., as very homogeneous compartments along a genome), as stated by the authors, is very misleading because such compartments cannot exist in natural DNA. Any DNA tract, indeed, is constrained by its genetic content (i.e., by its coding capacity as well as by the noncoding and regulatory elements present in it). Thus, complete homogeneity in an absolute sense is unrealistic; in particular, it is very difficult to identify isochore borders. However, if recursive segmentation methods are used (Li 2001), both prokaryotic and eukaryotic genomes can be partitioned in regions having a homogeneous GC content and the corresponding borders quite precisely mapped. These methods perform much better than do the usual window-based methods and do not need to assume randomness of natural DNA sequences, where, instead, short- and long-range correlations may exist between nucleotides. For example, the observation of an increasing compositional heterogeneity when moving from GC-poor to GC-rich isochores could be due to the requirement, in GC-rich regions, of AT-rich segments to stabilize the chromosome scaffold. Indeed, the compositional overview of human chromosomes shown in Fig. 7.3 clearly shows a mosaic organization of the human genome at the nucleotide level, corresponding to the isochore model earlier predicted by cesium chloride experiments (Bernardi 2001). However, to have a clear picture of genomic compositional compartmentalization, further investigations are required, comparing several genomes and using more reliable and suitable algorithms.

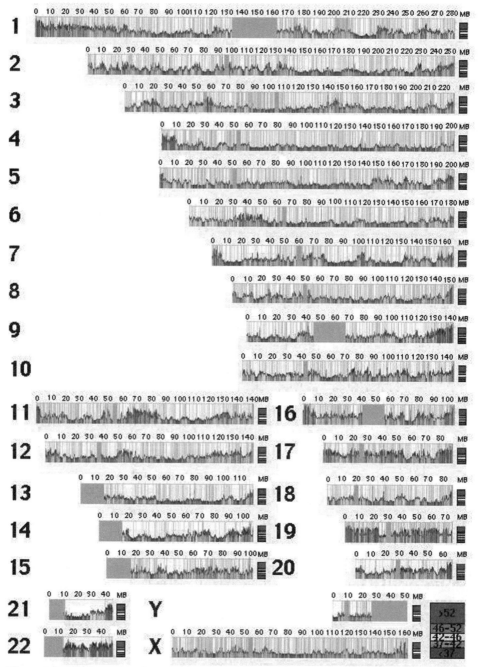

Figure 7.3. Window plot of the GC content of human chromosomes calculated by means of the draft genome sequence using nonoverlapping 100 kbp stretches. The 100 kbp windows were partitioned in five classes, corresponding to the five major isochore families (L1, <37% GC; L2, 37–42% GC; H1, 42–46% GC; H2, 46–52% GC; H3, >52% GC).

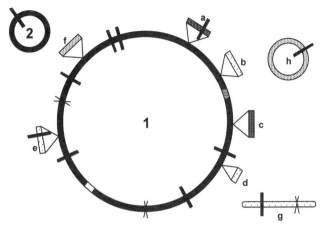

Figure 7.4. A generic dynamic bacterial genome is made up of an endogenome consisting of one or more endochromosomes shown in black (1, 2) and a number of accessory elements that may or may be not present (a–h). Components of the endogenome can be linear or circular and may or may not be undergoing segment shuffling or inversions (white section). Other systematic changes include slipped strand mispairing changes (indicated by crosses), nonorthologous replacements (crosshatched), and insertion of various kinds of mobile elements (thick black lines). Accessory elements include multiple types of plasmids that may exists as free replicons (h) or integrated (f), prophages or fragments thereof that may be integrated or in free plasmid forms (b,d,e,g), and pathogenicity islands (a,c). [Courtesy of S. Casjens; modified from Casjens (1998).]

7.4 EVOLUTION OF THE PROKARYOTIC GENOME

On the basis of the properties described in Chapter 1, the features of prokaryotic genomes can be summarized as follows: (1) compact size for an efficient and rapid response to environmental needs, (2) the ability to integrate the genetic material from the outside and thus to lose or acquire accessory elements quickly, and (3) strict dependence on the ecological niche of the organism. Therefore, it has a very dynamic structure, thanks to the cell capability to include new information at the level of the main genetic component, using other autoreplicating elements. According to the excellent review of Casjens (1998), prokaryotic genome has a core of genes, a structure universally present in all species, which can be called a *endogenome* and a number of other components, called *accessory elements* (Fig. 7.4). The latter, also called a *exogenome*, may both be integrated in the endogenome or be free in the cell as independent replicons. These elements may be functional or nonfunctional, useless or very important in particular environmental conditions. They consist of insertion sequences and transposons, prophages and phages, pathogenicity islands, or a combination of elements. Consider that in some cases the external origin of these elements can easily be recognized, particularly if they derive from a recent event; in others, instead, they cannot be easily distinguished from the remaining genomic part because of their evolutionary changes and adaptation to the entire genome.

The process of *lateral* or *horizontal genetic transfer* (LGT), that is, the acquisition of new genes from the outside, was described before the genomic era; however, genome sequencing has emphasized its importance by showing the extent of this phenomenon, which affects a very large portion of the genome in prokaryotes. Acquisition of new genes from the outside is very important from an evolutionary point of view. When new genes are imported from the outside, they have to spend much effort in the adaptation to the new environment. However, their fixation may take place through two mechanisms: (1) selection, since they confer an advantage to the living cell (e.g., acquisition of a new biosynthetic ability, antibiotic resistance, better adaptation to a new organism niche) and (2) neutral evolutionary processes.

In any case, new genes from the outside impose a new modality in tempo and mode of evolution. According to Lawrence and Ochman (1998) and Ochman, Lawrence et al. (2000), bacterial speciation, unlike eukaryotic speciation, is essentially driven by a high rate of horizontal transfer. This allows the genome, through its extremely dynamic structure, to adapt and exploit new environments rapidly. As a consequence, phylogeny of bacteria should not be a tree and their systematics cannot be based on monophyletic taxa. However, this point is questioned by Kurland (2000), who do not envisage such dramatic effects of LGT in genome evolution.

The bulk of information on microorganisms derives from data obtained from traditional bacterial cultures. However, our knowledge cannot be limited to these species because culturing provides poor access to many microbial species present in the environment. From an evolutionary perspective, not only should the DNA of the single species be considered but also the DNA present in an entire microbial community. This concept is illustrated by the term *metagenome*, which indicates collectively the genomes of all the microbiota found in a particular niche or environment (Rondon, Goodman et al. 1999; Rondon, August et al. 2000).

To know the full extent of microbial diversity, we need to develop methods allowing us to study the novel types of bacteria present in the environment. Metagenomic libraries can be constructed using BACs (bacterial artificial chromosomes). Such libraries already exist and represent a powerful tool for exploring microbial diversity. The data obtained from library scanning will increase enormously our knowledge of phylogenetic relationships and functional properties of the microbiota present in the environment, particularly those that cannot be cultivated. This in turn may open new perspectives in applied microbiology.

7.5   FROM PROKARYOTES TO EUKARYOTES

7.5.1   Origin of the Eukaryotic Cell

The main feature of the eukaryotic cell is the organization into compartments, formed by the expansion of membranes; therefore, the eukaryotic cell possesses a nucleus and cytoplasmic organelles. Among organelles, mitochondria and plastids were shown to possess their own genome and genetic system as early as the mid-1960s. It has long been known that the eukaryotic cell is the result of symbiotic events between primitive, prokaryotic organisms. The primitive hypotheses on the origin of the eukaryotes have recently been revised in the light of new information derived from studies at the molecular level, more particularly on genomics. These

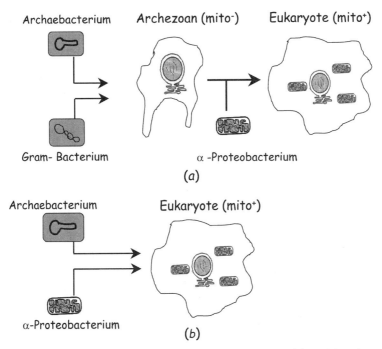

Figure 7.5. Current theories on the origin of the eukaryotic cell: (*a*) serial endosymbiontic hypothesis involving two endosymbiontic events, first creating an amitochondriate eukaryote; (*b*) hydrogen hypothesis involving a single endosymbiotic event.

theories include a hypothesis on the origin of mitochondria and plastids which are crucial to understanding the origin and essence of being an eukaryote.

More recently, two models, known as the *hydrogen hypothesis* (Martin and Muller 1998) and the *syntrophic hypothesis* (Moreira and Lopez-Garcia 1998), have been put forth for the origin of the eukaryotic cell. Both theories are a modification of the *primitive endosymbiontic `hypothesis* of Margulis (1970), also called the *serial endosymbiotic hypothesis*, which postulates that organelles originated from bacteria entering in symbiosis with a primitive eukaryotic cell. Both theories imply that the driving force for symbiosis was the requirement for hydrogen by a methanogenic archaeabacterium acting as the host.

Figure 7.5 illustrates major current theories on the evolution of the eukaryotic cell. The hydrogen hypothesis postulates a single fusion event of two primitive bacteria, an archaebacterium and an alpha-proteobacterium, which produced the eukaryotic cell: in other words, simultaneous appearance of the nucleus and the mitochondrion. This hypothesis is confirmed by the finding that not all amitochondriate taxa (e.g., Archaezoa) fall at the base of eukaryotic tree in phylogenetic analyses, as believed previously; since genes belonging to mitochondrial genome have been found in the nucleus of some Archaeozoa, these species originally did have mitochondria and lost them during evolution.

The syntrophic hypothesis is based on a primitive symbiosis between a methanogen archaeon and a delta-proteobacterium; this symbiosis was also related to the process of interspecies H_2 transfer. The bacterial partner produces H_2 as an end product of its metabolism, while the archaeal partner consumes H_2 and CO_2 taken from the environment to produce methane. The initial symbiosis led to the establishment of a permanent consortium in which the methanogen was entirely wrapped in the delta-proteobacterial cell. The subsequent step was an improvement of the H_2 transfer process, obtained by enlarging the physical contact among the symbiotic partners through the development of membrane invaginations, followed by the fusion of the bacterial membrane around the archeal cell, which would become the nucleus. One of the proofs in favor of this theory is the observation that such a mutualistic symbiosis is frequently found in anaerobic natural communities. It further proposes a third partner in symbiosis, an alpha-proteobacterium which once engulfed by the primitive cell would become the mitochondrion. The last stage was the loss of one of the plasma membranes and the beginning of a bidirectional, extensive gene flow and replacement which led to a complex, large nuclear genome and to a progressive reduction of the organellar genome. The main difference between the two hypotheses is that according to Martin and Muller (1998), mitochondria originated at essentially the same time as the nuclear component of the eukaryotic cell, while according to Moreira and Lopez-Garcia (1998), mitochondria originated from a later symbiosis with an alpha-proteobacterium.

Alternative hypotheses have been proposed. In the *ox-tox hypothesis*, suggested by Andersson and Kurland (1999), the acquisition of mitochondria would have occurred in two phases: initially, the symbiont functions in oxygen detoxification, and subsequently, it functions as a provider of ATP for the host. This detoxifying function has been proposed also by Vellai, Takacs et al. (1998). Karlin, Brocchieri et al. (1999) have proposed a polyphiletic origin of mitochondria based on genomic signatures: The symbiotic partners were a H_2-producing bacterium (similar to *Clostridium* spp.) and a H_2-consuming Archaeon (similar to the extant species *Sulfolobus*). The latter theory seems unlikely (Gray, Burger et al. 1999).

7.5.2   Evolution of the Mitochondrial Genome

Whatever the right evolutionary pathway, the presence of more than one genome seems to be against the economy of the eukaryotic cell. Indeed, to perform their function, mitochondria need more genes than those they are able to synthetize; the other products (more than 1000 proteins) are coded by the nuclear genetic system, translated by the protein-synthesizing machinery of the cytoplasm and then transferred to the organelle. The same applies to the genome of chloroplasts. Moreover, additional complicated intergenomic regulatory circuits must come into action to coordinate the replication and expression of each genome.

If so, why have organelles retained their genomes throughout evolution? Starting from the observation that organelles do not retain a random sample of their ancestral bacterial genomes but mainly specific genes (e.g., in mitochondria, those involved in respiration and gene expression machinery), some hypotheses have been put forward. Allen and Raven (1996) suggested that the redox potential of the organelle must regulate the expression of the gene involved in electron transfer itself; Race, Herrmann et al. (1999) emphasized the need to keep in a separate cell compartment some toxic mitochondrial components.

In any case, in the passage from an anaerobic to an aerobic metabolism, a progressive loss of genes from the primitive symbiontic alpha-proteobacterium occurred. These genes were transferred to the nucleus, which became a chimera incorporating contributions from different, both archaebacterial and eubacterial, progenitors. Such unidirectional transfer has been interpreted in different ways (Race, Herrmann et al. 1999; Blanchard and Lynch 2000; Dyall and Johnson 2000; Selosse, Albert et al. 2001).

The fact that with few exceptions in plants, unidirectional gene transfer occurred from organelles to nuclei, and not vice versa, may be explained easily, because in general a transfer from multicopy genome to single-copy genome is facilitated greatly. In a multicopy genome the adaptation of foreign material is much easier, due to the higher plasticity and flexibility of the endogenous genome. Probably, the organelle gene copy transferred to the nucleus should behave at first as a pseudogene in the new environment; then it could acquire all requisites to become an active nuclear gene. A possible primitive nuclear copy of the same gene could have then been lost. Among the features acquired by organellar proteins in the nucleus is the transit peptide (i.e., the presequence necessary to be imported into organelles). Interestingly, nuclear copies of mitochondrial DNA, apparently nontranscribed pseudogenes, are rather abundant in several eukaryotic species, including plants and animals (Bensasson, Zhang et al. 2001).

Another plausible explanation for a small genome in organelles could be positive selection. The advantages presented by small, compact genomes have often been discussed with regard to the genomes of prokaryotes. In the case of organelles, a compact genome should favor both its replication and expression. The case of yeast, in which rho⁻ mutants carrying a deleted mitochondrial genome can survive in nonrespiratory conditions, is often cited to support this thesis. In vitro and in vivo reduction of organelle DNA has been observed in other organisms. Indeed, progressive accumulation of organelle deletions is common in multicellular organisms, including humans. This phenomenon is observed in aging of human tissues, where deleted forms of mtDNA are found, although a real selective advantage for smaller molecules in human mitochondrial DNA replication has not yet been demonstrated clearly.

Finally, an explanation can be attempted in the light of the Muller ratchet principle. This is a fundamental rule of population genetics which postulates that asexual genomes (e.g., organellar genomes) would be doomed to extinction, due to the rapid accumulation of deleterious mutations which cannot be removed by recombination. However, in this context, we wonder why the cells did not eliminate the organellar DNA completely.

Indeed, there are many known mechanisms that might be utilized by mitochondria to maintain a functionally active organelle genome; whatever the mechanism, it has become evident that the mitochondrial genetic system is subject to strong selective pressure and to survive has to develop strategies to slow down or halt the ratchet. Among these we can mention the synergistic epistasis, the bottleneck of mitochondrial DNA (mtDNA) molecules during oocyte maturation, and the female germ cells atresia (Bergstrom and Pritchard 1998; Krakauer and Mira 1999).

The *epistasis hypothesis* suggests that the accumulation of deleterious mutations in a genome results in a decrease of individual fitness, to a point where no more mutations can be fixed, as they would be incompatible with fitness itself (this is not true, of course, for neutral mutations). For this reason, evolutionary rate slows over time, which is not the case, however, for animal mitochondria. In human and other metazoans, a severe bottleneck restricting the number of mtDNA molecules passing through the germline (Bergstrom and Pritchard 1998) is taking place. This mechanism is essential in maintaining mitochondrial genetic quality over evolution, as it is able to restore genetic homoplasmy among descendants; moreover, the bottleneck improves mitochondrial performance and, over a long time scale, acts to slow progression of the ratchet.

Another mechanism related to the mt genetic bottleneck is female germ cell atresia, which reduces the population of female germ cells to a small fraction of that present in early fetal life. Recent studies have demonstrated that this process not only assures the selection of high-quality mitochondria necessary for descendants but is also able to retard the Muller ratchet. Indeed, there is a significant positive correlation between the number of offspring and the number of mitochondria in germline cells and a negative correlation between the number of mitochondria in oocytes and the fraction of follicles undergoing atresia (Krakauer and Mira 1999). Species with small litters need to have high-quality mitochondria in oocytes to be sure that descendants are fit for survival; for this reason, the mitochondrial bottleneck needs to be stricter. Indeed, this mechanism decreases the number of mitochondria to a level that makes evident the difference in functionality among cells: only cells with healthy mitochondria will be able to escape the apoptotic process occurring during atresia. Based on the above, it appears clear that a central role is played by mitochondria in the fate and evolution of the eukaryotic cell.

It has recently been demonstrated that mitochondria are involved in many key cellular processes in addition to oxidative phosphorylation (which supplies up to 95% of cell energy). Mitochondria, indeed, contribute to apoptosis (Desagher and Martinou 2000). One of the first steps in this process is the loss of mitochondrial membrane potential, called *permeability transition*, which causes a fall in ATP production through disruption of the electron transport chain and the release of caspase activators such as apoptotic-inducing factors (AIF) and cytochrome *c*.

Mutations in mtDNA causing mitochondrial dysfunction have been associated with aging as well as to age-related degenerative diseases such as Huntington's, Parkinson's, and Alzheimer's diseases (Schon 2000; DiMauro and Schon 2001). Such mutations can result in severe disease phenotypes; the clinical manifestation of these diseases often involves muscles and the nervous system [e.g., in MELAS (mitochondrial encephalomyopathy, lactic acidosis, and stroke-like episodes), MERRF (myoclonic epilepsy and ragged red muscle fibers), and LHON (Leber hereditary optic neuropathy)], but they can also affect various organ systems, such as blood, colon, ear, eye, heart, kidney, liver, and pancreas.

Changes in the structure, function, and number of mitochondria play an important role in carcinogenesis, and several studies have shown an increase in mitochondrial gene expression in neoplasic cells. Somatic mutations in mtDNA have been identified in a variety of cancer (i.e., colorectal, gastric, papillary thyroid, and renal carcinomas) (Bianchi, Bianchi et al. 2001). mtDNA also seems to determine the cellular response to cancer therapeutic agents.

With regard to the evolutionary rate of mtDNA evolution, the data are very variable, since the rate is dependent on the properties and structure of the genome of that particular organism/lineage.

In plants, few studies reported analysis on the evolutionary rates of mtDNA; however, it has been shown that nuclear genes evolve faster than chloroplast genes, which evolve faster than mitochondrial genes (Wolfe, Li et al. 1987; Eyre-Walker and Gaut 1997). In animals, the most abundant data are those related to vertebrates, particularly to mammals. In vertebrate genes, owing to the rare occurrence of insertions and deletions, evolution measurement is based essentially on the analysis of nucleotide substitutions. In this case, the mathematical model used to calculate evolutionary rates is crucial to obtain correct results. Indeed, the compositional asymmetry of mammalian mtDNA, with high GC-skew values, implies that the probability that each nucleotide has be substituted by any of the others is not the same, and thus the two transitions and the four transversions are not equiprobable. Because of the foregoing characteristics of mtDNA evolution, more general stochastic models, such as the stationary Markov model (SMM; Lanave, Preparata et al. 1984; Saccone, Lanave et al. 1990), also known as the general time reversible (GTR) model, are advisable to use, to take into account the base composition of the homologous sequences under investigation and to treat differently the two transitional ($A \leftrightarrow G$, $C \leftrightarrow T$) and the four transversional ($A \leftrightarrow T, A \leftrightarrow C, C \leftrightarrow G, G \leftrightarrow T$) types.

Using the Markov model, the average evolutionary rate of the mitochondrial genes and of specific tracts of the D-loop region has been calculated (Pesole, Gissi et al. 1999). To obtain accurate estimates of the absolute nucleotide substitution rate, only pairs of organisms sharing a recent ancestor have been considered to avoid the problem of saturation of nucleotide substitutions and with divergence times known with sufficient accuracy from other molecular and nonmolecular sources. Saturation is essentially due to multiple mutational events affecting the same nucleotide position and occurring with an apparent reduction of the number of nucleotide differences observed for longer divergence rates. The overall effect of saturation is thus an underestimate of evolutionary distances, mainly between evolutionarily distant sequences. Figure 7.6 reports the absolute nucleotide substitution rates of the mitochondrial functional regions, calculated as mean rate values for six closely related mammalian species belonging to the orders Primates, Carnivora, Cetacea, and Perissodactyla. In general, two classes of functional regions can be identified: (1) slowly evolving regions, including nonsynonymous sites, tRNA and rRNA genes, and a D-loop central domain (Sbisà, Tanzariello et al. 1997); and (2) fast-evolving regions, including synonymous sites and CSB and ETAS D-loop domains, defined according to Sbisà, Tanzariello et al. (1997). The nonsynonymous sites are the most slowly evolving and the synonymous sites the most rapidly evolving, the latter being about 16-fold faster than the former. tRNA and rRNA genes evolve at a rate slightly higher than the nonsynonymous sites (about twofold), with 16S rRNA about 1.5-fold faster than 12S rRNA. All these regions are subject to strong functional constraints. However, different from nonsynonymous sites, which are forced to retain a specific primary sequence, tRNA and rRNA genes are mainly constrained to maintain specific secondary and tertiary structures, thus allowing slightly higher variability at the level of their primary structures.

Within the D-loop region, the central domain evolves about twofold faster than the nonsynonymous sites but much slower than the two peripheral D-loop domains.

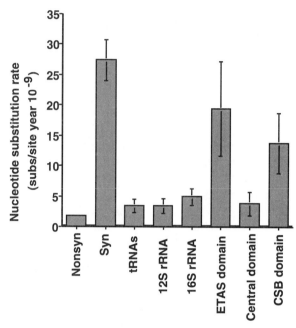

Figure 7.6. Absolute substitution rate of nonsynonymous sites (nonsyn), synonymous sites (syn), 12SrRNA, 16S rRNA, D-loop ETAS, and central and CSB domains of the mitochondrial genome calculated by comparing six closely related mammalian species. The standard deviation is also reported for each rate measured.

The central domain shows quite a homogeneous evolutionary rate among species, whereas in the CSB and ETAS domains strong rate heterogeneity has been found, as indicated by the high standard deviation (Fig. 7.6). The species-specific evolution of the D-loop region is due to its peculiar evolutionary pattern, since this region is prone to accept insertions/deletions as well as repeated sequences, particularly in ETAS and CSB domains (Saccone, Pesole et al. 1991; Sbisà, Tanzariello et al. 1997).

In the species considered, similar substitution rates have been found for tRNA, rRNAs, and protein-coding genes at the level of nonsynonymous sites. On the contrary, a rather high rate variability has been reported for the protein-coding genes of some mammalian orders (Honeycutt, Nebdal et al. 1995; Adkins, Honeycutt et al. 1996).

The mean synonymous and nonsynonymous absolute rates for each of the 13 protein-coding genes have been calculated for the same above-mentioned species pairs and are reported in Fig. 7.7. As expected, in each gene the synonymous rate is about one order of magnitude higher than the nonsynonymous one, and taking statistical fluctuations into account, it is approximately uniform between the various genes, with ND3 and ATP8 slightly slower and cytochrome *b* slightly faster. The nonsynonymous rate exhibits wide variations between genes, depending on the functional constraints, the cytochrome oxidase subunits (COI, COII, and COIII) being the slowest- and the ATP8 the fastest-evolving genes.

To compare the evolutionary dynamics of mtDNA and nDNA, nuclear- and mitochondrial-encoded tRNA, rRNA, and protein genes in the same species pairs

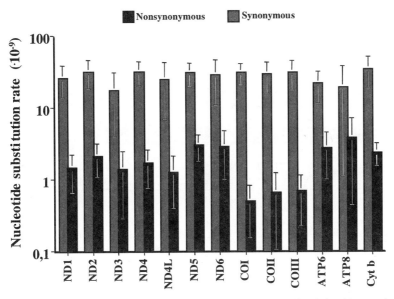

Figure 7.7. Mean synonymous and nonsynonymous rate for each of the 13 protein-coding genes of the mammalian mitochondrial genome. Rate values are plotted in log scale.

(human–chimps and rat–mouse) were analyzed (Pesole, Gissi et al. 1999). Figure 7.8 plots the nucleotide substitution rate ratios between the nuclear and the mitochondrial genome. The data show that, on average, the mitochondrial genome evolves faster that the nuclear genome but the fold difference is region-specific. In protein-coding genes, synonymous sites evolve about 22-fold faster in the mitochondrial than in the nuclear genome, whereas in nonsynonymous sites, a remarkable rate heterogeneity is observed in both genomes, as expected, due to the action of different selective constraints on the genes. The small and large rRNAs evolve about 19- and 4-fold faster, respectively, in the mitochondrial than in the nuclear genome. The highest rate difference has been found for mitochondrial tRNA genes, which evolve about 100-fold faster than their corresponding nuclear genes. For noncoding regions, the different structure of the mitochondrial and nuclear genomes makes the comparison quite meaningless.

To verify the existence of a molecular clock in the evolution of mammalian mtDNA, the relative rate test (RRT) of Muse and Gaut (1994) for protein-coding genes and of Muse and Weir (1992) for rRNA genes has been applied (Gissi, Reyes et al. 2000). RRT is able to detect substitution-rate heterogeneity along different gene lineages. The analyses were carried out on gap-free multialigned concatenated genes (supergenes) of the 12 H-stranded protein-coding genes and of the two ribosomal RNA genes. In the case of protein-coding supergenes (CDS), only the first and second codon positions were taken into account. In the case of ribosomal RNAs, ambiguously aligned sites, mostly adjacent to gaps, were excluded. In total, the number of sites analyzed was 7401 for the CDS and 2136 for the ribosomal supergene. The relative rate test (RRT) method is described in Sec. 6.11.2.

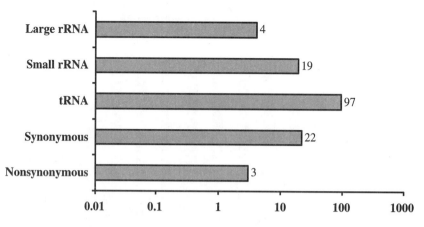

Figure 7.8. Ratio between the nucleotide substitution rates of the mitochondrial and the nuclear genome calculated by comparing homologous genes from human and chimp (synonymous and nonsynonymous rates) and mouse and rat (tRNA and rRNA rates).

Significant rate variations, not only between orders but even between closely related species of the same order, were found. Figure 7.9 plots the mean relative rates for mammalian interorder comparisons showing statistically significant rate differences. First and second codon positions (P12) of the 12 H-strand protein coding genes and rRNA genes have been taken into account for the analysis. A significant correlation can be observed between P12 and rRNA data, thus suggesting a high level of congruency between ribosomal and nonsynonymous sites. Primates and Proboscidea resulted in the fastest-evolving orders; Tubulidentata, Lagomorpha, and Perissodactyla were the slowest. Indeed, the rate variation observed did not exceed 1.8-fold between the fastest and the slowest order, thus supporting the suitability of mtDNA for drawing mammalian phylogeny (see Sec. 8.6.2).

7.5.3 Origin and Evolution of Plastids

The evolutionary history of plastids and their hosts has been unraveled by ultrastructural, biochemical, and molecular phylogenetics studies. In particular, the number and nature of plastid and tilacoid membranes, the types of pigments and of light-harvesting complexes, and more recently, the analysis of genome sequences have been used as the basis for drawing evolutionary relationships (Douglas 1998). The plastid-bearing organisms display an enormous variety in morphology and other properties; nevertheless, their origin can be traced back to a primary endosymbiosis event between a primitive eukaryote and a cyanobacterium-like ancestor (Fig. 7.10). After this event, two major lineages, rhodophyte and chlorophyte/methaphyte, and a minor group of glaucocystophytes have arisen. The process giving rise to plastids is called *primary endosymbiosis.*

Some eukaryotes appear to have acquired their plastids by secondary endosymbiosis; this is the case for Cryptomonads and Chlorarachniophytes, where soon after the symbiogenetic origin of chloroplasts from cyanobacteria, a chimaeric

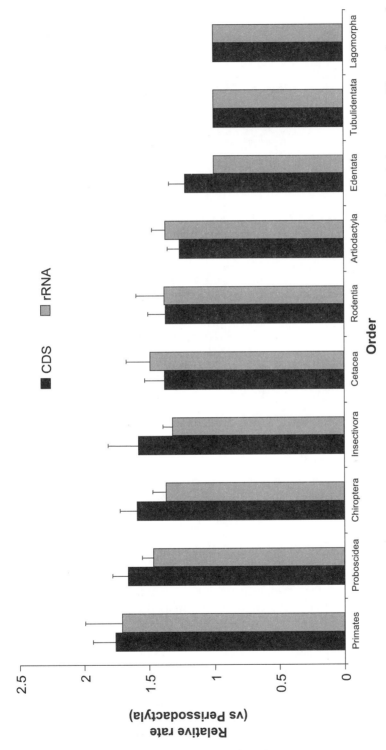

Figure 7.9. Mean relative rate for interorder comparisons with significant rate differences as defined by the relative rate test. First and second codon positions (P12) of the 12 H-strand protein-coding genes and rRNA genes have been taken into account for the analysis. Perissodactyla has been chosen as the reference order because of its lower evolutionary rate.

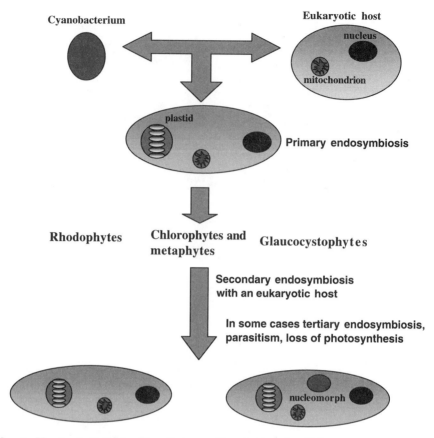

Cyanobacterium

Eukaryotic host

nucleus

mitochondrion

plastid

Primary endosymbiosis

Rhodophytes   Chlorophytes and
metaphytes   Glaucocystophytes

**Secondary endosymbiosis
with an eukaryotic host**

**In some cases tertiary endosymbiosis,
parasitism, loss of photosynthesis**

nucleomorph

**Heterokonts, Haptophytes, Dinoflagellates,   Cryptomonads, Chlorarachniophytes
Apiccomplexans, Euglenoids**

Figure 7.10. Possible model for the origin of exant plastids by primary, secondary, and tertiary endosymbioses.

integration of two evolutionarily distant eukaryotic cells took place. In the first taxon, the endosymbiont was a red alga, while Chlorarachniophytes acquired a green alga. In both of them a flagellate host contributed the nucleus, endomembranes, and mitochondria to the chimaera, whereas the photosynthetic endosymbiont provided its chloroplast, plasma membrane, and a second nucleus, called the *nucleomorph* (Maier, Fraunholz et al. 2000) (see Fig. 7.10). This secondary nucleus become miniaturized while keeping its envelope and three small chromosomes. As a result, these cells depend on four genomes, each encoding distinct protein synthesis machinery. Very recently, the three-chromosome nucleomorph of a cryptomonad, *Guillardia theta*, has been sequenced completely (Douglas, Zauner et al. 2001). This 551-kbp genome shows high gene density and short noncoding regions, and the GC content varies along the chromosomes, suggesting that there have been different mutational and selective pressures on base composition.

Analysis on this remnant of a nucleus suggested that the nucleus of the endosymbiont lost most of its genes over time, probably as the nucleomorph chromosomes cannot recombine; Many phytoplankton species undergo a similar process of secondary endosymbiosis, but all the genes of the nucleus of the engulfed symbiont were transferred to the host nucleus. So why have nucleomorphs been retained by cryptomonad and chlorarachniophyte cells? Some authors suggest that this relic has been kept as they encode something necessary for survival, probably proteins required to operate and maintain the chloroplast, and for some reason the gene transfer has been blocked in some way (Gilson and McFadden 2001).

The process of acquiring plastids is going on continuously through tertiary endosymbiosis events. In many cases, however, the extant organisms do not contain photosynthetic activity but only a vestigial form of plastid, whose DNA can reveal important steps in the story of that specific organism. Plastids surrounded by two membranes, such as Rhodophytes, probably arose by primary endosymbiosis, whereas those surrounded by three or more membranes may have arisen by multiple endosymbiontic events.

The study of plastids evolution accomplished so far on the basis of single genes has yielded many controversial results, probably because, as in many other molecular phylogenies, no great attention was given to the nature of genes being compared, most of which possibly derive from lateral gene transfer. Availability of the complete plastid genome sequence from a number of photosynthetic eukaryotes, as well as the progenitor cyanobacterium, Synechocystis, allows us to shed light on phylogenetic relationships (Douglas 1998). Among the organisms that contain vestigial nonphotosynthetic plastids, there are protozoan parasites, such as *Plasmodium*, *Toxoplasma*, and *Cryptosporidium*, belonging to the phylum Apicomplexa, whose name is derived from the former plastid, the apicoplast (Marechal and Cesbron-Delauw 2001). Figure 7.11 shows the gene content and organization of plastid-like DNA from *Toxoplasma gondii*.

The remaining apicoplast proteins are encoded by the nucleus. The function of apicoplasts is not completely understood, but a tight association with mitochondria suggests that there might be functional cooperation. Knowledge of apicoplasts provides a target for the development of drugs which act on the organelle but do not affect the mammalian host. Comparisons of available chloroplast DNAs, together with those based on mitochondrial sequences, support the clustering of green plants, red algae, and glaucophytes (Gilson, Maier et al. 1997).

7.6 FROM UNICELLULAR TO MULTICELLULAR STATE

In contrast to the single origin of other major transitions in evolution, the multicellular condition has arisen independently in the various life kingdoms, and several times also in the same phyla (Kaiser 2001). For example, prokaryotes, cyanobacteria (e.g., *Nostoc sp.*), and myxobacteria (e.g., *Myxococcus sp.*, *Stigmatella sp.*) adopted multicellularity in separate events. Interestingly, the genomes of multicellular prokaryotes are significantly larger than those of their closer unicellular relatives.

Also for eukaryotes, the multicellularity transition occurred independently in different taxa (e.g., sponges, animals, seed plants, fungi, brown and red algae). The fre-

Accession: NC_001799
Total Bases Sequenced: 34996 bp
Completed: Jun 28, 1999.

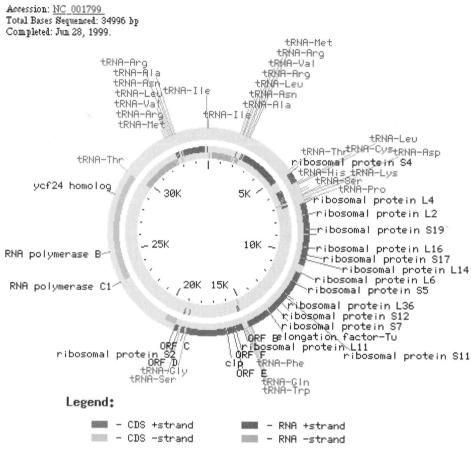

Figure 7.11. Structure of the 35-kb *T. gondii* apicoplast genome. The apicoplast organelle is *T. gondii* surrounded by four membranes, indicating that it was acquired by secondary endosymbiosis, probably from a green alga.

quent and independent transitions to multicellularity imply a large selective advantage, mainly concerning feeding and dispersion, although some price has to be paied like the restriction of regeneration capability only to germ cells.

Although a very thin line separates unicellular organisms from primitive multicellular forms, there is a concrete possibility of testing at the molecular level how multicellularity has been fixed in evolution: in other words, which and how many genes have been selected during the process giving rise to the differentiation responsible for a fitness increase in the multicellular state with respect to individually reproducing cells. A tentative search for such genes can be made using sequenced genomes such as those of *Nostoc* (9.76 Mbp) and Synechocystis (3.6 Mbp) as being representative of prokaryotes, *Drosophila* and *Caenorhabditis* as being representative of animals, and of yeast and *Arabidopsis* as representative of plants. The number of genes encoding components of signal transduction pathways is much higher in the multicellular cyanobacterium *Nostoc sp.* than in its close relative *Synechocystis sp.* The approach used by Aravind and Subramanian (1999) has been to look at the proteomes of uni- and multicellular organisms by using protein domain prediction methods (such as PSI-BLAST or HHMER; see Sec. 6.6) and databases of regulatory domains (such as SMART; see Sec. 5.9).

The results indicate that in both multicellular organisms, there were new—or an expansion of preexisting—regulatory domain families involved in extracellular adhesion or signal transduction as well as in DNA-binding transcription factors. In addition to the invention and proliferation of specific domains, new architectures created by juxtapositions of already existing domains were also detected in multicellular organisms. From a quantitative point of view, the recruitment of domains into new architectures was found to play the most prominent role (Copley, Schultz et al. 1999).

7.7 EVOLUTION OF THE NUCLEAR GENOME

The nuclear genome has two principal and peculiar features. The first is a genetic mosaicism that derives from its origin, as discussed previously. The second is genetic redundancy. Eukaryotic genomes have thousands of genes that have never been found in prokaryotes (Koonin, Aravind et al. 2000). Innovation can be generated through various mechanisms. It is observed that there is a general trend to increase genome size, particularly minimum size, proportional to organism complexity (see Table 2.3). As already stressed, genome size depends largely on expansion of the coding, and particularly of the noncoding, regions with repeated elements and/or repeated DNA regions. The scheme reported in Fig. 7.1 illustrates the major forces, both neutral and selective in nature, affecting genome size evolution (Petrov 2001).

7.7.1 Introns

It is well known that in eukaryotes, genes are interrupted by introns. These represent important noncoding elements that contribute to the increase in nuclear genome size. They are a peculiar feature of eukaryotic genes, although in some cases introns have been described in prokaryotes (e.g., in archaea). There are two major views about introns: the first hypothesis, called *introns early*, postulates that introns and RNA splicing are remnants of an RNA world and thus preceded the configuration of the intron-less genes we found in prokaryotes, which would have lost this feature for the sake of greater efficiency of gene expression.

The second hypothesis, called *introns late*, presumes that the prokaryotic state is the ancestral one and that intron insertion is a later process which is still going on. In other words, introns would have been acquired to favor functional diversity. To justify the additional weight on gene expression caused by the presence of introns, the advantages offered by this feature should be considered (Duret 2001). One of the most evident advantages consists of the fact that they enhance within-gene recombination and therefore increase selection efficiency (Hill-Robertson effects; Marais, Mouchiroud et al. 2001). Another important aspect is that introns often contain regulatory elements and, sometimes, other genes. Interestingly, it seems that the first intron(s) of a gene is generally the longest. It may also be assumed that the presence of introns decreases the possibility of recombination with pseudogenes,

TABLE 7.2. Coding Capacity of Completely Sequenced Eukaryotic Genomes[a]

Genome	Annotated Genes	Proteins with InterPro Matches
Fruitfly	23,964	9,815
Human	40,907	15,743
Weed	28,819	17,633
Worm	21,881	11,965
Yeast	7,235	3,979

[a] Counts of annotated genes and of InterPro matches are from the EuGENEs and InterPro resources, respectively (see Table 5.1).

particularly those based on retrotrasposition, and thus the risk of chromosome rearrangement. Intron evolution can be explained according to selectionist or neutralistic evolutionary models, which still require experimental demonstration. Recent studies on comparative genomics (Patthy 1999) have shown that the increase in genome size and gene number corresponds to an increase in intron length, thus creating a less compact genome structure.

Genome evolution implies an increased exon-shuffling process, producing modular multidomain proteins which became particularly relevant at the time of metazoan radiation. Alternative splicing is another fundamental mechanism to expand protein diversity, making possible a single gene to encode multiple proteins with opposing or subtle functional differences. As a result, the number of proteins in the proteome may exceed by orders of magnitude the total number of genes (Graveley 2001). A major challenge in the forthcoming proteomic era will be the structural and functional characterization of all possible transcripts potentially expressed by a gene.

7.7.2   Gene and Protein Number

Methods of gene prediction have allowed us to speculate and calculate roughly the number of genes in a genome. In Table 7.2 a comparative overview of completely sequenced genomes, in terms of coding capacity, is reported. Figures reported are not yet the true number of genes in these organisms. They just count gene records extracted from genome projects and following annotations and will hopefully converge on true counts as soon as the annotations become more accurate and validation experiments are carried out. Furthermore, the number of genes expressed by a genome does not correspond to the number of proteins. This is due mainly to alternative splicing that increases proteome complexity incredibly, but also to other posttranscriptional events, such as editing (Graveley 2001). For example, the ability to detect a wide range of sound frequencies in vertebrates derives from the multiple protein isoforms coded by the calcium-activated potassium channel (*slo*) gene produced by >500 alternatively spliced mRNA isoforms. More strikingly, the *Dscam* gene in *Drosophila* can generate about 38,000 different mRNA isoforms, that is, about twice the total number of genes in this organism.

For a better understanding of organismal life molecular requirements, it is advisable to look at the protein domains. Domain sharing among proteins is very common. In comparative studies between humans the two completely sequenced invertebrate genomes, and the yeast genomes, identical domain arrangements can be observed, mainly between human proteins but also between humans and other organisms. The proteins in which the largest number of domain types (nine) and the largest number of domains (130) are observed, so far are coded for by the human genome. Identical arrangements are found in many human proteins, and many are conserved in the various species. The largest domain arrangement shared by all four taxa is given by ornithine decarboxylase, containing 11 domains of two types.

A deeper investigation into protein similarities requires the implementation of clustering methods such as that proposed by Rost (1999) and modified by Li, Gu et al. (2001). According to the latter method, 3007 protein families were identified in the human proteome, 2041 of which were made by two members. The largest groups included L1 reverse trascriptase (139 members), immunoglobulin heavy chain (129 members), immunoglobulin light chain (124 members), zinc-finger proteins (104 members), and olfactory receptors (51 members). These investigations are clearly very important to an understanding of the gene makeup in different genomes. To meet this goal, we need better bioinformatic tools, including new specialized databases with reliable annotations.

7.7.3 Noncoding Elements

As we have already pointed out, only a small fraction of the genome, in humans about 1 to 2%, encodes proteins. The remaining part contains noncoding elements. In humans, about 44% corresponds to interspersed repetitive DNA, with the following major classes of interspersed repeats: SINES 13%, LINES 20%, LTR transposons 8%, and DNA transposons 3% (Lander, Linton et al. 2001). In addition, there are other classes of repetitive elements, including processed pseudogenes, simple sequence repeats (SSRs), segmental duplications, and tandemly repeated blocks (those at the centromeres and telomeres), amounting to about 50% of the genome. The distribution of repetitive elements varies across chromosomes, with some regions containing around 90% in intervals greater than 500 kbp, and in other regions, less than 2%.

Li, Gu et al. (2001) have recently analyzed human protein data applied to the International Protein Index (IPI; see the URL in Table 5.1) and found many occurrences of translated repetitive elements. Although most of them were found in "predicted" proteins, thus suggesting false positives in gene preditions, about 4% of known proteins were found to contain repetitive elements or truncated versions of them: in particular, L1 and Alu (Nekrutenko and Li 2001). The data indicate that the majority of repetitive elements have not been inserted directly into the open-reading frame but have become part of the protein by recruitment as novel exons through alternative splicing, which can sometimes extend or truncate the coding region. This *exon recruitment mechanism*, producing the insertion of repetitive elements in proteins, is not always deleterious, as expected, but in some cases may promote gene evolution and species differentiation.

7.7.4 Expansion of Gene Families

In the genome, the presence of strictly related genes, constituting gene families, is one of the aspects of genomic redundancy. Gene families can be expanded, and justification of this process builds up genome complexity; in other words, this is one of the strategies followed by the nuclear genome to enhance its information content. It takes place more frequently in complex organisms, where many family members are found and new interrelationships originate. For example, in metazoans, the increased genetic complexity of vertebrates may be considered as an "innovation" to enhance their morphological complexity, being the two complexity features strictly correlated. Vertebrates, however, are not the only organisms to employ gene family expansion, which, instead, takes place, although to a different extent, in all major eukaryotic lineages.

Gene families are generated by simple gene duplication, segmental genomic duplication or polyploidization of the genome, and by retrotrascription. The latter mechanism is very frequent and can often be recognized easily since it generates intron-less genes whose integration into DNA is witnessed by molecular events such as the generation of direct repeats in the flanking gene region. Gene expansion may generate different effects. In some cases, two or more duplicated genes have the same function, and this may have a well-defined meaning. As stressed in Sec. 7.2, gene redundancy is a device used by the cell in emergencies when a basic function fails. Indeed, for the preservation of a given function, even nonorthologous gene displacement can take place (see Table 7.1); this means that at the genome level, there are several mechanisms that assure functional reequilibrium in case of gene failure.

More often, new family members acquire new functions thus increasing genomic complexity because duplicated genes are freed from the constraining effects of natural selection by redundancy. This happens through molecular changes causing functional differences (in some cases it is possible to identify candidate changes; in other cases these changes are not so evident), or through changes in the regulation of specific genes, thus generating new regulatory circuits. The newly acquired functions may become lineage-specific and obviously make it difficult to find their strict functional equivalent in other lineages. In addition, depending on the specific constraints of the functions acquired, the evolutionary rate of a duplicated gene can also become lineage-specific. As a consequence, distance measurements between genes may become misleading, as they are dependent on the paralogous comparisons carried out (see Chapter 8).

It should be considered that the processes generating genetic redundancy can occur at specific branching times in phylogeny: for example, at Chordata radiation. This is particularly meaningful in evolution since the same type of redundancy may be transmitted to all derived lineages. The Hox gene family, described briefly below, is a clear example of this. However, additional gene-duplication events can occur within a given lineage, which contributes to generating further molecular redundancy and diversity. The acquisition of new genes and/or gene functions may be followed by gene loss. This also contributes to the nonequivalence between paralogous genes, even between close organisms, and represents an important issue to take into account for the use of model systems in experimental research.

From the observations above it is clear that the definition of gene family is not easy and unequivocal. Gene families can be classified on the basis of sequence similarity, functional similarity, and structural similarity; in many cases more than one criterion can be applied (Holm 1998). The concept of gene family can be amplified if specific functional domains (e.g., DNA binding domain, zinc-finger domain) are considered, which can be present in various, different genes/proteins. In this case, heterogeneous members will be grouped; thus it is preferable to use the term *superfamily*.

As a gene family is assumed to comprise a number of members derived from a common ancestor through one or more duplication events followed by sequence divergence and/or gene loss, the phylogenetic reconstruction of gene families overall is also very difficult. The availability of completely sequenced genomes such as those of *Saccharomyces cerevisiae*, *Drosophila melanogaster*, *Caenorhabditis elegans*, *Arabidopsis thaliana*, and *Homo sapiens* should allow us to compare their gene contents and calculate the number of gene families in each genome (Rubin, Yandell et al. 2000). In practice, this task has proved to be very difficult, since these organisms are distantly related, and many gene functions are unknown or only predicted. New family members are being identified continuously thanks to the powerful tools of postgenomics, which are opening new perspectives in gene family evolution.

Various attempts to list the principal gene families have been made and collected in specialized databases such as the COG database (Tatusov, Natale et al. 2001) (see Sec. 5.11). Here we present only a few examples of gene family expansion and evolution and refer the reader to the vast current literature for further examples. The Hox gene family is one of the best studied because it has a pivotal role in determining body plans and could explain the evolution of animal phyla (Meyer and Schartl 1999). The clustered organization of Hox genes, encoding a class of highly conserved transcription factors, provides an explicative example and a powerful opportunity to examine the events (e.g., gene duplication, gain and loss) which frequently take place in genomic evolution.

Hox genes are arranged in clusters and are expressed in a spatially collinear fashion. A phylogenetic survey of Hox genes shows an increase in gene number in some more recently evolved forms, particularly in vertebrates. Figure 7.12 shows a likely scenario for the evolution of Hox genes in vertebrate evolution (Meyer and Malaga-Trillo 1999). The increase in gene number has occurred through a two-step process involving first, gene expansion in cis of a primordial gene to produce 13 members forming a cluster, and second, cluster duplication in trans of the entire unit to form multiple clusters. As a consequence, Hox genes that occupy the same relative position along the chromosome are more similar in sequence and expression pattern than are adjacent genes on the same chromosome. Gene loss is also well substantiated in vertebrate Hox family evolution.

In mammals, 39 genes are located on four linkage groups on four different chromosome. Zebrafish has 42 genes in six clusters. For invertebrates, in the Cephalochordata *Amphioxus*, only a single Hox gene cluster has been found. This has been explained assuming a poliploidization of the genome: a possible two-round duplication [2R hypothesis of Ohno (1993)] in vertebrates and a three-round duplication in some fishes, as discussed later (see Sec. 7.8.5). However, this theory is still controversial and requires further investigation.

Other families that have attracted the interest of basic and applied researchers are those involved in oncogenesis, whose increasing availability of data at the transcriptomic and proteomic levels are providing new results and opening new fascinating scenarios. It has been reported that about 68% of human tumor genes have their counterparts in invertebrates. In this taxon the "equivalent" of the vertebrate genes composing a gene family is often present as a single product.

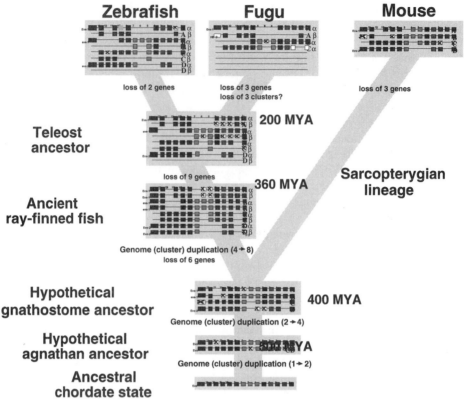

Figure 7.12. Evolution of the Hox gene cluster in vertebrates. Rectangles with crosses represent inferred gene losses, and white squares indicate pseudogenes. The datings of cluster duplications and gene losses are also shown on the tree. (Courtesy of Axel Meyer.)

It is well known that p53 is an oncosuppressor gene that plays multiple roles in many basic cellular functions connected to cell cycle regulation (DNA synthesis and repair, differentiation, programmed cell death). p53 is of great importance in basic science and in applied research for the development of new strategies in tumor diagnostics and therapy (Lane 1992; Levine 1997; Oren 1999; Oren and Rotter 1999; Deguin-Chambon, Vacher et al. 2000). p53 is a transcription factor that regulates several genes and is considered to be one of the most highly connected nodes in the cell, so is called the *hub* of the cell network system (Vogelstein, Lane et al. 2000). Yet, it is evident that an attack on p53 can interrupt basic cellular functions, and p53 divergence implies coevolution of target genes and regulatory pathways. For a long time, p53 was supposed to be a single-copy gene, different from other oncogenes and tumor suppressor genes. Great interest has therefore arisen in the scientific world following the quite recent discovery of two p53-related genes: p73 and p63 (Kaghad, Bonnet et al. 1997; Yang, Kaghad et al. 1998).

Obviously, such a discovery widens the scenario of cell cycle regulation and raises questions as to the functions of the new genes, their evolutionary origin and relationships, and in particular on whether these new genes could come into play when

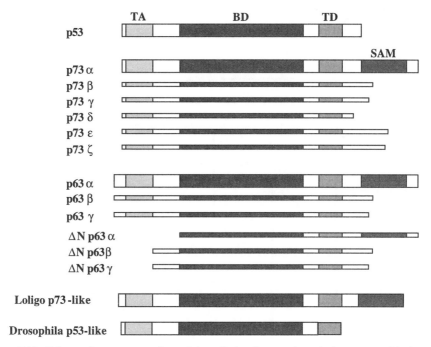

Figure 7.13. Schematic representation of the p53 family members in humans and in invertebrate homologs. p73 and p63 isoforms, derived from mechanisms of alternative splicing at the 3′ end of the gene, are also reported. ΔN-p63 isoforms are produced by a second internal promoter and initiation ATG codon in intron 3. TA, transactivation domain; BD, binding domain; TD, tetramerization domain; SAM, sterile alpha motif.

p53 function is lost. p73 and p63 show structural and functional resemblance to p53; nevertheless, to date, a complete functional redundancy has not been demonstrated (Levrero, De Laurenzi et al. 2000). The central domain, involved in DNA binding, and the tetramerization domain are structurally and functionally conserved throughout all family members; whereas other domains, such as the SAM domain involved in development, are randomly present. Figure 7.13 shows a schematic representation of the p53 family members in human and invertebrate homologs.

Because of p53 family member structural features such as gene length, close similarity, common presence of a carboxyterminal domain (SAM domain), and protein structural motifs, p63 and p73 are more closely related to one another than to p53. Thirty-two sequences of p53, three of p63, and four of p73 are presently available for vertebrates, and for invertebrates there are two p53-like sequences in mollusk and one sequence also identified recently in *D. melanogaster* and in *C. elegans*, respectively (D'Erchia, Pesole et al. 1999; D'Erchia, Tullo et al. 2001). The relatedness of the members of the p53 family can help us to understand the evolutionary history of this family.

Figure 7.14 shows the phylogenetic tree constructed with the NJ method (see Sec. 6.11.2) on amino acid similarity of the three proteins p53, p63, and p73. The pattern is very straightforward: It shows that p53, whose sequences are more numerous, although well conserved in vertebrates, evolves faster than p63 and p73. In the tree,

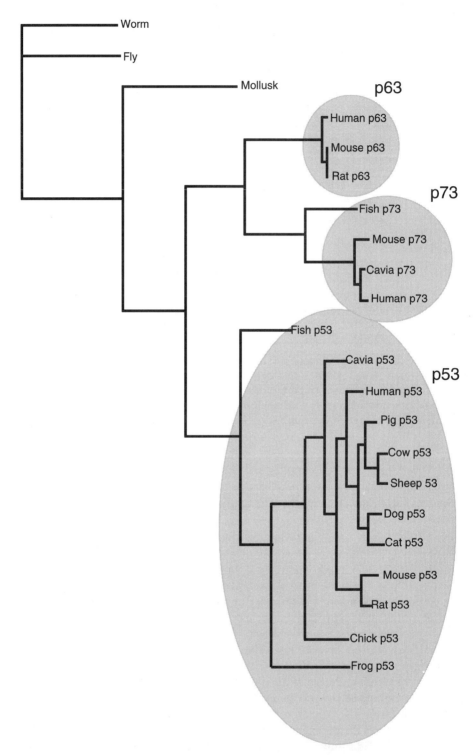

Figure 7.14. Phylogenetic tree of some representatives of the p53 gene family.

two main splits are observed. The common ancestor is the origin of p53 on one side and on the other of the precursor of p63 and p73. The closer relatedness of p63 and p73 may suggest that the second split is more recent and/or that these two proteins are evolving more slowly.

In invertebrates, the features of p53-like proteins are quite variable; some are more similar in length to p53, others to the two other members. This can easily be explained by the high divergence of the organisms. However, the functional role of the invertebrate genes and the relationships between vertebrate and invertebrate genes remain largely unknown. Obviously, to discover their evolutionary history, we need to know the sequences of many more p53-like genes in a broad range of animals. We also need data on their expression, which are at present available only in vertebrates. We know that in contrast to p53, both p73 and p63 show different transcripts, resulting from alternative splicing. Furthermore, the expression of these genes is differentially regulated and shows a different tissue distribution; there is evidence of a possible function of p63 and p73 in development and differentiation. p73 can, at least when overproduced, activate transcription from p53-responsive promoters but shows a degree of specificity for the promoters of target genes that is quantitatively distinct from the response mediated by p53 for p53 target genes. Furthermore, the roles of p63 and p73 in tumor suppression remain uncertain, as both genes appear not to be mutated in tumors and are not induced by any of the cellular stresses that cause DNA damage and activate p53.

The retinoblastoma (RB) is another oncosuppressor gene with a major role in cancer onset and progression. It is involved in cell growth, differentiation, and apoptosis by regulating cell cycle progression and gene expression. It belongs to a family consisting of three members: RB, RB2/p130, and p107. The function of retinoblastoma family proteins (pRB) is related to their ability to interact and modulate several proteins; main targets are the transcription factors E2F family and oncoprotein products of several DNA tumor viruses, such as E1A, SV40, and E7. In quiescent cells, pRBs are in their hypophosphorylated form and sequester E2Fs, blocking E2F-mediated cell cycle progression. During G1/S phase progression, the activation of cyclin-dependent kinases leads to the phosphorylation of pRB, the release of E2Fs, and the activation of E2F-dependent transcription. pRb/p105, pRb2/p130, and p107 are structurally and biochemically similar proteins carrying out analogous functional tasks. However, each exerts their growth-suppressive properties differentially in various cell lines, suggesting that although the proteins of members of the RB family proteins may complement one another, they are not fully functionally redundant. The RB carboxyl-terminal region contains the nuclear localization signal (NLS), and mutations at this site or deletion of the complete NLS can abrogate the protein-exclusive nuclear localization, yielding them confined to the cytoplasm and therefore determining the deregulation of cell cycle control (Cinti and Giordano 2000). pRBs are expressed in every tissue and genomic sequences are annotated for Rb1/p105: human, mouse, rat, and *D. melanogaster*; Rb2/p130: human, mouse, and rat; p107: human and mouse. In birds, expression of the three retinoblastoma family proteins has been reported (Kastner, Espanel et al. 1998).

An example of a superfamily is provided by Snail genes (Manzanares, Locascio et al. 2001). This superfamily groups zinc-finger transcription factors belonging to two independent, albeit closely related families, Snail and Scratch, involved in the embryonic development of the mesoderm and the nervous system of vertebrate and non-vertebrate chordates. The interest in this family arises since modifications in their functions could determine changes in body plan during evolution, and it has been shown to be involved in epithelial tumor progression. All the members share a conserved organization; they are composed of similar C terminals containing four to six zinc fingers, whose alignment highlights the presence of consensus sequences which serve as signatures allowing unambiguous assignment of any member to one of the families and construction of phylogenetic trees. It appears that an early duplication of a Snail gene at the base of metazoan radiation gave rise to two genes, Snail and Scratch, since identifiable members of both families can be found from nematodes to humans. *D. melanogaster* has three members of each family, Snail and Scratch, linked on chromosomes 2 and 3, respectively, suggesting intrachromosomal duplication events. In contrast, the Snail and Scratch genes of *C. elegans* and humans are located on different chromosomes, indicating that duplication in each family occurred independently in different groups of protostomes and in vertebrates and also that the clustered organization is not an evolutionarily conserved feature of this superfamily.

The transcription factor gene families represent in our opinion a clear example of the possible meaning of gene family expansion to increase the genetic complexity of higher organisms and thus to create new evolutionary opportunities. In addition, it demonstrates that evolutionary studies may be useful in understanding the specific role of the family members in basic physiological and pathological cellular processes such as cancer.

7.7.5 Genome Duplication

Genome duplication events play an important role in evolution since, as discussed above, they lead to novel functions and novel morphologies. It is well known that the concept of genome duplication was put forth for the first time by Ohno in 1970 (Ohno 1993), who suggested two rounds or complete duplication (2R) of the genome in the early evolution of vertebrates. It remains a very valid working hypothesis although current date do not allow to prove or disprove alternative hypotheses (Skrabanek and Wolfe, 1998).

Wang and Gu (2000) have dated a probable first and second genome duplication event analyzing 26 gene families as around 594 Mya for the first and 488 Mya for the second duplication in vertebrates. For some gene families (as for Hox genes in some fishes) the hypothesis of a third duplication could arise, but this remains rather unlikely and requires further investigation.

Hughes (1999) has questioned the 2R hypothesis on the basis of a phylogenetical reconstruction of nine protein families that are important in development. In particular, it was found that in only one case was topology of the type ((AB),(CD)), A being the sister group of B and C a sister group to D, in agreement with the 2R hypothesis, supported. In all other cases, topology of the type ((A),(BCD)), in which one paralog diverges prior to the others, was found.

Abi-Rached, Gilles et al. (2002), while investigating the major histocompatibility complex (MHC) paralogous regions, showed that duplications of this region occurred after the divergence of cephalochordata (lancelets) and vertebrates but before the gnathostomata (jawed vertebrate) radiation.

All these data and the relative controversies are based on the methods and assumptions used in the analyses and demonstrate clearly that further data and measurements are needed. In any case, it is evident that redundancy of genetic material is an important basic mechanism for creating novelties.

In addition to large-scale events (big-bang mode) implying complete genome duplication, continuous small-scale duplications are believed to occur during vertebrate evolution (continuous mode). Small-scale duplications imply gene family expansion by tandem of segmental duplication. An extensive analysis of 749 vertebrate gene families indicated that large- and small-scale gene duplications both make a significant contribution during the vertebrate evolution (Gu, Wang et al. 2002).

Segmental duplications are defined as the transfer of sequences of length 1 to 200 kbp from one genome region to one or more sites. Their detection is not very easy because of the presence of high numbers of copy repeats and many insertions and deletions in genomes. They indeed represent a serious impediment to accurate and complete sequencing of the human genome and other large and very redundant genomes.

Two types of segmental duplications can be distinguished: *intrachromosomal duplications and interchromosomal repeats*. The pericentromeric regions represent the preferential sites of location of segmental duplications, which are also located largely in subtelomeric regions. In the human genome it has been estimated that about 5% consists of interspersed duplications (Eichler 2001). The events of segmental duplication appear to be very recent (<35 Mya), as judged by the very high similarity between the segments duplicated (>95% identity). Many duplications contain possibly functional genes and thus may very well play a role in new gene evolution and in human diseases.

7.7.6 Conclusion

The linear primary sequence maps are inadequate for a thorough genome description (Bridger and Bickmore 1998). To reach a higher knowledge level as to the genome as a whole, we need to depict the spatial relationship between structural and functional properties: in other words, acquire a genome view in both space and time. Future genome maps should embody the richness of information we have not only from molecular genetics, but also genome topography, including new avenues of investigation by suitable large-scale experimental techniques, most of them still to be developed.

CHAPTER 8

MOLECULAR PHYLOGENY

8.1 INTRODUCTION

Another area that will be greatly affected by genomics is that related to molecular phylogeny and taxonomy. These disciplines have so far been based essentially on a comparison of sequences of single genes, RNA molecules, or proteins—thus on *microevolution*. Investigations of the molecular basis flourished in the 1970–1990s, due to the increasing availability of gene and protein sequences. In that period, research and debates also focused on the best methodology to measure genetic distances and build phylogenetic trees—arguments still not solved. Obviously, the detection of orthologs then became crucial, particularly in view of the use of nuclear genes in molecular phylogeny. For mitochondrial coded genes, this issue would be irrelevant, owing to their apparent monophyletic origin, particularly evident in metazoans. However, the problem is still centered on the validity of this genome for establishing species relatedness. Now we are dealing with complete genomes whose evolutionary characteristics have been discussed previously and whose comparisons are not easy to carry out. The scenario appears rather complex, but we cannot forget that it is time to change the focus of evolutionary studies, from microevolution to macroevolution, and gain a more complete vision.

In the paragraphs that follow we have selected for comment and discussion some topics we believe to be of central interest in guiding researchers in the new discipline of evolutionary genomics. In particular, we discuss briefly some general concepts on which molecular phylogeny is based: the molecular clock and the important definition of orthology and paralogy for sequences to be compared. The new perspectives opened up by genomics are illustrated here using two examples: the tree of life for the relationships between distantly related taxa, and the phylogeny of mammals as a concrete example of a problem still not completely solved on which many studies of molecular and nonmolecular nature already exist.

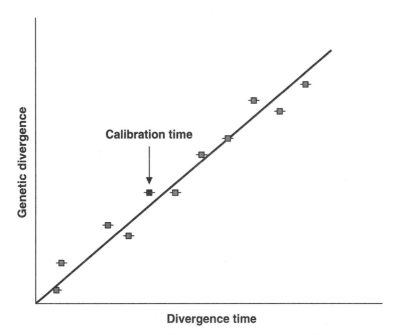

Figure 8.1. According to the molecular clock hypothesis, the genetic divergence between two homologous sequences, usually measured in terms of substitution/site, is linearly correlated with the corresponding time of divergence. Under this hypothesis, if a divergence time between two sequences is known from independent sources, all other times can be inferred from inferred genetic distances.

8.2 MOLECULAR CLOCK

It should be remembered that most molecular phylogeny has been, and still is, based on an important concept: the molecular clock hypothesis, which is worth discussing here. As a starting point, we should stress that it is usually assumed that molecular evolution is a divergent process, a notion which in some cases may be violated. In particular, the molecular clock concept postulates that biological molecules evolve, within statistical fluctuations, at a rather constant rate, thus allowing a measure of their genetic distances based on the divergence time between two markers generally inferred from other, nonmolecular sources (Fig. 8.1). Since its first enunciation by Zuckerkandl and Pauling (1965b), this hypothesis has received both strong support and ferocious criticism. In any case, as we have stressed, so far it is at the basis of most distance measurements, not only between molecules but also between organisms.

In the light of recent findings from complete genome sequencing, obviously this concept also needs to be revised. At first glance, from a genomics perspective, the molecular clock theory seems unrealistic both for bacteria, where extensive lateral gene transfer (LGT) takes place, and for the nuclear genome, because of the notions reported above: its mosaics origin and continuous gain and loss of genetic material. In addition, great difficulty rests in our ability to find true orthologs (see the next paragraph) and in the choice of the "right" molecule to measure distance in a given time span. The methodology used is an additional factor, together with the correct measurements of statistical fluctuations, since we should not expect biological molecules to demonstrate perfect, metronomic behavior.

As regards the choice of suitable molecules, rRNAs or rRNA genes might appear to be the best clock sequences because they are present in all cells; their import from the outside (e.g., through LGT) seems to be extremely improbable, and thus they should be inherited only vertically. However, trees constructed on ribosomal RNA genes are often weak or unreliable. An important reason is that ribosomal RNA genes, due to their highly developed secondary structure with several stems, need compensatory changes and thus use a kind of two-nucleotide language in the majority of their structure. If we consider that many regions of the rRNA genes would reach saturation of substitutions rapidly and that changes would depend on the nucleotide composition of the genome, we can easily understand how weak the trees constructed on rRNAs can be. In addition, base compositional bias of genes/genomes, multiple changes at the same site, and different variability among sites all represent additional factors interfering with molecular clock behavior.

In this scenario, the molecular clock hypothesis appears as a notion that should be rejected on theoretical grounds. Instead, in many instances, we have to admit that the molecular clock appears to be at work and data based on it are very frequently in agreement with other measurements using nonmolecular data. It has been suggested repeatedly we should consider the existence of a "local clock" instead of a "global clock." In this case, however, the definition of "global" and "local" in terms of time span, becomes even more difficult. Moreover, both clock types can be perturbed, for the reasons stated above.

It is generally true, particularly at the qualitative level, that when we have no convergent evolution, the more time that elapses, the more different the sequences become. But at the quantitative level, for measuring genetic distances more precisely, we have to take into account other problems as well as the new notions derived from genomics and postgenomics studies that we have treated above. From the latter, some new vision of molecular evolution mechanisms will shortly emerge. We wonder whether the relationships between changes and elapsed time are gradual, and whether variations can possibly have a remote influence or act on multiple sites simultaneously. It is too early to answer these questions; but genomics will certainly promote evolution of the way of thinking.

In conclusion, on the basis of the evidence available, a clocklike process seems to underlie the entire history of life, but the clock is perturbed continuously by various processes of great importance in evolution, not yet understood, which are time-dependent but not clocklike. All these processes may be driven by selection or randomly by events of a neutral nature.

8.3 SIMILARITY MEASURE: ORTHOLOGY VERSUS PARALOGY

Comparison is based on the possibility of measuring similarities between objects with respect to some of their properties. Sequence (or primary structure) similarity is one of the most important parameters when inferring relatedness between biological molecules. Comparisons can be done not only qualitatively but also quantitatively, the latter allowing the measurement of relative distance between molecules.

It should be stressed, however, that the measurement of genetic distances, particularly between distantly related sequences, could be very difficult. When sequence measures are extrapolated to distance between the organisms to which the sequences belong, the task is even more difficult. In the absence of data from other sources, measures at the molecular level may become very weak, lacking information on intermediate stages; thus we need to use arbitrary assumptions whose reliability often cannot be verified.

So far, comparison of nucleic acids or proteins having a certain degree of sequence similarity and thus presumably deriving from a common ancestor has been very popular not only for molecular taxonomy but also for gene prediction in newly sequenced genomes. Such studies, as stressed, belong to microevolution. However, it is clear today that the structure and complexity of a genome do not rely on coding regions alone. Repetitive elements, as well as gene organization and regulatory regions, have enormous importance for genome evolution, but their comparative analysis is not an easy task. We need to reach a more complex level of comparison, considering whole genomes—in other words, a *macroevolution* level.

Since comparison is based on similarity, to perform comparative analyses we need to define the parameters. Two sequences are said to be *homologous* if they share a common ancestor. Homology may be divided into subtypes. Sequences were defined by Fitch (1970) as *orthologous* when their divergence is due to species differentiation, while *paralogous* sequences would derive from gene duplication (Fig. 8.2). Such a definition is still commonly used not only by molecular evolutionists but also in genomics. In theory, orthologous genes should be unique in any given species; they should change by divergence and generally maintain the same function in the various lineages; whereas paralogous genes may be multiple and usually may acquire a new function or become pseudogenes (Gogarten and Olendzenski 1999).

Considering a very complicated subtype relationship between genes, Fitch (2000) argues that in a given set of organisms we can have more than one ortholog. For example, species A could have multiple orthologs compared with species B if in lineage A, but not in lineage B, gene duplication has occurred after the divergence of species A and B. Since after gene duplication, we cannot know how functions are conserved and what the rate of evolution is for each gene; in practice, it may be difficult to have a correct tree. The scenario depends essentially on the type of species (and gene/protein) considered and, more particularly, on the ability to detect the correct phylogenetic relationship. In practical terms, this approach may become highly subjective and strongly dependent on correct construction of the component tree, which is in itself not an easy task.

Actually, as is now being revealed by complete genome maps, in both prokaryotes and, more dramatically, in eukaryotes, it is extremely difficult to discriminate between orthologous and paralogous genes as defined previously. The criteria for such discrimination obviously cannot rely on the degree of similarity; if orthologous genes evolve very fast, in a given time span the degree of similarity may become very low in the short term. Conversely, for paralogous genes, if the duplication event is rather recent, the degree of similarity may be high. Yet another factor is that we cannot rely on gene function; in evolutionary time, after a duplication event in a given lineage, the old copy may be lost and the new one may retain the original function, whereas in another lineage, both copies might be conserved, with one of

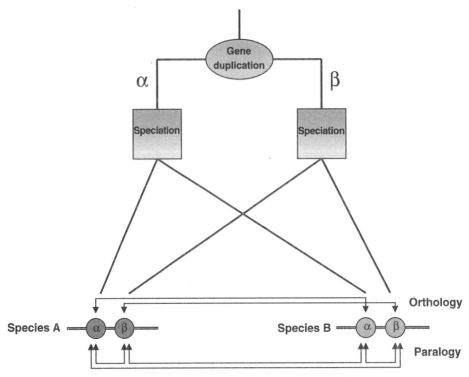

Figure 8.2. Orthologous and paralogous genes. Two genes, α and β, were derived from a duplication event. Then the ancestral species split into two species, A and B. The α and β genes in species A and B are orthologous, whereas the α genes are paralogous to all β genes, and vice versa.

them acquiring a new function. In addition, it is well known that genes having high sequence similarity can perform different functions. Various examples are offered by gene families whose members are very similar but often have different functions. In the p53 family, p63 and p73, which display a high degree of similarity with p53, regulate different metabolic pathways (Yang, Kaghad et al. 2002). Another striking example is offered by the enzymes/crystallines, a group of genes that code for proteins having enzyme functions but which can, at the same times, serve as structural proteins (Gonzalez, Hernandez-Calzadilla et al. 1994; Babiychuk, Kushnir et al. 1995). Conversely, the same function can be carried out by genes having very low or no significant similarity, as in gene displacement (see Sec. 7.2).

It is understandable how easily these situations can lead to wrong comparisons and cause erroneous results. In prokaryotes, for example, which have a very intensive LGT rate, we can have the worst impact on gene comparison if we assume orthology between genes deriving from different genetic sources (i.e., from different organisms). By mixing paralogs and orthologs, we have trees whose phylogenies cannot be correct not only for the sequences under consideration, but much more so for the taxa from which sequences derive.

Recently, the debate around the best definition—and the real meaning—of these terms has been opened repeatedly, particularly in connection with their frequent use in comparative genomic analysis. The concept of orthology and paralogy is in any case important for an understanding of the nature of forces acting on the evolution of the genome as a whole and for gene prediction. In this context, the function of a given gene comes into discussion and thus often the term *orthologous* is used with the meaning of *functionally orthologous*. This type of definition is clearly in contrast with that put forth by Fitch (1970). In relation to the very delicate problem of defining gene orthology for gene prediction, Attwood argues correctly that "in order to get the most from genomic data, we need . . . the adoption of a more holistic view of complex biological systems . . . since . . . Nature works by integration" (Attwood 2000a).

In conclusion, in light of the recent scenario emerging from complete genome sequencing, which is revealing the dynamic structure of genomes by continuous gains and losses of new genetic material, it is clear that it is very difficult to detect true orthologs. It is not a matter of best definition of the various types and subtypes of sequence relatedness, but simply that in most cases, the problem is the difficulty to detect correctly what has been "possibly precisely" defined. On the other hand, it is true that the terms *homology* and *orthology* are used with different meanings and with reference to different properties (e.g., functional or structural homology, etc.); thus we should be very cautious in assigning to them meanings out of practicality alone.

To sum up, since often, neither the evolutionary history of genes nor that of the organisms to which they belong is known, and probably will never be known, we cannot be sure of true gene orthology in most cases. In other words, because of differential gene duplication and selective gene loss in different evolutionary lineages, it is easy to see that we might be unable to recognize in two lineages genes whose divergence depends only on speciation, as postulated by the basic concept of evolutionary analysis, from genes born by duplication. It should be remembered that the original definitions of orthology and paralogy date to a time when little was known about the underlying processes in molecular evolution.

In the light of recent findings in comparative genomics, it would be better simply to use the term *homology* for genes sharing strong similarity and thus which are likely to have a common ancestor. We could even accept, in some very clearcut cases, classifying genes as *putative* or *bona fide* orthologous genes, but we cannot speculate further. For practical purposes we might tolerate the term *functionally* (or *structurally* or . . .) *orthologous* to mean that the genes have the same function (or structure or . . .) in different organisms. For the reasons stated above, we may expect a certain degree of paralogous or xenologous (i.e., genes acquired by LGT) contamination in specialized collections of orthologous genes, such as the COG database (Tatusov, Natale et al. 2001; see the URL in Table 5.1).

8.4 MOLECULAR PHYLOGENY IN THE GENOMICS ERA

In 1981, publication of the complete sequence of the first eukaryotic genome, the human mitochondrial genome (Anderson, Bankier et al. 1981), followed by that of many other mitochondrial DNAs, offered the first opportunity to compare genetic material using a complete genome instead of single genes. The mtDNA became suddenly a kind of model system for molecular evolutionary studies. With the advent of complete prokaryotic and nuclear genome sequencing and the explosion of comparative genomics, studies on molecular phylogeny have opened to a wider, global vision and are assuming a stronger validity. New concepts in macroevolution are emerging which can deeply affect several fundamental issues of molecular evolution. Some of these concepts were discussed earlier; here we emphasize the issues that have an impact in the field of molecular phylogeny.

One of the most important achievements in genomic evolution is the feature we have noted so often: genomes may be mosaics with traits derived from xenomic genetic material, as is the case with lateral gene transfer (LGT), particularly in prokaryotes. This process, together with the repeated origins of the same molecular trait as a result of reversals, parallelisms, or convergence (homoplasy) are the main factors that can make evolution not only divergent, but also convergent.

The other important issue directly affecting the estimate of molecular phylogeny from genomic data is a process called *nonorthologous gene displacement*, which is independent of the genetic distance between sequences (see Sec. 7.2). In addition to these processes, as discussed in Chapter 7, we can have duplication of single genes or genomic segments, polyploidization, and retrotransposition, all events generating new genetic material—thus a lot of paralogs—in a way that now appears unpredictable. But in evolution we can have a loss of genetic material as well, a very frequent process, which also affects the way molecular phylogeny is analyzed, in particular the recognition of orthologs (see also Chapter 7).

From the paradox of the C-value, and now from direct sequencing, we already know that genetic complexity and diversity depend largely on noncoding regions: namely, on the repeated elements that are generated by various mechanisms, as described above. This stresses the importance of the nature and suitability of the "characters" used in analyses of phylogenetic relationships. Obviously, at the molecular level the protein-coding genes remain the best tools. Instead, in the light of previous observations, the conservation of a short DNA tract in the genome or the comparison of regulatory regions (e.g., the D-loop region of vertebrate mt DNA) cannot provide reliable results.

Last but not least, in eukaryotes we have the problem of the coexistence, in the same cell, of more than a single genome: two genomes in all cells able to respire, the mitochondrial and the nuclear genomes; three genomes when plastids are present as in plant cells; and even more (see Chapter 3).

The presence of two or three genomes in the same organism, and thus the agreement between data generated from any single genome separately, is often a source of problems. On the one hand, it can be argued that more than one genome in a single cell can represent an advantage because we can compare two, sometimes three, genomes, instead of one. If data agree, the problem can be considered solved; but if there is disagreement, which data are correct? Can we assume that the evolutionary selective forces are the same for each cellular genome and the same for the species under comparison?

The diversity of organellar genomes among species may be as high as that of the corresponding nuclear genomes in the same organisms, but we cannot forget that within the cell, the various genomes must talk with one another, and to this end they have to establish regulatory circuits which are generally species-specific. In this scenario we wonder which molecular features best reflect the evolutionary history of the organisms under investigation.

These and other important implications derive from genomics analyses (i.e., when we consider a genome as a whole). It is clear that we cannot at present give precise answers because this field is in its infancy. However, these considerations should be taken into account in case of controversy and, more important, by authors when presenting results in their papers, to avoid the assertion of incontestability. The worst consequence of such an attitude is that data are accepted as such, without criticism, particularly by laypeople, almost as though they were in the Bible. Hence a sort of "frozen knowledge" is created, thanks to the rapid diffusion of data electronically and aided by the media. A way of thinking can easily be created that will be ever more difficult to modify or demolish.

8.5 INTERRELATIONSHIPS BETWEEN DISTANT TAXA: THE TREE OF LIFE

One of the major issues of molecular phylogeny is the division between Bacteria, Archaea, and Eukarya and their positions in the tree of life. The universal tree, based on ribosomal genes, supports the separated mono/holophyly of the three kingdoms. Can this notion continue to survive? More particularly, what are the relationships among the three kingdoms? According to several authors, recent analyses performed with more accurate methodologies would strongly support the monophyly of prokaryotes and cast serious doubts on the origin of eukaryotes and particularly on the generally accepted cluster of Archaea and Eukarya (Brinkmann and Philippe 1999).

By analyzing the data of only seven bacterial and three archaeal genomes, Koonin and Galperin (1997) concluded that in Bacteria and Archaea, protein sequences are overall highly conserved, which contrasts with the low conservation of genome organization (a few essential operons excepted). According to the authors, LGT and other nonvertical evolutionary mechanisms play a major role in the plasticity of the prokaryotic genome and in its evolution. The similarity between Archaea and Eukarya is limited to a set of products involved primarily in replication, transcription, and protein machinery, but this notion has also been disavowed by Philippe and Forterre (1999). On the basis of data collected so far, Koonin and Galperin (1997) depict the composition of the archaeal genome as shown in Fig. 8.3.

The recent hypotheses on the origin of the eukaryotic cell, discussed in Sec. 7.5, also seems to indicate a complicated pattern of relationships among the three kingdoms. Very recently, Brown, Douady et al. (2001) have analyzed 23 proteins from the genomic sequences of 45 species with the aim of extracting a credible history of extant life. The authors claim that some widely distributed genes have a common history that corresponds to the history of the organisms to which they belong. They essentially confirm the data obtained with rRNA.

As pointed out correctly by Olsen (2001), researchers can be classified as optimists and pessimists as to the possibility of deciphering the history of life using genomes or genome products. Indeed, those who believed in vertical evolution of the extant organisms and have, like Woese, introduced the idea of the tree of life, are now persuaded that life is probably more a network than a tree. In any case, we need further information. We should be prepared to revise the entire concept of evolutionary divergence and reach the conclusion that evolution has experienced

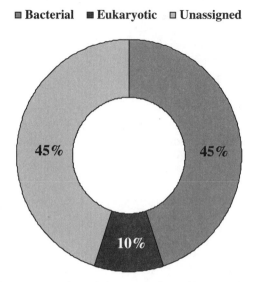

Figure 8.3. Schematic representation of the proportion of bacterial and eukaryotic proteins encoded in an archaeal genome. Proteins are classified as either bacterial or eukaryotic, depending on the greater similarity observed with bacterial or eukaryotic homologs. [Modified from Koonin and Galperin (1997).]

many pathways and the exchange of genetic material between organisms has taken place through a multiplicity of processes.

8.6 PHYLOGENY OF METAZOANS

There are many data of molecular and nonmolecular nature on animal organisms, particularly for vertebrates, mainly mammals. As stated earlier, the majority of molecular data rely on mtDNAs, which due to its reduced size, is now easy to sequence. However, the availability of nuclear genes, in some cases entire genomes, is rapidly increasing. This brings about the problems mentioned above, among which of prime importance is the agreement between data obtained by comparing mitochondrial and nuclear sequences. The latter topic is discussed briefly in the following section; the phylogeny of mammals has been chosen as an example because it is very likely the issue on which we possess the most abundant data sets.

8.6.1 Organellar versus Nuclear Taxonomy

The debate about the opportunity to use mitochondrial (mt) or nuclear genes for phylogenetic reconstruction has often been raised. If we consider the question of orthology, we must admit that mt genes, at least in some taxa (e.g., in metazoans, whose mtDNA size and gene content are constant), are largely superior to nuclear genes because they all share the same ancestry, and their sequences have diverged by speciation. Instead, nuclear genes are more tricky, and it is very difficult to identify true ortologs, for all the reasons discussed above (see also Chapter 7), in particular for the very frequent events of loss and gain of genetic material. Indeed, to date, mitochondrial genes are probably the most widely used for phylogenetic reconstruction, mainly in vertebrates. Saccone, Lanave et al. (1990) introduced the term *supergene* for mitochondrial genes to mean all genes concatenated to create a single product, which is then used in the analysis. In this way, the increased chain length gives better quantitative estimates of genetic distance with lower statistical fluctuations. The use of concatenated genes has been followed by other research groups. Both nuclear and mt genes have been linked together, in some cases even ribosomal genes and protein coding genes, a process that we do not find very accurate because it does not take into account the various constraints acting on the different products or in the different cellular compartments. Several authors have cast serious doubts about the use of complete mt genomes to infer a correct phylogeny. Naylor and Brown (1998) have claimed that mt genomes cannot resolve deep evolutionary divergences among vertebrates. According to the authors, the incorrect groupings would depend in part on base compositional bias among some taxa, creating an apparent case of convergent evolution. Takezaki and Gojobori (1999) have suggested that a change in amino acid composition in different lineages or homoplasies at sites with hydrophobic amino acids may lead to an incorrect phylogeny, which is in contrast with that commonly accepted by experts on the basis of other types of evidence.

Indeed, an incorrect alignment of the sequences, often due to subjective elimination of gaps, can easily generate wrong trees. Important controversies may arise from differences in the multialignment used to carry out the evolutionary analysis: for example, the monophyly and paraphyly of rodents obtained by comparing mtrRNAs by Frye and Hedges (1995) and Reyes, Pesole et al. (1998), respectively, were due only to a different starting multiple alignment of the sequences. We believe, however, that when all aspects discussed above are duly taken into account, the concatenated mt genes are capable of providing reliable results.

8.6.2 Phylogeny of Mammals

For obvious reasons, special attention has been paid to mammalian species that represent a good example of the state of the art in molecular phylogeny studies. For this group of organisms, advances made in the field of molecular biology have led to the availability of a huge amount of sequencing data from many species, which has added to previous knowledge based on morphology and fossil evidence.

Many hundreds of phylogenies of mammals have been published, some based on morphology and others based on molecular data (proteins and nucleic acids), yet there is no evidence that the different hypotheses of relationships among mammalian orders are converging on a single viewpoint. This uncertainty can be explained, at least partially, by the restricted and heterogeneous molecular data so far available for both nuclear and mitochondrial genomes, as well as from the limited knowledge of evolutionary processes at molecular level. We are optimistic that the use of more accurate methodologies and the availability of much more data, including several complete mitochondrial and nuclear genomes, will provide an increasing level of congruence.

The first classification of placental mammals was presented at the beginning of the last century (Gregory 1910) and was based on anatomical and paleontological data. Most of the morphology-based relationships among mammalian orders that have prevailed to date were established in this paper: (1) the grouping of Xenarthra and Pholidota; (2) the existence of a group called Archonta, which included Primates clustering with Scadentia and Chiroptera with Dermorptera; (3) the existence of the cohort Glires, which joined Rodentia with Lagomorpha; and (4) the split of Ungulata into two groups, the one containing Artiodactyla, Carnivora, and Cetacea, and the other containing Proboscidea, Sirenia, Hyracoidea, Tubulidentata, and Perissodactyla. Later, another important survey based on morphological data was published (Simpson 1945), and this has become a reference paper for the morphological classification of mammalian orders. Compared to the classification of Gregory, the classification of Simpson was less resolved, since he postulated that most orders appeared more or less simultaneously in the Cenozoic period as a consequence of a bushlike radiation process. Thus, he recognized only three groups: the cohort Glires as defined by Gregory; Unguiculata, comprising Insectivora, Chiroptera, Dermoptera, Primates, and Pholidota; and Ferungulata, comprising Carnivora, Tubulidentata, Perissodactyla, Artiodactyla, Proboscidea, Hyracoidea, and Sirenia. In the early 1990s, Novacek published a paper that summarized knowledge on mammalian phylogeny, based mainly on fossil and morphological data (Novacek 1992) as is presented in Fig. 8.4a. It can be observed that there is somehow a consensus tree between Gregory's and Simpson's ideas.

A completely different view is obtained when molecular data are used to infer mammalian phylogenies. In particular, recent surveys based primarily on the analysis of nuclear genes have produced new insight into the relationships among mammalian orders which is far from that described on the basis of morphological and fossil data (Madsen, Scally et al. 2001; Murphy, Eizirik et al. 2001). Phylogenetic analyses identify four primary superordinal clades, as shown in Fig. 8.4b: (1) Afrotheria, which comprises the orders Sirenia, Hyracoidea, Proboscidea, Macroscelidia, Tubulidentata, and some Insectivora; (2) Xenarthra; (3) Glires as a sister taxon to Scadentia, Dermoptera, and Primates; and (4) Laurasiatheria, which includes Carnivora, Pholidota, Perissodactyla, Cetacea + Artiodactyla, and Chiroptera. Compared to the morphological tree, the molecular tree shows changes in both the clustering of the various orders and the relative position of certain placental orders with regard to the base of the tree represented by Marsupialia. For example, Chiroptera are no longer included in the Archonta group along with Primates, Dermoptera, and Scadentia, but in the Laurasiatheria group; and Xenarthra and Pholidota are neither sister taxa nor placed at the base of the mammalian tree.

An important conclusion we can draw from these results is the lack of congruence between the different data sets (morphological versus molecular). Some of these differences may relate to real aspects of the data and their ability to reconstruct phylogeny, while others may relate to the analytical techniques employed. In particular, lack of a detailed and continuous fossil record and the use of maximum parsimony are the most frequent arguments against morphological trees. As to the poor quality of fossil records, this may be valid for the fossil record of some soft-bodied groups, such as worms or jellyfish, but there is no evidence for such a sceptical view of the fossil record of groups with hard parts, such as vertebrates in general and mammals in particular. The other controversial point is the use of maximum parsimony for tree reconstruction which assumes a very simplistic mode of evolution and can lead to misleading results. Despite this, fossils are extremely useful

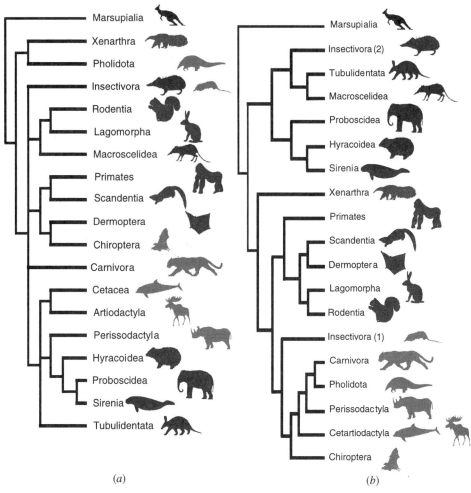

Figure 8.4. Phylogenetic tree of eutherian mammalian orders according to morphology (a) or molecular data (b). [Courtesy of Ole Madsen, as modified from (*a*) Novacek (1992) and (*b*) (Madsen, Scally et al. (2001).]

since they provide a unique view of the evolutionary history and allow the analysis of many primitive and derived characters that cannot be studied in extant taxa. As a consequence, morphological evidence from fossils may reveal patterns of evolution that could not be inferred using molecular data alone. With respect to molecular studies, it has been traditionally stated that their main drawback is the impossibility of studying extant species and the lack of representative sampling. Nevertheless, in the recent past it has been possible to obtain sequences from some extinct species such as Neanderthal man and different species of sloth and even the complete mitochondrial genome of two extinct moas. Moreover, nowadays sampling representation at the molecular level can be considered equivalent to that at the morphological level in terms of both sampled levels and number of species. In addition, molecular data have the advantage of including an ever-larger number of characters as well as the possibility of using complex and sophisticated models of evolution.

Thus, neither morphological nor protein or genome characters can be discarded as being consistently uninformative about phylogeny, since both of them have important roles in elucidating evolutionary history and phylogeny. Nevertheless, a better understanding of the underlying causes of the significant conflict between the molecular and morphological approaches is desirable for further resolution of relationships among mammalian orders. To achieve this goal, various strategies have been suggested:

- The revaluation of state assignments for both the molecular and morphological characters to correct errors in the original assessments of their homologies, especially in the case of fossil records, where these assignments were made a long time ago, when completely different views and methodological approaches were used.
- All phylogenetic methods are vulnerable to convergence to an incorrect phylogeny when their major assumptions about the evolutionary process are violated by mismatched data. Thus, one of the principal points is the reexamination of the methodology used currently to infer mammalian phylogeny at both the morphological and molecular levels.
- There is always a need to expand the databases of both molecular and morphological approaches. Efforts should also be made to broaden the representation of some mammalian orders in their existing molecular and morphological files, as in many cases each order is represented by a single species, lacking the obvious intraorder variability. Fulfilment of this line will provide the denser sampling of characters and taxa that is needed to test the generality of molecular versus morphological conflict and phylogenetic relationships among mammals.

Another important issue that we should consider at this point is the congruence between different molecular approaches (i.e., between nuclear and mitochondrial data sets). In the case of mammalian phylogeny, several studies have evidenced major differences compared to the results obtained with nuclear genes (see Reyes, Gissi et al., 2000, and references therein). However, it should be noted that the above mtDNA-based phylogenies were based on analyses including representatives from only 13 of 18 placental orders.

Remarkably, more recent analyses (Reyes et al., unpublished) using a much larger mtDNA dataset show a remarkable degree of congruence with the nuclear-based phylogeny reported by Murphy, Eizirik et al. (2001). This supports the idea that increasing the number of complete mitochondrial genomes in order to cover all orders, describe intraorder variability, and the use of more reliable algorithms for tree reconstruction, which would take into consideration the peculiar features of molecular evolution, will provide complete convergence between molecular data.

In conclusion, these issues, much as the problem of congruency between molecular and nonmolecular data, are not easy to settle. The result is that too many controversial data exist in the literature. However, the switch of focus from the gene/protein to the genome should help greatly to increase our knowledge of the relations between genotype and phenotype, and then on the genetic relationships among organisms.

APPENDIX

URLs CITED IN THE TEXT

Web Site	URL
3D-PSSM	www.sbg.bio.ic.ac.uk/~3dpssm
Arachne	www-genome.wi.mit.edu/wga
CATH Database	www.biochem.ucl.ac.uk/bsm/cath_new/index.html
CMR (Comprehensive Microbial Resource)	www.tigr.org/tigr-scripts/CMR2/CMRHomePage.spl
CodonW	www.molbiol.ox.ac.uk/cu/codonW.html
COILS	www.york.ac.uk/depts/biol/units/coils/coilcoil.html
Comparative RNA	www.rna.icmb.utexas.edu
CrossMatch	www.phrap.org
DPInteract Database	arep.med.harvard.edu/dpinteract
EMBOSS (European Molecular Biology Open Software Suite)	www.emboss.org
FMGP (Fungal Mitochondrial Genome Project)	megasun.bch.umontreal.ca/People/lang/FMGP/FMGP.html
Fugu Genomics Project	fugu.hgmp.mrc.ac.uk
GadFly	www.fruitfly.org/annot/index.html
GAP4	www.mrc-lmb.cam.ac.uk/pubseq
GeneExpress	www.cshl.org/genomere/supplement/houlgautte
GeneThreader	bioinf.cs.ucl.ac.uk/psiform.html
Insight II	www.msi.com/materials/insight
LBSN (List of Bacterial Names with Standing in Nomenclature)	www.bacterio.cict.fr
MarFinder	www.ncgr.org/Mar-Search
Markov Method	bighost.area.ba.cnr.it/BIG/Markov
MarWiz	www.futuresoft.org
Mascot	www.matrixscience.com/cgi/index.pl?page=/search_form_select.html
MatrixPlot	www.cbs.dtu.dk/services/MatrixPlot
MFold	bioinfo.math.rpi.edu/~mfold/rna/form1.cgi

Web Site	URL
MGED (Microarray Gene Expression Database) Group	www.mged.org
Modeller	guitar.rockefeller.edu/modeller/modeller.html
Mosquito Genome Draft	www.ensembl.org/Anopheles_gambiae
Mouse Genome Draft	mouse.ensembl.org
MOWSE	www.hgmp.mrc.ac.uk/Bioinformatics/Webapp/mowse
MrBayes	morphbank.ebc.uu.se/mrbayes
Multicoil	nightingale.lcs.mit.edu/cgi-bin/multicoil
OGMP (Organelle Genome Megasequencing Program)	megasun.bch.umontreal.ca/ogmp
ORF Finder	www.ncbi.nlm.nih.gov/gorf/gorf.html
ParCoil	nightingale.lcs.mit.edu/cgi-bin/score
PAUP* (Phylogenetic Analysis Using Parsimony)	www.lms.si.edu/PAUP/about.html
PHRED/PHRAP/CONSED	www.phrap.org
PHYLIP	evolution.genetics.washington.edu/phylip.html
PipMaker	bio.cse.psu.edu/pipmaker
RepeatMasker	ftp.genome.washington.edu/cgi-bin/RepeatMasker
RDP (Ribosomal Database Project)	rdp.cme.msu.edu/html
Rosetta (*ab initio* Fold Predictor)	depts.washington.edu/bakerpg
rRNA WWW Server	rrna.uia.ac.be
SAGEmap	www.ncbi.nlm.nih.gov/SAGE
SCOP (Structural Classification of Protein)	scop.mrc-lmb.cam.ac.uk/scop
STACK	www.sanbi.ac.za/Dbases.html
Swiss-Model	www.expasy.ch/swissmod/SWISS-MODEL.html
TIGR Assembler	www.tigr.org/softlab/assembler
TIGR Gene Indices	www.tigr.org/tdb/tgi.shtml
Tree-Puzzle	www.tree-puzzle.de
Unigene Build Procedure	www.ncbi.nlm.nih.gov/UniGene/build.html
VecScreen	www.ncbi.nlm.nih.gov/VecScreen/VecScreen_docs.html
Vista	www-gsd.lbl.gov/vista/index.html
Wentian Li Gene Finding Page	linkage.rockefeller.edu/wli/gene
YID (Yeast Intron Database)	www.cse.ucsc.edu/research/compbio/yeast_introns.html

REFERENCES

Aach, J., W. Rindone et al. (2000). Systematic management and analysis of yeast gene expression data. *Genome Res.* **10**(4): 431–45.

Aaronson, J. S., B. Eckman et al. (1996). Toward the development of a gene index to the human genome: an assessment of the nature of high-throughput EST sequence data. *Genome Res.* **6**(9): 829–45.

Abi-Rached, L., A. Gilles et al. (2002). Evidence of en bloc duplication in vertebrate genomes. *Nat. Genet.* **31**(1): 100–5.

Adachi, J. and M. Hasegawa (1996). Model of amino acid substitution in proteins encoded by mitochondrial DNA. *J. Mol. Evol.* **42**(4): 459–68.

Adachi, J., P. J. Waddell et al. (2000). Plastid genome phylogeny and a model of amino acid substitution for proteins encoded by chloroplast DNA. *J. Mol. Evol.* **50**(4): 348–58.

Adams, K. L., M. J. Clements et al. (1998). The *Peperomia* mitochondrial *coxI* group I intron: timing of horizontal transfer and subsequent evolution of the intron. *J. Mol. Evol.* **46**(6): 689–96.

Adams, K. L., D. O. Daley et al. (2000). Repeated, recent and diverse transfers of a mitochondrial gene to the nucleus in flowering plants. *Nature* **408**(6810): 354–57.

Adams, K. L., K. Song et al. (1999). Intracellular gene transfer in action: dual transcription and multiple silencings of nuclear and mitochondrial *cox2* genes in legumes. *Proc. Natl. Acad. Sci. USA* **96**(24): 13863–68.

Adams, M. D., S. E. Celniker et al. (2000). The genome sequence of *Drosophila melanogaster*. *Science* **287**(5461): 2185–95.

Adams, M. D., J. M. Kelley et al. (1991). Complementary DNA sequencing: expressed sequence tags and human genome project. *Science* **252**(5013): 1651–56.

Adams, M. D., A. R. Kerlavage et al. (1995). Initial assessment of human gene diversity and expression patterns based upon 83 million nucleotides of cDNA sequence. *Nature* **377**(6547 Suppl.): 3–174.

Adkins, R. M., R. L. Honeycutt et al. (1996). Evolution of eutherian cytochrome *c* oxidase subunit II: heterogeneous rates of protein evolution and altered interaction with cytochrome *c*. *Mol. Biol. Evol.* **13**: 1393–1404.

Allen, J. F. and J. A. Raven (1996). Free-radical-induced mutation vs. redox regulation: costs and benefits of genes in organelles. *J. Mol. Evol.* **42**(5): 482–92.

Alm, R. A., L. S. Ling et al. (1999). Genomic-sequence comparison of two unrelated isolates of the human gastric pathogen *Helicobacter pylori. Nature* **397**(6715): 176–80.

Almagor, H. (1983). A Markov analysis of DNA sequences. *J. Theor. Biol.* **104**: 633–45.

Altman, S., L. Kirsebom et al. (1993). Recent studies of ribonuclease P. *FASEB J.* **7**(1): 7–14.

Altschul, S. F. (1991). Amino acid substitution matrix from information theoretic perspective. *J. Mol. Biol.* **219**: 555–65.

Altschul, S. F. (1993). A protein alignment scoring system sensitive at all evolutionary distances. *J. Mol. Evol.* **36**(3): 290–300.

Altschul, S. F., M. S. Boguski et al. (1994). Issues in searching molecular sequence databases. *Nat. Genet.* **6**: 119–29.

Altschul, S. F. and W. Gish (1996). Local alignment statistics. *Methods Enzymol.* **266**: 460–80.

Altschul, S. F., W. Gish et al. (1990). Basic Local Alignment Search Tool. *J. Mol. Biol.* **215**: 403–10.

Altschul, S. F., T. L. Madden et al. (1997). Gapped BLAST and PSI-BLAST: a new generation of protein database search programs. *Nucleic Acids Res.* **25**(17): 3389–3402.

Alwine, J. C., D. J. Kemp et al. (1977). Method for detection of specific RNAs in agarose gels by transfer to diazobenzyloxymethyl-paper and hybridization with DNA probes. *Proc. Natl. Acad. Sci. USA* **74**(12): 5350–54.

Ames, B. N., M. K. Shigenaga et al. (1995). Mitochondrial decay in aging. *Biochim. Biophys. Acta* **1271**(1): 165–70.

Anderson, I. and A. Brass (1998). Searching DNA databases for similarities to DNA sequences: when is a match significant? *Bioinformatics* **14**(4): 349–56.

Anderson, J. B., C. Wickens et al. (2001). Infrequent genetic exchange and recombination in the mitochondrial genome of *Candida albicans. J. Bacteriol.* **183**(3): 865–72.

Anderson, N. L., A. D. Matheson et al. (2000). Proteomics: applications in basic and applied biology. *Curr. Opin. Biotechnol.* **11**(4): 408–12.

Anderson, S., A. T. Bankier et al. (1981). Sequence and organization of the human mitochondrial genome. *Nature* **290**: 457–65.

Andersson, J. O. and S. G. Andersson (1999). Insights into the evolutionary process of genome degradation. *Curr. Opin. Genet. Dev.* **9**(6): 664–71.

Andersson, S. G. and C. G. Kurland (1999). Origins of mitochondria and hydrogenosomes. *Curr. Opin. Microbiol.* **2**(5): 535–41.

Andersson, S. G., A. Zomorodipour et al. (1998). The genome sequence of *Rickettsia prowazekii* and the origin of mitochondria. *Nature* **396**(6707): 133–40.

Anfinsen, C. B. (1973). Principles that govern the folding of protein chains. *Science* **181**: 223–30.

Apweiler, R., T. K. Attwood et al. (2001). The InterPro database, an integrated documentation resource for protein families, domains and functional sites. *Nucleic Acids Res.* **29**(1): 37–40.

Apweiler, R., M. Biswas et al. (2001). Proteome Analysis Database: online application of InterPro and CluSTr for the functional classification of proteins in whole genomes. *Nucleic Acids Res.* **29**(1): 44–48.

Arabidopsis Genome Initiative (2000). Analysis of the genome sequence of the flowering plant *Arabidopsis thaliana. Nature* **408**(6814): 796–815.

Aravind, L. and G. Subramanian (1999). Origin of multicellular eukaryotes—insights from proteome comparisons. *Curr. Opin. Genet. Dev.* **9**(6): 688–94.

Ashburner, M., C. A. Ball et al. (2000). Gene ontology: tool for the unification of biology. The Gene Ontology Consortium. *Nat. Genet.* **25**(1): 25–29.

Attardi, G., A. Chomyn et al. (1990). Regulation of mitochondrial gene expression in mammalian cells. *Biochem. Soc. Trans.* **18**(4): 509–13.

Attimonelli, M., J. M. Cooper et al. (1999). Update of the Human MitBASE database. *Nucleic Acids Res.* **27**(1): 143–46.

Attwood, T. K. (2000a). Genomics: the Babel of bioinformatics. *Science* **290**(5491): 471–73.

Attwood, T. K. (2000b). The role of pattern databases in sequence analysis. *Brief. Bioinf.* **1**(1): 45–59.

Attwood, T. K., M. D. Croning et al. (2000). PRINTS-S: the database formerly known as PRINTS. *Nucleic Acids Res.* **28**(1): 225–27.

Babiychuk, E., S. Kushnir et al. (1995). *Arabidopsis thaliana* NADPH oxidoreductase homologs confer tolerance of yeasts toward the thiol-oxidizing drug diamide. *J. Biol. Chem.* **270**(44): 26224–31.

Bachtrog, D., S. Weiss et al. (1999). Distribution of dinucleotide microsatellites in the *Drosophila melanogaster* genome. *Mol. Biol. Evol.* **16**(5): 602–10.

Backert, S., B. L. Nielsen et al. (1997). The mystery of the rings: structure and replication of mitochondrial genomes from higher plants. *Trends Plant Sci.* **2**(2): 477–83.

Bailey, T. L. and C. Elkan (1994). Fitting a mixture model by expectation maximization to discover motifs in biopolymers. *Proc. Int. Conf. Intell. Syst. Mol. Biol.* **2**: 28–36.

Bairoch, A. and R. Apweiler (2000). The SWISS-PROT protein sequence database and its supplement TrEMBL in 2000. *Nucleic Acids Res.* **28**(1): 45–48.

Bao, Q., Y. Tian et al. (2002). A complete sequence of the T. tengcongensis genome. *Genome Res.* **12**(5): 689–700.

Barbour, A. G. (1988). Plasmid analysis of *Borrelia burgdorferi*, the Lyme disease agent. *J. Clin. Microbiol.* **26**(3): 475–78.

Barker, W. C., J. S. Garavelli et al. (2001). Protein Information Resource: a community resource for expert annotation of protein data. *Nucleic Acids Res.* **29**(1): 29–32.

Bashiardes, S. and M. Lovett (2001). cDNA detection and analysis. *Curr. Opin. Chem. Biol.* **5**(1): 15–20.

Bateman, A., E. Birney et al. (2000). The Pfam protein families database. *Nucleic Acids Res.* **28**(1): 263–66.

Batzoglou, S., D. B. Jaffe et al. (2002). ARACHNE: a whole-genome shotgun assembler. *Genome Res.* **12**(1): 177–89.

Beagley, C. T., J. L. Macfarlane et al. (1995). Mitochondrial genomes of Anthozoa (Cnidaria). In *Progress in Cell Research*, Vol. 5. F. Palmieri, S. Papa, C. Saccone and M. N. Gadaleta, eds. Amsterdam: Elsevier, pp. 149–53.

Beagley, C. T., N. A. Okada et al. (1996). Two mitochondrial group I introns in a metazoan, the sea anemone *Metridim senile*: one intron contains genes for subunits 1 and 3 of NADH dehydrogenase. *Proc. Natl. Acad. Sci. USA* **93**: 5619–23.

Beavis, R. and D. Fenyo (2000). Database searching with mass-spectrometric information. *Proteom. Trends Guide* **2000**(1): 22–27.

Beckman, K. B. and B. N. Ames (1998). The free radical theory of aging matures. *Physiol. Rev.* **78**(2): 547–81.

Begu, D., P. V. Graves et al. (1990). RNA editing of wheat mitochondrial ATP synthase subunit 9: direct protein and cDNA sequencing. *Plant Cell* **2**(12): 1283–90.

Bell, S. D. and S. P. Jackson (1998). Transcription and translation in Archaea: a mosaic of eukaryal and bacterial features. *Trends Microbiol.* **6**(6): 222–28.

Belmont, A. S., S. Dietzel et al. (1999). Large-scale chromatin structure and function. *Curr. Opin. Cell Biol.* **11**(3): 307–11.

Benne, R. (1985). Mitochondrial genes in trypanosomes. *Trends Genet.* **1**: 117–21.

Benne, R., J. Van den Burg et al. (1986). Major transcript of the frameshifted *coxII* gene from trypanosome mitochondria contains four nucleotides that are not encoded in the DNA. *Cell* **46**(6): 819–26.

Bennett, S. T. and I. J. Leitch (1995). Nuclear DNA amounts in angiosperms. *Ann. Bot.* **76**: 113–76.

Bennetzen, J. L. and B. D. Hall (1982). Codon selection in yeast. *J. Biol. Chem.* **252**: 3026–31.

Bensasson, D., D. Zhang et al. (2001). Mitochondrial pseudogenes: evolution's misplaced witnesses. *Trends Ecol. Evol.* **16**(6): 314–21.

Benson, G. (1999). Tandem repeats finder: a program to analyze DNA sequences. *Nucleic Acids Res.* **27**(2): 573–80.

Bentley, S. D., K. F. Chater et al. (2002). Complete genome sequence of the model actinomycete *Streptomyces coelicolor* A3(2). *Nature* **417**(6885): 141–7.

Berg, O. G. and P. H. von Hippel (1988). Selection of DNA binding sites by regulatory proteins. *Trends Biochem. Sci.* **13**(6): 207–11.

Berger, B., D. B. Wilson et al. (1995). Predicting coiled coils by use of pairwise residue correlations. *Proc. Natl. Acad. Sci. USA* **92**(18): 8259–63.

Bergstrom, C. T. and J. Pritchard (1998). Germline bottlenecks and the evolutionary maintenance of mitochondrial genomes. *Genetics* **149**(4): 2135–46.

Berk, A. J. and P. A. Sharp (1977). Sizing and mapping of early adenovirus mRNAs by gel electrophoresis of S1 endonuclease-digested hybrids. *Cell* **12**(3): 721–32.

Berman, H. M., J. Westbrook et al. (2000). The Protein Data Bank. *Nucleic Acids Res.* **28**(1): 235–42.

Bernander, R. (2000). Chromosome replication, nucleoid segregation and cell division in Archaea. *Trends Microbiol.* **8**(6): 278–83.

Bernardi, G. (2000a). The compositional evolution of vertebrate genomes. *Gene* **259**(1–2): 31–43.

Bernardi, G. (2000b). Isochores and the evolutionary genomics of vertebrates. *Gene* **241**(1): 3–17.

Bernardi, G. (2001). Misunderstandings about isochores. Part 1. *Gene* **276**(1–2): 3–13.

Bernardi, G., B. Olofsson et al. (1985). The mosaic genome of warm-blooded vertebrates. *Science* **228**(4702): 953–58.

Bernstein, H. J. (2000). Recent changes to RasMol, recombining the variants. *Trends Biochem. Sci.* **25**(9): 453–55.

Beroud, C. and T. Soussi (1998). p53 gene mutation: software and database. *Nucleic Acids Res.* **26**(1): 200–204.

Bianchi, N. O., M. S. Bianchi et al. (2001). Mitochondrial genome instability in human cancers. *Mutat. Res.* **488**(1): 9–23.

Binder, S., F. Hatzack et al. (1995). A novel pea mitochondrial in vitro transcription system recognizes homologous and heterologous mRNA and tRNA promoters. *J. Biol. Chem.* **270**(38): 22182–89.

Binder, S., A. Marchfelder et al. (1996). Regulation of gene expression in plant mitochondria. *Plant Mol. Biol.* **32**(1–2): 303–14.

Bittner-Eddy, P., A. F. Monroy et al. (1994). Expression of mitochondrial genes in the germinating conidia of *Neurospora crassa*. *J. Mol. Biol.* **235**(3): 881–97.

Black, W. C. T. and R. L. Roehrdanz (1998). Mitochondrial gene order is not conserved in arthropods: prostriate and metastriate tick mitochondrial genomes. *Mol. Biol. Evol.* **15**(12): 1772–85.

Blake, J. A., J. T. Eppig et al. (2001). The Mouse Genome Database (MGD): integration nexus for the laboratory mouse. *Nucleic Acids Res.* **29**(1): 91–94.

Blanchard, J. L. and M. Lynch (2000). Organellar genes: why do they end up in the nucleus? *Trends Genet.* **16**(7): 315–20.

Blanchette, M., T. Kunisawa et al. (1999). Gene order breakpoint evidence in animal mitochondrial phylogeny. *J. Mol. Evol.* **49**(2): 193–203.

Blattner, F. R., G. Plunkett III et al. (1997). The complete genome sequence of *Escherichia coli* K-12. *Science* **277**(5331): 1453–74.

Blom, N., S. Gammeltoft et al. (1999). Sequence and structure-based prediction of eukaryotic protein phosphorylation sites. *J. Mol. Biol.* **294**(5): 1351–62.

Blumenthal, T. (1995). Trans-splicing and polycistronic transcription in *Caenorhabditis elegans*. *Trends Genet.* **11**(4): 132–36.

Blumenthal, T. (1998). Gene clusters and polycistronic transcription in eukaryotes. *Bioessays* **20**(6): 480–87.

Bock, R. (2001). RNA editing in plant mitochondria and chloroplasts. In *RNA Editing*. B. L. Bass, ed. Oxford: Oxford University Press, pp. 38–60.

Bode, J., C. Benham et al. (2000). Transcriptional augmentation: modulation of gene expression by scaffold/matrix-attached regions (S/MAR elements). *Crit. Rev. Eukaryot. Gene Expr.* **10**(1): 73–90.

Boguski, M. S., T. M. Lowe et al. (1993). dbEST—database for "expressed sequence tags". *Nat. Genet.* **4**(4): 332–33.

Boguski, M. S. and G. D. Schuler (1995). ESTablishing a human transcript map. *Nat. Genet.* **10**(4): 369–71.

Bolotin, A., P. Wincker et al. (2001). The complete genome sequence of the lactic acid bacterium *Lactococcus lactis* ssp. *lactis* IL1403. *Genome Res.* **11**(5): 731–53.

Bonaldo, M., P. Jelenc et al. (1996). Identification and characterization of three genes and two pseudogenes on chromosome 13. *Hum. Genet.* **97**(4): 441–52.

Bonfield, J. K., K. Smith et al. (1995). A new DNA sequence assembly program. *Nucleic Acids Res.* **23**(24): 4992–99.

Boore, J. L. (1999). Animal mitochondrial genomes. *Nucleic Acids Res.* **27**(8): 1767–80.

Boore, J. L. and W. M. Brown (1994). Complete DNA sequence of the mitochondrial genome of the black chiton, *Katharina tunicata*. *Genetics* **138**(2): 423–43.

Boore, J. L. and W. M. Brown (1995). Complete sequence of the mitochondrial DNA of the annelid worm *Lumbricus terrestris*. *Genetics* **141**(1): 305–19.

Boore, J. L. and W. M. Brown (1998). Big trees from little genomes: mitochondrial gene order as a phylogenetic tool. *Curr. Opin. Genet. Dev.* **8**: 668–74.

Borner, G. V., M. Morl et al. (1996). RNA editing changes the identity of a mitochondrial tRNA in marsupials. *EMBO J.* **15**(21): 5949–57.

Bowler, L. D., M. Hubank et al. (1999). Representational difference analysis of cDNA for the detection of differential gene expression in bacteria: development using a model of iron-regulated gene expression in *Neisseria meningitidis*. *Microbiology* **145**(Pt. 12): 3529–37.

Bowman, S., C. Churcher et al. (1997). The nucleotide sequence of *Saccharomyces cerevisiae* chromosome XIII. *Nature* **387**(6632 Suppl.): 90–93.

Bowman, S., D. Lawson et al. (1999). The complete nucleotide sequence of chromosome 3 of *Plasmodium falciparum*. *Nature* **400**(6744): 532–38.

Bowtell, D. D. (1999). Options available—from start to finish—for obtaining expression data by microarray. *Nat. Genet.* **21**(1 Suppl.): 25–32.

Bradnam, K. R., C. Seoighe et al. (1999). G + C content variation along and among *Saccharomyces cerevisiae* chromosomes. *Mol. Biol. Evol.* **16**(5): 666–75.

Brazma, A., I. Jonassen et al. (1998). Predicting gene regulatory elements *in silico* on a genomic scale. *Genome Res.* **8**(11): 1202–15.

Brazma, A., A. Robinson et al. (2000). One-stop shop for microarray data. *Nature* **403**: 699–700.

Brazma, A. and J. Vilo (2000). Gene expression data analysis. *FEBS Lett.* **480**(1): 17–24.

Brendel, V., J. S. Beckmann et al. (1986). Linguistics of nucleotide sequences: morphology and comparison of vocabularies. *J. Biomol. Struct. Dyn.* **4**(1): 11–21.

Brendel, V., P. Bucher et al. (1992). Methods and algorithms for statistical analysis of protein sequences. *Proc. Natl. Acad. Sci. USA* **89**(6): 2002–6.

Brenner, S. E. (1999). Errors in genome annotation. *Trends Genet.* **15**(4): 132–33.

Brennicke, A., A. Marchfelder et al. (1999). RNA editing. *FEMS Microbiol. Rev.* **23**(3): 297–316.

Brent, R. (2000). Genomic biology. *Cell* **100**(1): 169–83.

Bridger, J. M. and W. A. Bickmore (1998). Putting the genome on the map. *Trends Genet.* **14**(10): 403–9.

Brinkmann, H. and H. Philippe (1999). Archaea sister group of bacteria? Indications from tree reconstruction artifacts in ancient phylogenies. *Mol. Biol. Evol.* **16**(6): 817–25.

Brochier, C., E. Bapteste et al. (2002). Eubacterial phylogeny based on translational apparatus proteins. *Trends Genet.* **18**(1): 1–5.

Brookes, A. J., H. Lehvaslaiho et al. (2000). HGBASE: a database of SNPs and other variations in and around human genes. *Nucleic Acids Res.* **28**(1): 356–60.

Brown, J. R., C. J. Douady et al. (2001). Universal trees based on large combined protein sequence data sets. *Nat. Genet.* **28**(3): 281–85.

Brown, M. P., W. N. Grundy et al. (2000). Knowledge-based analysis of microarray gene expression data by using support vector machines. *Proc. Natl. Acad. Sci. USA* **97**(1): 262–67.

Brown, W. M., E. M. Prager et al. (1982). Mitochondrial DNA sequences of primates: tempo and mode of evolution. *J. Mol. Evol.* **18**: 225–39.

Bruce, B. D. (2000). Chloroplast transit peptides: structure, function and evolution. *Trends Cell Biol* **10**(10): 440–47.

Brunak, S., J. Engelbrecht et al. (1991). Prediction of human mRNA donor and acceptor sites from the DNA sequences. *J. Mol. Biol.* **220**: 49–65.

Bucher, P. (1990). Weight matrix descriptions of four eukaryotic RNA polymerase II promoter elements derived from 502 unrelated promoter sequences. *J. Mol. Biol.* **212**(4): 563–78.

Bucher, P. (1999). Regulatory elements and expression profiles. *Curr. Opin. Struct. Biol.* **9**(3): 400–407.

Budd, M. E. and J. L. Campbell (1997). The roles of the eukaryotic DNA polymerases in DNA repair synthesis. *Mutat. Res.* **384**(3): 157–67.

Budd, M. E. and J. L. Campbell (2000). Interrelationships between DNA repair and DNA replication. *Mutat. Res.* **451**(1–2): 241–55.

Bulmer, M. (1990). The effect of context on synonymous codon usage in genes with low codon usage bias. *Nucleic Acids Res.* **18**(10): 2869–73.

Bult, C. J., O. White et al. (1996). Complete genome sequence of the methanogenic archaeon, *Methanococcus jannaschii*. *Science* **273**(5278): 1058–73.

Burckhardt, F., A. von Haeseler et al. (1999). HvrBase: compilation of mtDNA control region sequences from primates. *Nucleic Acids Res.* **27**(1): 138–42.

Burge, C. and S. Karlin (1997). Prediction of complete gene structures in human genomic DNA. *J. Mol. Biol.* **268**(1): 78–94.

Burger, G., D. Saint-Louis et al. (1999). Complete sequence of the mitochondrial DNA of the red alga *Porphyra purpurea*. Cyanobacterial introns and shared ancestry of red and green algae. *Plant Cell* **11**(9): 1675–94.

Burger, G., Y. Zhu et al. (2000). Complete sequence of the mitochondrial genome of *Tetrahymena pyriformis* and comparison with *Paramecium aurelia* mitochondrial DNA. *J. Mol. Biol.* **297**(2): 365–80.

Burke, J., D. Davison et al. (1999). d2••cluster: a validated method for clustering EST and full-length cDNA sequences. *Genome Res.* **9**(11): 1135–42.

Burke, J., H. Wang et al. (1998). Alternative gene form discovery and candidate gene selection from gene indexing projects. *Genome Res.* **8**(3): 276–90.

Campbell, A., J. Mrazek et al. (1999). Genome signature comparisons among prokaryote, plasmid, and mitochondrial DNA. *Proc. Natl. Acad. Sci. USA* **96**(16): 9184–89.

Cantatore, P., M. N. Gadaleta, M. Roberti, C. Saccone, A. C. Wilson (1987). Duplication and remoulding of tRNA genes during the evolutionary rearrangement of mitochondrial genomes. *Nature* **329**(6142): 853–5.

Cantatore, P., M. Roberti et al. (1990). Mapping and characterization of *Paracentrotus lividus* mitochondrial transcripts: multiple and overlapping transcription units. *Curr. Genet.* **17**(3): 235–45.

Cantatore, P., M. Roberti et al. (1989). The complete nucleotide sequence, gene organization, and genetic code of the mitochondrial genome of *Paracentrotus lividus*. *J. Biol. Chem.* **264**(19): 10965–75.

Cantatore, P. and C. Saccone (1987). Organization, structure, and evolution of mammalian mitochondrial genes. *Int. Rev. Cytol.* **108**: 149–208.

Carrodeguas, J. A., K. Theis et al. (2001). Crystal structure and deletion analysis show that the accessory subunit of mammalian DNA polymerase gamma, pol gamma B, functions as a homodimer. *Mol. Cell* **7**(1): 43–54.

Casjens, S. (1998). The diverse and dynamic structure of bacterial genomes. *Annu. Rev. Genet.* **32**: 339–77.

Casjens, S., M. Murphy et al. (1997). Telomeres of the linear chromosomes of Lyme disease spirochaetes: nucleotide sequence and possible exchange with linear plasmid telomeres. *Mol. Microbiol.* **26**(3): 581–96.

Casjens, S., N. Palmer et al. (2000). A bacterial genome in flux: the twelve linear and nine circular extrachromosomal DNAs in an infectious isolate of the Lyme disease spirochete *Borrelia burgdorferi*. *Mol. Microbiol.* **35**(3): 490–516.

Castresana, J., G. Feldmaier-Fuchs et al. (1998). The mitochondrial genome of the hemichordate *Balanoglossus carnosus* and the evolution of deuterostome mitochondria. *Genetics* **150**(3): 1115–23.

Catalano, D., F. Licciulli et al. (2000). Update of KEYnet: a gene and protein names database for biosequences functional organisation. *Nucleic Acids Res.* **28**(1): 372–73.

Cavalier-Smith, T. (1993). Kingdom protozoa and its 18 phyla. *Microbiol. Rev.* **57**(4): 953–94.

Celis, J. E., M. Kruhoffer et al. (2000). Gene expression profiling: monitoring transcription and translation products using DNA microarrays and proteomics. *FEBS Lett.* **480**(1): 2–16.

Chaitin, G. (1969). On the length of programs for computing finite binary sequences. *J. Assoc. Comput. Mach.* **16**: 407–22.

Chalfie, M. and E. M. Jorgensen (1998). *C. elegans* neuroscience: genetics to genome. *Trends Genet.* **14**(12): 506–12.

Chambaud, I., R. Heilig et al. (2001). The complete genome sequence of the murine respiratory pathogen *Mycoplasma pulmonis*. *Nucleic Acids Res.* **29**(10): 2145–53.

Chao, K.-M., W. R. Pearson et al. (1992). Aligning two sequences within a specified diagonal band. *CABIOS* **8**(5): 481–87.

Chappell, S. A., G. M. Edelman et al. (2000). A 9-nt segment of a cellular mRNA can function as an internal ribosome entry site (IRES) and when present in linked multiple copies greatly enhances IRES activity. *Proc. Natl. Acad. Sci. USA* **97**(4): 1536–41.

Chargaff, E. (1950). Chemical specificity of nucleic acids and mechanism of their enzymic degradation. *Experientia* **6**: 201–9.

Chen, C. W. (1996). Complications and implications of linear bacterial chromosomes. *Trends Genet.* **12**(5): 192–96.

Chen, J. J., J. D. Rowley et al. (2000). Generation of longer cDNA fragments from serial analysis of gene expression tags for gene identification. *Proc. Natl. Acad. Sci. USA* **97**(1): 349–53.

Cherry, J. M., C. Ball et al. (1997). Genetic and physical maps of *Saccharomyces cerevisiae*. *Nature* **387**(6632 Suppl.): 67–73.

Chesnokov, I., M. Gossen et al. (1999). Assembly of functionally active *Drosophila* origin recognition complex from recombinant proteins. *Genes Dev.* **13**(10): 1289–96.

Cho, Y. and J. D. Palmer (1999). Multiple acquisitions via horizontal transfer of a group I intron in the mitochondrial *cox1* gene during evolution of the Araceae family. *Mol. Biol. Evol.* **16**(9): 1155–65.

Cho, Y., Y. L. Qiu et al. (1998). Explosive invasion of plant mitochondria by a group I intron. *Proc. Natl. Acad. Sci. USA* **95**(24): 14244–49.

Choo, K. H. (2000). Centromerization. *Trends Cell Biol.* **10**(5): 182–88.

Chou, A. and J. Burke (1999). CRAWview: for viewing splicing variation, gene families, and polymorphism in clusters of ESTs and full-length sequences. *Bioinformatics* **15**(5): 376–81.

Chou, P. Y. and G. D. Fasman (1978a). Empirical predictions of protein conformations. *Annu. Rev. Biochem.* **47**: 251–76.

Chou, P. Y. and G. D. Fasman (1978b). Prediction of the secondary structure of proteins from their amino acid sequence. *Adv. Enzymol.* **47**: 45–148.

Christoffels, A., A. van Gelder et al. (2001). STACK: Sequence Tag Alignment and Consensus Knowledgebase. *Nucleic Acids Res.* **29**(1): 234–38.

Churcher, C., S. Bowman et al. (1997). The nucleotide sequence of *Saccharomyces cerevisiae* chromosome IX. *Nature* **387**(6632 Suppl.): 84–87.

Cinti, C. and A. Giordano (2000). The retinoblastoma gene family: its role in cancer onset and progression. Newly emerging therapeutic targets. *Emerg. Ther. Targets* **6**: 765–83.

Claros, M. G., S. Brunak et al. (1997). Prediction of N-terminal protein sorting signals. *Curr. Opin. Struct. Biol.* **7**(3): 394–98.

Claros, M. G. and P. Vincens (1996). Computational method to predict mitochondrially imported proteins and their targeting sequences. *Eur. J. Biochem.* **241**(3): 779–86.

Claros, M. G. and G. von Heijne (1994). TopPred II: an improved software for membrane protein structure predictions. *Comput. Appl. Biosci.* **10**(6): 685–86.

Clary, D. O. and D. R. Wolstenholme (1985). The mitochondrial DNA molecular of *Drosophila yakuba*: nucleotide sequence, gene organization, and genetic code. *J. Mol. Evol.* **22**(3): 252–71.

Claverie, J. M. (1999). Computational methods for the identification of differential and coordinated gene expression. *Hum. Mol. Genet.* **8**(10): 1821–32.

Clayton, D. A. (1982). Replication of animal mitochondrial DNA. *Cell* **28**: 693–705.

Clayton, D. A. (1991). Replication and transcription of vertebrate mitochondrial DNA. *Annu. Rev. Cell Biol.* **7**: 453–78.

Clayton, D. A. (1992). Transcription and replication of animal mitochondrial DNAs. *Int. Rev. Cytol.* **141**: 217–32.

Clayton, D. A. (2000). Vertebrate mitochondrial DNA—a circle of surprises. *Exp. Cell Res.* **255**(1): 4–9.

Clegg, M. T., B. S. Gaut et al. (1994). Rates and patterns of chloroplast DNA evolution. *Proc. Natl. Acad. Sci. USA* **91**(15): 6795–6801.

Cole, S. T., R. Brosch et al. (1998). Deciphering the biology of *Mycobacterium tuberculosis* from the complete genome sequence. *Nature* **393**(6685): 537–44.

Cole, S. T., K. Eiglmeier et al. (2001). Massive gene decay in the leprosy bacillus. *Nature* **409**(6823): 1007–11.

Comay, E., R. Nussinov et al. (1984). An accelerated algorithm for calculating the secondary structure of single stranded RNAs. *Nucleic Acids Res.* **12**(1, Pt. 1): 53–66.

Copley, R. R., J. Schultz et al. (1999). Protein families in multicellular organisms. *Curr. Opin. Struct. Biol.* **9**(3): 408–15.

Corpet, F., F. Servant et al. (2000). ProDom and ProDom-CG: tools for protein domain analysis and whole genome comparisons. *Nucleic Acids Res.* **28**(1): 267–69.

Cortopassi, G. A. and A. Wong (1999). Mitochondria in organismal aging and degeneration. *Biochim. Biophys. Acta* **1410**(2): 183–93.

Costanzo, M. C., M. E. Crawford et al. (2001). YPD, PombePD and WormPD: model organism volumes of the BioKnowledge library, an integrated resource for protein information. *Nucleic Acids Res.* **29**(1): 75–79.

Costanzo, M. C., J. D. Hogan et al. (2000). The yeast proteome database (YPD) and *Caenorhabditis elegans* proteome database (WormPD): comprehensive resources for the organization and comparison of model organism protein information. *Nucleic Acids Res.* **28**(1): 73–76.

Crozier, R. H. and Y. C. Crozier (1993). The mitochondrial genome of the honeybee *Apis mellifera*: complete sequence and genome organization. *Genetics* **133**(1): 97–117.

Cserzo, M., E. Wallin et al. (1997). Prediction of transmembrane alpha-helices in prokaryotic membrane proteins: the dense alignment surface method. *Protein Eng.* **10**(6): 673–76.

Cuff, J. A. and G. J. Barton (2000). Application of multiple sequence alignment profiles to improve protein secondary structure prediction. *Proteins* **40**(3): 502–11.

Cuff, J. A., M. E. Clamp et al. (1998). JPred: a consensus secondary structure prediction server. *Bioinformatics* **14**(10): 892–93.

Cummings, D. J., K. L. McNally et al. (1990). The complete DNA sequence of the mitochondrial genome of *Podospora anserina*. *Curr. Genet.* **17**(5): 375–402.

Dai, H., Y. S. Lo et al. (1998). Population heterogeneity of higher-plant mitochondria in structure and function. *Eur. J. Cell. Biol.* **75**(2): 198–209.

Darr, S. C., J. W. Brown et al. (1992). The varieties of ribonuclease P. *Trends Biochem. Sci.* **17**(5): 178–82.

da Silva, A. C., J. A. Ferro et al. (2002). Comparison of the genomes of two *Xanthomonas* pathogens with differing host specificities. *Nature* **417**(6887): 459–63.

Datson, N. A., J. van der Perk-de Jong et al. (1999). MicroSAGE: a modified procedure for serial analysis of gene expression in limited amounts of tissue. *Nucleic Acids Res.* **27**(5): 1300–1307.

Davies, H. A. (2000). The ProteinChip System from Ciphergen: a new technique for rapid, micro-scale protein biology. *J. Mol. Med.* **78**(7): B29.

Dayhoff, M. O., R. M. Schwartz et al. (1978). A model of evolutionary change in proteins. In *Atlas of Protein Sequence and Structure*, Vol. 5. M. O. Dayhoff, ed. Washington D.C.: National Biomedical Research Foundation, pp. 345–52.

Debouck, C. (1995). Differential display or differential dismay? *Curr. Opin. Biotechnol.* **6**(5): 597–99.

Deckert, G., P. V. Warren et al. (1998). The complete genome of the hyperthermophilic bacterium *Aquifex aeolicus*. *Nature* **392**(6674): 353–58.

De Giorgi, C., A. Martiradonna et al. (1996a). Complete sequence of the mitochondrial DNA in the sea urchin *Arbacia lixula*: conserved features of the echinoid mitochondrial genome. *Mol. Phylogenet. Evol.* **5**(2): 323–32.

De Giorgi, C., A. Martiradonna et al. (1996b). Evolutionary analysis of sea urchin mitochondrial tRNAs: folding of the molecules as suggested by the non-random occurrence of nucleotides. *Curr. Genet.* **30**(3): 191–99.

Deguin-Chambon, V., M. Vacher et al. (2000). Direct transactivation of c-Ha-Ras gene by p53: evidence for its involvement in p53 transactivation activity and p53-mediated apoptosis. *Oncogene* **19**(51): 5831–41.

Delcher, A. L., D. Harmon et al. (1999). Improved microbial gene identification with GLIMMER. *Nucleic Acids Res.* **27**(23): 4636–41.

Delcher, A. L., S. Kasif et al. (1999). Alignment of whole genomes. *Nucleic Acids Res.* **27**(11): 2369–76.

Deleage, G. and B. Roux (1987). An algorithm for protein secondary structure prediction based on class prediction. *Protein Eng.* **1**(4): 289–94.

DelVecchio, V. G., V. Kapatral et al. (2002). The genome sequence of the facultative intracellular pathogen *Brucella melitensis*. *Proc. Natl. Acad. Sci. USA* **99**(1): 443–48.

de Massy, B., V. Rocco et al. (1995). The nucleotide mapping of DNA double-strand breaks at the CY53 initiation site of meiotic recombination in *Saccharomyces cerevisiae*. *EMBO J.* **14**(18): 4589–98.

Dennis, P. P. (1997). Ancient ciphers: translation in Archaea. *Cell* **89**(7): 1007–10.

DePamphilis, M. L. (2000). Review: nuclear structure and DNA replication. *J. Struct. Biol.* **129**(2–3): 186–97.

D'Erchia A. M., G. Pesole et al. (1999). Guinea pig p53 mRNA: identification of new elements in coding and untranslated regions and their functional and evolutionary implications. *Genomics* **58**(1): 50–64.

D'Erchia, A., A. Tullo et al. (2002). p53 gene family: structural, functional and evolutionary features. *Curr. Genom.* (in press).

Deppenmeier, U., A. Johann et al. (2002). The genome of *Methanosarcina mazei*: evidence for lateral gene transfer between bacteria and archaea. *J. Mol. Microbiol. Biotechnol.* **4**(4): 453–61.

Desagher, S. and J. C. Martinou (2000). Mitochondria as the central control point of apoptosis. *Trends Cell Biol.* **10**(9): 369–77.

Desjardins, P. and R. Morais (1990). Sequence and gene organization of the chicken mitochondrial genome: a novel gene order in higher vertebrates. *J. Mol. Biol.* **212**: 599–634.

de Zamaroczy, M. and G. Bernardi (1985). Sequence organization of the mitochondrial genome of yeast: a review. *Gene* **37**(1–3): 1–17.

Diachenko, L. B., J. Ledesma et al. (1996). Combining the technique of RNA fingerprinting and differential display to obtain differentially expressed mRNA. *Biochem. Biophys. Res. Commun.* **219**(3): 824–28.

Dietrich, A., I. Small et al. (1996). Editing and import: strategies for providing plant mitochondria with a complete set of functional transfer RNAs. *Biochimie* **78**(6): 518–29.

Dietrich, A., J. H. Weil et al. (1992). Nuclear-encoded transfer RNAs in plant mitochondria. *Annu. Rev. Cell Biol.* **8**: 115–31.

DiMauro, S. and E. A. Schon (2001). Mitochondrial DNA mutations in human disease. *Am. J. Med. Genet.* **106**(1): 18–26.

Discala, C., X. Benigni et al. (2000). DBcat: a catalog of 500 biological databases. *Nucleic Acids Res.* **28**(1): 8–9.

Dmitrova, M., G. Younes-Cauet et al. (1998). A new LexA-based genetic system for monitoring and analyzing protein heterodimerization in *Escherichia coli*. *Mol. Gen. Genet.* **257**(2): 205–12.

Dodd, I. B. and J. B. Egan (1990). Improved detection of helix-turn-helix DNA-binding motifs in protein sequences. *Nucleic Acids Res.* **18**(17): 5019–26.

Dodge, C., R. Schneider et al. (1998). The HSSP database of protein structure-sequence alignments and family profiles. *Nucleic Acids Res.* **26**(1): 313–15.

Dombrowski, S., A. Brennicke et al. (1997). 3′-Inverted repeats in plant mitochondrial mRNAs are processing signals rather than transcription terminators. *EMBO J.* **16**(16): 5069–76.

Douglas, S., S. Zauner et al. (2001). The highly reduced genome of an enslaved algal nucleus. *Nature* **410**(6832): 1091–96.

Douglas, S. E. (1998). Plastid evolution: origins, diversity, trends. *Curr. Opin. Genet. Dev.* **8**(6): 655–61.

Dujon, B., K. Albermann et al. (1997). The nucleotide sequence of *Saccharomyces cerevisiae* chromosome XV. *Nature* **387**(6632 Suppl.): 98–102.

Dujon, B., D. Alexandraki et al. (1994). Complete DNA sequence of yeast chromosome XI. *Nature* **369**(6479): 371–78.

Durbin, R. and J. Thierry-Mieg (1994). *The ACEDB Genome Database*. New York: Plenum Press.

Duret, L. (2001). Why do genes have introns? Recombination might add a new piece to the puzzle. *Trends Genet.* **17**(4): 172–75.

Dyall, S. D. and P. J. Johnson (2000). Origins of hydrogenosomes and mitochondria: evolution and organelle biogenesis. *Curr. Opin. Microbiol.* **3**(4): 404–11.

Ebeling, W. and M. A. Jimenez-Montano (1980). On grammars, complexity and informational measures of biological macromolecules. *Math. Biosci.* **52**: 53–71.

Eddy, S. R. (1998). Profile hidden Markov models. *Bioinformatics* **14**(9): 755–63.

Edgell, D. R. and W. F. Doolittle (1997). Archaea and the origin(s) of DNA replication proteins. *Cell* **89**(7): 995–98.

Eichler, E. E. (2001). Recent duplication, domain accretion and the dynamic mutation of the human genome. *Trends Genet.* **17**(11): 661–69.

Eisenberg, D., E. Schwarz et al. (1984). Analysis of membrane and surface protein sequences with the hydrophobic moment plot. *J. Mol. Biol.* **179**(1): 125–42.

Ellis, J., P. Dodds et al. (2000). Structure, function and evolution of plant disease resistance genes. *Curr. Opin. Plant Biol.* **3**(4): 278–84.

Emanuelsson, O., H. Nielsen et al. (2000). Predicting subcellular localization of proteins based on their N-terminal amino acid sequence. *J. Mol. Biol.* **300**(4): 1005–16.

Emanuelsson, O., G. von Heijne et al. (2001). Analysis and prediction of mitochondrial targeting peptides. *Methods Cell Biol.* **65**: 175–87.

Epstein, C. B., J. A. Waddle et al. (2001). Genome-wide responses to mitochondrial dysfunction. *Mol. Biol. Cell* **12**(2): 297–308.

Esnouf, R. M. (1997). An extensively modified version of MolScript that includes greatly enhanced coloring capabilities. *J. Mol. Graph. Model.* **15**(2): 132–34, 112–13.

Estevez, A. M. and L. Simpson (1999). Uridine insertion/deletion RNA editing in trypanosome mitochondria—a review. *Gene* **240**(2): 247–60.

Etzold, T. and P. Argos (1993). SRS—an indexing and retrieval tool for flat file data libraries. *Comput. Appl. Biosci.* **9**: 49–57.

Etzold, T., A. Ulyanov et al. (1996). SRS: information retrieval system for molecular biology data banks. *Methods Enzymol.* **266**: 114–28.

Euzeby, J. P. (1997). List of Bacterial Names with Standing in Nomenclature: a folder available on the Internet. *Int. J. Syst. Bacteriol.* **47**(2): 590–92.

Ewing, B. and P. Green (1998). Base-calling of automated sequencer traces using phred. II. Error probabilities. *Genome Res.* **8**(3): 186–94.

Ewing, B., L. Hillier et al. (1998). Base-calling of automated sequencer traces using phred. I. Accuracy assessment. *Genome Res.* **8**(3): 175–85.

Eyre-Walker, A. and B. S. Gaut (1997). Correlated rates of synonymous site evolution across plant genomes. *Mol. Biol. Evol.* **14**(4): 455–60.

Eyre-Walker, A. and L. D. Hurst (2001). The evolution of isochores. *Nat. Rev. Genet.* **2**(7): 549–55.

Fagegaltier, D., A. Lescure et al. (2000). Structural analysis of new local features in SECIS RNA hairpins. *Nucleic Acids Res.* **28**(14): 2679–89.

Fan, L. and L. S. Kaguni (2001). Multiple regions of subunit interaction in *Drosophila* mitochondrial DNA polymerase: three functional domains in the accessory subunit. *Biochemistry* **40**(15): 4780–91.

Fekete, M., I. L. Hofacker et al. (2000). Prediction of RNA base pairing probabilities on massively parallel computers. *J. Comput. Biol.* **7**(1–2): 171–82.

Feldmann, H., M. Aigle et al. (1994). Complete DNA sequence of yeast chromosome II. *EMBO J.* **13**(24): 5795–5809.

Felsenstein, J. (1978). Cases in which parsimony or compatibility methods will be positively misleading. *Syst. Zool.* **27**: 401–10.

Felsenstein, J. (1981). Evolutionary trees from DNA sequences: a maximum likelihood approach. *J. Mol. Evol.* **17**: 368–76.

Felsenstein, J. (1985). Confidence limits on phylogenies: an approach using the bootstrap. *Evolution* **39**: 783–91.

Feng, D. F. and R. F. Doolittle (1987). Progressive sequence alignment as a prerequisite to correct phylogenetic trees. *J. Mol. Evol.* **25**(4): 351–60.

Feng, D. F., M. S. Johnson et al. (1984). Aligning amino acid sequences: comparison of commonly used methods. *J. Mol. Evol.* **21**(2): 112–25.

Fenyo, D. (2000). Identifying the proteome: software tools. *Curr. Opin. Biotechnol.* **11**(4): 391–95.

Fernandez-Silva, P., F. Martinez-Azorin et al. (1997). The human mitochondrial transcription termination factor (mTERF) is a multizipper protein but binds to DNA as a monomer, with evidence pointing to intramolecular leucine zipper interactions. *EMBO J.* **16**(5): 1066–79.

Ferretti, J. J., W. M. McShan et al. (2001). Complete genome sequence of an M1 strain of *Streptococcus pyogenes*. *Proc. Natl. Acad. Sci. USA* **98**(8): 4658–63.

Fickett, J. W. (1982). Recognition of protein coding regions in DNA sequences. *Nucleic Acids Res.* **10**: 5303–18.

Fickett, J. W. and A. G. Hatzigeorgiou (1997). Eukaryotic promoter recognition. *Genome Res.* **7**(9): 861–78.

Fields, S. and O. Song (1989). A novel genetic system to detect protein–protein interactions. *Nature* **340**(6230): 245–46.

Figueroa, P., I. Gomez et al. (1999). Transfer of rps14 from the mitochondrion to the nucleus in maize implied integration within a gene encoding the iron–sulphur subunit of succinate dehydrogenase and expression by alternative splicing. *Plant J.* **18**(6): 601–9.

Filipski, J. (1987). Correlation between molecular clock ticking, codon usage fidelity of DNA repair, chromosome banding and chromatin compactness in germline cells. *FEBS Lett.* **217**(2): 184–86.

Finley, R. L., Jr. and R. Brent (1994). Interaction mating reveals binary and ternary connections between *Drosophila* cell cycle regulators. *Proc. Natl. Acad. Sci. USA* **91**(26): 12980–84.

Fisher, A., Y. Shi et al. (2000). Functional correlation in amino acid residue mutations of yeast iso-2-cytochrome *c* that is consistent with the prediction of the concomitantly variable codon theory in cytochrome *c* evolution. *Biochem. Genet.* **38**(5–6): 181–200.

Fitch, W. M. (1970). Distinguishing homologous from analogous proteins. *Syst. Zool.* **19**(2): 99–113.

Fitch, W. M. (2000). Homology: a personal view on some of the problems. *Trends Genet.* **16**(5): 227–31.

Fitz-Gibbon, S. T., H. Ladner et al. (2002). Genome sequence of the hyperthermophilic crenarchaeon *Pyrobaculum aerophilum*. *Proc. Natl. Acad. Sci. USA* **99**(2): 984–89.

Fleischmann, R. D., M. D. Adams et al. (1995). Whole-genome random sequencing and assembly of *Haemophilus influenzae* Rd. *Science* **269**(5223): 496–512.

Florea, L., G. Hartzell et al. (1998). A computer program for aligning a cDNA sequence with a genomic DNA sequence. *Genome Res.* **8**(9): 967–74.

Flores-Rozas, H. and R. D. Kolodner (2000). Links between replication, recombination and genome instability in eukaryotes. *Trends Biochem. Sci.* **25**(4): 196–200.

Flybase Consortium (1996). FlyBase—the *Drosophila* database. *Nucleic Acids Res.* **24**: 53–56.

Fontaine, J. M., S. Rousvoal et al. (1995). The mitochondrial LSU rDNA of the brown alga *Pylaiella littoralis* reveals alpha-proteobacterial features and is split by four group IIB introns with an atypical phylogeny. *J. Mol. Biol.* **251**(3): 378–89.

Forterre, P. and C. Elie (1993). *The Biochemistry of Archaea (Archaeabacteria)*. Amsterdam: Elsevier.

Foury, F., T. Roganti et al. (1998). The complete sequence of the mitochondrial genome of *Saccharomyces cerevisiae*. *FEBS Lett.* **440**(3): 325–31.

Fraser, C. M., S. Casjens et al. (1997). Genomic sequence of a Lyme disease spirochaete, *Borrelia burgdorferi*. *Nature* **390**(6660): 580–86.

Fraser, C. M., J. D. Gocayne et al. (1995). The minimal gene complement of *Mycoplasma genitalium*. *Science* **270**(5235): 397–403.

Fraser, C. M., S. J. Norris et al. (1998). Complete genome sequence of *Treponema pallidum*, the syphilis spirochete. *Science* **281**(5375): 375–88.

Freeman, W. M., S. J. Walker et al. (1999). Quantitative RT-PCR: pitfalls and potential. *Biotechniques* **26**(1): 112–22, 124–25.

Freier, S. M., R. Kierzek et al. (1986). Improved free-energy parameters for predictions of RNA duplex stability. *Proc. Natl. Acad. Sci. USA* **83**(24): 9373–77.

Freist, W., D. H. Gauss et al. (1997). Glutaminyl-tRNA synthetase. *Biol. Chem.* **378**(10): 1103–17.

Frishman, D. and P. Argos (1995). Knowledge-based protein secondary structure assignment. *Proteins* **23**(4): 566–79.

Frishman, D. and P. Argos (1997). Seventy-five percent accuracy in protein secondary structure prediction. *Proteins* **27**(3): 329–35.

Fromont-Racine, M., J. C. Rain et al. (1997). Toward a functional analysis of the yeast genome through exhaustive two-hybrid screens. *Nat. Genet.* **16**(3): 277–82.

Frye, M. S. and S. B. Hedges (1995). Monophyly of the order Rodentia inferred from mitochondrial DNA sequences of the genes for 12S rRNA, 16S rRNA, and tRNA-valine. *Mol. Biol. Evol.* **12**(1): 168–76.

Fryxell, K. J. and E. Zuckerkandl (2000). Cytosine deamination plays a primary role in the evolution of mammalian isochores. *Mol. Biol. Evol.* **17**(9): 1371–83.

Fung, E. T., V. Thulasiraman et al. (2001). Protein biochips for differential profiling. *Curr. Opin. Biotechnol.* **12**(1): 65–69.

Gaasterland, T. and M. Oprea (2001). Whole-genome analysis: annotations and updates. *Curr. Opin. Struct. Biol.* **11**(3): 377–81.

Gadaleta, M. N., A. Cormio et al. (1998). Aging and mitochondria. *Biochimie* **80**(10): 863–70.

Galagan, J. E., C. Nusbaum et al. (2002). The genome of *M. acetivorans* reveals extensive metabolic and physiological diversity. *Genome Res.* **12**(4): 532–42.

Galibert, F., D. Alexandraki et al. (1996). Complete nucleotide sequence of *Saccharomyces cerevisiae* chromosome X. *EMBO J.* **15**(9): 2031–49.

Galibert, F., T. M. Finan et al. (2001). The composite genome of the legume symbiont *Sinorhizobium meliloti*. *Science* **293**(5530): 668–72.

Gardiner-Garden, M. and M. Frommer (1987). CpG islands in vertebrate genomes. *J. Mol. Biol.* **196**(2): 261–82.

Gardiner-Garden, M. and T. G. Littlejohn (2001). A comparison of microarray databases. *Brief. Bioinf.* **2**(2): 143–58.

Gardner, M. J., H. Tettelin et al. (1998). Chromosome 2 sequence of the human malaria parasite *Plasmodium falciparum*. *Science* **282**(5391): 1126–32.

Garesse, R. and C. G. Vallejo (2001). Animal mitochondrial biogenesis and function: a regulatory cross-talk between two genomes. *Gene* **263**(1–2): 1–16.

Garnier, J., J. F. Gibrat et al. (1996). GOR method for predicting protein secondary structure from amino acid sequence. *Methods Enzymol.* **266**: 540–53.

Garnier, J., D. J. Osguthorpe et al. (1978). Analysis of the accuracy and implications of simple methods for predicting the secondary structure of globular proteins. *J. Mol. Biol.* **120**: 97–120.

Gatlin, L. L. (1966). The information content of DNA. *J. Theor. Biol.* **10**: 281–300.

Gatlin, L. L. (1968). The information content of DNA. II. *J. Theor. Biol.* **18**: 181–94.

Gautier, C. (2000). Compositional bias in DNA. *Curr. Opin. Genet. Dev.* **10**(6): 656–61.

Genetics Computer Group (1998). *GCG Wisconsin Package Version 9.0*. Madison, Wis.: GCG.

Geourjon, C. and G. Deleage (1994). SOPM: a self-optimized method for protein secondary structure prediction. *Protein Eng.* **7**(2): 157–64.

Geourjon, C. and G. Deleage (1995). SOPMA: significant improvements in protein secondary structure prediction by consensus prediction from multiple alignments. *Comput. Appl. Biosci.* **11**(6): 681–84.

Gerstein, M. and R. Jansen (2000). The current excitement in bioinformatics-analysis of whole-genome expression data: how does it relate to protein structure and function? *Curr. Opin. Struct. Biol.* **10**(5): 574–84.

Gibbons, N. and R. Murray (1978). Proposals concerning the higher taxa of bacteria. *Int. J. Syst. Bacteriol.* **28**: 1–6.

Gibrat, J. F., J. Garnier et al. (1987). Further developments of protein secondary structure prediction using information theory. New parameters and consideration of residue pairs. *J. Mol. Biol.* **198**(3): 425–43.

Gilbert, D. M. (1998). Replication origins in yeast versus metazoa: separation of the haves and the have nots. *Curr. Opin. Genet. Dev.* **8**(2): 194–99.

Gilson, P. R., U. G. Maier et al. (1997). Size isn't everything: lessons in genetic miniaturisation from nucleomorphs. *Curr. Opin. Genet. Dev.* **7**(6): 800–806.

Gilson, P. R. and G. I. McFadden (2001). Genome sequencing. A grin without a cat. *Nature* **410**(6832): 1040–41.

Gissi, C., A. Reyes et al. (2000). Lineage-specific evolutionary rate in mammalian mtDNA. *Mol. Biol. Evol.* **17**(7): 1022–31.

Glaser, P., L. Frangeul et al. (2001). Comparative genomics of *Listeria* species. *Science* **294**(5543): 849–52.

Glass, J. I., E. J. Lefkowitz et al. (2000). The complete sequence of the mucosal pathogen *Ureaplasma urealyticum*. *Nature* **407**(6805): 757–62.

Glover, K. E., D. F. Spencer et al. (2001). Identification and structural characterization of nucleus-encoded transfer RNAs imported into wheat mitochondria. *J. Biol. Chem.* **276**(1): 639–48.

Gockel, G. and W. Hachtel (2000). Complete gene map of the plastid genome of the nonphotosynthetic euglenoid flagellate *Astasia longa*. *Protist* **151**(4): 347–51.

Goddard, J. M. and D. J. Cummings (1975). Structure and replication of mitochondrial DNA from *Paramecium aurelia*. *J. Mol. Biol.* **97**(4): 593–609.

Goffeau, A., B. G. Barrell et al. (1996). Life with 6000 genes. *Science* **274**(5287): 546, 563–67.

Gogarten, J. P. and L. Olendzenski (1999). Orthologs, paralogs and genome comparisons. *Curr. Opin. Genet. Dev.* **9**(6): 630–36.

Goldman, N., J. P. Anderson et al. (2000). Likelihood-based tests of topologies in phylogenetics. *Syst. Biol.* **49**: 652–70.

Golenberg, E. M., M. T. Clegg et al. (1993). Evolution of a noncoding region of the chloroplast genome. *Mol. Phylogenet. Evol.* **2**(1): 52–64.

Gonzalez, P., C. Hernandez-Calzadilla et al. (1994). Comparative analysis of the zetacrystallin/quinone reductase gene in guinea pig and mouse. *Mol. Biol. Evol.* **11**(2): 305–15.

Goff, S. A., D. Ricke et al. (2002). A draft sequence of the rice genome (*Oryza sativa L. ssp. japonica*). *Science* **296**(5565): 92–100.

Goodner, B., G. Hinkle et al. (2001). Genome sequence of the plant pathogen and biotechnology agent *Agrobacterium tumefaciens* C58. *Science* **294**(5550): 2323–28.

Gordon, D., C. Abajian et al. (1998). Consed: a graphical tool for sequence finishing. *Genome Res.* **8**(3): 195–202.

Gouy, M. (1987). Codon contexts in enterobacterial and coliphage genes. *Mol. Biol. Evol.* **4**(4): 426–44.

Gouy, M. and C. Gautier (1982). Codon usage in bacteria: correlation with gene expressivity. *Nucleic Acids Res.* **10**(22): 7055–74.

Gouy, M., C. Gautier et al. (1985). ACNUC—a portable retrieval system for nucleic acid sequence databases: logical and physical designs and usage. *CABIOS* **1**(3): 167–72.

Grantcharova, V., E. J. Alm et al. (2001). Mechanisms of protein folding. *Curr. Opin. Struct. Biol.* **11**(1): 70–82.

Grantham, R. (1974). Amino acid difference formula to help explain protein evolution. *Science* **185**(154): 862–64.

Grantham, R., C. Gautier et al. (1980). Codon catalog usage and the genome hypothesis. *Nucleic Acids Res.* **8**: r49–r62.

Graveley, B. R. (2001). Alternative splicing: increasing diversity in the proteomic world. *Trends Genet.* **17**(2): 100–107.

Gray, M., W. Cedergren et al. (1989). On the evolutionary origin of the plant mitochondrion and its genome. *Proc. Natl. Acad. Sci. USA* **86**: 2267–71.

Gray, M. W. (1992). The endosymbiont hypothesis revisited. *Int. Rev. Cytol.* **141**: 233–357.

Gray, M. W. (1993). Origin and evolution of organelle genomes. *Curr. Opin. Genet. Dev.* **3**(6): 884–90.

Gray, M. W., G. Burger et al. (1999). Mitochondrial evolution. *Science* **283**(5407): 1476–81.

Gray, M. W., B. F. Lang et al. (1998). Genome structure and gene content in protist mitochondrial DNAs. *Nucleic Acids Res.* **26**(4): 865–78.

Gray, S. A. and M. E. Konkel (1999). Codon usage in the A/T-rich bacterium *Campylobacter jejuni*. *Adv. Exp. Med. Biol.* **473**: 231–35.

Green, E. D. (2001). Strategies for the systematic sequencing of complex genomes. *Nat. Rev. Genet.* **2**(8): 573–83.

Greenblatt, J. (1997). RNA polymerase II holoenzyme and transcriptional regulation. *Curr. Opin. Cell. Biol.* **9**(3): 310–19.

Gregory, T. R. (2001). Coincidence, coevolution, or causation? DNA content, cell size, and the *C*-value enigma. *Biol. Rev. Camb. Philos. Soc.* **76**(1): 65–101.

Gregory, W. K. (1910). The order of mammals. *Bull. Am. Mus. Nat. Hist.* **27**: 1–542.

Gribskov, M., A. D. McLachlan et al. (1987). Profile analysis: detection of distantly related proteins. *Proc. Natl. Acad. Sci. USA* **84**: 4355–58.

Gribskov, M. and S. Veretnik (1996). Identification of sequence pattern with profile analysis. *Methods Enzymol.* **266**: 198–212.

Grillo, G., M. Attimonelli et al. (1996). CLEANUP: a fast computer program for cleaning nucleic acids sequence databases from redundancies. *CABIOS* **12**: 1–8.

Grosskopf, D. and R. M. Mulligan (1996). Developmental- and tissue-specificity of RNA editing in mitochondria of suspension-cultured maize cells and seedlings. *Curr. Genet.* **29**(6): 556–63.

Gu, X. and W. H. Li (1995). The size distribution of insertions and deletions in human and rodent pseudogenes suggests the logarithmic gap penalty for sequence alignment. *J. Mol. Evol.* **40**(4): 464–73.

Gu, X., Y. Wang et al. (2002). Age distribution of human gene families shows significant roles of both large- and small-scale duplications in vertebrate evolution. *Nat. Genet.* **31**(2): 205–9.

Guermeur, Y., C. Geourjon et al. (1999). Improved performance in protein secondary structure prediction by inhomogeneous score combination. *Bioinformatics* **15**(5): 413–21.

Guex, N. and M. C. Peitsch (1997). Swiss-Model and the Swiss-PdbViewer: an environment for comparative protein modeling. *Electrophoresis* **18**(15): 2714–23.

Gusev, V. D., L. A. Nemytikova et al. (1999). On the complexity measures of genetic sequences. *Bioinformatics* **15**(12): 994–99.

Gusfield, D. (1997). *Algorithms on Strings, Trees and Sequences: Computer Science and Computational Biology*. New York: Cambridge University Press.

Gutell, R. R. (1994). Collection of small subunit (16S- and 16S-like) ribosomal RNA structures: 1994. *Nucleic Acids Res.* **22**(17): 3502–7.

Gygi, S. P., Y. Rochon et al. (1999). Correlation between protein and mRNA abundance in yeast. *Mol. Cell. Biol.* **19**(3): 1720–30.

Hammer, J. (1995). New methods to predict MHC-binding sequences within protein antigens. *Curr. Opin. Immunol.* **7**(2): 263–69.

Hampsey, M. and D. Reinberg (1999). RNA polymerase II as a control panel for multiple coactivator complexes. *Curr. Opin. Genet. Dev.* **9**(2): 132–39.

Hancock, K. and S. L. Hajduk (1990). The mitochondrial tRNAs of *Trypanosoma brucei* are nuclear encoded. *J. Biol. Chem.* **265**(31): 19208–15.

Hansen, J. E., O. Lund et al. (1995). Prediction of *O*-glycosylation of mammalian proteins: specificity patterns of UDP-GalNAc:polypeptide *N*-acetylgalactosaminyltransferase. *Biochem. J.* **308**(Pt. 3): 801–13.

Hasegawa, M., H. Kishino et al. (1985). Dating of the human-ape splitting by a molecular clock of mitochondrial DNA. *J. Mol. Evol.* **22**(2): 160–74.

Hayashi, T., K. Makino et al. (2001). Complete genome sequence of enterohemorrhagic *Escherichia coli* O157:H7 and genomic comparison with a laboratory strain K-12. *DNA Res.* **8**(1): 11–22.

Haynes, P., I. Miller et al. (1998). Proteins of rat serum: I. Establishing a reference two-dimensional electrophoresis map by immunodetection and microbore high performance liquid chromatography-electrospray mass spectrometry. *Electrophoresis* **19**(8–9): 1484–92.

He, Q., H. Chen et al. (1994). A physical map of the *Myxococcus xanthus* chromosome. *Proc. Natl. Acad. Sci. USA* **91**(20): 9584–87.

Hedtke, B., T. Borner et al. (1997). Mitochondrial and chloroplast phage-type RNA polymerases in *Arabidopsis*. *Science* **277**(5327): 809–11.

Heidelberg, J. F., J. A. Eisen et al. (2000). DNA sequence of both chromosomes of the cholera pathogen *Vibrio cholerae*. *Nature* **406**(6795): 477–83.

Henderson, S. L. and B. Sollner-Webb (1990). The mouse ribosomal DNA promoter has more stringent requirements in vivo than in vitro. *Mol. Cell Biol.* **10**(9): 4970–73.

Hendy, M. D. and P. D. Hendy (1982). Branch and bound algorithms to determine minimal evolutionary trees. *Math. Biosci.* **59**: 277–90.

Henikoff, J. G., E. A. Greene et al. (2000). Increased coverage of protein families with the blocks database servers. *Nucleic Acids Res.* **28**(1): 228–30.

Henikoff, S., K. Ahmad et al. (2001). The centromere paradox: stable inheritance with rapidly evolving DNA. *Science* **293**(5532): 1098–1102.

Henikoff, S. and J. G. Henikoff (1992). Amino acid substitution matrices from protein blocks. *Proc. Natl. Acad. Sci. USA* **89**: 10915–19.

Henikoff, S. and J. G. Henikoff (1993). Performance evaluation of amino acid substitution matrices. *Proteins* **17**(1): 49–61.

Hentze, M. W. and L. C. Kuhn (1996). Molecular control of vertebrate iron metabolism: mRNA-based regulatory circuits operated by iron, nitric oxide, and oxidative stress. *Proc. Natl. Acad. Sci. USA* **93**: 8175–82.

Hernould, M., S. Suharsono et al. (1993). Male-sterility induction in transgenic tobacco plants with an unedited atp9 mitochondrial gene from wheat. *Proc. Natl. Acad. Sci. USA* **90**(6): 2370–74.

Herrmann, J. M. and W. Neupert (2000). Protein transport into mitochondria. *Curr. Opin. Microbiol.* **3**(2): 210–14.

Hertz, G. Z., G. W. Hartzell et al. (1990). Identification of consensus patterns in unaligned DNA sequences known to be functionally related. *CABIOS* **6**(2): 81–92.

Hertz, G. Z. and G. D. Stormo (1999). Identifying DNA and protein patterns with statistically significant alignments of multiple sequences. *Bioinformatics* **15**(7–8): 563–77.

Hess, J. F., M. A. Parisi et al. (1991). Impairment of mitochondrial transcription termination by a point mutation associated with the MELAS subgroup of mitochondrial encephalomyopathies. *Nature* **351**(6323): 236–39.

Hide, W., J. Burke et al. (1994). Biological evaluation od d2, an algorithm for high-performance sequence comparison. *J. Comput. Biol.* **1**: 199–215.

Higgins, D. G., J. D. Thompson et al. (1996). Using CLUSTAL for multiple sequence alignments. *Methods Enzymol.* **266**: 383–402.

Himmelreich, R., H. Hilbert et al. (1996). Complete sequence analysis of the genome of the bacterium *Mycoplasma pneumoniae*. *Nucleic Acids Res.* **24**(22): 4420–49.

Hinnebusch, J. and A. G. Barbour (1991). Linear plasmids of *Borrelia burgdorferi* have a telomeric structure and sequence similar to those of a eukaryotic virus. *J. Bacteriol.* **173**(22): 7233–39.

Hinnebusch, J., S. Bergstrom et al. (1990). Cloning and sequence analysis of linear plasmid telomeres of the bacterium *Borrelia burgdorferi*. *Mol. Microbiol.* **4**(5): 811–20.

Hirokawa, T., S. Boon-Chieng et al. (1998). SOSUI: classification and secondary structure prediction system for membrane proteins. *Bioinformatics* **14**(4): 378–79.

Hixson, J. E. and D. A. Clayton (1985). Initiation of transcription from each of the two human mitochondrial promoters requires unique nucleotides at the transcriptional start sites. *Proc. Natl. Acad. Sci. USA* **82**(9): 2660–64.

Hobbs, A. E., M. Srinivasan et al. (2001). Mmm1p, a mitochondrial outer membrane protein, is connected to mitochondrial DNA (mtDNA) nucleoids and required for mtDNA stability. *J. Cell Biol.* **152**(2): 401–10.

Hoffmann, R. J., J. L. Boore et al. (1992). A novel mitochondrial genome organization for the blue mussel, *Mytilus edulis*. *Genetics* **131**: 397–412.

Hofmann, K., P. Bucher et al. (1999). The ProSite database, its status in 1999. *Nucleic Acids Res.* **27**(1): 215–19.

Hogenesch, J. B., K. A. Ching et al. (2001). A comparison of the Celera and Ensembl predicted gene sets reveals little overlap in novel genes. *Cell* **106**(4): 413–15.

Hogue, C. W., H. Ohkawa et al. (1996). A dynamic look at structures: WWW-Entrez and the Molecular Modeling Database. *Trends Biochem. Sci.* **21**(6): 226–29.

Holm, L. (1998). Unification of protein families. *Curr. Opin. Struct. Biol.* **8**(3): 372–79.

Holm, L. and C. Sander (1998). Touring protein fold space with Dali/FSSP. *Nucleic Acids Res.* **26**(1): 316–19.

Holt, I. J., H. E. Lorimer et al. (2000). Coupled leading- and lagging-strand synthesis of mammalian mitochondrial DNA. *Cell* **100**(5): 515–24.

Honeycutt, R. J., M. McClelland et al. (1993). Physical map of the genome of *Rhizobium meliloti* 1021. *J. Bacteriol.* **175**(21): 6945–52.

Honeycutt, R. L., M. A. Nebdal et al. (1995). Mammalian mitochondrial DNA evolution: a comparison of the cytochrome *b* and cytochrome *c* oxidase II genes. *J. Mol. Evol.* **40**: 260–72.

Hopp, T. P. (1986). Protein surface analysis. Methods for identifying antigenic determinants and other interaction sites. *J. Immunol. Methods* **88**(1): 1–18.

Hori, H. and S. Osawa (1986). Evolutionary change in 5S rRNA secondary structure and a phylogenic tree of 352 5S rRNA species. *Biosystems* **19**(3): 163–72.

Hoskins, J., W. E. Alborn, Jr. et al. (2001). Genome of the bacterium *Streptococcus pneumoniae* strain R6. *J. Bacteriol.* **183**(19): 5709–17.

Houlgatte, R., R. Mariage-Samson et al. (1995). The Genexpress Index: a resource for gene discovery and the genic map of the human genome. *Genome Res.* **5**(3): 272–304.

Hu, J. C., E. K. O'Shea et al. (1990). Sequence requirements for coiled-coils: analysis with lambda repressor-GCN4 leucine zipper fusions. *Science* **250**(4986): 1400–1403.

Huang, J. Y. and D. L. Brutlag (2001). The EMOTIF database. *Nucleic Acids Res.* **29**(1): 202–4.

Huang, X. (1992). A contig assembly program based on sensitive detection of fragment overlaps. *Genomics* **14**: 18–25.

Huang, X. (1996). An improved sequence assembly program. *Genomics* **33**(1): 21–31.

Huang, X. and A. Madan (1999). CAP3: a DNA sequence assembly program. *Genome Res.* **9**(9): 868–77.

Huelsenbeck, J. P. and J. J. Bull (1996). A likelihood ratio test to detect conflicting phylogenetic signal. *Syst. Biol.* **45**: 92–98.

Huelsenbeck, J. P. and B. Rannala (1997). Phylogenetic methods come of age: testing hypotheses in an evolutionary context. *Science* **276**(5310): 227–32.

Huelsenbeck, J. P., B. Rannala et al. (1997). Statistical tests of host–parasite cospeciation. *Evolution* **51**: 410–19.

Huelsenbeck, J. P., F. Ronquist et al. (2001). Bayesian inference of phylogeny and its impact on evolutionary biology. *Science* **294**(5550): 2310–14.

Hughes, A. L. (1999). Phylogenies of developmentally important proteins do not support the hypothesis of two rounds of genome duplication early in vertebrate history. *J. Mol. Evol.* **48**(5): 565–76.

Hughes, S., D. Zelus et al. (1999). Warm-blooded isochore structure in Nile crocodile and turtle. *Mol. Biol. Evol.* **16**(11): 1521–27.

Hughes, T. R., M. J. Marton et al. (2000). Functional discovery via a compendium of expression profiles. *Cell* **102**(1): 109–26.

Ikeda, T. M. and M. W. Gray (1999). Characterization of a DNA-binding protein implicated in transcription in wheat mitochondria. *Mol. Cell Biol.* **19**(12): 8113–22.

Insdorf, N. F. and D. F. Bogenhagen (1989). DNA polymerase gamma from *Xenopus laevis.* II. A 3′–5′ exonuclease is tightly associated with the DNA polymerase activity. *J. Biol. Chem.* **264**(36): 21498–503.

Ito, T. and Y. Sakaki (1997). Fluorescent differential display. *Methods Mol. Biol.* **85**: 37–44.

Ito, T., K. Tashiro et al. (2000). Toward a protein–protein interaction map of the budding yeast: a comprehensive system to examine two-hybrid interactions in all possible combinations between the yeast proteins. *Proc. Natl. Acad. Sci. USA* **97**(3): 1143–47.

Iwabe, N., K. Kuma et al. (1989). Evolutionary relationship of archaebacteria, eubacteria, and eukaryotes inferred from phylogenetic trees of duplicated genes. *Proc. Natl. Acad. Sci. USA* **86**(23): 9355–59.

Iwata, S., J. W. Lee et al. (1998). Complete structure of the 11-subunit bovine mitochondrial cytochrome bc1 complex. *Science* **281**(5373): 64–71.

Jacob, F. and J. Monod (1961). On the regulation of gene activity. *Cold Spring Harbor Symp. Quant. Biol.* **26**: 193–211.

Jacobs, H. T. and D. J. Elliot (1988). Nucleotide sequence and gene organization of sea urchin mitochondrial DNA. *J. Mol. Biol.* **202**: 185–217.

Jacq, C., J. Alt-Morbe et al. (1997). The nucleotide sequence of *Saccharomyces cerevisiae* chromosome IV. *Nature* **387**(6632 Suppl.): 75–78.

Jacquier, A. (1996). Group II introns: elaborate ribozymes. *Biochimie* **78**(6): 474–87.

Jameson, B. A. and H. Wolf (1988). The antigenic index: a novel algorithm for predicting antigenic determinants. *Comput. Appl. BioSci.* **4**: 181–86.

Jamet-Vierny, C., O. Begel et al. (1980). Senescence in *Podospora anserina*: amplification of a mitochondrial DNA sequence. *Cell* **21**(1): 189–94.

Janin, J. (1979). Surface and inside volumes in globular proteins. *Nature* **277**(5696): 491–92.

Janke, A. and U. Arnason (1997). The complete mitochondrial genome of *Alligator mississipiensis* and the separation between recent Archosauria (birds and crocodiles). *Mol. Biol. Evol.* **14**: 1266–72.

Janke, A., X. Xu et al. (1997). The complete mitochondrial genome of the wallaroo (*Macropus robustus*) and the phylogenetic relationship among Monotremata, Marsupialia and Eutheria. *Proc. Natl. Acad. Sci. USA* **94**: 1276–81.

Jazwinski, S. M. (2000). Metabolic control and ageing. *Trends Genet.* **16**(11): 506–11.

Jazwinski, S. M. (2001). New clues to old yeast. *Mech. Ageing Dev.* **122**(9): 865–82.

Johnston, M. (2000). The yeast genome: on the road to the Golden Age. *Curr. Opin. Genet. Dev.* **10**(6): 617–23.

Johnston, M., S. Andrews et al. (1994). Complete nucleotide sequence of *Saccharomyces cerevisiae* chromosome VIII. *Science* **265**(5181): 2077–82.

Jones, D. T. (1999). Protein secondary structure prediction based on position-specific scoring matrices. *J. Mol. Biol.* **292**(2): 195–202.

Jones, D. T. (2000). Protein structure prediction in the postgenomic era. *Curr. Opin. Struct. Biol.* **10**(3): 371–79.

Jones, D. T., W. R. Taylor et al. (1992). The rapid generation of mutation data matrices from protein sequences. *Comput. Appl. BioSci.* **8**(3): 275–82.

Jones, D. T., W. R. Taylor et al. (1994). A model recognition approach to the prediction of all-helical membrane protein structure and topology. *Biochemistry* **33**(10): 3038–49.

Jongeneel, C. V. (2000). Searching the expressed sequence tag (EST) databases: panning for genes. *Brief. Bioinform.* **1**(1): 76–92.

Jordan, I. K. and J. F. McDonald (1999). Comparative genomics and evolutionary dynamics of *Saccharomyces cerevisiae* Ty elements. *Genetica* **107**(1–3): 3–13.

Jukes, T. H. and C. R. Cantor (1969). Evolution of protein molecules. In *Mammalian Protein Metabolism*. H. N. Munro, ed. New York: Academic Press, pp. 21–132.

Jurecic, R., R. G. Nachtman et al. (1998). Identification and cloning of differentially expressed genes by long-distance differential display. *Anal. Biochem.* **259**(2): 235–44.

Jurka, J. (2000). Repbase update: a database and an electronic journal of repetitive elements. *Trends Genet.* **16**(9): 418–20.

Kaback, D. B. (1989). Meiotic segregation of circular plasmid-minichromosomes from intact chromosomes in *Saccharomyces cerevisiae*. *Curr. Genet.* **15**(6): 385–92.

Kabsch, W. and C. Sander (1983). Dictionary of protein secondary structure: pattern recognition of hydrogen bonded and geometrical features. *Biopolymers* **22**: 2577–2637.

Kadowaki, K., N. Kubo et al. (1996). Targeting presequence acquisition after mitochondrial gene transfer to the nucleus occurs by duplication of existing targeting signals. *EMBO J.* **15**(23): 6652–61.

Kaghad, M., H. Bonnet et al. (1997). Monoallelically expressed gene related to p53 at 1p36, a region frequently deleted in neuroblastoma and other human cancers. *Cell* **90**(4): 809–19.

Kaiser, D. (2001). Building a multicellular organism. *Annu. Rev. Genet.* **35**: 103–23.

Kalman, S., W. Mitchell et al. (1999). Comparative genomes of *Chlamydia pneumoniae* and *C. trachomatis*. *Nat. Genet.* **21**(4): 385–89.

Kanaya, S., Y. Yamada et al. (1999). Studies of codon usage and tRNA genes of 18 unicellular organisms and quantification of *Bacillus subtilis* tRNAs: gene expression level and species-specific diversity of codon usage based on multivariate analysis. *Gene* **238**(1): 143–55.

Kanehisa, M. and S. Goto (2000). KEGG: Kyoto encyclopedia of genes and genomes. *Nucleic Acids Res.* **28**(1): 27–30.

Kaneko, T., Y. Nakamura et al. (2000). Complete genome structure of the nitrogen-fixing symbiotic bacterium *Mesorhizobium loti*. *DNA Res.* **7**(6): 331–38.

Kaneko, T., Y. Nakamura et al. (2001). Complete genomic sequence of the filamentous nitrogen-fixing cyanobacterium *Anabaena* sp. strain PCC 7120. *DNA Res.* **8**: 205–13.

Kaneko, T., S. Sato et al. (1996). Sequence analysis of the genome of the unicellular cyanobacterium *Synechocystis* sp. strain PCC6803. II. Sequence determination of the entire genome and assignment of potential protein-coding regions. *DNA Res.* **3**(3): 109–36.

Kapatral, V., I. Anderson et al. (2002). Genome sequence and analysis of the oral bacterium *Fusobacterium nucleatum* strain ATCC 25586. *J. Bacteriol.* **184**(7): 2005–18.

Karlin, S. and S. F. Altschul (1990). Methods for assessing the statistical significance of molecular sequence features by using general scoring schemes. *Proc. Natl. Acad. Sci. USA* **87**: 2264–68.

Karlin, S. and S. F. Altschul (1993). Application and statistics for multiple high-scoring segments in molecular sequences. *Proc. Natl. Acad. Sci. USA* **90**: 5873–77.

Karlin, S., L. Brocchieri et al. (1999). A chimeric prokaryotic ancestry of mitochondria and primitive eukaryotes. *Proc. Natl. Acad. Sci. USA* **96**(16): 9190–95.

Karlin, S., J. Mrazek et al. (1997). Compositional biases of bacterial genomes and evolutionary implications. *J. Bacteriol.* **179**(12): 3899–913.

Karp, P. D. (2000). An ontology for biological function based on molecular interactions. *Bioinformatics* **16**(3): 269–85.

Karp, P. D., M. Riley et al. (2000). The EcoCyc and MetaCyc databases. *Nucleic Acids Res.* **28**(1): 56–59.

Karwan, R. (1993). RNase MRP/RNase P: a structure–function relation conserved in evolution? *FEBS Lett.* **319**(1–2): 1–4.

Karwan, R., J. L. Bennett et al. (1991). Nuclear RNase MRP processes RNA at multiple discrete sites: interaction with an upstream G box is required for subsequent downstream cleavages. *Genes Dev.* **5**(7): 1264–76.

Kastner, A., X. Espanel et al. (1998). Transient accumulation of retinoblastoma/E2F-1 protein complexes correlates with the onset of neuronal differentiation in the developing quail neural retina. *Cell Growth Differ.* **9**(10): 857–67.

Katinka, M. D., S. Duprat et al. (2001). Genome sequence and gene compaction of the eukaryote parasite *Encephalitozoon cuniculi*. *Nature* **414**(6862): 450–53.

Kawai, J., A. Shinagawa et al. (2001). Functional annotation of a full-length mouse cDNA collection. *Nature* **409**(6821): 685–90.

Kawarabayasi, Y., Y. Hino et al. (1999). Complete genome sequence of an aerobic hyper-thermophilic crenarchaeon, *Aeropyrum pernix* K1. *DNA Res.* **6**(2): 83–101, 145–52.

Kawarabayasi, Y., Y. Hino et al. (2001). Complete genome sequence of an aerobic thermo-acidophilic crenarchaeon, *Sulfolobus tokodaii* strain7. *DNA Res.* **8**(4): 123–40.

Kawarabayasi, Y., M. Sawada et al. (1998). Complete sequence and gene organization of the genome of a hyper-thermophilic archaebacterium, *Pyrococcus horikoshii* OT3. *DNA Res.* **5**(2): 55–76.

Kawashima, T., N. Amano et al. (2000). Archaeal adaptation to higher temperatures revealed by genomic sequence of *Thermoplasma volcanium*. *Proc. Natl. Acad. Sci. USA* **97**(26): 14257–62.

Kelley, S. (2000). Getting started with Acedb. *Brief. Bioinf.* **1**(2): 131–37.

Kent, W. J. (2002). BLAT—the BLAST-like alignment tool. Genome Res. **12**(4): 656–64.

Ketting, R. F., S. E. Fischer et al. (1997). Target choice determinants of the Tc1 transposon of *Caenorhabditis elegans*. *Nucleic Acids Res.* **25**(20): 4041–47.

Kimura, M. (1980). A simple method for estimating evolutionary rates of base substitution through comparative studies of nucleotide sequences. *J. Mol. Evol.* **16**: 111–20.

Kimura, M. (1983). *The Neutral Theory of Molecular Evolution.* Cambridge: Cambridge University Press.

King, R. D. and M. J. Sternberg (1996). Identification and application of the concepts important for accurate and reliable protein secondary structure prediction. *Protein Sci.* **5**(11): 2298–2310.

Kingston, R. E., C. A. Bunker et al. (1996). Repression and activation by multiprotein complexes that alter chromatin structure. *Genes Dev.* **10**(8): 905–20.

Kirchman, P. A., S. Kim et al. (1999). Interorganelle signaling is a determinant of longevity in *Saccharomyces cerevisiae*. *Genetics* **152**(1): 179–90.

Kishino, H. and M. Hasegawa (1989). Evaluation of the maximum likelihood estimate of the evolutionary tree topologies from DNA sequence data, and the branching order in hominoidea. *J. Mol. Evol.* **29**(2): 170–79.

Kiss, T. and W. Filipowicz (1992). Evidence against a mitochondrial location of the 7-2/MRP RNA in mammalian cells. *Cell* **70**(1): 11–16.

Klappenbach, J. A., J. M. Dunbar et al. (2000). rRNA operon copy number reflects ecological strategies of bacteria. *Appl. Environ. Microbiol.* **66**(4): 1328–33.

Klenk, H. P., R. A. Clayton et al. (1997). The complete genome sequence of the hyperthermophilic, sulphate-reducing archaeon *Archaeoglobus fulgidus*. *Nature* **390**(6658): 364–70.

Kneller, D. G., F. E. Cohen et al. (1990). Improvements in protein secondary structure prediction by an enhanced neural network. *J. Mol. Biol.* **214**(1): 171–82.

Knoll, A. H. (1992). The early evolution of eukaryotes: a geological perspective. *Science* **256**(5057): 622–27.

Knoop, V. and A. Brennicke (1991). A mitochondrial intron sequence in the 5′-flanking region of a plant nuclear lectin gene. *Curr. Genet.* **20**(5): 423–25.

Knoop, V. and A. Brennicke (1994). Promiscuous mitochondrial group II intron sequences in plant nuclear genomes. *J. Mol. Evol.* **39**(2): 144–50.

Knoop, V., M. Unseld et al. (1996). Copia-, gypsy- and LINE-like retrotransposon fragments in the mitochondrial genome of *Arabidopsis thaliana*. *Genetics* **142**(2): 579–85.

Koch, A. L. (1996). What size should a bacterium be? A question of scale. *Annu. Rev. Microbiol.* **50**: 317–48.

Kogelnik, A. M., M. T. Lott et al. (1998). MITOMAP: a human mitochondrial genome database—1998 update. *Nucleic Acids Res.* **26**(1): 112–15.

Kohler, S., C. F. Delwiche et al. (1997). A plastid of probable green algal origin in *Apicomplexan* parasites. *Science* **275**(5305): 1485–89.

Kolaskar, A. S. and P. C. Tongaonkar (1990). A semi-empirical method for prediction of antigenic determinants on protein antigens. *FEBS Lett.* **276**(1–2): 172–74.

Kolesnikova, O. A., N. S. Entelis et al. (2000). Suppression of mutations in mitochondrial DNA by tRNAs imported from the cytoplasm. *Science* **289**(5486): 1931–33.

Kolmogorov, A. N. (1965). Three approaches to the definition of the concept "quantity of information." *IEEE Trans. Inf. Theory* **14**: 662–69.

Koonin, E. V., L. Aravind et al. (2000). The impact of comparative genomics on our understanding of evolution. *Cell* **101**(6): 573–76.

Koonin, E. V. and M. Y. Galperin (1997). Prokaryotic genomes: the emerging paradigm of genome-based microbiology. *Curr. Opin. Genet. Dev.* **7**(6): 757–63.

Kozak, M. (1989). The scanning model for translation: an update. *J. Cell Biol.* **108**: 229–41.

Kozak, M. (1999). Initiation of translation in prokaryotes and eukaryotes. *Gene* **234**(2): 187–208.

Kozak, M. (2001). New ways of initiating translation in eukaryotes? *Mol. Cell Biol.* **21**(6): 1899–1907.

Kozhukhin, C. A. and P. A. Pevzner (1991). Genome inhomogeneity is determined mainly by WW and SS dinucleotides. *CABIOS* **7**(1): 39–49.

Krakauer, D. C. and A. Mira (1999). Mitochondria and germ-cell death. *Nature* **400**(6740): 125–26.

Krawczak, M., N. A. Chuzhanova et al. (2000). Changes in primary DNA sequence complexity influence the phenotypic consequences of mutations in human gene regulatory regions. *Hum. Genet.* **107**(4): 362–65.

Kriventseva, E. V., W. Fleischmann et al. (2001). CluSTr: a database of clusters of SWISS-PROT + TrEMBL proteins. *Nucleic Acids Res.* **29**(1): 33–36.

Krogh, A. (1997). Two methods for improving performance of an HMM and their application for gene finding. *Proc. Int. Conf. Intell. Syst. Mol. Biol.* **5**: 179–86.

Krogh, A. (2000). Using database matches with for HMMGene for automated gene detection in *Drosophila*. *Genome Res.* **10**(4): 523–28.

Krogh, A., M. Brown et al. (1994). Hidden Markov models in computational biology. Applications to protein modeling. *J. Mol. Biol.* **235**(5): 1501–31.

Krogh, A., B. Larsson et al. (2001). Predicting transmembrane protein topology with a hidden Markov model: application to complete genomes. *J. Mol. Biol.* **305**(3): 567–80.

Ku, H. M., T. Vision et al. (2000). Comparing sequenced segments of the tomato and *Arabidopsis* genomes: large-scale duplication followed by selective gene loss creates a network of synteny. *Proc. Natl. Acad. Sci. USA* **97**(16): 9121–26.

Kubo, N., K. Harada et al. (1999). A single nuclear transcript encoding mitochondrial RPS14 and SDHB of rice is processed by alternative splicing: common use of the same mitochondrial targeting signal for different proteins. *Proc. Natl. Acad. Sci. USA* **96**(16): 9207–11.

Kubo, T., S. Nishizawa et al. (2000). The complete nucleotide sequence of the mitochondrial genome of sugar beet (*Beta vulgaris* L.) reveals a novel gene for tRNA(Cys)(GCA). *Nucleic Acids Res.* **28**(13): 2571–76.

Kumazawa, Y., H. Ota et al. (1998). The complete nucleotide sequence of a snake (*Dinodon semicarinatus*) mitochondrial genome with two identical control regions. *Genetics* **150**(1): 313–29.

Kunkel, T. A. and P. S. Alexander (1986). The base substitution fidelity of eukaryotic DNA polymerases: misparing frequencies, site preferences, insertion preferences, and base substitution by dislocation. *J. Biol. Chem.* **261**: 160–66.

Kunkel, T. A. and A. Soni (1988). Exonucleolytic proofreading enhances the fidelity of DNA synthesis by chick embryo DNA polymerase-gamma. *J. Biol. Chem.* **263**: 4450–59.

Kunst, F., N. Ogasawara et al. (1997). The complete genome sequence of the gram-positive bacterium *Bacillus subtilis*. *Nature* **390**(6657): 249–56.

Kurabayashi, A. and R. Ueshima (2000). Complete sequence of the mitochondrial DNA of the primitive opisthobranch gastropod *Pupa strigosa*: systematic implication of the genome organization. *Mol. Biol. Evol.* **17**(2): 266–77.

Kurland, C. G. (2000). Something for everyone. Horizontal gene transfer in evolution. *EMBO Rep.* **1**(2): 92–95.

Kuroda, M., T. Ohta et al. (2001). Whole genome sequencing of meticillin-resistant *Staphylococcus aureus*. *Lancet* **357**(9264): 1225–40.

Kurtz, S. and C. Schleiermacher (1999). REPuter: fast computation of maximal repeats in complete genomes. *Bioinformatics* **15**(5): 426–27.

Kyrpides, N. C. and C. A. Ouzounis (1999). Transcription in archaea. *Proc. Natl. Acad. Sci. USA* **96**(15): 8545–50.

Kyte, J. and R. F. Doolittle (1982). A simple method for displaying the hydropathic character of a protein. *J. Mol. Biol.* **157**: 105–32.

Lafay, B., J. C. Atherton et al. (2000). Absence of translationally selected synonymous codon usage bias in *Helicobacter pylori*. *Microbiology* **146**(Pt. 4): 851–60.

Laforest, M. J., I. Roewer et al. (1997). Mitochondrial tRNAs in the lower fungus *Spizellomyces punctatus*: tRNA editing and UAG "stop" codons recognized as leucine. *Nucleic Acids Res.* **25**(3): 626–32.

Lanave, C., S. Liuni et al. (2000). Update of AMmtDB: a database of multi-aligned metazoa mitochondrial DNA sequences. *Nucleic Acids Res.* **28**(1): 153–54.

Lanave, C., G. Preparata et al. (1984). A new method for calculating evolutionary substitution rates. *J. Mol. Evol.* **20**: 86–93.

Lander, E. S., L. M. Linton et al. (2001). Initial sequencing and analysis of the human genome. *Nature* **409**(6822): 860–921.

Lane, D. P. (1992). Cancer. p53, guardian of the genome. *Nature* **358**(6381): 15–16.

Lang, B. F., G. Burger et al. (1997). An ancestral mitochondrial DNA resembling a eubacterial genome in miniature. *Nature* **387**(6632): 493–97.

Laroche, J. and J. Bousquet (1999). Evolution of the mitochondrial rps3 intron in perennial and annual angiosperms and homology to nad5 intron 1. *Mol. Biol. Evol.* **16**(4): 441–52.

Larsson, N. G., J. Wang et al. (1998). Mitochondrial transcription factor A is necessary for mtDNA maintenance and embryogenesis in mice. *Nat. Genet.* **18**(3): 231–36.

Lawrence, C. E., S. F. Altschul et al. (1993). Detecting subtle sequence signals: a Gibbs sampling strategy for multiple alignment. *Science* **262**(5131): 208–14.

Lawrence, J. G. and H. Ochman (1998). Molecular archaeology of the *Escherichia coli* genome. *Proc. Natl. Acad. Sci. USA* **95**(16): 9413–17.

Le, Q. H., S. Wright et al. (2000). Transposon diversity in *Arabidopsis thaliana*. *Proc. Natl. Acad. Sci. USA* **97**(13): 7376–81.

Le, S. Y. and J. V. Maizel (1997). A common RNA structural motif involved in the internal initiation of translation of cellular mRNAs. *Nucleic Acids Res.* **25**(2): 362–69.

Leatherwood, J. (1998). Emerging mechanisms of eukaryotic DNA replication initiation. *Curr. Opin. Cell Biol.* **10**(6): 742–48.

Lecompte, O., R. Ripp et al. (2001). Genome evolution at the genus level: comparison of three complete genomes of hyperthermophilic archaea. *Genome Res.* **11**(6): 981–93.

Lecrenier, N. and F. Foury (2000). New features of mitochondrial DNA replication system in yeast and man. *Gene* **246**(1–2): 37–48.

Lee, K. H. (2001). Proteomics: a technology-driven and technology-limited discovery science. *Trends Biotechnol.* **19**(6): 217–22.

Lee, W. J. and T. D. Kocher (1995). Complete sequence of a sea lamprey (*Petromyzon marinus*) mitochondrial genome: early establishment of the vertebrate genome organization. *Genetics* **139**(2): 873–87.

Lehman, N. (2001). Molecular evolution: Please release me, genetic code. *Curr. Biol.* **11**(2): R63–R66.

Lehvaslaiho, H. (2000). Human sequence variation and mutation databases. *Brief. Bioinf.* **1**(2): 161–66.

Lemer, C. M., M. J. Rooman et al. (1995). Protein structure prediction by threading methods: evaluation of current techniques. *Proteins* **23**(3): 337–55.

Levine, A. J. (1997). p53, the cellular gatekeeper for growth and division. *Cell* **88**(3): 323–31.

Levings, C. S., III and G. G. Brown (1989). Molecular biology of plant mitochondria. *Cell* **56**(2): 171–79.

Levitt, M. (1978). Conformational preferences of amino acids in globular proteins. *Biochemistry* **17**: 4277–85.

Levrero, M., V. De Laurenzi et al. (2000). The p53/p63/p73 family of transcription factors: overlapping and distinct functions. *J. Cell Sci.* **113**: 1661–70.

Lewis, D. L., C. L. Farr et al. (1995). *Drosophila melanogaster* mitochondrial DNA: completion of the nucleotide sequence and evolutionary comparisons. *Insect Mol. Biol.* **4**(4): 263–78.

Li, W. (2001). Delineating relative homogeneous G + C domains in DNA sequences. *Gene* **276**(1–2): 57–72.

Li, W. H., Z. Gu et al. (2001). Evolutionary analyses of the human genome. *Nature* **409**(6822): 847–49.

Liang, F., I. Holt et al. (2000a). Gene index analysis of the human genome estimates approximately 120,000 genes. *Nat. Genet.* **25**(2): 239–40.

Liang, F., I. Holt et al. (2000b). An optimized protocol for analysis of EST sequences. *Nucleic Acids Res.* **28**(18): 3657–65.

Liang, P. and A. B. Pardee (1992). Differential display of eukaryotic messenger RNA by means of the polymerase chain reaction. *Science* **257**(5072): 967–71.

Liang, P. and A. B. Pardee (1995). Recent advances in differential display. *Curr. Opin. Immunol.* **7**(2): 274–80.

Lilly, J. W. and M. J. Havey (2001). Small, repetitive DNAs contribute significantly to the expanded mitochondrial genome of cucumber. *Genetics* **159**(1): 317–28.

Lim, V. I. (1974). Algorithms for prediction of alpha-helical and beta-structural regions in globular proteins. *J. Mol. Biol.* **88**(4): 873–94.

Lindahl, T. (1993). Instability and decay of primary structure of DNA. *Nature* **362**: 709–15.

Lingner, J. and T. R. Cech (1998). Telomerase and chromosome end maintenance. *Curr. Opin. Genet. Dev.* **8**(2): 226–32.

Lipman, D. J., S. F. Altschul et al. (1989). A tool for multiple sequence alignment. *Proc. Natl. Acad. Sci. USA* **86**(12): 4412–15.

Lipshutz, R. J., S. P. Fodor et al. (1999). High density synthetic oligonucleotide arrays. *Nat. Genet.* **21**(1 Suppl.): 20–24.

Lisitsyn, N. and M. Wigler (1993). Cloning the differences between two complex genomes. *Science* **259**(5097): 946–51.

Liu, S. L. and K. E. Sanderson (1995). Rearrangements in the genome of the bacterium *Salmonella typhi*. *Proc. Natl. Acad. Sci. USA* **92**(4): 1018–22.

Liu, Z. and R. A. Butow (1999). A transcriptional switch in the expression of yeast tricarboxylic acid cycle genes in response to a reduction or loss of respiratory function. *Mol. Cell Biol.* **19**(10): 6720–28.

Lobry, J. R. (1996). Asymmetric substitution patterns in the two DNA strands of bacteria. *Mol. Biol. Evol.* **13**(5): 660–65.

Lobry, J. R. (1997). Influence of genomic G + C content on average amino-acid composition of proteins from 59 bacterial species. *Gene* **205**(1–2): 309–16.

Lockhart, D. J., H. Dong et al. (1996). Expression monitoring by hybridization to high-density oligonucleotide arrays. *Nat. Biotechnol.* **14**(13): 1675–80.

Lockhart, D. J. and E. A. Winzeler (2000). Genomics, gene expression and DNA arrays. *Nature* **405**(6788): 827–36.

Lo Conte, L., B. Ailey et al. (2000). SCOP: a structural classification of proteins database. *Nucleic Acids Res.* **28**(1): 257–59.

Logsdon, J. M., Jr. (1998). The recent origins of spliceosomal introns revisited. *Curr. Opin. Genet. Dev.* **8**(6): 637–48.

Lonergan, K. M. and M. W. Gray (1993). Editing of transfer RNAs in *Acanthamoeba castellanii* mitochondria. *Science* **259**(5096): 812–16.

Lowe, T. M. and S. R. Eddy (1997). tRNAscan-SE: a program for improved detection of transfer RNA genes in genomic sequence. *Nucleic Acids Res.* **25**(5): 955–64.

Lozovskaya, E. R., D. L. Hartl et al. (1995). Genomic regulation of transposable elements in *Drosophila*. *Curr. Opin. Genet. Dev.* **5**(6): 768–73.

Lu, B., R. K. Wilson et al. (1996). Protein polymorphism generated by differential RNA editing of a plant mitochondrial rps12 gene. *Mol. Cell Biol.* **16**(4): 1543–49.

Lunt, D. H. and B. C. Hyman (1997). Animal mitochondrial DNA recombination. *Nature* **387**: 247.

Lupas, A. (1997). Predicting coiled-coil regions in proteins. *Curr. Opin. Struct. Biol.* **7**(3): 388–93.

Lupold, D. S., A. G. Caoile et al. (1999). The maize mitochondrial *cox2* gene has five promoters in two genomic regions, including a complex promoter consisting of seven overlapping units. *J. Biol. Chem.* **274**(6): 3897–3903.

Lupski, J. R. and G. M. Weinstock (1992). Short, interspersed repetitive DNA sequences in prokaryotic genomes. *J. Bacteriol.* **174**(14): 4525–29.

Lynn, A., C. Kashuk et al. (2000). Patterns of meiotic recombination on the long arm of human chromosome 21. *Genome Res.* **10**(9): 1319–32.

Macasev, D., E. Newbigin et al. (2000). How do plant mitochondria avoid importing chloroplast proteins? Components of the import apparatus Tom20 and Tom22 from *Arabidopsis* differ from their fungal counterparts. *Plant Physiol.* **123**(3): 811–16.

Madsen, O., M. Scally et al. (2001). Parallel adaptive radiations in two major clades of placental mammals. *Nature* **409**(6820): 610–14.

Maier, D., C. L. Farr et al. (2001). Mitochondrial single-stranded DNA-binding protein is required for mitochondrial DNA replication and development in *Drosophila melanogaster*. *Mol. Biol. Cell* **12**(4): 821–30.

Maier, R. M., P. Zeltz et al. (1996). RNA editing in plant mitochondria and chloroplasts. *Plant Mol. Biol.* **32**(1–2): 343–65.

Maier, U. G., M. Fraunholz et al. (2000). A nucleomorph-encoded CbbX and the phylogeny of RuBisCo regulators. *Mol. Biol. Evol.* **17**(4): 576–83.

Maizel, J. V., Jr. and R. P. Lenk (1981). Enhanced graphic matrix analysis of nucleic acid and protein sequences. *Proc. Natl. Acad. Sci. USA* **78**(12): 7665–69.

Maizels, N. and A. M. Weiner (1995). Phylogeny from function: the origin of tRNA is replication, not translation. In *Tempo and Mode in Evolution*. W. M. Fitch and F. J. Ayala, eds. Washington, D.C.: National Academy Press.

Malek, O., A. Brennicke et al. (1997). Evolution of trans-splicing plant mitochondrial introns in pre-Permian times. *Proc. Natl. Acad. Sci. USA* **94**(2): 553–58.

Manzanares, M., A. Locascio et al. (2001). The increasing complexity of the Snail gene super-family in metazoan evolution. *Trends Genet.* **17**(4): 178–81.

Marais, G., D. Mouchiroud et al. (2001). Does recombination improve selection on codon usage? Lessons from nematode and fly complete genomes. *Proc. Natl. Acad. Sci. USA* **98**(10): 5688–92.

Marchfelder, A., S. Binder et al. (1998). RNA editing by base conversion in plant organellar RNAs. In *Modification and Editing of RNA*. H. Grosjean and R. Benne, eds. Washington, D.C.: ASM Press, pp. 307–323.

Marchler-Bauer, A., A. R. Panchenko et al. (2002). CDD: a database of conserved domain alignments with links to domain three-dimensional structure. *Nucleic Acids Res.* **30**(1): 281–83.

Marcotte, E. M. (2000). Computational genetics: finding protein function by nonhomology methods. *Curr. Opin. Struct. Biol.* **10**(3): 359–65.

Marechal, E. and M. F. Cesbron-Delauw (2001). The apicoplast: a new member of the plastid family. *Trends Plant Sci.* **6**(5): 200–205.

Marechal-Drouard, L., D. Ramamonjisoa et al. (1993). Editing corrects mispairing in the acceptor stem of bean and potato mitochondrial phenylalanine transfer RNAs. *Nucleic Acids Res.* **21**(21): 4909–14.

Margulis, L. (1970). *Origin of Eukaryotic Cells*. New Haven, Conn.: Yale University Press.

Martin, W. and R. G. Herrmann (1998). Gene transfer from organelles to the nucleus: how much, what happens, and why? *Plant Physiol.* **118**(1): 9–17.

Martin, W. and M. Muller (1998). The hydrogen hypothesis for the first eukaryote. *Nature* **392**(6671): 37–41.

Masters, B. S., L. L. Stohl et al. (1987). Yeast mitochondrial RNA polymerase is homologous to those encoded by bacteriophages T3 and T7. *Cell* **51**(1): 89–99.

Mathews, D. H., J. Sabina et al. (1999). Expanded sequence dependence of thermodynamic parameters improves prediction of RNA secondary structure. *J. Mol. Biol.* **288**(5): 911–40.

Maxam, A. M. and W. Gilbert (1977). A new method for sequencing DNA. *Proc. Natl. Acad. Sci. USA* **74**(2): 560–64.

May, B. J., Q. Zhang et al. (2001). Complete genomic sequence of *Pasteurella multocida*, Pm70. *Proc. Natl. Acad. Sci. USA* **98**(6): 3460–65.

McCabe, P. F. and C. J. Leaver (2000). Programmed cell death in cell cultures. *Plant Mol. Biol.* **44**(3): 359–68.

McClelland, M., F. Mathieu-Daude et al. (1995). RNA fingerprinting and differential display using arbitrarily primed PCR. *Trends Genet.* **11**(6): 242–46.

McClelland, M., K. E. Sanderson et al. (2001). Complete genome sequence of *Salmonella enterica* serovar *typhimurium* LT2. *Nature* **413**(6858): 852–56.

McCulloch, V., B. L. Seidel-Rogol et al. (2002). A human mitochondrial transcription factor is related to RNA adenine methyltransferases and binds s-adenosylmethionine. *Mol. Cell Biol.* **22**(4): 1116–25.

McInerney, J. O. (1998). Replicational and transcriptional selection on codon usage in *Borrelia burgdorferi*. *Proc. Natl. Acad. Sci. USA* **95**(18): 10698–703.

McKeown, M. (1992). Alternative mRNA splicing. *Annu. Rev. Cell Biol.* **8**: 133–55.

McLean, M. J., K. H. Wolfe et al. (1998). Base composition skews, replication orientation, and gene orientation in 12 prokaryote genomes. *J. Mol. Evol.* **47**(6): 691–96.

McVean, G. A. and G. D. Hurst (2000). Evolutionary lability of context-dependent codon bias in bacteria. *J. Mol. Evol.* **50**(3): 264–75.

Medigue, C., A. Viari et al. (1991). *Escherichia colis* molecular genetic map (1500 kbp): update II. *Mol. Microbiol.* **5**(11): 2629–40.

Mengeritsky, G. and T. F. Smith (1987). Recognition of characteristic patterns in sets of functionally equivalent DNA sequences. *CABIOS* **3**(3): 223–27.

Meyer, A. and E. Malaga-Trillo (1999). Vertebrate genomics: more fishy tales about Hox genes. *Curr. Biol.* **9**(6): R210–R213.

Meyer, A. and M. Schartl (1999). Gene and genome duplications in vertebrates: the one-to-four (-to-eight in fish) rule and the evolution of novel gene functions. *Curr. Opin. Cell Biol.* **11**(6): 699–704.

Michel, F., A. Jacquier et al. (1982). Comparison of fungal mitochondrial introns reveals extensive homologies in RNA secondary structure. *Biochimie* **64**(10): 867–81.

Mignone, F., C. Gissi et al. (2002). Untranslated regions of mRNAs. *Genome Biol.* **3**(3): Reviews0004.

Miller, I., P. Haynes et al. (1998). Proteins of rat serum: II. Influence of some biological parameters of the two-dimensional electrophoresis pattern. *Electrophoresis* **19**(8–9): 1493–1500.

Miller, R. T., A. G. Christoffels et al. (1999). A comprehensive approach to clustering of expressed human gene sequence: the sequence tag alignment and consensus knowledge base. *Genome Res.* **9**(11): 1143–55.

Mindell, D., M. Sorenson et al. (1998). Multiple independent origins of mitochondrial gene order in birds. *Proc. Natl. Acad. Sci. USA* **95**: 10693–97.

Mireau, H., A. Cosset et al. (2000). Expression of *Arabidopsis thaliana* mitochondrial alanyl-tRNA synthetase is not sufficient to trigger mitochondrial import of tRNAAla in yeast. *J. Biol. Chem.* **275**(18): 13291–96.

Miret, J. J., L. Pessoa-Brandao et al. (1998). Orientation-dependent and sequence-specific expansions of CTG/CAG trinucleotide repeats in *Saccharomyces cerevisiae*. *Proc. Natl. Acad. Sci. USA* **95**(21): 12438–43.

Moraes, C. T., L. Kenyon et al. (1999). Mechanisms of human mitochondrial DNA maintenance: the determining role of primary sequence and length over function. *Mol. Biol. Cell.* **10**(10): 3345–56.

Moreira, D. and P. Lopez-Garcia (1998). Symbiosis between methanogenic archaea and delta-proteobacteria as the origin of eukaryotes: the syntrophic hypothesis. *J. Mol. Evol.* **47**(5): 517–30.

Moritz, C., T. Dowling et al. (1987). Evolution of animal mitochondrial DNA: relevance for population and systematics. *Annu. Rev. Ecol. Syst.* vol. 18: 269–92.

Morton, B. R. (1999). Strand asymmetry and codon usage bias in the chloroplast genome of *Euglena gracilis*. *Proc. Natl. Acad. Sci. USA* **96**(9): 5123–28.

Moszer, I., E. P. Rocha et al. (1999). Codon usage and lateral gene transfer in *Bacillus subtilis*. *Curr. Opin. Microbiol.* **2**(5): 524–28.

Moult, J., K. Fidelis et al. (2001). Critical assessment of methods of protein structure prediction (CASP): round IV. *Proteins* **45**(Suppl. 5): 2–7.

Mrazek, J. and S. Karlin (1998). Strand compositional asymmetry in bacterial and large viral genomes. *Proc. Natl. Acad. Sci. USA* **95**(7): 3720–5.

Muller, C. W. (2001). Transcription factors: global and detailed views. *Curr. Opin. Struct. Biol.* **11**(1): 26–32.

Mulligan, R. M., M. A. Williams et al. (1999). RNA editing site recognition in higher plant mitochondria. *J. Hered.* **90**(3): 338–44.

Mullis, K., F. Faloona et al. (1986). Specific enzymatic amplification of DNA in vitro: the polymerase chain reaction. *Cold Spring Harbor Symp. Quant. Biol.* **51**(Pt. 1): 263–73.

Munoz, V. and L. Serrano (1997). Development of the multiple sequence approximation within the AGADIR model of alpha-helix formation: comparison with Zimm–Bragg and Lifson–Roig formalisms. *Biopolymers* **41**(5): 495–509.

Murphy, W. J., E. Eizirik et al. (2001). Molecular phylogenetics and the origins of placental mammals. *Nature* **409**(6820): 614–18.

Murray, R. (1984). The higher taxa, or, a place for everything. In *Bergey's Manual of Systematic Bacteriology*, Vol. 1. N. Krieg and J. Holt, eds. Baltimore: Williams & Wilkins, pp. 31–34.

Muse, S. V. and B. S. Gaut (1994). A likelihood approach for comparing synonymous and nonsynonymous nucleotide substitution rates, with application to the chloroplast genome. *Mol. Biol. Evol.* **11**(5): 715–24.

Muse, S. V. and B. S. Weir (1992). Testing for equality of evolutionary rates. *Genetics* **132**: 269–76.

Mushegian, A. (1999). The minimal genome concept. *Curr. Opin. Genet. Dev.* **9**(6): 709–14.

Mushegian, A. R. and E. V. Koonin (1996). A minimal gene set for cellular life derived by comparison of complete bacterial genomes. *Proc. Natl. Acad. Sci. USA* **93**(19): 10268–73.

Muto, A. and S. Osawa (1987). The guanine and cytosine content of genomic DNA and bacterial evolution. *Proc. Natl. Acad. Sci. USA* **84**(1): 166–69.

Myler, P. J., L. Audleman et al. (1999). *Leishmania* major Friedlin chromosome 1 has an unusual distribution of protein-coding genes. *Proc. Natl. Acad. Sci. USA* **96**(6): 2902–6.

Myllykallio, H., P. Lopez et al. (2000). Bacterial mode of replication with eukaryotic-like machinery in a hyperthermophilic archaeon. *Science* **288**(5474): 2212–15.

Nakai, K. and P. Horton (1999). PSORT: a program for detecting sorting signals in proteins and predicting their subcellular localization. *Trends Biochem. Sci.* **24**(1): 34–36.

Nakamura, Y., T. Gojobori et al. (2000). Codon usage tabulated from international DNA sequence databases: status for the year 2000. *Nucleic Acids Res.* **28**(1): 292.

Nardi, F., A. Carapelli et al. (2001). The complete mitochondrial DNA sequence of the basal hexapod *Tetrodontophora bielanensis*: evidence for heteroplasmy and tRNA translocations. *Mol. Biol. Evol.* **18**(7): 1293–1304.

Naylor, G. J. P. and W. M. Brown (1998). Amphioxus mitochondrial DNA, chordate phylogeny, and the limits of inference based on comparisons of sequences. *Syst. Biol.* **47**(1): 61–76.

Needleman, S. B. and C. D. Wunsch (1970). A general method applicable to the search for similarities in the amino acid sequences of two proteins. *J. Mol. Biol.* **48**: 443–53.

Neilson, L., A. Andalibi et al. (2000). Molecular phenotype of the human oocyte by PCR-SAGE. *Genomics* **63**(1): 13–24.

Nekrutenko, A. and W. H. Li (2001). Transposable elements are found in a large number of human protein-coding genes. *Trends Genet.* **17**(11): 619–21.

Nelson, K. E., R. A. Clayton et al. (1999). Evidence for lateral gene transfer between archaea and bacteria from genome sequence of *Thermotoga maritima*. *Nature* **399**(6734): 323–29.

Neuwald, A. F., J. S. Liu et al. (1995). Gibbs motif sampling: detection of bacterial outer membrane protein repeats. *Protein Sci.* **4**(8): 1618–32.

Neuwald, A. F., J. S. Liu et al. (1997). Extracting protein alignment models from the sequence database. *Nucleic Acids Res.* **25**(9): 1665–77.

Ng, W. V., S. P. Kennedy et al. (2000). Genome sequence of *Halobacterium* species NRC-1. *Proc. Natl. Acad. Sci. USA* **97**(22): 12176–81.

Nielsen, H., J. Engelbrecht et al. (1997). Identification of prokaryotic and eukaryotic signal peptides and prediction of their cleavage sites. *Protein Eng.* **10**(1): 1–6.

Nierman, W. C., T. V. Feldblyum et al. (2001). Complete genome sequence of *Caulobacter crescentus*. *Proc. Natl. Acad. Sci. USA* **98**(7): 4136–41.

Ning, Z., A. J. Cox, and J. C. Mullikin (2001). SSAHA: a fast search method for large DNA databases. *Genome Res.* **11**(10): 1725–29.

Noguchi, Y., K. Endo et al. (2000). The mitochondrial genome of the brachiopod *Laqueus rubellus*. *Genetics* **155**(1): 245–59.

Noller, H. F., R. Green et al. (1995). Structure and function of ribosomal RNA. *Biochem. Cell Biol.* **73**(11–12): 997–1009.

Nolling, J., G. Breton et al. (2001). Genome sequence and comparative analysis of the solvent-producing bacterium *Clostridium acetobutylicum*. *J. Bacteriol.* **183**(16): 4823–38.

Nosek, J., L. Tomaska et al. (1998). Linear mitochondrial genomes: 30 years down the line. *Trends Genet.* **14**(5): 184–88.

Notredame, C. and D. G. Higgins (1996). SAGA: sequence alignment by genetic algorithm. *Nucleic Acids Res.* **24**(8): 1515–24.

Notredame, C., E. A. O'Brien et al. (1997). RAGA: RNA sequence alignment by genetic algorithm. *Nucleic Acids Res.* **25**(22): 4570–80.

Novacek, N. J. (1992). Mammalian phylogeny: shaking the tree. *Nature* **379**: 333–35.

Nowak, M. A., M. C. Boerlijst et al. (1997). Evolution of genetic redundancy. *Nature* **388**(6638): 167–71.

Nunnari, J., W. F. Marshall et al. (1997). Mitochondrial transmission during mating in *Saccharomyces cerevisiae* is determined by mitochondrial fusion and fission and the intramitochondrial segregation of mitochondrial DNA. *Mol. Biol. Cell* **8**(7): 1233–42.

Nussinov, R. (1980). Some rules in the ordering of nucleotides in the DNA. *Nucleic Acids Res.* **8**: 4545–62.

Nussinov, R. (1981). Nearest neighbor nucleotide patterns: structural and biological implications. *J. Biol. Chem.* **256**: 8458–62.

Nussinov, R. (1987). Theoretical molecular biology: prospectives and perspectives. *J. Theor. Biol.* **125**: 219–35.

Ochman, H., J. G. Lawrence et al. (2000). Lateral gene transfer and the nature of bacterial innovation. *Nature* **405**(6784): 299–304.

Oda, K., K. Yamato et al. (1992). Gene organization deduced from the complete sequence of liverwort *Marchantia polymorpha* mitochondrial DNA. A primitive form of plant mitochondrial genome. *J. Mol. Biol.* **223**(1): 1–7.

Ogata, H., S. Audic et al. (2001). Mechanisms of evolution in *Rickettsia conorii* and *R. prowazekii*. *Science* **293**(5537): 2093–98.

Ohno, S. (1993). Patterns in genome evolution. *Curr. Opin. Genet. Dev.* **3**(6): 911–14.

Ojala, D., C. Merkel et al. (1980). The tRNA genes punctuate the reading of genetic information in human mitochondrial DNA. *Cell* **22**(2, Pt. 2): 393–403.

Okimoto, R., H. M. Chamberlin et al. (1991). Repeated sequence sets in mitochondrial DNA molecules of root knot nematodes (*Meloidogyne*): nucleotide sequences, genome location and potential for host-race identification. *Nucleic Acids Res.* **19**(7): 1619–26.

Okimoto, R., J. L. Macfarlane et al. (1992). The mitochondrial genome of two nematodes, *Caenorhabditis elegans* and *Ascaris suum*. *Genetics* **130**: 471–98.

Oldenburg, D. J. and A. J. Bendich (2001). Mitochondrial DNA from the liverwort *Marchantia polymorpha*: circularly permuted linear molecules, head-to-tail concatemers, and a 5′ protein. *J. Mol. Biol.* **310**(3): 549–62.

Olsen, G. J. (2001). The history of life. *Nat. Genet.* **28**(3): 197–98.

Olson, A. and J. Stenlid (2001). Plant pathogens. Mitochondrial control of fungal hybrid virulence. *Nature* **411**(6836): 438.

O'Neill, M. J. and A. H. Sinclair (1997). Isolation of rare transcripts by representational difference analysis. *Nucleic Acids Res.* **25**(13): 2681–82.

Oren, M. (1999). Regulation of the p53 tumor suppressor protein. *J. Biol. Chem.* **274**(51): 36031–34.

Oren, M. and V. Rotter (1999). Introduction: p53—the first twenty years. *Cell. Mol. Life Sci.* **55**(1): 9–11.

Osguthorpe, D. J. (2000). Ab initio protein folding. *Curr. Opin. Struct. Biol.* **10**(2): 146–52.

Otsuga, D., B. R. Keegan et al. (1998). The dynamin-related GTPase, Dnm1p, controls mitochondrial morphology in yeast. *J. Cell Biol.* **143**(2): 333–49.

Ouali, M. and R. D. King (2000). Cascaded multiple classifiers for secondary structure prediction. *Protein Sci.* **9**(6): 1162–76.

Page, R. D. M. and E. C. Holmes (1998). *Molecular Evolution: A Phylogenetic Approach.* Oxford: Blackwell Science.

Palleschi, C., S. Francisci et al. (1984). Expression of the clustered mitochondrial tRNA genes in *Saccharomyces cerevisiae*: transcription and processing of transcripts. *EMBO J.* **3**(6): 1389–95.

Palmer, J. D. (1990). Contrasting modes and tempos of genome evolution in land plant organelles. *Trends Genet.* **6**(4): 115–20.

Pan, A., C. Dutta et al. (1998). Codon usage in highly expressed genes of *Haemophillus influenzae* and *Mycobacterium tuberculosis*: translational selection versus mutational bias. *Gene* **215**(2): 405–13.

Pappin, D. D. J., P. Højrup et al. (1993). Rapid identification of proteins by peptide-mass finger printing. *Curr. Biol.* **3**: 327–32.

Paques, F. and J. E. Haber (1999). Multiple pathways of recombination induced by double-strand breaks in *Saccharomyces cerevisiae*. *Microbiol. Mol. Biol. Rev.* **63**(2): 349–404.

Paquin, B., M. J. Laforest et al. (1997). The fungal mitochondrial genome project: evolution of fungal mitochondrial genomes and their Gene Expression. *Curr. Genet.* **31**(5): 380–95.

Paquin, B., M. J. Laforest et al. (2000). Double-hairpin elements in the mitochondrial DNA of allomyces: evidence for mobility. *Mol. Biol. Evol.* **17**(11): 1760–68.

Parikh, V. S., M. M. Morgan et al. (1987). The mitochondrial genotype can influence nuclear Gene Expression in yeast. *Science* **235**(4788): 576–80.

Parisi, M. A., B. Xu et al. (1993). A human mitochondrial transcriptional activator can functionally replace a yeast mitochondrial HMG-box protein both in vivo and in vitro. *Mol. Cell Biol.* **13**(3): 1951–61.

Parkhill, J., M. Achtman et al. (2000). Complete DNA sequence of a serogroup A strain of *Neisseria meningitidis* Z2491. *Nature* **404**(6777): 502–6.

Parkhill, J., G. Dougan et al. (2001). Complete genome sequence of a multiple drug resistant *Salmonella enterica* serovar *typhi* CT18. *Nature* **413**(6858): 848–52.

Parkhill, J., B. W. Wren et al. (2000). The genome sequence of the food-borne pathogen Campylobacter jejuni reveals hypervariable sequences. *Nature* **403**(6770): 665–68.

Parkhill, J., B. W. Wren et al. (2001). Genome sequence of *Yersinia pestis*, the causative agent of plague. *Nature* **413**(6855): 523–27.

Parry-Smith, D. J., A. W. Payne et al. (1998). CINEMA—a novel colour INteractive editor for multiple alignments. *Gene* **221**(1): GC57–GC63.

Patterson, D. J. (1999). The diversity of eukaryotes. *Am. Nat.* **154**(S4): S96–S124.

Patthy, L. (1999). Genome evolution and the evolution of exon-shuffling: a review. *Gene* **238**(1): 103–14.

Paule, M. R. and R. J. White (2000). Survey and summary: transcription by RNA polymerases I and III. *Nucleic Acids Res.* **28**(6): 1283–98.

Pavesi, A., F. Conterio et al. (1994). Identification of new eukaryotic tRNA genes in genomic DNA databases by a multistep weight matrix analysis of transcriptional control regions. *Nucleic Acids Res.* **22**(7): 1247–56.

Pavlicek, A., K. Jabbari et al. (2001). Similar integration but different stability of Alus and LINEs in the human genome. *Gene* **276**(1–2): 39–45.

Pearl, F., A. E. Todd et al. (2000). Using the CATH domain database to assign structures and functions to the genome sequences. *Biochem. Soc. Trans.* **28**(2): 269–75.

Pearson, H. (2001). Biology's name game. *Nature* **411**(6838): 631–32.

Pearson, H. (2002). Surviving a knockout blow. *Nature* **415**(6867): 8–9.

Pearson, W. R. (1990). Rapid and sensitive sequence comparison with FASTP and FASTA. *Methods Enzymol.* **183**: 63–98.

Pearson, W. R. (1996). Effective protein sequence comparison. *Methods Enzymol.* **266**: 227–58.

Pearson, W. R. (1998). Empirical statistical estimates for sequence similarity searches. *J. Mol. Biol.* **276**(1): 71–84.

Pearson, W. R. (2000). Flexible sequence similarity searching with the FASTA3 program package. *Methods Mol. Biol.* **132**: 185–219.

Pearson, W. R. and D. J. Lipman (1988). Improved tools for biological sequence comparison. *Proc. Natl. Acad. Sci. USA* **85**: 2444–48.

Pellegrini, M. (2001). Computational methods for protein function analysis. *Curr. Opin. Chem. Biol.* **5**(1): 46–50.

Pellegrini, M., E. M. Marcotte et al. (1999). Assigning protein functions by comparative genome analysis: protein phylogenetic profiles. *Proc. Natl. Acad. Sci. USA* **96**(8): 4285–88.

Perez-Jannotti, R. M., S. M. Klein et al. (2001). Two forms of mitochondrial DNA ligase III are produced in *Xenopus laevis* oocytes. *J. Biol. Chem.* **276**(52): 48978–87.

Perkins, D. N., D. J. Pappin et al. (1999). Probability-based protein identification by searching sequence databases using mass spectrometry data. *Electrophoresis* **20**(18): 3551–67.

Perna, N. T., G. Plunkett III et al. (2001). Genome sequence of enterohaemorrhagic *Escherichia coli* O157:H7. *Nature* **409**(6819): 529–33.

Persson, B. and P. Argos (1994). Prediction of transmembrane segments in proteins utilising multiple sequence alignments. *J. Mol. Biol.* **237**(2): 182–92.

Pesole, G., M. Attimonelli et al. (1996). Linguistic analysis of nucleotide sequences: algorithms for pattern recognition and analysis of codon strategy. *Methods Enzymol.* **266**: 281–94.

Pesole, G., C. Gissi et al. (1999). Nucleotide substitution rate of mammalian mitochondrial genomes. *J. Mol. Evol.* **48**(4): 427–34.

Pesole, G., C. Gissi et al. (2000a). MitoNuc and MitoAln: two related databases of nuclear genes coding for mitochondrial proteins. *Nucleic Acids Res.* **28**(1): 163–65.

Pesole, G., C. Gissi et al. (2000b). Analysis of oligonucleotide AUG start codon context in eukariotic mRNAs. *Gene* **261**: 85–91.

Pesole, G., G. Grillo et al. (2000). The untranslated regions of eukaryotic mRNAs: structure, function, evolution and bioinformatics tools for their analysis. *Brief. Bioinf.* **1**: 236–49.

Pesole, G., S. Liuni et al. (2000a). PatSearch: a pattern matcher software that finds functional elements in nucleotide and protein sequences and assesses their statistical significance. *Bioinformatics* **16**(5): 439–50.

Pesole, G., S. Liuni et al. (2000b). UTRdb and UTRsite: specialized databases of sequences and functional elements of 5′ and 3′ untranslated regions of eukaryotic mRNAs. *Nucleic Acids Res.* **28**(1): 193–96.

Pesole, G., N. Prunella et al. (1992). WORDUP: an efficient algorithm for discovering statistically significant patterns in DNA sequences. *Nucleic Acids Res.* **20**(11): 2871–75.

Peters, D. G., A. B. Kassam et al. (1999). Comprehensive transcript analysis in small quantities of mRNA by SAGE-lite. *Nucleic Acids Res.* **27**(24): e39.

Peterson, J. D., L. A. Umayam et al. (2001). The Comprehensive Microbial Resource. *Nucleic Acids Res.* **29**(1): 123–25.

Petrov, D. A. (2001). Evolution of genome size: new approaches to an old problem. *Trends Genet.* **17**(1): 23–28.

Petsko, G. A. (2002). Grain of truth. *Genome Biol.* **3**(5): comment1007.

Pevzner, P. A. (1992). Statistical distance between texts and filtration methods in sequence comparison. *Comput. Appl. BioSci.* **8**(2): 121–27.

Pevzner, P. A. and S. H. Sze (2000). Combinatorial approaches to finding subtle signals in DNA sequences. *Proc. Int. Conf. Intell. Syst. Mol. Biol.* **8**: 269–78.

Philippe, H. and P. Forterre (1999). The rooting of the universal tree of life is not reliable. *J. Mol. Evol.* **49**(4): 509–23.

Phreaner, C. G., M. A. Williams et al. (1996). Incomplete editing of rps12 transcripts results in the synthesis of polymorphic polypeptides in plant mitochondria. *Plant Cell* **8**(1): 107–17.

Pietrokovski, S. (2001). Intein spread and extinction in evolution. *Trends Genet.* **17**(8): 465–72.

Pietrokovski, S., J. Hirshon et al. (1990). Linguistic measure of taxonomic and functional relatedness of nucleotide sequences. *J. Biomol. Struct. Dyn.* **7**: 1251–68.

Pinz, K. G., S. Shibutani et al. (1995). Action of mitochondrial DNA polymerase gamma at sites of base loss or oxidative damage. *J. Biol. Chem.* **270**: 9202–6.

Pittman, K. F., C. A. Walczak et al. (1991). Codes and abbreviations for approved of effectively published names of genera of bacteria published from January 1980 to December 1990. *Int. J. Syst. Bacteriol.* **41**(4): 571–79.

Pont-Kingdon, G. A., N. A. Okada et al. (1995). A coral mitochondrial MutS gene. *Nature* **375**: 109–11.

Pont-Kingdon, G. A., N. A. Okada et al. (1998). Mitochondrial DNA of the coral *Sarcophyton glaucum* contains a gene for a homologue bacterial MutS: a possible case of gene transfer from the nucleus to the mitochondrion. *J. Mol. Evol.* **46**: 419–31.

Popov, O., D. M. Segal et al. (1996). Linguistic complexity of protein sequences as compared to texts of human languages. *Biosystems* **38**(1): 65–74.

Prlic, A., F. S. Domingues et al. (2000). Structure-derived substitution matrices for alignment of distantly related sequences. *Protein Eng.* **13**(8): 545–50.

Proudfoot, N. (2000). Connecting transcription to messenger RNA processing. *Trends Biochem. Sci.* **25**(6): 290–93.

Pruitt, K. D., K. S. Katz et al. (2000). Introducing RefSeq and LocusLink: curated human genome resources at the NCBI. *Trends Genet.* **16**(1): 44–47.

Pyronnet, S. and N. Sonenberg (2001). Cell-cycle-dependent translational control. *Curr. Opin. Genet. Dev.* **11**(1): 13–18.

Quackenbush, J. (2001). Computational analysis of microarray data. *Nat. Rev. Genet.* **2**(6): 418–27.

Quackenbush, J., F. Liang et al. (2000). The TIGR gene indices: reconstruction and representation of expressed gene sequences. *Nucleic Acids Res.* **28**(1): 141–45.

Quandt, K., K. Frech et al. (1995). MatInd and MatInspector: new fast and versatile tools for detection of consensus matches in nucleotide sequence data. *Nucleic Acids Res.* **23**(23): 4878–84.

Race, H. L., R. G. Herrmann et al. (1999). Why have organelles retained genomes? *Trends Genet.* **15**(9): 364–70.

Ragnini, A. and H. Fukuhara (1988). Mitochondrial DNA of the yeast *Kluyveromyces*: guanine–cytosine rich sequence clusters. *Nucleic Acids Res.* **16**(17): 8433–42.

Raha, S. and B. H. Robinson (2000). Mitochondria, oxygen free radicals, disease and ageing. *Trends Biochem. Sci.* **25**(10): 502–8.

Ramamonjisoa, D., S. Kauffmann et al. (1998). Structure and expression of several bean (*Phaseolus vulgaris*) nuclear transfer RNA genes: relevance to the process of tRNA import into plant mitochondria. *Plant Mol. Biol.* **36**(4): 613–25.

Rambaut, A. and N. C. Grassly (1997). Seq-Gen: an application for the Monte Carlo simulation of DNA sequence evolution along phylogenetic trees. *Comput. Appl. Biosci.* **13**(3): 235–38.

Ranwez, V. and O. Gascuel (2001). Quartet-based phylogenetic inference: improvements and limits. *Mol. Biol. Evol.* **18**(6): 1103–16.

Rapp, W. D., D. S. Lupold et al. (1993). Architecture of the maize mitochondrial atp1 promoter as determined by linker-scanning and point mutagenesis. *Mol. Cell Biol.* **13**(12): 7232–38.

Read, T. D., R. C. Brunham et al. (2000). Genome sequences of *Chlamydia trachomatis* MoPn and *Chlamydia pneumoniae* AR39. *Nucleic Acids Res.* **28**(6): 1397–406.

Reeck, G. R., C. de Haen et al. (1987). "Homology" in proteins and nucleic acids: a terminology muddle and a way out of it [letter]. *Cell* **50**(5): 667.

Reese, M. G., D. Kulp et al. (2000). Genie—gene finding in *Drosophila melanogaster*. *Genome Res.* **10**(4): 529–38.

Reyes, A., C. Gissi et al. (1998). Asymmetrical directional mutation pressure in the mitochondrial genome of mammals. *Mol. Biol. Evol.* **15**(8): 957–66.

Reyes, A., C. Gissi et al. (2000). Where do rodents fit? Evidence from the complete mitochondrial genome of *Sciurus vulgaris*. *Mol. Biol. Evol.* **17**(6): 979–83.

Reyes, A., G. Pesole et al. (1998). Complete mitochondrial DNA sequence of the fat dormouse, *Glis glis:* further evidence of rodent paraphyly. *Mol. Biol. Evol.* **15**(5): 499–505.

Rice, P., I. Longden et al. (2000). EMBOSS: the European Molecular Biology Open Software Suite. *Trends Genet.* **16**(6): 276–77.

Ridanpaa, M., H. van Eenennaam et al. (2001). Mutations in the RNA component of RNase MRP cause a pleiotropic human disease, cartilage-hair hypoplasia. *Cell* **104**(2): 195–203.

Ringwald, M., J. T. Eppig et al. (2000). GXD: integrated access to Gene Expression data for the laboratory mouse. *Trends Genet.* **16**(4): 188–90.

Ringwald, M., J. T. Eppig et al. (2001). The Mouse Gene Expression Database (GXD). *Nucleic Acids Res.* **29**(1): 98–101.

Robb, F. T., D. L. Maeder et al. (2001). Genomic sequence of hyperthermophile, *Pyrococcus furiosus*: implications for physiology and enzymology. *Methods Enzymol.* **330**: 134–57.

Robison, K., A. M. McGuire et al. (1998). A comprehensive library of DNA-binding site matrices for 55 proteins applied to the complete *Escherichia coli* K-12 genome. *J. Mol. Biol.* **284**(2): 241–54.

Rodriguez-Tome, P. and P. Lijnzaad (2001). RHdb: the radiation hybrid database. *Nucleic Acids Res.* **29**(1): 165–66.

Rodriguez-Trelles, F., R. Tarrio et al. (1999). Switch in codon bias and increased rates of amino acid substitution in the *Drosophila saltans* species group. *Genetics* **153**(1): 339–50.

Rodriguez-Trelles, F., R. Tarrio et al. (2000). Evidence for a high ancestral GC content in *Drosophila*. *Mol. Biol. Evol.* **17**(11): 1710–17.

Roeder, R. G. (1996). The role of general initiation factors in transcription by RNA polymerase II. *Trends Biochem. Sci.* **21**(9): 327–35.

Rogic, S., A. K. Mackworth et al. (2001). Evaluation of gene-finding programs on mammalian sequences. *Genome Res.* **11**(5): 817–32.

Rohou, H., S. Francisci et al. (2001). Reintroduction of a characterized Mit tRNA glycine mutation into yeast mitochondria provides a new tool for the study of human neurodegenerative diseases. *Yeast* **18**(3): 219–27.

Romero, D., J. Martinez-Salazar et al. (1999). Repeated sequences in bacterial chromosomes and plasmids: a glimpse from sequenced genomes. *Res. Microbiol.* **150**(9–10): 735–43.

Romero, H., A. Zavala et al. (2000). Codon usage in *Chlamydia trachomatis* is the result of strand-specific mutational biases and a complex pattern of selective forces. *Nucleic Acids Res.* **28**(10): 2084–90.

Rondon, M. R., P. R. August et al. (2000). Cloning the soil metagenome: a strategy for accessing the genetic and functional diversity of uncultured microorganisms. *Appl. Environ. Microbiol.* **66**(6): 2541–47.

Rondon, M. R., R. M. Goodman et al. (1999). The Earth's bounty: assessing and accessing soil microbial diversity. *Trends Biotechnol.* **17**(10): 403–9.

Rossello-Mora, R. and R. Amann (2001). The species concept for prokaryotes. *FEMS Microbiol. Rev.* **25**(1): 39–67.

Rossmanith, W., A. Tullo et al. (1995). Human mitochondrial tRNA processing. *J. Biol. Chem.* **270**(21): 12885–91.

Rost, B. (1999). Twilight zone of protein sequence alignments. *Protein Eng.* **12**(2): 85–94.

Rost, B., R. Casadio et al. (1995). Transmembrane helices predicted at 95% accuracy. *Protein Sci.* **4**(3): 521–33.

Rost, B., P. Fariselli et al. (1996). Topology prediction for helical transmembrane proteins at 86% accuracy. *Protein Sci.* **5**(8): 1704–18.

Rost, B. and C. Sander (1994). Conservation and prediction of solvent accessibility in protein families. *Proteins* **20**(3): 216–26.

Rost, B., C. Sander et al. (1994). PHD—an automatic mail server for protein secondary structure prediction. *Comput. Appl. Biosci.* **10**: 53–60.

Rubin, G. M. and E. B. Lewis (2000). A brief history of *Drosophila*'s contributions to genome research. *Science* **287**(5461): 2216–18.

Rubin, G. M., M. D. Yandell et al. (2000). Comparative genomics of the eukaryotes. *Science* **287**(5461): 2204–15.

Ruepp, A., W. Graml et al. (2000). The genome sequence of the thermoacidophilic scavenger *Thermoplasma acidophilum*. *Nature* **407**(6803): 508–13.

Ruiz De Mena, I., E. Lefai et al. (2000). Regulation of mitochondrial single-stranded DNA-binding protein Gene Expression links nuclear and mitochondrial DNA replication in *Drosophila*. *J. Biol. Chem.* **275**(18): 13628–36.

Saccone, C. (1999). Structure and evolution of the metazoan mitochondrial genome. In *Frontiers of Cellular Bioenergetics*. S. Papa, F. Guerrieri and J. M. Tager, eds. New York: Kluwer Academic, pp. 521–551.

Saccone, C., M. Attimonelli et al. (1990). The role of tRNA genes in the evolution of animal mitochondrial DNA. In *Structure, Function and Biogenesis of Energy Transfer Systems*. E. Quagliariello, S. Papa, F. Palmieri and C. Saccone, eds. Amsterdam, Elsevier, pp. 9–96.

Saccone, C., C. Lanave et al. (1990). Influence of base composition on quantitative estimates of gene evolution. *Methods Enzymol.* **183**: 570–83.

Saccone, C., G. Pesole et al. (1991). The main regulatory region of mammalian mitochondrial DNA: structure–function model and evolutionary pattern. *J. Mol. Evol.* **33**: 83–91.

Saccone, C. and E. Sbisà (1994). The evolution of the mitochondrial genome. In *Principles of Medical Biology*. E. E. Bittar, ed. Greenwich, Conn.: JAI Press, pp. 39–72.

Saccone, S., A. De Sario et al. (1993). Correlations between isochores and chromosomal bands in the human genome. *Proc. Natl. Acad. Sci. USA* **90**(24): 11929–33.

Saitou, N. and M. Nei (1987). The neighbor-joining method: a new method for reconstructing phylogeneti trees. *Mol. Biol. Evol.* **4**: 406–25.

Sakaguchi, K. (1990). Invertrons, a class of structurally and functionally related genetic elements that includes linear DNA plasmids, transposable elements, and genomes of adeno-type viruses. *Microbiol. Rev.* **54**(1): 66–74.

Salamov, A. A. and V. V. Solovyev (1995). Prediction of protein secondary structure by combining nearest-neighbor algorithms and multiple sequence alignments. *J. Mol. Biol.* **247**(1): 11–15.

Salanoubat, M., S. Genin et al. (2002). Genome sequence of the plant pathogen *Ralstonia solanacearum*. *Nature* **415**(6871):497–502.

Sali, A. and T. L. Blundell (1993). Comparative protein modelling by satisfaction of spatial restraints. *J. Mol. Biol.* **234**(3): 779–815.

Salzberg, S. L., O. White et al. (2001). Microbial genes in the human genome: lateral transfer or gene loss? *Science* **292**(5523): 1903–6.

Sanchez, H., T. Fester et al. (1996). Transfer of rps19 to the nucleus involves the gain of an RNP-binding motif which may functionally replace RPS13 in *Arabidopsis mitochondria*. *EMBO J.* **15**(9): 2138–49.

Sanchirico, M. E., T. D. Fox et al. (1998). Accumulation of mitochondrially synthesized *Saccharomyces cerevisiae cox2p* and *cox3p* depends on targeting information in untranslated portions of their mRNAs. *EMBO J.* **17**(19): 5796–804.

Sanger, F., S. Nicklen et al. (1977). DNA sequencing with chain-terminating inhibitors. *Proc. Natl. Acad. Sci. USA* **74**(12): 5463–67.

Santos, M. A., C. Cheesman et al. (1999). Selective advantages created by codon ambiguity allowed for the evolution of an alternative genetic code in *Candida* spp. *Mol. Microbiol.* **31**(3): 937–47.

Sato, N. (2001). Was the evolution of plastid genetic machinery discontinuous? *Trends Plant Sci.* **6**(4): 151–55.

Sauer, F. and R. Tjian (1997). Mechanisms of transcriptional activation: differences and similarities between yeast, *Drosophila*, and man. *Curr. Opin. Genet. Dev.* **7**(2): 176–81.

Sayle, R. A. and E. J. Milner-White (1995). RASMOL: biomolecular graphics for all. *Trends Biochem. Sci.* **20**(9): 374.

Sbisà, E., M. Nardelli et al. (1990). The complete and symmetric transcription of the main noncoding region of rat mitochondrial genome: in vivo mapping of heavy and light transcripts. *Curr. Genet.* **17**(3): 247–53.

Sbisà, E., F. Tanzariello et al. (1997). Mammalian mitochondrial D-loop region structural analysis: identification of new conserved sequences and their functional and evolutionary implications. *Gene* **205**(1–2): 125–40.

Sbisà, E., A. Tullo et al. (1992). Transcription mapping of the Ori L region reveals novel precursors of mature RNA species and antisense RNAs in rat mitochondrial genome. *FEBS Lett.* **296**(3): 311–16.

Schaffer, A. A., Y. I. Wolf et al. (1999). IMPALA: matching a protein sequence against a collection of PSI-BLAST-constructed position-specific score matrices. *Bioinformatics* **15**(12): 1000–1011.

Scharfe, C., P. Zaccaria et al. (2000). MITOP, the mitochondrial proteome database: 2000 update. *Nucleic Acids Res.* **28**(1): 155–58.

Schatz, G. and B. Dobberstein (1996). Common principles of protein translocation across membranes. *Science* **271**(5255): 1519–26.

Schena, M., D. Shalon et al. (1995). Quantitative monitoring of gene expression patterns with a complementary DNA microarray. *Science* **270**(5235): 467–70.

Schena, M., D. Shalon et al. (1996). Parallel human genome analysis: microarray-based expression monitoring of 1000 genes. *Proc. Natl. Acad. Sci. USA* **93**(20): 10614–19.

Scherf, M., A. Klingenhoff et al. (2000). Highly specific localization of promoter regions in large genomic sequences by PromoterInspector: a novel context analysis approach. *J. Mol. Biol.* **297**(3): 599–606.

Schmitt, M. E., J. L. Bennett et al. (1993). Secondary structure of RNase MRP RNA as predicted by phylogenetic comparison. *FASEB J.* **7**(1): 208–13.

Schnable, P. S. and R. P. Wise (1998). The molecular basis of cytoplasmic male sterility and fertility restoration. *Trends Plant Sci.* **3**: 175–80.

Schneider, T. D. and R. M. Stephens (1990). Sequence logos: a new way to display consensus sequences. *Nucleic Acids Res.* **18**(20): 6097–6100.

Schon, E. A. (2000). Mitochondrial genetics and disease. *Trends Biochem. Sci.* **25**(11): 555–60.

Schonbach, C., P. Kowalski-Saunders et al. (2000). Data warehousing in molecular biology. *Brief. Bioinf.* **1**(2): 190–98.

Schuler, G. D. (1997a). Pieces of the puzzle: expressed sequence tags and the catalog of human genes. *J. Mol. Med.* **75**(10): 694–98.

Schuler, G. D. (1997b). Sequence mapping by electronic PCR. *Genome Res.* **7**(5): 541–50.

Schuler, G. D. (1998). Electronic PCR: bridging the gap between genome mapping and genome sequencing. *Trends Biotechnol.* **16**(11): 456–59.

Schuler, G. D., M. S. Boguski et al. (1996). A gene map of the human genome. *Science* **274**(5287): 540–46.

Schultz, J., R. R. Copley et al. (2000). SMART: a Web-based tool for the study of genetically mobile domains. *Nucleic Acids Res.* **28**(1): 231–34.

Schuster, W. and A. Brennicke (1994). The plant mitochondrial genome: physical structure, information content, RNA editing and gene migration to the nucleus. *Annu. Rev. Plant Physiol. Plant Mol. Biol.* **45**: 61–78.

Schwartz, S., Z. Zhang et al. (2000). PipMaker—a Web server for aligning two genomic DNA sequences. *Genome Res.* **10**(4): 577–86.

Scissum-Gunn, K., M. Gandhi et al. (1998). Separation of different confomations of plant mitochondrial DNA molecules by field inversion gen electrophoresis. *Plant Mol. Biol. Rep.* **16**: 219–29.

Scott, W. G. (1998). RNA catalysis. *Curr. Opin. Struct. Biol.* **8**(6): 720–26.

Seilacher, A., P. K. Bose et al. (1998). Triploblastic animals more than 1 billion years ago: trace fossil evidence from India. *Science* **282**(5386): 80–83.

Selosse, M.-A., B. Albert et al. (2001). Reducing the genome size of organelles favours gene transfer to the nucleus. *Trends Ecol. Evol.* **16**: 135–41.

Senapathy, P., M. B. Shapiro et al. (1990). Splice junctions, branch point sites, and exons: sequence statistics, identification, and applications to genome project. *Methods Enzymol.* **183**: 252–78.

Seoighe, C., N. Federspiel et al. (2000). Prevalence of small inversions in yeast gene order evolution. *Proc. Natl. Acad. Sci. USA* **97**(26): 14433–37.

Seoighe, C. and K. H. Wolfe (1998). Extent of genomic rearrangement after genome duplication in yeast. *Proc. Natl. Acad. Sci. USA* **95**(8): 4447–52.

Seraphin, B., A. Boulet et al. (1987). Construction of a yeast strain devoid of mitochondrial introns and its use to screen nuclear genes involved in mitochondrial splicing. *Proc. Natl. Acad. Sci. USA* **84**(19): 6810–14.

Shang, J. and D. A. Clayton (1994). Human mitochondrial transcription termination exhibits RNA polymerase independence and biased bipolarity in vitro. *J. Biol. Chem.* **269**(46): 29112–20.

Shao, R., N. J. Campbell et al. (2001). Numerous gene rearrangements in the mitochondrial genome of the wallaby louse, *Heterodoxus macropus* (Phthiraptera). *Mol. Biol. Evol.* **18**(5): 858–65.

Sharp, P. M. and W. H. Li (1987). The codon adaptation index—a measure of directional synonymous codon usage bias, and its potential applications. *Nucleic Acids Res.* **15**(3): 1281–95.

Sharp, P. M., M. Stenico et al. (1993). Codon usage: mutational bias, translational selection, or both? *Biochem. Soc. Trans.* **21**(4): 835–41.

Sharp, P. M., T. M. Tuohy et al. (1986). Codon usage in yeast: cluster analysis clearly differentiates highly and lowly expressed genes. *Nucleic Acids Res.* **14**(13): 5125–43.

She, Q., R. K. Singh et al. (2001). The complete genome of the crenarchaeon *Sulfolobus solfataricus* P2. *Proc. Natl. Acad. Sci. USA* **98**(14): 7835–40.

Shepard, K. A. and M. P. Yaffe (1999). The yeast dynamin-like protein, Mgm1p, functions on the mitochondrial outer membrane to mediate mitochondrial inheritance. *J. Cell. Biol.* **144**(4): 711–20.

Sherry, S. T., M. H. Ward et al. (2001). dbSNP: the NCBI database of genetic variation. *Nucleic Acids Res.* **29**(1): 308–11.

Shigenobu, S., H. Watanabe et al. (2000). Genome sequence of the endocellular bacterial symbiont of aphids *Buchnera* sp. APS. *Nature* **407**(6800): 81–86.

Shimizu, T., K. Ohtani et al. (2002). Complete genome sequence of *Clostridium perfringens*, an anaerobic flesh-eater. *Proc. Natl. Acad. Sci. USA* **99**(2): 996–1001.

Shimkets, L. (1998). Structure and sizes of the genomes of Archaea and Bacteria. In *Bacterial Genomes: Physical Structure and Analysis.* L. J. W. G. de Bruijn, ed. New York: Chapman & Hall, pp. 5–11.

Shimko, N., L. Liu et al. (2001). GOBASE: the organelle genome database. *Nucleic Acids Res.* **29**(1): 128–32.

Shimodaira, H. and M. Hasegawa (1999). Multiple comparisons of log-likelihoods with applications to phylogenetic inference. *Mol. Biol. Evol.* **16**(8): 1114–16.

Shirai, M., H. Hirakawa et al. (2000). Comparison of whole genome sequences of *Chlamydia pneumoniae* J138 from Japan and CWL029 from USA. *Nucleic Acids Res.* **28**(12): 2311–14.

Shmatkov, A. M., A. A. Melikyan et al. (1999). Finding prokaryotic genes by the "frame-by-frame" algorithm: targeting gene starts and overlapping genes. *Bioinformatics* **15**(11): 874–86.

Shpaer, E. G. (1986). Constraints on codon context in *Escherichia coli* genes. Their possible role in modulating the efficiency of translation. *J. Mol. Biol.* **188**(4): 555–64.

Silke, J. (1997). The majority of long non-stop reading frames on the antisense strand can be explained by biased codon usage. *Gene* **194**(1): 143–55.

Simpson, A. J., F. C. Reinach et al. (2000). The genome sequence of the plant pathogen *Xylella fastidiosa*. The *Xylella fastidiosa* Consortium of the Organization for Nucleotide Sequencing and Analysis. *Nature* **406**(6792): 151–57.

Simpson, A. M., Y. Suyama et al. (1989). Kinetoplastid mitochondria contain functional tRNAs which are encoded in nuclear DNA and also contain small minicircle and maxicircle transcripts of unknown function. *Nucleic Acids Res.* **17**(14): 5427–45.

Simpson, G. G. (1945). Classification of mammals. *Bull. Am. Mus. Nat. Hist.* **85**: 1–350.

Sinclair, D. A., K. Mills et al. (1998). Molecular mechanisms of yeast aging. *Trends Biochem. Sci.* **23**(4): 131–34.

Singh, G. B., J. A. Kramer et al. (1997). Mathematical model to predict regions of chromatin attachment to the nuclear matrix. *Nucleic Acids Res.* **25**(7): 1419–25.

Siomi, H. and G. Dreyfuss (1997). RNA-binding proteins as regulators of Gene expression. *Curr. Opin. Genet. Dev.* **7**(3): 345–53.

Sipos, L. and G. von Heijne (1993). Predicting the topology of eukaryotic membrane proteins. *Eur. J. Biochem.* **213**(3): 1333–40.

Sippl, M. J. and H. Flockner (1996). Threading thrills and threats. *Structure* **4**(1): 15–19.

Skrabanek, L. and K. H. Wolfe (1998). Eukaryote genome duplication—where's the evidence? *Cur. Opin. Genet. Dev.* **8**(6): 694–700.

Slesarev, A. I., K. V. Mezhevaya et al. (2002). The complete genome of hyperthermophile *Methanopyrus kandleri* AV19 and monophyly of archaeal methanogens. *Proc. Natl. Acad. Sci. USA* **99**(7): 4644–49.

Smith, C. W. and J. Valcarcel (2000). Alternative pre-mRNA splicing: the logic of combinatorial control. *Trends Biochem. Sci.* **25**(8): 381–88.

Smith, D. R., L. A. Doucette-Stamm et al. (1997). Complete genome sequence of *Methanobacterium thermoautotrophicum* deltaH: functional analysis and comparative genomics. *J. Bacteriol.* **179**(22): 7135–55.

Smith, J. M. and N. H. Smith (1996). Site-specific codon bias in bacteria. *Genetics* **142**(3): 1037–43.

Smith, S. D., D. J. O'Mahony et al. (1993). Transcription from the rat 45S ribosomal DNA promoter does not require the factor UBF. *Gene Expr.* **3**(3): 229–36.

Smith, T. F. and M. S. Waterman (1981). Identification of common molecular subsequences. *J. Mol. Biol.* **147**: 195–97.

Smith, T. F., M. S. Waterman et al. (1985). The statistical distribution of nucleic acid similarities. *Nucleic Acids Res.* **13**(2): 645–56.

Snel, B., P. Bork et al. (1999). Genome phylogeny based on gene content. *Nat. Genet.* **21**(1): 108–10.

Soares, M. B., M. F. Bonaldo et al. (1994). Construction and characterization of a normalized cDNA library. *Proc. Natl. Acad. Sci. USA* **91**(20): 9228–32.

Sokal, R. R. and C. D. Michener (1958). A statistical method for evaluating systematic relationships. *Univ. Kans. Sci. Bull.* **28**: 1409–38.

Somanchi, A. and S. P. Mayfield (1999). Nuclear-chloroplast signalling. *Curr. Opin. Plant Biol.* **2**(5): 404–9.

Spellman, P. T., G. Sherlock et al. (1998). Comprehensive identification of cell cycle-regulated genes of the yeast *Saccharomyces cerevisiae* by microarray hybridization. *Mol. Biol. Cell* **9**(12): 3273–97.

Spinella, D. G., A. K. Bernardino et al. (1999). Tandem arrayed ligation of expressed sequence tags (TALEST): a new method for generating global Gene expression profiles. *Nucleic Acids Res.* **27**(18): e22.

Spruyt, N., C. Delarbre et al. (1998). Complete sequence of the amphioxus (*Branchiostoma lanceolatum*) mitochondrial genome: relations to vertebrates. *Nucleic Acids Res.* **26**(13): 3279–85.

Staden, R. (1996). The Staden sequence analysis pckage. *Mol. Biotechnol.* **5**: 233–41.

Stanhope, M. J., A. Lupas et al. (2001). Phylogenetic analyses do not support horizontal gene transfers from bacteria to vertebrates. *Nature* **411**(6840): 940–44.

Stein, L., P. Sternberg et al. (2001). WormBase: network access to the genome and biology of *Caenorhabditis elegans*. *Nucleic Acids Res.* **29**(1): 82–86.

Steinhauser, S., S. Beckert et al. (1999). Plant mitochondrial RNA editing. *J. Mol. Evol.* **48**(3): 303–12.

Stephens, R. S., S. Kalman et al. (1998). Genome sequence of an obligate intracellular pathogen of humans: *Chlamydia trachomatis*. *Science* **282**(5389): 754–59.

Stevens, R. and C. Miller (2000). Wrapping and interoperating bioinformatics resources using CORBA. *Brief. Bioinf.* **1**(1): 9–21.

Stoebe, B. and K. V. Kowallik (1999). Gene-cluster analysis in chloroplast genomics. *Trends Genet.* **15**(9): 344–47.

Stoesser, G., W. Baker et al. (2001). The EMBL nucleotide sequence database. *Nucleic Acids Res.* **29**(1): 17–21.

Stollberg, J., J. Urschitz et al. (2000). A quantitative evaluation of SAGE. *Genome Res.* **10**(8): 1241–48.

Stormo, G. D. (2000). Gene-finding approaches for eukaryotes. *Genome Res.* **10**(4): 394–97.

Stover, C. K., X. Q. Pham et al. (2000). Complete genome sequence of *Pseudomonas aeruginosa* PA01, an opportunistic pathogen. *Nature* **406**(6799): 959–64.

Strimmer, K. and A. Von Haeseler (1996). Quartet puzzling: a quartet maximum-likelihood method for reconstructing tree topologies. *Mol. Biol. Evol.* **13**(7): 964–69.

Strivens, M. (2000). Not who you know, but what you know. *Trends Biotechnol.* **18**(6): 227–28.

Stroud, R. M. and P. Walter (1999). Signal sequence recognition and protein targeting. *Curr. Opin. Struct. Biol.* **9**(6): 754–59.

Strunnikov, A. V. (1998). SMC proteins and chromosome structure. *Trends Cell Biol.* **8**(11): 454–59.

Sturniolo, T., E. Bono et al. (1999). Generation of tissue-specific and promiscuous HLA ligand databases using DNA microarrays and virtual HLA class II matrices. *Nat. Biotechnol.* **17**(6): 555–61.

Sugiura, M. (1992). The chloroplast genome. *Plant Mol. Biol.* **19**(1): 149–68.

Sutcliffe, J. G., P. E. Foye et al. (2000). TOGA: an automated parsing technology for analyzing expression of nearly all genes. *Proc. Natl. Acad. Sci. USA* **97**(5): 1976–81.

Sutton, G., O. White et al. (1995). TIGR assembler: a new tool for assembling large shotgun sequencing project. *Genome Sci. Technol.* **1**: 9–18.

Suyama, Y. and K. Miura (1968). Size and structural variations of mitochondrial DNA. *Proc. Natl. Acad. Sci. USA* **60**: 235–42.

Suzuki, Y. and S. Sugano (2001). Construction of full-length-enriched cDNA libraries. The oligo-capping method. *Methods Mol. Biol.* **175**: 143–53.

Swofford, D. L. (2000). *PAUP*: Phylogenetic Analysis Using Parsimony (and Other Methods), Version 4.* Sunderland, Mass.: Sinauer Associates.

Swofford, D. L., G. J. Olsen et al. (1996). Phylogenetic inference. In *Molecular Systematics.* D. M. Hillis, C. Moritz and K. Mable, eds. Sunderland, Mass.: Sinauer Associates, pp. 407–514.

Taanman, J. W. (1999). The mitochondrial genome: structure, transcription, translation and replication. *Biochim. Biophys. Acta* **1410**(2): 103–23.

Takami, H., K. Nakasone et al. (2000). Complete genome sequence of the alkaliphilic bacterium *Bacillus halodurans* and genomic sequence comparison with *Bacillus subtilis. Nucleic Acids Res.* **28**(21): 4317–31.

Takezaki, N. and T. Gojobori (1999). Correct and incorrect vertebrate phylogenies obtained by the entire mitochondrial DNA sequences. *Mol. Biol. Evol.* **16**(5): 590–601.

Tamayo, P., D. Slonim et al. (1999). Interpreting patterns of Gene expression with self-organizing maps: methods and application to hematopoietic differentiation. *Proc. Natl. Acad. Sci. USA* **96**(6): 2907–12.

Tamura, K. (1992). Estimation of the number of nucleotide substitutions when there are strong transition-transversion and G + C-content biases. *Mol. Biol. Evol.* **9**(4): 678–87.

Tateno, Y., S. Miyazaki et al. (2000). DNA data bank of Japan (DDBJ) in collaboration with mass sequencing teams. *Nucleic Acids Res.* **28**(1): 24–26.

Tatusov, R. L., E. V. Koonin et al. (1997). A genomic perspective on protein families. *Science* **278**(5338): 631–37.

Tatusov, R. L., A. R. Mushegian et al. (1996). Metabolism and evolution of *Haemophilus influenzae* deduced from a whole-genome comparison with *Escherichia coli. Curr. Biol.* **6**(3): 279–91.

Tatusov, R. L., D. A. Natale et al. (2001). The COG database: new developments in phylogenetic classification of proteins from complete genomes. *Nucleic Acids Res.* **29**(1): 22–28.

Tauch, A., I. Homann et al. (2002). Strategy to sequence the genome of *Corynebacterium glutamicum* ATCC 13032: use of a cosmid and a bacterial artificial chromosome library. *J. Biotechnol.* **95**(1): 25–38.

Tavazoie, S., J. D. Hughes et al. (1999). Systematic determination of genetic network architecture. *Nat. Genet.* **22**(3): 281–85.

T.C.e.S. Consortium (1998). "Genome sequence of the nematode *C. elegans:* a platform for investigating biology. The *C. elegans* Sequencing Consortium. *Science* **282**(5396): 2012–18.

Tenzen, T., T. Yamagata et al. (1997). Precise switching of DNA replication timing in the GC content transition area in the human major histocompatibility complex. *Mol. Cell Biol.* **17**(7): 4043–50.

Tettelin, H., K. E. Nelson et al. (2001). Complete genome sequence of a virulent isolate of *Streptococcus pneumoniae. Science* **293**(5529): 498–506.

Tettelin, H., N. J. Saunders et al. (2000). Complete genome sequence of *Neisseria meningitidis* serogroup B strain MC58. *Science* **287**(5459): 1809–15.

Thieffry, D., H. Salgado et al. (1998). Prediction of transcriptional regulatory sites in the complete genome sequence of *Escherichia coli* K-12. *Bioinformatics* **14**: 391–400.

Thommes, P., C. L. Farr et al. (1995). Mitochondrial single-stranded DNA-binding protein from *Drosophila* embryos. Physical and biochemical characterization. *J. Biol. Chem.* **270**(36): 21137–43.

Tian, G. L., C. Macadre et al. (1991). Incipient mitochondrial evolution in yeasts. I. The physical map and gene order of *Saccharomyces douglasii* mitochondrial DNA discloses a translocation of a segment of 15,000 base-pairs and the presence of new introns in comparison with *Saccharomyces cerevisiae*. *J. Mol. Biol.* **218**(4): 735–46.

Tillier, E. R. and R. A. Collins (2000). The contributions of replication orientation, gene direction, and signal sequences to base-composition asymmetries in bacterial genomes. *J. Mol. Evol.* **50**(3): 249–57.

Tinoco, I. and C. Bustamante (1999). How RNA folds. *J. Mol. Biol.* **293**(2): 271–81.

Tomb, J. F., O. White et al. (1997). The complete genome sequence of the gastric pathogen *Helicobacter pylori*. *Nature* **388**(6642): 539–47.

Tomilin, N. V. (1999). Control of genes by mammalian retroposons. *Int. Rev. Cytol.* **186**: 1–48.

Topal, M. D. and J. R. Fresco (1976). Complementary base pairing and the origin of substitution mutations. *Nature* **263**(5575): 285–89.

Topper, J. N., J. L. Bennett et al. (1992). A role for RNAase MRP in mitochondrial RNA processing. *Cell* **70**(1): 16–20.

Toronen, P., M. Kolehmainen et al. (1999). Analysis of Gene expression data using self-organizing maps. *FEBS Lett.* **451**(2): 142–46.

Toth, G., Z. Gaspari et al. (2000). Microsatellites in different eukaryotic genomes: survey and analysis. *Genome Res.* **10**(7): 967–81.

Trifonov, E. N. (1989). The multiple codes of nucleotide sequences. *Bull. Math. Biol.* **51**(4): 417–32.

Trifonov, E. N. (1999). Elucidating sequence codes: three codes for evolution. *Ann. N. Y. Acad. Sci.* **870**: 330–38.

Turmel, M., C. Otis et al. (1999). The complete chloroplast DNA sequence of the green alga *Nephroselmis divaceal*: insights into the architecture of ancestral chloroplast genomes. *Proc. Natl. Acad. Sci. USA* **96**(18): 10248–53.

Tusnady, G. E. and I. Simon (1998). Principles governing amino acid composition of integral membrane proteins: application to topology prediction. *J. Mol. Biol.* **283**(2): 489–506.

Uetz, P., L. Giot et al. (2000). A comprehensive analysis of protein–protein interactions in *Saccharomyces cerevisiae*. *Nature* **403**(6770): 623–27.

Unseld, M., J. R. Marienfeld et al. (1997). The mitochondrial genome of *Arabidopsis thaliana* contains 57 genes in 366,924 nucleotides. *Nat. Genet.* **15**(1): 57–61.

Valverde, J., B. Batuecas et al. (1994). The complete mitochondrial sequence of the crustacean *Artemia franciscana*. *J. Mol. Evol.* **39**: 400–408.

van Belkum, A. (1999). Short sequence repeats in microbial pathogenesis and evolution. *Cell. Mol. Life Sci.* **56**(9–10): 729–34.

van Belkum, A., S. Scherer et al. (1998). Short-sequence DNA repeats in prokaryotic genomes. *Microbiol. Mol. Biol. Rev.* **62**(2): 275–93.

van den Boogaart, P., J. Samallo et al. (1982). Similar genes for a mitochondrial ATPase subunit in the nuclear and mitochondrial genomes of *Neurospora crassa*. *Nature* **298**(5870): 187–89.

van der Velden, A. W. and A. A. Thomas (1999). The role of the 5′ untranslated region of an mRNA in translation regulation during development. *Int. J. Biochem. Cell. Biol.* **31**(1): 87–106.

van Helden, J., B. Andre et al. (1998). Extracting regulatory sites from the upstream region of yeast genes by computational analysis of oligonucleotide frequencies. *J. Mol. Biol.* **281**(5): 827–42.

van Helden, J., A. F. Rios et al. (2000). Discovering regulatory elements in non-coding sequences by analysis of spaced dyads. *Nucleic Acids Res.* **28**(8): 1808–18.

Velculescu, V. E., B. Vogelstein et al. (2000). Analysing uncharted transcriptomes with SAGE. *Trends Genet.* **16**(10): 423–25.

Velculescu, V. E., L. Zhang et al. (1995). Serial analysis of Gene expression. *Science* **270**(5235): 484–87.

Velculescu, V. E., L. Zhang et al. (1997). Characterization of the yeast transcriptome. *Cell* **88**(2): 243–51.

Vellai, T., K. Takacs et al. (1998). A new aspect to the origin and evolution of eukaryotes. *J. Mol. Evol.* **46**(5): 499–507.

Venclovas, A. Zemla et al. (2001). Comparison of performance in successive CASP experiments. *Proteins* **45**(Suppl. 5): 163–70.

Venkatesh, B., P. Gilligan et al. (2000). Fugu: a compact vertebrate reference genome. *FEBS Lett.* **476**(1–2): 3–7.

Venter, J. C., M. D. Adams et al. (2001). The sequence of the human genome. *Science* **291**(5507): 1304–51.

Vingron, M. and M. S. Waterman (1994). Sequence alignment and penalty choice. Review of concepts, case studies and implications. *J. Mol. Biol.* **235**: 1–12.

Vogelstein, B., D. Lane et al. (2000). Surfing the p53 network. *Nature* **408**(6810): 307–10.

Volff, J. N. and J. Altenbuchner (1998). Genetic instability of the *Streptomyces* chromosome. *Mol. Microbiol.* **27**(2): 239–46.

Volpetti, V., R. Gallerani et al. (2000). PLMItRNA, a database for tRNAs and tRNA genes in plant mitochondria: enlargement and updating. *Nucleic Acids Res.* **28**(1): 159–62.

von Nickisch-Rosenegk, M., W. M. Brown et al. (2001). Complete sequence of the mitochondrial genome of the tapeworm *Hymenolepis diminuta*: gene arrangements indicate that platyhelminths are eutrochozoans. *Mol. Biol. Evol.* **18**(5): 721–30.

Vriend, G. (1990). WHAT IF: a molecular modeling and drug design program. *J. Mol. Graph.* **8**(1): 52–56, 29.

Wahl, M. C. and M. Sundaralingam (1995). New crystal structures of nucleic acids and their complexes. *Curr. Opin. Struct. Biol.* **5**(3): 282–95.

Wahle, E. and W. Keller (1996). The biochemistry of polyadenylation. *Trends Biochem. Sci.* **21**(7): 247–50.

Wako, H. and T. L. Blundell (1994). Use of amino acid environment-dependent substitution tables and conformational propensities in structure prediction from aligned sequences of homologous proteins. I. Solvent accessibility classes. *J. Mol. Biol.* **238**(5): 682–92.

Walbot, V. (2000). *Arabidopsis thaliana* genome. a green chapter in the book of life. *Nature* **408**(6814): 794–95.

Walshaw, J. and D. N. Woolfson (2001). Socket: a program for identifying and analysing coiled-coil motifs within protein structures. *J. Mol. Biol.* **307**(5): 1427–50.

Wang, D. G., J. B. Fan et al. (1998). Large-scale identification, mapping, and genotyping of single-nucleotide polymorphisms in the human genome. *Science* **280**(5366): 1077–82.

Wang, E., L. D. Miller et al. (2000). High-fidelity mRNA amplification for gene profiling. *Nat. Biotechnol.* **18**(4): 457–59.

Wang, J. and J. L. Manley (1997). Regulation of pre-mRNA splicing in metazoa. *Curr. Opin. Genet. Dev.* **7**(2): 205–11.

Wang, Y. and X. Gu (2000). Evolutionary patterns of gene families generated in the early stage of vertebrates. *J. Mol. Evol.* **51**(1): 88–96.

Washburn, M. and J. Yates III (2000). New methods of proteome analysis: multidimensional chromatography and mass spectrometry. *Proteom. Trends Guide* **1**: 27–30.

Watson, S. (2001). Editorial overview: new technologies in the post-genomic era. *Curr. Opin. Pharmacol.* **1**(5): 511–12.

Weiller, G. F., H. Bruckner et al. (1991). A GC cluster repeat is a hotspot for mit- macro-deletions in yeast mitochondrial DNA. *Mol. Gen. Genet.* **226**(1–2): 233–40.

Werner, T. (2000). Identification and functional modelling of DNA sequence elements of transcription. *Brief. Bioinf.* **1**(4): 372–80.

Wheelan, S. J., D. M. Church et al. (2001). Spidey: a tool for mRNA-to-genomic alignments. *Genome Res.* **11**(11): 1952–57.

Wheeler, D. L., D. M. Church et al. (2001). Database resources of the National Center for Biotechnology Information. *Nucleic Acids Res.* **29**(1): 11–16.

White, J. V., C. M. Stultz et al. (1994). Protein classification by stochastic modeling and optimal filtering of amino-acid sequences. *Math. BioSci.* **119**(1): 35–75.

White, O., J. A. Eisen et al. (1999). Genome sequence of the radioresistant bacterium *Deinococcus radiodurans* R1. *Science* **286**(5444): 1571–77.

White, O. and A. R. Kerlavage (1996). TDB: new databases for biological discovery. *Methods Enzymol.* **266**: 27–40.

Whitman, W. B., D. C. Coleman et al. (1998). Prokaryotes: the unseen majority. *Proc. Natl. Acad. Sci. USA* **95**(12): 6578–83.

Wickens, M., P. Anderson et al. (1997). Life and death in the cytoplasm: messages from the 3′ end. *Curr. Opin. Genet. Dev.* **7**(2): 220–32.

Wicky, C., A. M. Villeneuve et al. (1996). Telomeric repeats (TTAGGC)n are sufficient for chromosome capping function in *Caenorhabditis elegans*. *Proc. Natl. Acad. Sci. USA* **93**(17): 8983–88.

Will, C. L. and R. Luhrmann (2001). Spliceosomal UsnRNP biogenesis, structure and function. *Curr. Opin. Cell. Biol.* **13**(3): 290–301.

Williams, M. A., Y. Johzuka et al. (2000). Addition of non-genomically encoded nucleotides to the 3-terminus of maize mitochondrial mRNAs: truncated rps12 mRNAs frequently terminate with CCA. *Nucleic Acids Res.* **28**(22): 4444–51.

Wilson, M., J. DeRisi et al. (1999). Exploring drug-induced alterations in Gene Expr.ession in *Mycobacterium tuberculosis* by microarray hybridization. *Proc. Natl. Acad. Sci. USA* **96**(22): 12833–38.

Wilson, R. J., P. W. Denny et al. (1996). Complete gene map of the plastid-like DNA of the malaria parasite *Plasmodium falciparum*. *J. Mol. Biol.* **261**(2): 155–72.

Wodicka, L., H. Dong et al. (1997). Genome-wide expression monitoring in *Saccharomyces cerevisiae*. *Nat. Biotechnol.* **15**(13): 1359–67.

Woese, C. R. and G. E. Fox (1977). Phylogenetic structure of the prokaryotic domain: the primary kingdoms. *Proc. Natl. Acad. Sci. USA* **74**(11): 5088–90.

Woese, C. R., O. Kandler et al. (1990). Towards a natural system of organisms: proposal for the domains Archaea, Bacteria, and Eucarya. *Proc. Natl. Acad. Sci. USA* **87**(12): 4576–79.

Wolf, E., P. S. Kim et al. (1997). MultiCoil: a program for predicting two- and three-stranded coiled coils. *Protein Sci.* **6**(6): 1179–89.

Wolfe, K. H., W. H. Li et al. (1987). Rates of nucleotide substitution vary greatly among plant mitochondrial, chloroplast, and nuclear DNAs. *Proc. Natl. Acad. Sci. USA* **84**(24): 9054–58.

Wolfe, K. H., P. M. Sharp et al. (1989). Mutation rates differ among regions of the mammalian genome. *Nature* **337**(6204): 283–85.

Wolfe, K. H. and D. C. Shields (1997). Molecular evidence for an ancient duplication of the entire yeast genome. *Nature* **387**(6634): 708–13.

Wolff, G., G. Burger et al. (1993). Mitochondrial genes in the colourless alga *Prototheca wickerhamii* resemble plant genes in their exons but fungal genes in their introns. *Nucleic Acids Res.* **21**(3): 719–26.

Wolstenholme, D. R. (1992). Animal mitochondrial DNA: structure and evolution. *Int. Rev. Cytol.* **141**: 173–216.

Womble, D. D. (2000). GCG: The Wisconsin Package of sequence analysis programs. *Methods Mol. Biol.* **132**: 3–22.

Wood, D. W., J. C. Setubal et al. (2001). The genome of the natural genetic engineer *Agrobacterium tumefaciens* C58. *Science* **294**(5550): 2317–23.

Wood, V., R. Gwilliam et al. (2002). The genome sequence of *Schizosaccharomyces pombe*. *Nature* **415**(6874): 871–80.

Woodcock, C. L. and S. Dimitrov (2001). Higher-order structure of chromatin and chromosomes. *Curr. Opin. Genet. Dev.* **11**(2): 130–35.

Wootton, J. C. (1994). Non-globular domains in protein sequences: automated segmentation using complexity measures. *Comput. Chem.* **18**(3): 269–85.

Wootton, J. C. and S. Federhen (1996). Analysis of compositionally biased regions in sequence databases. *Methods Enzymol.* **266**: 554–71.

Wright, F. (1990). The "effective number of codons" used in a gene. *Gene* **87**(1): 23–29.

Wright, F. and M. J. Bibb (1992). Codon usage in the G + C-rich *Streptomyces* genome. *Gene* **113**(1): 55–65.

Wright, F., W. Lemon et al. (2001). A draft annotation and overview of the human genome. *Genome Biol.* **2**(7): research0025.1–0025.18.

Xie, T., D. Ding et al. (1998). The relationship between synonymous codon usage and protein structure. *FEBS Lett.* **434**(1–2): 93–6.

Yamazaki, N., R. Ueshima et al. (1997). Evolution of pulmonate gastropod mitochondrial genomes: comparisons of complete gene organization of *Euhadra*, *Cepaea* and *Albinaria* and implications of unusual tRNA secondary structures. *Genetics* **145**: 749–58.

Yang, A., M. Kaghad et al. (1998). p63, a p53 homolog at 3q27–29, encodes multiple products with transactivating, death-inducing, and dominant-negative activities. *Mol. Cell* **2**(3): 305–16.

Yang, A., M. Kaghad et al. (2002). On the shoulders of giants: p63, p73 and the rise of p53. *Trends Genet.* **18**(2): 90–95.

Yang, Z. (1994). Estimating the pattern of nucleotide substitution. *J. Mol. Evol.* **39**(1): 105–11.

Yeh, R. F., L. P. Lim et al. (2001). Computational inference of homologous gene structures in the human genome. *Genome Res.* **11**(5): 803–16.

Yu, J., S. Hu et al. (2002). A draft sequence of the rice genome (*Oryza sativa* L. ssp. *indica*). *Science* **296**(5565): 79–92.

Zannis-Hadjopoulos, M. and G. B. Price (1999). Eukaryotic DNA replication. *J. Cell Biochem.* (Suppl. 32–33): 1–14.

Zardoya, R. and A. Meyer (1998). Complete mitochondrial genome suggests diapsid affinities of turtles. *Proc. Natl. Acad. Sci. USA* **95**: 14226–31.

Zeeberg, B. (2002). Shannon information theoretic computation of synonymous codon usage biases in coding regions of human and mouse genomes. *Genome Res.* **12**(6): 944–55.

Zenvirth, D., T. Arbel et al. (1992). Multiple sites for double-strand breaks in whole meiotic chromosomes of *Saccharomyces cerevisiae*. *EMBO J.* **11**(9): 3441–47.

Zhang, Z., S. Schwartz et al. (2000). A greedy algorithm for aligning DNA sequences. *J. Comput. Biol.* **7**: 203–14.

Zhu, H., M. Bilgin et al. (2001). Global analysis of protein activities using proteome chips. *Science* **293**(5537): 2101–5.

Zuckerkandl, E. and L. Pauling (1965a). Evolutionary divergence and convergence in proteins. In *Evolving Genes and Proteins*. V. Bryson and H. J. Vogel, eds. New York: Academic Press, pp. 97–166.

Zuckerkandl, E. and L. Pauling (1965b). Molecules as documents of evolutionary history. *J. Theor. Biol.* **8**(2): 357–66.

Zuker, M. (1989a). Computer prediction of RNA structure. *Methods Enzymol.* **180**: 262–88.

Zuker, M. (1989b). On finding all suboptimal foldings of an RNA molecule. *Science* **244**: 48–52.

Zuker, M. (2000). Calculating nucleic acid secondary structure. *Curr. Opin. Struct. Biol.* **10**(3): 303–10.

Zuker, M. and A. B. Jacobson (1998). Using reliability information to annotate RNA secondary structures. *RNA* **4**(6): 669–79.

Zuker, M. and P. Stiegler (1981). Optimal computer folding of large RNA sequences using thermodynamics and auxiliary information. *Nucleic Acids Res.* **9**: 133–48.

Zuther, E., J. J. Johnson et al. (1999). Growth of *Toxoplasma gondii* is inhibited by aryloxyphenoxypropionate herbicides targeting acetyl-CoA carboxylase. *Proc. Natl. Acad. Sci. USA* **96**(23): 13387–92.

INDEX